Frontispiece

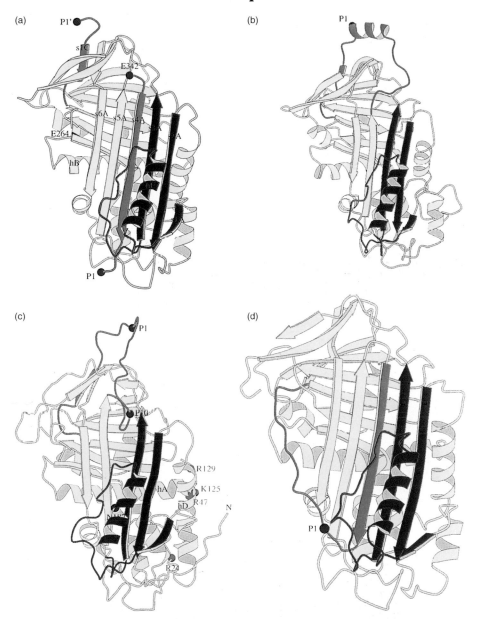

Figure 4.1. Schematic pictures of serpins illustrating different conformations of the sequence equivalent to the bait loop and strand s1C (light blue). The bait loop is identifiable by the position of the P1 residue, and the proximal hinge of the loop by residues P10–P14. The changes in tertiary structure involve opening and closing of sheet A owing to the shift of a small fragment of the protein (pink) in relation to the rest of the structure (yellow). (a) Cleaved α1-proteinase inhibitor showing the sites of the common Z (E342) and S (E264) mutations and the site of the rare Siiyama mutation (S53). (b) Intact ovalbumin. (c) Intact ATIII showing important heparin-binding residues (green) and the site of the Rouen VI variant (N187). In the crystal structure, the bait loop interacts with sheet C of another intact antithrombin molecule which has a conformation similar to latent PAI-1. (d) Latent PAI-1. Reproduced from Stein and Carrell (1995) with permission from *Nature Structural Biology* and Professor R.W. Carrell.

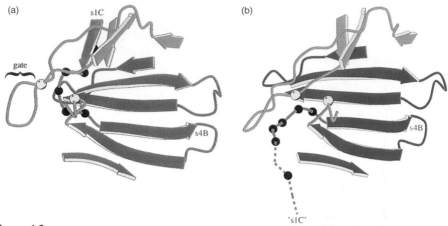

Figure 4.2.

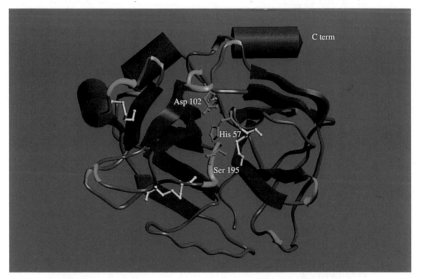

Figure 5.5.

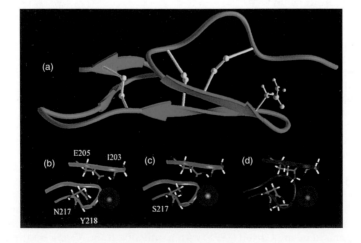

Figure 6.3.

Figure 4.2. Schematic pictures showing selected regions of (a) active and (b) latent forms of intact ATIII in a dimeric crystal structure. These forms illustrate two extremes of the different conformations that may be adopted by the distal hinge (red) formed by strand s1C and the s1C–s4B turn, in association with changes in the conformation of the bait loop. In the latent ATIII molecule (b), the sequence equivalent to strand s1C (red dashed line) is disordered in the crystal structure. Sheet C (except s1C) is coloured light blue, sheet B dark blue, and the C terminus green. $C\alpha$ atoms of the cysteines forming the disulphide bond linking the C terminus and the 'gate' region are shown as yellow spheres. Sites are shown (black spheres) of a number of mutations in the distal hinge region that cause loss of inhibitory activity. Reproduced from Stein and Carrell (1995) with permission from *Nature Structural Biology*.

Figure 5.5. Molecular model of the β-barrel-like fold of the serine protease domain of activated (human) protein C. α Helices are shown as red barrels, β sheets as red arrows, β turns as light green arrows and random coil as purple ribbons. The heavy atoms of Cys residues are represented as yellow sticks and the catalytic site residues (Asp102, His57 and Ser195; chymotrypsin numbering) are shown as dark green sticks. The location of the C-terminal α helix is also indicated. Grateful thanks to Adam Wacey for provision of the picture.

Figure 6.3. Molecular model of the fourth EGF-like domain of protein S and its putative Ca^{2+} binding pocket. (a) The secondary structure of the fourth EGF domain of protein S. β-Pleated sheets are shown as green arrows, random coil as a blue ribbon, disulphide bridges in gold and cysteine atoms in yellow. The side chain of residue Asn217 is a ball-and-stick representation. (b) Close-up view of the wild-type protein S Ca^{2+}-binding pocket. The backbone is represented by a cyan ribbon; residues are shown as stick and are coloured by atom. (c) Close-up view of the mutant protein S Ca^{2+}-binding pocket. The backbone is represented by a blue ribbon; residues are shown as stick and are coloured by atom. (d) Close-up view of the wild-type factor X Ca^{2+}-binding pocket. The backbone is represented by a purple ribbon; residues are shown as stick and are coloured by atom. The calcium-binding geometry of the fourth EGF-like domain was inferred from nuclear magnetic resonance (NMR) analysis of the first EGF domain of factor X. The backbone carbonyl of Gly47 and Gly64 and the side-chain carbonyl oxygens of Hya63 and Gln49 together form a pocket capable of binding calcium (d). The replacement of Glu49 and Hya63 of factor X by the analogous Glu205 and Asn217 residues of protein S preserves the location of the apical reactive groups of the side chain while rotating their geometry through 180° (b). The orientation of the reactive side chains of Asn217 and Glu205 may bring the carbonyl group of Asn217 into closer contact with the calcium ion than is observed in the NMR structure of factor X. Substitution of Asn217 by Ser results in the loss of a γ carbon and the replacement of the apical amide and reactive carbonyl group by a less reactive hydroxyl group (c). The resultant reduction in both steric and polaric contact between the mutated residue and calcium ion would be expected to lead to a loss of calcium binding affinity in this pocket consistent with the dysfunctional phenotype observed. Taken from Formstone *et al.* (1995) by kind permission of W.B. Saunders Co. and with grateful thanks to Adam Wacey.

Venous Thrombosis: from genes to clinical medicine

The molecular genetics of an archetypal multigene disorder

The HUMAN MOLECULAR GENETICS series

Series Advisors

D.N. Cooper, *Institute of Medical Genetics, University of Wales College of Medicine, Cardiff, UK*

S.E. Humphries, *Division of Cardiovascular Genetics, University College London Medical School, London, UK*

T. Strachan, *Department of Human Genetics, University of Newcastle upon Tyne, Newcastle upon Tyne, UK*

Human Gene Mutation
From Genotype to Phenotype
Functional Analysis of the Human Genome
Molecular Genetics of Cancer
Environmental Mutagenesis
HLA and MHC: Genes, Molecules and Function
Human Genome Evolution
Gene Therapy
Molecular Endocrinology
Venous Thrombosis: from Genes to Clinical Medicine

Forthcoming titles
Protein Dysfunction and Human Genetic Disease
Molecular Genetics of Early Human Development

Venous Thrombosis: from genes to clinical medicine

The molecular genetics of an archetypal multigene disorder

D.N. Cooper and M. Krawczak
Institute of Medical Genetics, University of Wales College of Medicine, Cardiff, UK

βIOS
SCIENTIFIC
PUBLISHERS

© BIOS Scientific Publishers Limited, 1997

First published in 1997

A CIP catalogue record for this book is available from the British Library.

ISBN 1 872748 94 5

BIOS Scientific Publishers Ltd
9 Newtec Place, Magdalen Road, Oxford OX4 1RE, UK.
Tel. +44 (0) 1865 726286. Fax. +44 (0) 1865 246823
World-Wide Web home page: http://www.Bookshop.co.uk/BIOS/

DISTRIBUTORS

Australia and New Zealand
    DA Information Services
    648 Whitehorse Road, Mitcham
    Victoria 3132

India
    Viva Books Private Limited
    4346/4C Ansari Road
    New Delhi 110002

Singapore and South East Asia
    Toppan Company (S) PTE Ltd
    38 Liu Fang Road, Jurong
    Singapore 2262

USA and Canada
    BIOS Scientific Publishers
    PO Box 605, Herndon,
    VA 20172-0605

Truth is not that which is demonstrable but that which is ineluctable.

Antoine de Saint Exupéry (1939), *Wind, Sand and Stars*

Typeset by Saxon Graphics Ltd, Derby, UK.
Printed by Biddles Ltd, Guildford, UK.

Contents

Abbreviations

5mC	5-methylcytosine
AMP	adenosine monophosphate
APC	activated protein C
APTT	activated partial thromboplastin time
ATIII	antithrombin III
C4bBP	C4b-binding protein
CAT	chloramphenicol acetyltransferase
CBS	cystathionine β-synthase
cDNA	complementary DNA
C/EBP	CCAAT/enhancer-binding protein site
CI	confidence interval
C1I	C1 inhibitor
CTR	clotting time ratio
DDAVP	desamino-D-arginylvasopressin
DGGE	denaturing gradient gel electrophoresis
DNase	deoxyribonuclease
DVT	deep vein thrombosis
EACA	6-aminohexanoic acid
EGF	epidermal growth factor
ELISA	enzyme-linked immunosorbent assay
EPCR	endothelial cell protein C receptor
FDP	fibrinogen degradation product
FGA	α-fibrinogen
FGB	β-fibrinogen
FGG	γ-fibrinogen
FpA	fibrinopeptide A
FpB	fibrinopeptide B
Gla	γ-carboxyglutamic acid
GLA	α-galactosidase A
GRP	glucose-regulating protein
HBS	heparin-binding site
HCF2	heparin cofactor II
HMWK	high-molecular-weight kininogen
HNF	hepatocyte nuclear factor
HRG	histidine-rich glycoprotein
IHD	ischaemic heart disease
IL-6	interleukin-6
LDLR	low-density lipoprotein receptor
LINE	long interspersed nuclear elements
MLE	minisatellite sequence
MLTF	major late transcription factor

mRNA	messenger RNA
MTHFR	5,10-methylenetetrahydrofolate reductase
Myr	million years
NMR	nuclear magnetic resonance
PAI	plasminogen activator inhibitor
PAR	percentage attributable risk
PCI	protein C inhibitor
PCR	polymerase chain reaction
PE	pulmonary embolism
PFGE	pulsed field gel electrophoresis
POAD	premature occlusion arterial disease
PTCA	percutaneous transluminal coronary angioplasty
RFLP	restriction fragment length polymorphism
RR	relative risk
RS	reactive site
RT-PCR	reverse transcriptase polymerase chain reaction
SCR	structurally conserved region
SCRe	short consensus repeat
SD	standard deviation
SDS-PAGE	sodium dodecyl sulphate polyacrylamide gel electrophoresis
SHBG	sex hormone-binding globulin
snRNA	small nuclear RNA
SSCP	single-strand conformational polymorphism
STS	steroid sulphatase
TAFI	thrombin-activatable fibrinolysis inhibitor
TAT	thrombin/antithrombin
TFPI	tissue factor pathway inhibitor
TGF	transforming growth factor
TNFα	tumour necrosis factor α
TP53	tumour protein p53
tPA	tissue plasminogen activator
uPA	urokinase-type plasminogen activator
USF	upstream stimulatory factor
UV	ultraviolet
VNTR	variable number tandem repeat

Preface

Venous thrombosis has a good claim to be the best characterized of all human multifactorial diseases. It is now clear that hereditary predisposition to venous thrombosis may arise as a consequence of a lesion in any one of at least 12 different genes. These genes encode proteins with either an enzymatic, structural or regulatory role in the clotting cascade. Some 400 different lesions have so far been reported at these loci in patients with a family history of venous thrombosis.

Inheritance of mutation(s) causing a deficiency state does not automatically imply that the individual concerned will come to clinical attention. Although homozygous or compound heterozygous carriers almost invariably experience some thrombotic symptoms (from relatively mild to neonatally fatal), most heterozygotes probably go undetected. Recently, it has become clear that the co-inheritance of different deficiency states is not uncommon and, when it occurs, greatly increases the risk that an individual will suffer thrombotic episodes. It has also been recognized that polymorphic variation within the non-coding portions of genes may affect their expression and, by tilting haemostasis in either a procoagulant or antifibrinolytic direction, influence the likelihood of venous thrombosis. Other non-genetic predisposing factors such as pregnancy, immobility, surgery, trauma and contraceptive pill usage are clearly also important in precipitating thrombotic events and interact in combinatorial fashion with genetic factors.

The study of gene mutations underlying venous thrombosis has proven extremely important for studies of protein structure and function and has provided new insights into the genotype–phenotype relationship in the different deficiency states. In the diagnostic context, mutation detection is important for proper risk assessment and retrospectively to 'explain' the clinical phenotypes occurring in a given family. However, owing to the multifactorial nature of the disease, molecular genetic data alone will continue to have limited predictive potential. It will be some considerable time before our understanding at the molecular level is translated into improvements in the specificity of individual therapeutic regimes.

David N. Cooper (*Cardiff*)
Michael Krawczak (*Cardiff*)

Acknowledgements

The authors would like to thank Adam Wacey for molecular modelling. DNC wishes to thank Vijay Kakkar for introducing him to the study of venous thrombosis, David Millar, Lutz-Peter Berg, Caroline Formstone, Debbie Scopes, Paula Hallam, Catherine Grundy, Ted Tuddenham, Margaret McLaughlin and Annie Procter for their many and varied contributions down the years, and Peter Harper for his continuing enthusiasm and encouragement. MK is grateful to Jörg Schmidtke and Regina Krawczak for their unceasing support. This work was supported in part by the Deutsche Forschungsgemeinschaft through a Heisenberg grant (Kr 1093/5–1) to MK.

Foreword

Advancing knowledge of blood coagulation over the past 30 years has gradually changed our model of the process from a simple linear enzymatic amplification cascade to an intricate network containing both positive and negative feedback loops. Such is the complexity of the multiple interactions involved that is often not possible to guess intuitively the outcome of raising or lowering the level of the 20 or so components. However, levels do vary quite widely in normal individuals as well as in disease, both hereditary and acquired. When functioning normally, haemostasis fulfils the physiological requirement that blood within vessels is always fluid but rapidly solidifies upon contact with extravascular tissue. Furthermore, the coagulant stimulus is tightly localized to wound sites and is prevented from propagating back into the circulation. The regulatory molecules and feedback loops are essential for maintaining the stability of this complex system. Loss of stability due to even partial defects of regulation can lead to catastrophic thrombosis. Concurrent with our advancing knowledge of the system, an increasing number of genetic thrombophilia syndromes have been characterized and these are described lucidly in the present work.

A particular feature of thrombophilia is its multifactorial nature and it could be said to be a paradigm for a disease in which multiple epistatic genetic effects interact with several well-established environmental risk factors to precipitate acute events. Arguably, the environmental genetic interaction is now better understood for venous thromboembolism than for any other disease complex. Nevertheless there remain some areas of uncertainty. About 30% of attenders at thrombophilia clinics have no abnormality detectable by present assays. A precise estimate of future risk cannot be given based on test results although prospective studies may lead to useful clinical algorithms. The contribution of the vessel wall to thrombotic risk is only just beginning to be studied in thrombophilia and the third member of Virchow's triad, blood flow, has been little studied since the decline of interest in blood rheology and is only thought of nowadays in relation to obvious stasis-producing events.

A satisfactory synthesis of all these multiple inputs to venous thromboembolism is not yet within our grasp but may be glimpsed from the masterly overview provided in the present volume.

E.G.D. Tuddenham (*London*)

Introduction

A starlit or a moonlit dome disdains
All that man is,
All mere complexities,
The fury and the mire of human veins.

W.B. Yeats, 'Byzantium'.

1.1 Venous thrombosis: causes and clinical significance

1.1.1 Thrombotic disease: an overview

Vascular disease is the major cause of disability and death in Western societies (Beaglehole, 1990): coronary thrombosis, peripheral vascular disease and pulmonary embolism (PE) alone are collectively implicated in the death of over half of all individuals. A prominent factor in this high morbidity and mortality is thrombosis, which forms at the site of already diseased or atheromatous blood vessels. Thrombosis, the inappropriate clotting of blood in an artery or vein with partial or complete occlusion of the vessel, may be regarded as a consequence of defective haemostasis. Tissue necrosis may result if alternative vessels are not able to maintain the circulation in the area. The patient is also at risk of embolism, the occlusion of a vessel by clot fragments which have broken free from a remote location and come to rest in a critical site (e.g. lung).

Disabling or fatal thrombotic disease may stem from the formation of thrombi within *arteries* as a consequence of a platelet or vessel wall abnormality (Schafer and Kroll, 1993) or at sites of endothelial injury (e.g. atheroma; Fuster *et al.*, 1992a, b; Nachman, 1992; Ross, 1993; Davies, 1996) or alternatively within *veins* as a result of venous stasis and/or hypercoagulability. Thrombi can also form within the heart chambers, on prosthetic valves or within the microcirculation. Venous thrombi are composed predominantly of fibrin and red cells whereas arterial thrombi are composed predominantly of platelets.

Intriguingly, there is an as yet unexplained tendency for thrombi to affect certain anatomical regions of the vasculature disproportionately and in a given individual to affect some areas far more severely than others (Weinmann and Salzman, 1994). Furthermore, specific risk factors appear to be selectively more important in the development or progression of thrombi in particular blood vessels. For example, hypertension is associated particularly with a risk of cerebrovascular arterial disease, whilst heavy cigarette smoking is associated with both arterial and venous peripheral vascular disease as well as with coronary artery thrombosis. Clearly the pathogenesis of thrombosis is complex and usually results from the interaction of multiple factors, including a genetic predisposition.

Objective diagnostic testing for venous thrombosis usually involves one or both of two non-invasive methods: Doppler ultrasonography and serial impedance plethysmography. These diagnostic methods do, however, vary in terms of their accuracy,

Venous Thrombosis: from Genes to Clinical Medicine, D.N. Cooper and M. Krawczak.
© 1997 BIOS Scientific Publishers, Oxford.

reliability and cost-effectiveness (Weinmann and Salzman, 1994; Hull *et al.*, 1995; Wells *et al.*, 1995; De Stefano *et al.*, 1996). Venous thrombosis is currently treated with either heparin or oral anticoagulants such as the coumarins (British Committee for Standards in Haematology, 1990; Greaves and Preston, 1991; Verstraete, 1993; Hyers *et al.*, 1995; ten Cate *et al.*, 1995; De Stefano *et al.*, 1996). Treatment depends upon whether the patients manifest temporary/reversible risk factors (e.g. surgery), in which case short-term treatment is normally recommended, or whether they have permanent risk factors such as an inherited deficiency state, in which case long-term/life-long therapy may have to be considered (Ginsberg, 1996). Thrombolytic therapy (Marder, 1995) may, however, be employed in cases of massive PE. These treatments have proved remarkably effective and it is clear that preventive medicine in the form of early therapeutic (or even prophylactic) intervention would save many lives at the same time as dramatically cutting the costs incurred in surgery, patient recovery and rehabilitation.

Clearly prophylactic measures require the early detection of those most at risk, but accurate tests to predict thrombotic disease are still lacking. Although a number of risk factors (e.g. age, diet, lifestyle, viral infection, haemostatic variables, etc.) have been identified in arterial thrombosis (Fuster *et al.*, 1992b; Heinrich *et al.*, 1994: Vercellotti, 1995; Catto and Grant, 1995), the clinical variability manifested by this complex disease group, together with its probabilistic nature, has severely hampered universal risk assessment. Similarly, in venous thrombosis, certain groups (e.g. post-surgical patients, patients with major trauma, individuals with hyperlipidaemia, pregnant women and cases of the inherited deficiency of any one of a number of coagulation factors) are known to be particularly at risk (Weinmann and Salzman, 1994; Bauer, 1995; Geerts *et al.*, 1995; Kawasaki *et al.*, 1995; De Stefano *et al.*, 1996; Lane *et al.*, 1996c, d). The various inherited deficiency states, which are clinically indistinguishable from each other, are now becoming increasingly well understood in terms of their physiology, biochemistry, molecular biology and genetics. Although all thrombotic disease is by its very nature multifactorial, it is our contention that we are most likely to be able to gain a mechanistic understanding of the disease process in those cases of thrombosis that result from the deficiency of a single gene/gene product. For this reason, this volume is devoted to the study of those inherited disorders which confer a predisposition to venous thrombosis.

Previous volumes on venous thrombosis (e.g. Pitney, 1981; Ogston, 1987) have concentrated on the clinical manifestations and pathophysiology of the disease, the potential role of haemostatic or environmental factors in its causation, or its diagnosis, treatment and prevention. Here we have attempted to study the disease process from a genetic standpoint. Sufficient information is also provided to allow the reader to make sense of new genetic studies as they appear in the literature or, alternatively, to design their own. The focus of this book is venous thrombotic disease and its association with abnormalities of the haemostatic system. This is for two reasons: firstly, because most of the known deficiency states responsible for conferring a familial predisposition to venous thrombosis are associated with a reduced ability to regulate haemostasis, these disorders comprise a natural grouping; secondly, since the mode of inheritance of the different predisposing conditions causing venous thrombosis is fairly well defined, they are amenable to molecular genetic analysis. Genetic defects may also underlie arterial thrombosis (as in the hyperlipidaemias and hypertension) but the mode of inheritance is often less well defined and the contribution of genetic variation at various loci to the pathophysiology of the condition is less well understood (MacCluer and Kammerer, 1991).

In many ways, the study of venous thrombosis provides a paradigm for those interested in the molecular genetics of multifactorial disease. In no other inherited condition are so many different genes known to be involved, either singly or in concert. By focusing on venous thrombosis, we can take a detailed look at the influence of such genetic factors as the mode of inheritance, allelic heterogeneity, locus heterogeneity, variable penetrance and epistasis (as well as various environmental factors) on the clinical phenotype in this common multi-gene disorder. We can also assess the importance of these factors for the accurate prediction of thrombotic disease and effective genetic counselling.

1.1.2 Venous thrombosis: risk factors and pathogenesis

Virchow (1856) originally proposed that three main factors determine the location, nature and extent of thrombus formation: blood flow, the constituents of the blood and the vessel wall. This view is still valid today and indeed provides a useful starting point in any attempt to understand the aetiology of venous thrombosis. In addition to the risk factors outlined in Section 1.1.1, numerous clinical disorders (e.g. cancer, diabetes, etc.; Ogston, 1987; Murray, 1991) are known to be associated with thrombosis. This has led to the emergence of the concept of the hypercoagulable or prethrombotic state (Kitchens, 1985; Greaves and Preston, 1991; Schafer, 1994) in which the haemostatic equilibrium is tilted in a procoagulant direction. Attempts to define this state in molecular terms have followed Virchow in so far as they have focused upon changes in the vessel wall (Rodgers, 1988; Vanhoutte and Scott-Burden, 1994), disturbance of blood flow (Carter, 1994a; Grabowski and Lam, 1995) or abnormalities in the composition of the blood in patients with thrombotic disease (Bauer and Rosenberg, 1987; Mannucci and Giangrande, 1992; Lee et al., 1994). The venous circulation is particularly sensitive to stasis and decreased blood flow, even in otherwise healthy individuals. Thus, prolonged immobilization (e.g. after surgery) is an important risk factor for venous thrombosis.

Damage to the endothelium and subsequent exposure of sub-endothelial collagen may be an important trigger in thrombus formation (Ross, 1993). The endothelial cell plays a dynamic role in vascular haemostasis (Pearson, 1993): it not only synthesises a number of important anticoagulant (e.g. protein C, protein S, thrombomodulin, tissue factor pathway inhibitor) and fibrinolytic [e.g. tissue plasminogen activator (tPA) and its inhibitor, PAI-1] molecules but also provides a suitable surface upon which the various anticoagulant and antifibrinolytic proteins may be appropriately positioned (Nawroth et al., 1985; Pearson, 1993). It is the major physiological site for thrombin inhibition by virtue of the membrane-anchored heparan sulphate proteoglycans which bind antithrombin III (ATIII), thereby dramatically increasing the serpin's affinity for thrombin. Thus, under normal conditions, the lumen of the endothelium provides a potent antithrombotic surface. However, following a haemostatic challenge or as a consequence of atherosclerotic disease, myocardial infarction or ischaemia, the properties of the endothelial cell may be altered dramatically (Gerritsen and Bloor, 1993; Lüscher et al., 1993; Lüscher, 1994). These changes, which take place as a result of exposure to cytokines and other mediators of the inflammatory response, serve both to render the endothelial surface more procoagulant and to inhibit the fibrinolytic pathway (e.g. through down-regulation of the synthesis and release of PAI-1).

The components of the blood also play an important role in thrombus formation. The conversion of soluble fibrinogen into an insoluble fibrin clot (the structural framework of the thrombus) is mediated by thrombin whose own conversion from its inactive zymogen, prothrombin, is brought about by the reactions of the clotting cascades (Ofosu et al., 1996). Elevated concentrations of either procoagulant factors or the inhibitors of fibrinolysis, or reduced concentrations of the coagulation factors or the plasminogen activators, will therefore lead to the emergence of a prethrombotic state and an increased risk of thrombosis, whether arterial or venous (Thompson et al., 1995). Not surprisingly, predisposition to venous thrombosis may be caused by either an inherited or an acquired defect in the mechanism of haemostasis. Both can disturb the delicate balance of pro- and anti-coagulant forces and allow a comparatively minor intravascular injury to proceed to major thrombosis. Many of the protein factors involved in platelet activation, clot formation and fibrinolysis have now been identified (see Section 1.2); the detection of altered levels of these factors in clinical studies is already of proven importance in risk assessment (Bauer and Rosenberg, 1987). Metabolites (e.g. activation peptides) generated as a consequence of these processes can be assayed biochemically and their predictive potential evaluated (Boisclair et al., 1990; Bauer and Rosenberg, 1994). Such tests represent valuable indicators of thrombotic disease and promise to be of enormous benefit in screening programmes to detect the early signs of this common yet eminently preventable cause of early death and disablement.

In venous thrombosis, the highest known risk is conferred by inherited defects in the genes encoding the haemostatic regulatory factors as well as, to a lesser extent, the fibrinogen chains and plasminogen (Miletich et al., 1993; De Stefano et al., 1996). A considerable number of genetic lesions causing venous thrombotic disease have now been described (see Appendix and individual chapters) and result either in the increased deposition or decreased dissolution of fibrin. Between them, these inherited defects may account for the majority of all patients presenting with recurrent venous thrombosis.

Even though most cases of disease predisposition in affected families appear to arise largely through the influence of a single defective gene, it would be misleading to describe the inherited thrombophilias simply as single-gene disorders. This important point is illustrated by the fact that many apparent carriers of a disease gene lesion are asymptomatic both in affected families and in the general population. Much more information on haemostatic variables, genetic variation and environmental risk factors and 'triggers' will be needed to resolve the tangled issue of the pathogenesis of 'thrombophilia'.

1.1.3 Epidemiology: the problem in its context

The epidemiology of venous thrombosis has been well reviewed (Alpert and Dalen, 1994; Carter, 1994b). Clinically, the most significant complication of venous thrombosis is PE which usually arises from thrombi in the deep veins of the leg (Bell and Smith, 1982; Rubinstein et al., 1988). It certainly poses the greatest risk to life; it is estimated that deep vein thromboses (DVTs) result in 300 000–600 000 hospitalizations per year in the United States (Bernstein, 1986), of whom 10% die from PE within a few hours (Bell and Smith, 1982). In 30% of the remainder, a diagnosis is rapidly established and mortality kept to around 8%. However, the 70% of patients who remain undiagnosed, and therefore untreated, exhibit a much higher mortality of around 30%. Early, preferably presymptomatic, diagnosis and accurate risk assessment are therefore major goals in thrombosis research.

The prevalence of DVT in the general population has been estimated at between 2 and 5% (Rodeghiero and Tosetto, 1993). The incidence of venous thrombosis in some specific circumstances, such as the post-operative period, has been defined with some degree of precision, but a meaningful assessment of the overall prevalence in the community has been difficult to obtain. From a longitudinal community study, it has been estimated that the total annual incidence of clinically recognized DVT in the United States is over 250 000, whereas that of superficial thrombophlebitis is over, 123 000 (Coon et al., 1973). Clinical diagnosis of venous thrombosis is, however, notoriously unreliable (Hull et al., 1981; Weinmann and Salzman, 1994) and it is likely that the frequency of venous thrombosis is considerably underestimated when its detection is based on physical examination alone.

One area in which the mortality associated with heritable thrombophilia is likely to have been underestimated is the antenatal period. Recently, Preston et al. (1996) have shown that the risk of fetal loss was higher in women with inherited thrombophilia [odds ratio 1.35; 95% confidence interval (CI), 1.01–1.82]. The risk was higher for stillbirth than miscarriage (odds ratios 3.6 vs. 1.27; 95% CI, respectively, 1.4–9.4 and 0.94–1.71).

In recent years, an increasing number of congenital alterations in blood components have been recognised to endow the affected individual with a marked predisposition to thrombosis, often at an early age. A recent review of the prevalence of hereditary thrombotic disease in patients under 45 years of age with proven venous thrombosis have concluded that at least 20% can be expected to have ATIII, protein C or protein S deficiency (Hirsh et al., 1989). In some cases, double or triple deficiency states may occur in the same individual thereby greatly increasing the likelihood that a given individual will come to clinical attention (Wolf et al., 1990; Jobin et al., 1991). A detailed assessment of the genetic and environmental contributions to the epidemiology of venous thrombosis is to be found in Chapter 15.

1.1.4 Molecular genetics of familial venous thrombosis: the inheritance of risk

An hereditary predisposition to venous thrombosis may arise as a consequence of a lesion in any one of at least 12 different genes encoding ATIII (AT3), factor V (F5), factor XII (F12), the three chains of fibrinogen (FGA, FGB, FGG), heparin cofactor II (HCF2), plasminogen (PLG), protein C (PROC), protein S (PROS), thrombin (F2) and thrombomodulin (THBD) (Tuddenham and Cooper, 1994). In excess of 400 different gene lesions (single base-pair substitutions, deletions and insertions) have so far been reported at these loci in patients with thrombotic disease (see Appendix and subsequent chapters). One of the main topics addressed in this volume is how different gene mutations at different loci give rise to what is a broadly similar phenotype.

Possession of an inherited deficiency state does not automatically imply that the individual concerned will come to clinical attention. Although homozygous or doubly heterozygous individuals will almost certainly experience some thrombotic symptoms (from relatively mild to neonatally fatal), most heterozygotes probably go undetected. Thus, clinically symptomatic ATIII deficiency in the general population occurs at an 80-fold lower frequency ($f = 2.1 \times 10^{-5}$) than does the clinically asymptomatic deficiency state ($f = 1.65 \times 10^{-3}$; Tait *et al.*, 1994). Similarly, clinically symptomatic protein C (PROC) deficiency occurs at a 23- to 52-fold lower frequency ($f = 2.8 \times 10^{-5} - 6.2 \times 10^{-5}$) than its asymptomatic counterpart ($f = 1.4 \times 10^{-3}$) (Miletich *et al.*, 1987; Gladson *et al.*, 1988; Tait *et al.*, 1995). One gene lesion (factor V Leiden, Arg506→Gln), implicated in the etiology of perhaps 40% of all cases of familial thrombotic disease, occurs with a prevalence of between 2 and 7% in the general population (Beauchamp *et al.*, 1994; Bertina *et al.*, 1994; Svensson and Dahlbäck, 1994) but only a small fraction of carriers are clinically affected. Thus, a defect in any one of the above-mentioned genes is on its own probably insufficient to cause thrombosis. A second gene lesion at another locus might be involved. Other environmentally based predisposing factors such as pregnancy, surgery, trauma, hormone replacement therapy and contraceptive pill usage may also contribute to the likelihood of a thrombotic episode (Allaart *et al.*, 1993; Henkens *et al.*, 1993; Geerts *et al.*, 1995; World Health Organization Collaborative Study of Cardiovascular Disease and Steroid Hormone Contraception, 1995; Daly *et al.*, 1996b; Jick *et al.*, 1996).

A number of other genetic factors may also be involved in determining the likelihood that a given individual will come to clinical attention. For example, *genotypic variation* in the *PROC* gene promoter influences transcriptional efficiency of the *PROC* gene, the plasma protein C concentration and ultimately thrombotic risk (Scopes *et al.*, 1995; Spek *et al.*, 1995a). *Epistasis*, the influence of other gene loci, may also play a role. Thus in the Dutch population, co-inheritance of the factor V Leiden mutation with protein C deficiency has been shown to increase the risk of an individual coming to clinical attention by some six-fold (Koeleman *et al.*, 1994; Gandrille *et al.*, 1995b) although this risk may vary between different populations (Hallam *et al.*, 1995a). Similarly, co-inheritance of the factor V Leiden mutation with protein S or ATIII deficiency also increases the risk of an individual coming to clinical attention (Zöller *et al.*, 1995a; van Boven *et al.*, 1996). Such combinatorial effects of different loci are unlikely to be rare since many of the deficiency states giving rise to thrombosis are individually fairly frequent in the general population (e.g. *AT3*, 0.0016; *PROC*, 0.0014; *F5*, 0.02–0.07; *F12*, 0.023; *HCF2*, 0.009).

ATIII deficiency and protein C deficiency are unusual in that they are recessive disorders with respect to both mortality and reproductive fitness but dominant disorders with respect to morbidity, albeit with reduced penetrance (Rosendaal *et al.*, 1991; Allaart *et al.*, 1993, 1995). Prophylactic anticoagulation is an effective treatment for those who have already exhibited thrombotic manifestations. However, it is not normally recommended for relatives of the clinically affected individual, even if they are proven carriers of a predisposing gene lesion, unless there are clear mitigating circumstances (e.g. trauma, surgery). This is because, as indicated earlier, the penetrance of predisposing gene lesions is usually low and many other factors contribute to the eventual likelihood of a thrombotic episode. It may, therefore, be that the detection of clinically asymptomatic heterozygous individuals by molecular genetic means will be of limited predictive utility. The study of the molecular genetics of venous thrombotic disease nevertheless promises to greatly increase our understanding of how genetic and environmental risk factors can combine to increase the likelihood of disease.

1.2 An introduction to haemostasis

Blood coagulation is a system which has evolved to meet two conflicting requirements. Firstly, blood should be able to flow freely at all times within the vasculature, reaching all tissues. Secondly, at the site of any leakage, whether large or small, blood must transform itself rapidly into a solid plug in order to stem its own outflow. This second requirement becomes more demanding the higher the pressure of blood and the more dynamic the circulation. However, a system that permits the rapid clotting of blood in response to a small stimulus must be tightly controlled since inappropriate clotting could obstruct vital areas of the circulation. The haemostatic mechanism in mammals meets these requirements by means of a complex multipathway interactive system with multiple fail-safe and back-up circuits, and both positive- and negative-feedback loops. The complexities of the system cannot be conveyed in a simple diagram. However, dissecting the system into discrete entities greatly simplifies its presentation. Although the nature of the mechanisms which permit balance and fine-tuning of haemostasis are still debatable, and even although the role of some factors is not yet clear, the primary structures of the proteins of haemostasis are now fully defined. The clinical effect of deficiencies of some of these factors may be clearly observed in thrombophilia and in various bleeding disorders. Other factors appear to be dispensable for haemostasis and are of uncertain physiological importance. The following brief overview is intended to provide an introductory perspective on current theories of haemostasis. More detailed information may be found in recent specialist volumes (Bloom *et al.*, 1994; Colman *et al.*, 1994) and reviews (e.g. Davie *et al.*, 1991; Esmon, 1993a; Davie, 1995).

1.2.1 The coagulation cascade

A series of enzyme interactions accelerated by cofactors converts inactive precursors, zymogens or procofactors to activated forms, culminating in the production of thrombin which converts soluble fibrinogen to an insoluble fibrin clot. Although the mechanism of initiation of clotting *in vivo* is somewhat controversial, two potential initiators have been studied extensively *in vitro*: the contact phase and tissue factor.

1.2.2 Contact phase of coagulation

Human blood taken out of the circulation and spilled on to any artificial surface forms a solid clot within minutes. Surfaces which possess a negative charge or which are 'wettable' such as glass, promote especially rapid clot formation. The basis of this effect, surface contact activation, lies in the ability of three zymogens and a cofactor to bind to negatively charged surfaces and to activate factor XI. Factor XII binds to surfaces, thereupon undergoing a structural change making it more susceptible to activation by kallikrein. Kallikrein cleaves factor XII at two sites; the first cleavage yields αXIIa, the second βXIIa. Kallikrein also circulates as a zymogen, pre-kallikrein, which is cleaved by activated factor XII (factor XIIa) to convert it to its active two-chain form, kallikrein. High-molecular-weight kininogen acts as a cofactor in this reaction by binding to pre-kallikrein and kallikrein, and promotes their binding to negatively charged surfaces. These reactions may be of physiological importance but the precise *in vivo* mechanism of initiation of this 'intrinsic pathway' of coagulation is unclear. Whatever the explanation, αXIIa once generated cleaves factor XI (a homodimer) at one site in each chain, yielding factor XIa which comprises two heavy chains and two light chains held together by disulphide bonds. This reaction is also facilitated by high-molecular-weight kininogen which binds factor XI and promotes its binding to surfaces. One way in which the contact factors may be bypassed to initiate coagulation is via the action of thrombin: factor XI is activated by thrombin in the presence of a negatively charged surface and factor XIa is then capable of autocatalytically activating factor XI.

1.2.3 Intermediate stages of coagulation

Factor XIa leaves the surface and, in the only free solution reaction in the clotting cascade other than the terminal reaction of thrombin with fibrinogen, activates factor IX in the presence of calcium. This step involves two cleavages, resulting in the release of a small activation peptide from factor IX. In the next step, a trimolecular ('tenase') complex is assembled on the phospholipid surface of the activated platelet. The presence of an activated cofactor, factor VIIIa, is essential for this assembly to occur efficiently. Factor VIII circulates in a complex with its carrier protein, von Willebrand factor, from which it is released and activated by limited cleavage by either thrombin or factor Xa. In the presence of calcium and a phospholipid surface, factor VIIIa dramatically accelerates the activation of factor X by factor IXa in the 'tenase' complex (*Figure 1.1*). Factor X circulates as a two-chain molecule which is further cleaved by factor IXa, releasing an activation peptide to yield the active enzyme, factor Xa. Factor Xa then goes on to promote the formation of the 'prothrombinase' complex which comprises factors Va and Xa and prothrombin (*Figure 1.1*). Factor V circulates as an inactive procofactor which is activated by thrombin or factor Xa. Factor Va then binds to the activated platelet surface and, in the presence of calcium, accelerates the activation of prothrombin by factor Xa. Cleavage of prothrombin by factor Xa in the prothrombinase complex then generates the serine proteinase thrombin, leaving the γ-carboxyglutamic acid (Gla) and kringle containing portion of the molecule (fragment 1.2) attached to the phospholipid membrane.

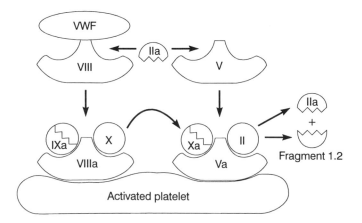

Figure 1.1. Formation of the 'tenase' (left) and 'prothrombinase' (right) complexes.

1.2.4 The extrinsic pathway of coagulation: tissue factor-dependent initiation

Much more important than the intrinsic pathway, the activation of the extrinsic pathway requires tissue factor, an integral membrane glycoprotein found in many cell types (Rapaport and Rao, 1995).

Normally, tissue factor is not exposed to blood but if blood comes into contact with tissue after vascular injury, the activation sequence diagrammed in *Figure 1.2* occurs. Tissue factor functions as a cofactor in the activation of factors VII and X, accelerating the reactions by causing conformational changes in the proteinases. Factors VIIa and Xa are capable of reciprocal activation and this process may be triggered by the presence of minute amounts of factor Xa or factor VIIa. Factor VIIa cleaves factor X in the same fashion as factor IXa in the tenase complex, thereby bringing about convergence of the intrinsic and extrinsic pathways. Factor Xa generated from these interactions then enters the prothrombinase complex (*Figure 1.1*). Factor IX is also activated by tissue factor/factor VIIa (*Figure 1.2*), a reaction which may be as important as factor X activation when the amount of tissue factor is limiting. Factor IXa then enters the tenase complex. The revised representation of the coagulation cascade shown in *Figure 1.3* is a more accurate representation of how coagulation is initiated.

1.2.5 Terminal stage of coagulation

Fibrinogen is a very abundant soluble protein which consists of three pairs of chains (α, β and γ) with the molecular formula $\alpha_2\beta_2\gamma_2$. Thrombin cleaves fibrinogen to release two small peptides, fibrinopeptides A and B. These peptides are the N termini of the α and β chains of fibrinogen. The resulting fibrin monomer rapidly and spontaneously polymerizes in a staggered array. Fibrin strands form the physical meshwork of the insoluble clot but a clot is relatively unstable unless and until it is converted to a covalently linked and stable form by the action of factor XIII. Factor XIII, a transamidase, circulates as an inactive precursor but is activated by thrombin, a process accelerated by fibrin formation. Factor XIIIa then covalently cross-links the α and γ chains of fibrin.

1.2.6 Control mechanisms in coagulation

The purpose of the coagulation cascade is the rapid generation of thrombin. However, in order to limit thrombin generation and to prevent inappropriate activation, regulatory control systems must operate. These controls are vital to ensure the maintenance of the fluidity of blood and to prevent the consequences, inappropriate

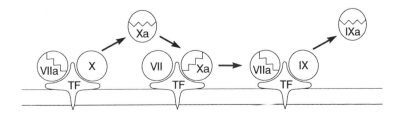

Figure 1.2. Initial reactions of the extrinsic pathway.

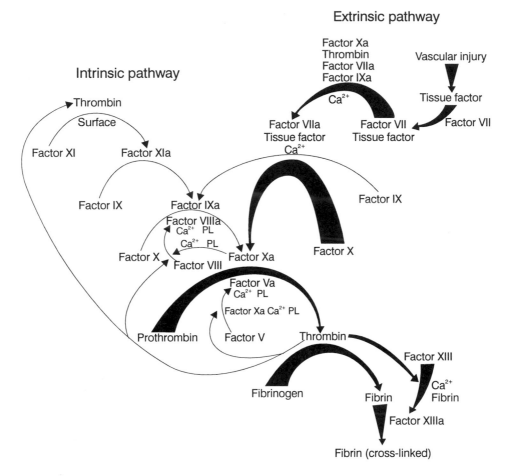

Figure 1.3. Coagulation cascade and fibrin formation by the intrinsic and extrinsic pathways. Reprinted after Davie (1991) with permission. © 1991 American Chemical Society.

intravascular coagulation and thrombosis. These regulatory systems will be discussed in some detail in subsequent chapters because defects in key molecules in these systems commonly lead to an inherited thrombotic disorder.

Plasma contains a family of serine proteinase inhibitors (serpins) whose role is to neutralize the activated serine proteinases of coagulation and fibrinolysis. The most important serpin of coagulation is ATIII which forms a one-to-one stoichiometric complex with thrombin as well as factors VIIa, XIa, IXa and Xa. Once formed, such complexes are stable and totally inactive, and are rapidly removed from the circulation. The rate of serpin:proteinase complex formation is enormously enhanced by the sulphated mucopolysaccharide heparin. Although heparin does not occur naturally in the circulation, it possesses a a physiological counterpart in the glycosaminoglycans on the luminal surface of intact endothelium. Heparin binds to a specific region of ATIII, resulting in a conformational change which enhances interaction of ATIII with its target proteinases. Heparin also binds to, and may be important for, the approximation of the proteinase to the serpin in a 'ternary complex'.

Thrombin initiates a negative feedback control pathway through the activation of protein C. Activation of protein C requires an endothelial cell-surface receptor, the glycoprotein thrombomodulin, which binds to thrombin thereby inhibiting the proteinase's various procoagulant functions. Activated protein C (APC) requires a cofactor for its action on factors V and VIII, namely protein S. In the presence of phospholipid and calcium ions, APC/protein S inactivates factors Va and VIIIa. Protein S also fulfils two other anticoagulant roles: it inhibits the activity of the prothrombinase complex by interacting with factors Va and Xa and it inhibits factor X activation by interacting with factor VIII.

Plasma also contains an inhibitor of tissue factor activation, tissue factor pathway inhibitor (TFPI). TFPI is related in structure to the Kunitz inhibitor and it forms an inhibitory complex with factor Xa. This complex then binds tissue factor/factor VIIa in a quaternary inhibitor complex. Two of the three Kunitz domains in TFPI are involved in this complex, one each for factors Xa and VIIa. TFPI does not inhibit factor IX activation by factor VII/tissue factor so that factor X activation may continue via the intrinsic pathway. Indeed, TFPI inhibition of factor X activation may be reduced by factors IX and VIII.

Fibrin clot formed *in vitro* or *in vivo* dissolves over a period of hours due to digestion by the proteinase plasmin. Plasmin circulates as an inactive zymogen, plasminogen, whose activation is localized to and initiated by the presence of fibrin, to which it binds together with an enzyme synthesised by endothelium, tPA. Fibrin-specific localization of activator and zymogen ensures that active plasmin is released in the presence of its substrate, fibrin. Fibrin is cleaved by plasmin at specific sites, releasing fibrin fragments (X, Y, D, E, D dimer). Free plasmin that diffuses away from the clot is neutralized exceedingly rapidly by the serpin α_2-antiplasmin, whilst tPA is inactivated by PAI-1. Regulation of fibrinolysis involves other activator and inhibitor proteins (e.g. urokinase, and complement inhibitor), and may be initiated by contact phase interactions, but the plasminogen/tPA system is probably the most important.

1.2.7 Platelets and haemostasis

Platelets are essential for effective haemostasis and have multiple interactions with plasma coagulation factors. Platelets play an integral role in blood coagulation through the provision of phospholipid surface receptors, local release of procoagulants and structural interaction with the fibrin clot. Under most circumstances of actual injury to blood vessels, both the platelet response and the coagulation cascade are needed for effective haemostasis.

Platelets circulate until they encounter extravascular surfaces such as collagen or basement membranes generated by vascular injury. The platelet will adhere to all such 'foreign' surfaces, and will then spread out in contact, and finally aggregate to other platelets, thus building up a haemostatic plug that stems the outflow of blood from the wound. This process depends on platelet membrane glycoprotein receptors and their interaction with plasma and cell-surface adhesive proteins (i.e. fibrinogen, fibronectin and von Willebrand factor). Adhesion at high shear rate (as found in the microcirculation) is mediated by von Willebrand factor, a polymeric protein that possesses binding sites for both platelet membrane glycoprotein I and for collagen. Adherent platelets then spread to gain intimate contact, a step involving both platelet membrane glycoprotein IIb/IIIa and von Willebrand factor. During this process of attachment and spreading, the platelet changes dramatically in shape and releases the contents of its dense bodies and α granules. This coincides with the aggregation of further free platelets to the adherent platelets, a step that requires the platelet membrane glycoprotein complex GPIIb/IIIa and the adhesive proteins fibrinogen, fibronectin and thrombospondin. These proteins are released from platelet α granules along with von Willebrand factor, factor V, platelet factor 4 and fibrinogen, among others.

Platelets also contribute directly to the intrinsic pathway in terms of supporting the 'tenase' and 'prothrombinase' complexes and protecting them from inactivation by heparin–ATIII. The phospholipid contributed by activated platelets serves the vital function of restricting coagulation to the site of vascular injury and ensuring that it does not trigger inappropriate coagulation elsewhere in the circulation.

1.2.8 Protein domains and evolution

As we shall see, different groups of coagulation factors betray their evolutionary relatedness through their shared possession of certain types of protein domain. Take for example the six vitamin K-dependent factors of coagulation: factors VII, IX and X, protein C, protein S and prothrombin. All six contain Gla domains whose role is to potentiate a calcium-mediated interaction between the proteinases and negatively charged membrane phospholipids. The common possession of the Gla domain is best explained by the recruitment of a Gla domain into a vitamin K-dependent factor ancestral protein early on in vertebrate evolution. Two epidermal growth factor (EGF) domains (which play a role in substrate/cofactor binding and in establishing the correct Gla/EGF conformation) are present in four of these factors, the exceptions being protein S (which has four) and prothrombin (which has replaced its EGF domains with kringles). A process of gene duplication and divergence also accounts for five of the six vitamin K-dependent factors possessing very similar catalytic domains. The sixth factor, protein S, has replaced its catalytic domain with one of unknown function which is homologous to that found in sex hormone-binding globulin. Such gains and losses of domains are believed to occur by a process of exon shuffling in which different exons encoding different domains have been transferred to and between different proteins. The evolution of the vitamin K-dependent factors is explored in much more detail in Chapter 3.

Pathological mutations: DNA sequence and protein structure

2.1 Introduction

Mutation denotes both the process and the outcome of changes in the cellular genetic material, deoxyribonucleic acid (DNA). Mutational changes can either be induced by external agents, such as ionizing/ultraviolet (UV) irradiation and chemical mutagens, or have internal causes, the most important one being the less than perfect fidelity of the intracellular DNA replication and repair machinery. Since all biochemical processes in a living organism are indirectly controlled by its genes, alteration of the function of one or more gene products through mutation inevitably results in an incongruity between the cell or organism and the environment to which it is otherwise well adapted. In most cases, this incongruity will not be to the benefit of the bearer and will manifest itself in malfunction or death. However, if no changes were ever to occur in DNA, the evolution of complex organisms such as humans would have scarcely been possible. Inherited mutations thus represent two sides to the same coin: pathological lesions, on the one hand, threaten those carrying them with severe disease; their advantageous counterparts, on the other hand, represent the driving force or 'fuel' of evolution. Not surprisingly, however, the molecular mechanisms underlying the two categories of mutation appear to be the same (Krawczak and Cooper, 1996).

Since the advent of modern DNA technology, many different types of inherited gene mutations (as distinct from large-scale 'chromosomal' or 'genome' mutations) have been characterized in considerable detail in the context of genetic disease. These comprise single base-pair substitutions, which make up the bulk of disease-associated variations, deletions, insertions, inversions, and complex rearrangements involving combinations thereof. What all these lesions have in common is that their location and prevalence is highly specific, and that this specificity is determined to a considerable extent by the local DNA sequence environment (Cooper and Krawczak, 1993). Early evidence for such an association emerged from the observation of mutational 'hotspots' in phage T4 (Benzer, 1961). In eukaryotes, the efficiency of chemical (Coulondre et al., 1978; Loeb and Preston, 1986) and enzymatic (Loeb and Kunkel, 1982; Modrich, 1987) DNA modification, thought today to play an important role in the endogenous process of single nucleotide substitution, was shown to be sequence-dependent. Studies of gene deletions provided further evidence for contextual effects.

For example, spontaneous deletions in the *lac*I gene of *Escherichia coli* were frequently found to be flanked by direct repeats, an observation which was suggestive of DNA 'slippage' as a mutational mechanism (Schaaper *et al.*, 1986). Finally, mutation rates calculated for human genetic diseases varied quite widely (Vogel and Motulsky, 1986; Kondrashov and Crow, 1993), suggesting that the germline mutability of the underlying gene sequences was highly variable.

In the following sections, a brief introduction will be given to our current knowledge of *in vivo* mechanisms of human germline mutagenesis, insights that stem almost exclusively from the biases prevalent in the mutational spectra associated with inherited diseases. Not only has the study of thrombotic disorders contributed substantially to this knowledge, but basic rules for assessing and predicting the relative frequency and location of lesions in the genes involved, if and where they can be established, promise to help in improving the efficacy of mutation search strategies and, therefore, of the predictive diagnosis of venous thrombosis.

2.2 Single base-pair substitutions

2.2.1 The hypermutable CpG dinucleotide

The most abundant chemical modification of eukaryotic DNA is 5-methylcytosine (5mC). It occurs predominantly in CpG dinucleotides and, owing to its propensity to undergo deamination to form thymidine (*Figure 2.1*), renders this doublet particularly

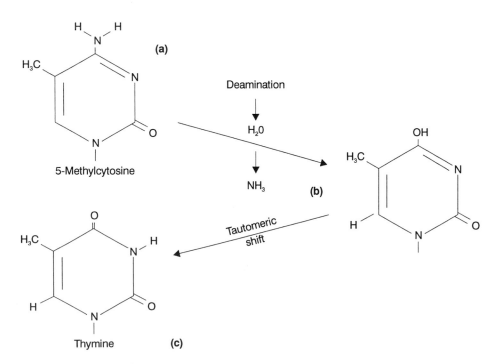

Figure 2.1. Spontaneous deamination of 5mC. In aqueous solution, a spontaneous hydrophilic attack at C-4 of 5mC (a) results in a replacement of the amino by a hydroxyl group. The intermediate *enol*-type product (b) undergoes a tautomeric transition into the more stable *keto*-type product, thymine (c). Reprinted from Cooper and Krawczak (1993).

mutable. One consequence of the hypermutability of CpG is the striking paucity of CpG in vertebrate DNA ('CpG suppression'; Bird, 1980). Only 1% of vertebrate genomes consists of regions which are comparatively rich in CpG, and these so-called 'CpG islands' probably represent the last remnants of the unmethylated domains that once predominated. CpG islands often coincide with transcribed regions (Bird, 1986). At the chromosomal level, they are primarily found in early replicating (R band) as opposed to late replicating (G band) sections (Craig and Bickmore, 1994).

An estimate of the *in vivo* rate at which 5mC is deaminated and fixed as thymidine was arrived at by extrapolation from *in vitro* data ($1.66 \times 10^{-16}\,sec^{-1}$; Cooper and Krawczak, 1989). This estimate, later confirmed using a different mathematical approach (Cooper and Krawczak, 1993), turned out to be consistent with the evolutionary pattern of CpG substitution in the α-globin gene and pseudogene sequences of human, chimpanzee and macaque (Cooper and Krawczak, 1989). Mathematical modelling of the dynamics underlying vertebrate CpG suppression revealed that the time span required in order to create the currently observed CpG frequency (0.01) was 50–100 million years (Myr) but that CpG loss must have been going on for approximately 450 Myr in order for the mononucleotide (C and G) frequencies to attain their current level (Cooper and Krawczak, 1989). Remarkably, this time span corresponds closely to the estimated time since the emergence and adaptive radiation of vertebrates and thus coincides with the probable advent of their heavily methylated genomes.

At the level of the gene, a high rate of C→T transitions was first reported by Vogel and Röhrborn (1965) in their analysis of human haemoglobin variants. Later studies (Vogel and Kopun, 1977) confirmed the existence of this phenomenon. Since then, compelling evidence has accumulated that CpG is indeed a hotspot for mutation in human genes. Of the 5864 non-identical single base-pair substitutions logged in the Human Gene Mutation Database (http://www.cf.ac.uk/uwcm/mg/hgmd0.html) by October 1996, 29.8% involved a CpG dinucleotide, and 79.1% of these (23.6% of the total) were either C→T or G→A transitions (the latter arise as a result of C→T transitions on the antisense DNA strand and subsequent miscorrection, G→A, on the sense strand). These data imply that the rate at which CpG mutates to either TpG or CpA is approximately six times higher than the base mutation rate. It should be remembered, however, that CpG hypermutability on its own represents only indirect evidence for CpG methylation. That 5mC deamination is itself directly responsible for mutational events is evidenced by the fact that several cytosine residues known to have undergone germline mutation in the human low-density lipoprotein receptor (*LDLR*) and tumour protein p53 (*TP53*) genes are indeed methylated in this tissue (Rideout *et al.*, 1990).

The proportion of human gene mutations compatible with a model of methylation-mediated deamination is very much an average figure and provides no information on individual genes. When recalculated on a gene-specific basis, considerable variation becomes apparent. Inspection of the Human Gene Mutation Database reveals that the frequency of CG→TG and CG→CA mutations may be less than 10% [*HBB* (3%), *HPRT* (5%), *TTR* (9%)] or more than 50% [*MYH7* (54%), *ADA* (59%)]. In the assumed absence of a detection bias, it may be surmised that this variation is due either to differences in germline DNA methylation and/or relative intragenic CpG frequency. However, Green *et al.* (1990) found a disproportionately high number of CpG mutations at conserved residues of the *F9* gene causing haemophilia B. This

excess was explained in terms of selection retaining CpG dinucleotides at these locations despite the attendant high risk of deleterious mutation. The authors concluded
that consideration of the ratio of CpG to non-CpG mutation might lead to an overes-
timation of the intrinsic mutability of CpG dinucleotides. However, the fact that similar estimates for the relative mutability of CpG emerged from clinical and evolutionary studies (Cooper and Krawczak, 1989) argues strongly against the validity of this
postulate. Furthermore, Koeberl *et al*. (1990) and Wacey *et al*. (1994) have shown that
F9 missense mutations are over-represented at evolutionarily conserved residues in
general, and not only at those containing CpG.

A possible ethnic difference in mutability of specific CpG sites has been mooted for
the *F8* gene by Pattinson *et al*. (1990) who noted a disproportionate number of mutations at these sites in individuals of Indo-Pakistani origin as compared with European
Caucasians. By contrast, the pattern of germline CpG mutation in the *F9* gene appears
to be indistinguishable between Asians, mostly of Korean origin, and Caucasians
(Bottema *et al*., 1990). Consistent with the latter finding, analysis of methylation patterns exhibited by a variety of DNA sequences gave no indication of frequent differences in methylation between individuals from different ethnic backgrounds (Behn-
Krappa *et al*., 1991).

2.2.2. Other mechanisms of germline single base-pair substitution

DNA replication is an accurate yet error-prone multistep process. Its final precision
depends very much upon both the initial fidelity of the replicative step and the efficiency of subsequent error correction. Since DNA polymerases are involved in replication, recombination and repair, their accuracy is likely to be a critical factor in
determining the rate of cellular mutation. Comparison of the relative single base-pair
substitution rates underlying human inherited disease with *in vitro* measured error
rates of vertebrate DNA polymerases revealed a significant positive correlation for
polymerase β, but not for polymerases α and δ (Cooper and Krawczak, 1993). It
should be noted that the efficacy of proofreading and post-replicative mismatch repair
has been neglected in this comparison since purified polymerases *in vitro* lack the 3'
and 5' exonuclease activities thought to be responsible for proofreading *in vivo*.
Nevertheless, the result obtained for polymerase β is consistent with the view that a
substantial proportion of germline nucleotide substitutions results from misincorporation during DNA replication.

An alternative model of single base-pair mutagenesis has been suggested which
seeks to explain nucleotide misincorporation through transient misalignment of the
primer–template complex caused by the looping-out of a template base ('slipped mispairing'; Kunkel, 1985; Kunkel and Soni, 1988). During replication synthesis, the
template strand slips back one base, resulting in the misincorporation of the next
nucleotide on the primer strand. After realignment of both primer and template
strand, the mismatch may be corrected in favour of the misincorporated base. If misincorporation mediated by one-base slippage was an important mutational mechanism
in vivo; however, a substantial proportion of single base-pair substitutions should
exhibit identity of the newly introduced base to one of the bases flanking the mutation
site. Since this phenomenon was not observed in a large sample of mutations associated with human genetic disease (Cooper *et al*., 1995), and is still not apparent employing data from the Human Gene Mutation Database, one-base slippage is unlikely to
play a prominent role in germline single base-pair mutagenesis in humans.

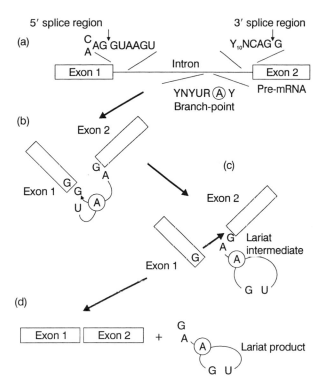

Figure 2.2. Basic model of mRNA splicing in higher eukaryotes. (a) pre-mRNA from an archetypal gene with two exons and intervening intron. Consensus sequences of the splice sites and branch region are given. (b) The 2′ hydroxyl group of the branch-point A nucleotide (circled) attacks the 5′ phosphoryl end of the first intron nucleotide. (c) The 3′ hydroxyl group of the last nucleotide of exon 1 attacks the 5′ phosphoryl end of the first nucleotide of exon 2. (d) Spliced mRNA plus lariat product. Reprinted from Cooper and Krawczak (1993).

Several other explanations for base misincorporation may be considered. Depurination, the cleavage of the N-glycosylic bond connecting a purine to the DNA 'backbone', occurs spontaneously through hydrolysis, may be caused by chemical agents, or can occur during DNA repair. Since DNA polymerases misincorporate bases opposite abasic sites during replication, this could in principle represent an important cause of spontaneous mutation. The most frequently incorporated base at abasic sites is deoxyadenosine (Kunkel, 1984; Takeshita et al., 1987). Thus, if depurination was a major cause of germline mutation in human genes, the observed mutational spectrum would be expected to be heavily biased towards G/A→T and C/T→A substitutions. Analysis of the data logged in the Human Gene Mutation Database reveals, however, that this is not the case. Another chemical mutation mechanism is through damage by oxygen free radicals, a major product of which is 8-hydroxyguanine. Cheng et al. (1992) demonstrated that, depending upon whether this derivative serves as a template or a substrate during DNA replication, 8-hydroxyguanine induces G→T and A→C transitions in E. coli. Since neither type of mutation is particularly over-represented in the Human Gene Mutation Database, however, oxidative damage is probably not a major mechanism of germline mutation in man.

A variety of DNA sequence motifs are known to play an important role in the breakage and rejoining of DNA, and these sequence elements could represent potential determinants of single base-pair mutagenesis. Only one triplet, however, was found to be over-represented in the immediate vicinity of single base-pair substitutions associated with human inherited disease: CTT, the topoisomerase I cleavage site (Cooper and Krawczak, 1993). The only two tetranucleotides found to be significantly frequent at mutational sites were TCGA (*Taq*I restriction site) and TGGA. The latter motif comprises the first four nucleotides of the deletion hotspot consensus sequence (Cooper and Krawczak, 1993) which in turn resembles the putative arrest site for DNA polymerase α. Thus the arrest or pausing of the polymerase at the replication fork may dispose the replication complex to misincorporation of nucleotides.

2.2.3 Single base-pair substitutions affecting mRNA splicing

Most eukaryotic genes are templates for the production of RNAs with large amounts of sequence (introns) interupting the actual coding sequence (exons). Introns are removed from the RNA prior to translation into protein in a process called 'splicing' (*Figure 2.2*). Splicing occurs in two stages: first, RNA is cleaved at the 5' end ('5' splice site') of the intron involving the so-called 'branch-point' sequence. This motif is located some 20–40 nucleotides upstream of the 3' end of the intron and bonds to the 5' end to drag it into a lariat-like structure. Then, RNA is cleaved at the 3' splice site and the free ends of the two flanking exons ligate. Both 5' and 3' splice sites are characterized by sequence motifs with considerable inter-intronic homology. The intronic dinucleotides immediately abutting the two neighbouring exons are virtually invariant (GT in all 5' splice sites, AG in all 3' splice sites).

Mutations which affect mRNA splicing fall into three main categories: (i) mutations within 5' or 3' splice sites which reduce the amount of correctly processed mature RNA and/or activate an alternative ('cryptic') splice site in the vicinity, (ii) mutations outwith actual splice sites creating cryptic splice sites, and (iii) mutations in the branch-point sequence. The vast majority of lesions reported so far have been single base-pair substitutions within splice sites, not only because these are comparatively frequent but also because they are both readily detectable and highly likely to result in a severe clinical phenotype. It was estimated that point mutations causing an mRNA splicing defect represent some 15% of all point mutations associated with human genetic disease (Krawczak *et al.*, 1992). This estimate has only slightly changed (12% in the Human Gene Mutation Database) and compares well with its theoretical expectation (Krawczak *et al.*, 1992).

Mutations affecting 5' splice sites are approximately twice as frequent as mutations at 3' splice sites (Krawczak *et al.*, 1992). This discrepancy coincides with a much higher level of sequence conservation at 5' splice sites and is likely to reflect a strong requirement of U1 small nuclear ribonucleic acid (snRNA) binding at 5' splice sites to promote alignment and cleavage. The 5' end of U1 snRNA is Watson–Crick homologous to the 5' splice site consensus sequence (Krainer and Maniatis, 1985). A mutation within the latter would destroy the homology and interfere with proper binding. A pattern consistent with this supposition emerged when the numbers of mutations observed at individual nucleotide positions within splice sites were compared with their expectation, based solely upon nearest neighbour-dependent substitution rates. The ratio of observed over expected was highest for the invariant dinucleotides and correlated well with the level of sequence conservation (Krawczak *et al.*, 1992). Regarding the phenotypic consequences of mutations affecting mRNA splicing, the faulty exclusion of one or more exons from the end-product ('exon skipping') was observed more frequently than the usage of cryptic splice sites. Some evidence exists, at least for 5' splice sites, that cryptic splice site usage is favoured under conditions where a number of potential sites are present in the vicinity of the mutated splice site and where these potential splice sites exhibit sufficient homology to the consensus sequence (Krawczak *et al.*, 1992).

Mutations creating new cryptic splice sites are more difficult to detect than mutations at actual splice sites. This is because comparatively few sequence data exist for introns and investigation of RNA (instead of DNA) would be a prerequisite for assessing the phenotypic consequences of a base change outside a coding region or splice site. Nevertheless, the available data indicate that, in most cases, the activating mutation improves the similarity between the cryptic site and the splice site consensus sequence. At 3' splice sites, the amount of mRNA product utilizing the cryptic splice site appears to be correlated with the level of similarity to the consensus sequences; at 5' splice sites, the distance to the nearest wild-type splice site may also play a role (Krawczak *et al.*, 1992).

2.2.4 Single base-pair substitution in human gene pathology and evolution

Single base-pair substitutions in human gene pathology and evolution may be viewed as two sides to the same coin. This appealing supposition has been corroborated by an in-depth comparison of mutations causing inherited disease with mutations in non-coding DNA that have become fixed during the evolutionary divergence of human and other mammals (Krawczak and Cooper, 1996). Whilst the latter mutations may date back many million years and their survival has been independent of natural selection, disease-associated substitutions are of fairly recent origin in comparison with the evolutionary timescale. This difference notwithstanding, the relative base substitution rates associated with human genetic disease were found to be remarkably similar to those arrived at using data from human gene/pseudogene alignments (Krawczak and Cooper, 1996). The only notable discrepancy was a slight under-representation among pseudogene mutations of C→T and G→A substitutions. This paucity is explicable in terms of either a lower level of germline methylation or, more likely, a more pronounced CpG suppression in non-coding DNA sequences.

That relative mutation rates were correlated between disease-associated and evolutionary mutations might appear to contradict experimental results. Hanawalt (1990) has shown, for example, that *in vitro*-induced pyrimidine dimers and interstrand DNA crosslinks are repaired with a substantially higher efficiency in active genes than in non-coding regions. Although these lesions were specific to particular exogenous mutagens, the idea of a generally more effective removal of endogenous mutations from coding DNA as opposed to non-coding DNA is appealing. This is because efficient DNA repair should only have conferred a substantial selective advantage in coding regions. However, comparison of evolutionary and disease data (Krawczak and Cooper, 1996) has revealed that the relative contribution (via variable efficiency) of different DNA repair pathways to the generation of mutations is unlikely to differ much between intragenic and intergenic sequences.

When codon substitutions were categorized according to whether they are neutral or whether they change the hydropathy or polarity of the encoded amino acid residue, it emerged that neutral substitutions have higher likelihoods of generation through mutation than non-neutral substitutions (Krawczak and Cooper, 1996). This disparity prompted the hypothesis that selection has operated on the cellular DNA repair machinery in such a way as to avoid disadvantageous mutations. Further evidence for such a bias came from the notion that the likelihood of generation of an amino acid substitution is negatively correlated with its likelihood of coming to clinical attention. Clinical observation likelihoods of amino acid substitutions, in turn, were found to be negatively correlated with a quantity called 'relative evolutionary acceptability', calculated from more than 300 human–rodent complementary DNA (cDNA) sequence alignments. Thus, the more likely a given amino acid substitution is to result in a disease phenotype, the less often has it been tolerated during the evolution of human and rodent protein sequences. Taken together, these observations strongly suggest that the evolutionary requirement to avoid a deleterious phenotype has left its footprints in the mechanisms of mutation generation. Of course, the most promising target for selection would appear to be the intracellular DNA repair mechanisms. If the efficiency of mutation removal were directed by the immediate DNA sequence context of a lesion, it may be that this has facilitated the avoidance during evolution of deleterious amino acid replacements by 'consideration' of the genetic code.

2.3 Deletions

2.3.1 Gross gene deletions

Gross gene deletions may arise through a number of different mechanisms, including unequal recombination between homologous DNA sequences and non-homologous recombination between sequences with little or no homology. Perhaps the best understood example of repetitive DNA-mediated gene deletion is provided by *Alu* sequences. *Alu* sequences are primate-specific DNA sequence elements approximately 300 bp in length of which up to 900 000 copies are thought to exist in the human genome at an average spacing of 4 kb (Hwu *et al.*, 1986; Kariya *et al.*, 1987). Deletion breakpoints have been found to be flanked by *Alu* sequences in a considerable number of human genetic conditions, involving defects of the genes for α- and β-globin (Henthorn *et al.*, 1986; Nicholls *et al.*, 1987), apolipoprotein B (Huang *et al.*, 1989), β-hexosaminidase (Myerowitz and Hogikyan, 1987), and many others. *Alu* sequence-mediated deletion can be of either of three types: (i) recombination between an *Alu* repeat and a non-repetitive DNA sequence with or without homology to the *Alu* repeat, (ii) recombination between *Alu* sequences oriented in opposite directions, or (iii) recombination between *Alu* sequences oriented in the same direction.

The most dramatic examples of the involvement of *Alu* sequences in deletion events occur in the *LDLR* and C1 inhibitor (*C1I*) genes. Almost all known deletion breakpoints in the *LDLR* gene occur within an *Alu* repeat sequence. In the vast majority of cases, the *Alu* repeats occur in the same orientation. In the *C1I* gene, *Alu* repeats represent more than a third of the intronic sequence (Carter *et al.*, 1991). Deletions and partial deletions account for 15–20% of lesions in this gene, causing type 1 hereditary angioneurotic oedema, and a much higher than expected proportion of deletions occur within *Alu* sequences (Stoppa-Lyonnet *et al.*, 1991). There are, however, examples which suggest that the extent of association between deletion breakpoints and *Alu* sequences is dependent upon the gene under study. Henthorn *et al.* (1990) collated data on over 30 deletions in the β-globin cluster, but noted the presence of *Alu* sequences at only four breakpoints. In the case of β-globin, it could be argued that this was due to the relative paucity of *Alu* sequences (only eight copies in 60 kb) within the corresponding gene cluster. Of 130 patients with Fabry disease, however, only five possessed a partial deletion of the associated α-galactosidase A (*GLA*) gene (Kornreich *et al.*, 1990); three breakpoints occurred within an *Alu* sequence. This is a rather small figure considering that some 30% of the *GLA* gene comprises *Alu* repeat sequence.

Most gene deletions are not associated with *Alu* repeat sequences and other types of repetitive sequence element are also thought to be capable of mediating homologous unequal recombination. For example, Yen *et al.* (1990) reported that in 24 of 26 ichthyosis patients with a deletion in their steroid sulphatase (*STS*) gene, breakpoints clustered at the S232-type repetitive sequences flanking the gene. This finding suggested that the high frequency of deletion at the *STS* locus (90% of ichthyosis patients) may be due to recombination between S232 sequences. The long terminal repeats of the RTV_L-H family have also been found to mediate homologous unequal recombination events (Mager and Goodchild, 1989). Finally, sequence analysis of deletion breakpoints located within the intron 43 deletion hotspot in the dystrophin genes of two unrelated Duchenne muscular dystrophy patients has revealed the presence of a transposon-like element belonging to the THE-1 family (Pizzuti *et al.*, 1992), indicating a possible involvement in deletion mediation.

2.3.2 Short gene deletions

Short human gene deletions (less than 20 bp) have been analysed by Cooper and Krawczak (1993) in a search for DNA sequence motifs in the vicinity of these lesions, trying to discern mechanisms responsible for their generation. Many mechanistic models for deletion mutagenesis have been proposed but no one model alone appears to be sufficient to account for the spectrum of small deletions observed in human genes. In many cases, slipped mispairing at the replication fork between homologous sequences in close proximity to one another on complementary DNA strands appears to be the causative mechanism (*Figure 2.3*). Slipped mispairing is thought to be potentiated by the presence of either direct repeats, inverted repeats or symmetric elements, and all three sequence features are indeed found to be associated with small deletions (Krawczak and Cooper, 1991; Cooper and Krawczak, 1993).

A consensus sequence, TGRRKM ($R=^A/_G$, $K=^T/_G$, $M=^A/_C$), has been identified which appears to be common to micro-deletion 'hotspots' in human genes (Krawczak and Cooper, 1991). This consensus sequence is strikingly similar to the core motifs, TGGGG and TGAGC, of the tandemly repeated immunoglobulin switch regions (Gritzmacher, 1989) and to putative arrest sites for DNA polymerase α, which often contain a GAG motif (Weaver and DePamphilis, 1982). Polymerase arrest sites may fulfil one or more important biological functions such as the initial localization of the replication fork, the synchronization or termination of DNA replication, or the promotion of recombination at the replication fork. All these factors render the involvement of DNA polymerase α arrest sites in the generation of small human gene deletions a biologically meaningful scenario. A more elaborate search for associated DNA sequence motifs revealed that, in addition to the TGRRKM motif, only polypyrimidine runs of at least 5 bp (YYYYY) appear to be over-represented in the vicinity of short human gene deletions. In contrast to these findings, however,

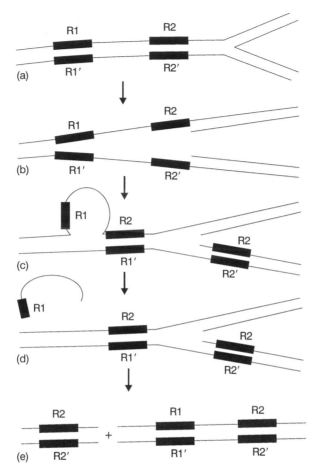

Figure 2.3. The slipped mispairing model for the generation of deletions during DNA replication. (a) Duplex DNA containing direct repeat sequences. (b) Duplex becomes single-stranded at replication fork. (c) R2 repeat base-pairs with complementary R1′ repeat producing a single-stranded loop. (d) Loop excised and rejoined by DNA repair enzymes. (e) Daughter duplexes, one of which contains only one of the two repeats and lacks the intervening sequence between R1 and R2. Reprinted from Cooper and Krawczak (1993).

Monnat *et al.* (1992) observed a significant association with hypoxanthine guanine phosphoribosyltransferase gene deletion breakpoints only for CTY vertebrate topo-isomerase I cleavage sites.

2.4 Other mechanisms of mutation

2.4.1 Gene conversion

Gene conversion is the 'modification of one of two alleles by the other' (Vogel and Motulsky, 1986). The end result is very similar to that consequent to a double unequal crossing-over. The difference between the two processes is, however, that the 'correction' of one allele brought about by gene conversion is non-reciprocal, leaving the other allele physically unchanged. The process of gene conversion may involve the whole or only a part of a gene and occurs not only between allelic genes but also between highly homologous but non-allelic genes. Examples of the latter include the G-γ and A-γ globin genes (Shen *et al.*, 1981; Stoeckert *et al.*, 1984; Powers and Smithies, 1986) and the α1- and α2-globin genes (Liebhaber *et al.*, 1981). The mechanism underlying gene conversion remains elusive but presumably entails the close physical interaction between the homologous DNA sequences. Gene conversion has been invoked in instances where it is necessary to account for the association of the same disease-causing mutation with two or more different haplotypes [e.g. the β^E-globin alleles in south-east Asian populations (Antonarakis *et al.*, 1982) and the β^s-globin mutation in African populations (Antonarakis *et al.*, 1984)]. Gene conversion has also had important effects upon the evolution of multigene families. It may act so as to homogenize the sequences of repeated genes, such as the primate immunoglobulin Cα gene segments (Kawamura *et al.*, 1992), or may alternatively promote diversity by introducing sequence changes into different members of gene families (e.g. in the major histocompatibility complex; Ohta, 1991a)

2.4.2 Insertion

A substantial proportion of mutations causing human genetic disease involve short sequence duplications or insertions, usually resulting in an alteration of the reading frame of the encoded protein and most often causing termination of translation at some distance downstream. That insertional mutagenesis might be as intrinsically specific as point mutations and gene deletions was strongly suggested by the findings of Fearon *et al.* (1990), who reported 10 independent examples of DNA insertion within the same 170 bp intronic region of the *DCC* gene (a locus which has been proposed to play an important role in human colorectal neoplasia). Cooper and Krawczak (1991) studied the nature of short insertions in human genes causing inherited disease. Although their sample size was too small to identify mutational hotspots, two broad conclusions nevertheless emerged: (i) the likelihood of insertional mutagenesis is highly dependent upon the local DNA sequence context, and (ii) mechanistic models which have explanatory value in the context of short gene deletions, such as slipped mispairing mediated by repetitive sequence motifs, also appear to account for the nature of small gene insertions.

At the level of gross lesions, a number of examples of gene inactivation through the insertion of *Alu* or long interspersed nuclear elements (LINEs) into gene coding sequences are known (Cooper and Krawczak, 1993). Both *Alu* sequences (Kariya *et al.*, 1987) and LINEs (Kazazian *et al.*, 1988) exhibit a preference for integration at AT-rich sequences, an observation which was reminiscent of the AT-rich insertional target sequences of retroviruses (Shih *et al.*, 1988; Umlauf and Cox, 1988). Indeed, the two LINE element target sites in the *F8* gene (Kazazian *et al.*, 1988) are 80% homologous to a 10 nucleotide motif (GAAGACATAC) present in one of the highly favoured retroviral insertion target sequences reported by Shih *et al.* (1988). If the generality of this observation is supported by further data, then gross insertions are as sequence-specific as the insertions of a few base pairs.

2.4.3 Gene duplication

The duplication of either whole genes or their constituent exons contributes significantly to human genome pathology (Hu and Worton, 1992). Two distinct mechanisms are currently envisaged: (i) homologous unequal recombination either between homologous chromosomes or sister chromatids, and (ii) non-homologous recombination at sites with minimal homology. Several possible topoisomerase I cleavage sites have been noted in the vicinity of the *F8* gene duplication described by Casula *et al.* (1990). This observation is potentially interesting since topoisomerase activity has been implicated in several cases of non-homologous recombination (Bullock *et al.*, 1985). Other examples of topoisomerase cleavage sites have been reported to be associated with gene duplications (Kornreich *et al.*, 1990; Hu *et al.*, 1991); potential sites for topoisomerases I and II were found exactly coincident with the breakpoints of a duplication in the dystrophin gene (Hu *et al.*, 1991). The significance of these findings remains to be elucidated.

2.4.4 Mutation of repetitive DNA

Although not yet directly relevant to venous thrombosis, a mutational mechanism of particular clinical importance should be mentioned here which involves the instability of certain trinucleotide repeats (reviewed by Caskey *et al.*, 1992; Rousseau *et al.*, 1992; Willems, 1994; Sutherland and Richards, 1995). This mechanism was first reported in the context of fragile X syndrome, the most frequent cause of inherited mental retardation. In the 5' untranslated region of the gene underlying fragile X syndrome (*FMR1*; Verkerk *et al.*, 1991), a highly unusual CGG repeat array was found to exhibit length variation which correlates with the severity of the disease phenotype (Oberlé *et al.*, 1991; Yu *et al.*, 1991). Whilst the bulk of the normal healthy population exhibits 25–35 repeat copies, affected males have more than 300 (Dobkin *et al.*, 1991).

A total of 10 human disorders/phenotypes have so far been found to be associated with triplet repeat expansion. These include four fragile sites, all of which are due to the expansion of CGG tracts to more than 200 repeats. A second category relates to five neurodegenerative disorders (Huntington disease, spinocerebellar ataxia type 1, dentatorubral pallidoluysian atrophy, Machado-Joseph disease, and Kennedy disease) in which a CAG trinucleotide cluster is expanded within the open reading frame of a gene. All five diseases are progressive, autosomal, have late onset and are characterized by similar neuropathological features. The repeat number above which meiotic instability becomes apparent is approximately 35 in all five diseases; the upper limit for expansion appears to be 100 repeats. Myotonic dystrophy, a progressive disorder of muscle weakness inherited as an autosomal dominant trait, is as yet the only representative of a third group of triplet disorders. Expansion affects a CTG repeat sequence within the 3' untranslated region of a putative serine–threonine protein kinase gene (Brook *et al.*, 1992; Harley *et al.*, 1992). The repeat number varies from 5 to 37 repeats in controls (Brunner *et al.*, 1992; Davies *et al.*, 1992) and is more than 50 in patients.

The process of triplet repeat expansion has been termed 'dynamic mutation' (Richards *et al.*, 1992), since the probability of mutation increases with repeat number and transition of a normal allele to a clinically significant copy number can involve a number of steps. As to the *in vivo* mechanism of instability manifested by the repeats, slipped mispairing is one possibility. This idea is supported by the observed expansion of the repeat in multiples of 3 bp. Nevertheless, it has become clear that the precise mechanism of change in repeat copy number is complex and may differ from one repeat to another. Richards and Sutherland (1994) suggested that relatively small changes in the normal range of triplet repeat clusters result from replication slippage of Okazaki fragments still anchored at their 5' end by unique sequence flanking the repeat array. If the number of (uninterrupted) repeats exceeds a certain threshold, however, strand breakage becomes likely to occur at both ends within the repeat sequence. Then, the Okazaki fragment is no longer anchored but can slip and slide freely during polymerization, thereby giving rise to much larger expansions. Kuryavyi and Jovin (1995) proposed a novel structural model of DNA (and possibly RNA), called 'triad-DNA', which consists of an antiparallel double helix of base triads instead of the base pairs of conventional B-DNA. Such a structure could provide a stable self-structured conformation for the extruded strand in a slippage event, serving to stabilize the intermediate in the pathological expansion of a triple repeat sequence.

2.5 The study of pathological mutations at the protein level: human factor IX

The nature, frequency and location of gene lesions causing human genetic disease are highly specific and, as outlined in previous sections of this chapter, are determined in part by the local DNA sequence environment. Once a given mutation has arisen, however, the likelihood that it will come to clinical attention is a complex function of the nature of the resulting amino acid substitution, its precise location and immediate environment within the protein molecule, and its effects upon protein structure and function. Whilst in-depth investigations of the *in vivo* effects of missense mutations upon specific human proteins are generally rare, two such studies have nevertheless been performed for human factor IX (Bottema *et al.*, 1991; Wacey *et al.*, 1994), the liver expressed zymogen of a vitamin K-dependent serine protease which activates factor X in the presence of factor VIIIa (Reiner and Davie, 1995). These studies are described in detail here not only because they reveal some of the effects mutations can have on proteins but also because they might anticipate future results for the lesions described in Chapters 5 and 6, which deal with the homologous proteins C and S.

The vast majority of known lesions in the *F9* gene causing haemophilia B are missense mutations (Giannelli *et al.*, 1996), causing approximately 59% of severe (less than 1% FIX: C) and moderate (1–5% FIX: C) haemophilia, and perhaps as much as 97% of mild (more than 5% FIX: C) haemophilia (Sommer *et al.*, 1992). On the basis of 95 independent missense mutations, Bottema *et al.* (1991) concluded that substitutions of 'generic' factor IX residues (conserved in factor IX of other mammals and in related human serine proteases) almost invariably cause haemophilia B. Mutations at factor IX-specific residues (conserved only in mammalian factor IX) and non-conserved residues, by contrast, were found to be some six- to 33-fold less likely to result in a disease phenotype. Even though the study of Bottema *et al.* (1991) provided new insights into the identity of amino acid residues of structural or functional importance to factor IX, the authors did not employ models of the tertiary structure of the protein or its constituent domains. The significance of the location of specific amino acid residues within the structure of the factor IX molecule to the consequences of mutation could thus not be assessed. Moreover, neither the variable propensity of different regions of the *F9* gene to mutate nor the nature of the resulting amino acid exchanges were considered.

Factor IX in many ways represented an ideal system in which to assess the influence of positional determinants upon the disease-associated mutational spectrum of a single protein. Firstly, the number of known *F9* missense mutations has always been among the highest of all human genes (347 different; current Human Gene Mutation Database). Secondly, the amino acid sequences of numerous other vertebrate factor IX proteins and evolutionarily related serine proteases have been available for direct comparison (Sarkar *et al.*, 1990; Bottema *et al.*, 1991). Finally, although the structure of factor IX has only recently been determined by X-ray crystallography (Brandstetter *et al.*, 1995), the three-dimensional (3D) structures of a number of homologous serine proteases were known for quite some time. These conditions prompted Wacey *et al.* (1994) to construct, by comparative methods (Swindells and Thornton, 1991), a multi-domain model of the quaternary structure of activated factor IX (FIXa) and to use this model in the analysis of the expression pathway of *F9* gene lesions from geno-type to clinical phenotype.

A total of 277 different single base-pair substitutions in the *F9* gene, comprising 241 missense mutations and 36 nonsense mutations, were analysed by Wacey *et al.* (1994). Comparison of the relative nearest neighbour-dependent single base-pair sub-stitution rates in the *F9* gene with estimates from a wide range of other human genes (Cooper and Krawczak, 1993) revealed similar profiles, suggesting that similar muta-tional mechanisms were operating at the DNA level. In particular, CpG dinucleotides turned out to be hotspots for *F9* gene mutation, too (see Section 2.2.1).

Wacey *et al.* (1994) classified missense mutations as either *conservative* or *non-conservative* on the basis of the chemical difference between the wild-type and the mutant amino acid residue. Chemical difference is a measure originally devised by Grantham (1974), combining the three interdependent properties of composition, polarity and molecular volume of an amino acid residue. When the magnitude of bio-chemical change upon substitution was measured in this way and related to the clini-cal severity of the resulting disease phenotype, non-conservative substitutions (i.e. those characterized by large chemical differences) were found to result in severe rather than mild or moderate haemophilia B approximately 1.7 times more often than conservative substitutions. In turn, changes of the latter type were three to four times more likely to cause a moderate or mild rather than a severe haemophilia B phenotype than their non-conservative counterparts.

The possibility was considered that conservative amino acid substitutions might be more likely to come to clinical attention in the tightly packed core of the protein and were more readily tolerated on the surface of the molecule. However, since mutations from all chemical difference classes appeared to be scattered over all domains of the protein and since no one chemical difference class was found to be associated solely with surface or buried residues, there appears to be no relationship between the mag-nitude of the amino acid exchange and the location of the affected residue.

The extent of evolutionary sequence conservation exhibited by an amino acid residue in factor IX was also found to correlate with disease severity. Whilst 71% of mutations at residues conserved in all mammalian factor IX proteins and in three other serine proteases ('highly conserved') caused severe rather than mild or moder-ate haemophilia B, this was the case for only 50% of mutations at less conserved residues. Furthermore, Wacey *et al.* (1994) estimated that missense mutations at non-conserved residues are 15–20 times less likely than mutations at conserved residues to result in a disease phenotype at all. Although this implies that many missense muta-tions of evolutionarily unconserved residues are tolerated by the molecule and do not come to clinical attention, the relative importance of such residues was found to be greater than previously perceived (Bottema *et al.*, 1991). Several explanations were suggested for this discrepancy. Firstly, the sample of mutations used by Wacey *et al.* (1994) was three times larger than that of Bottema *et al.* (1991). Secondly, these authors allowed for determinants neglected by Bottema *et al.* (1991), that is, the actual *F9* gene sequence and the redundancy of the genetic code. Finally, the two studies were not directly comparable since Wacey *et al.* (1994) confined their estimation of clinical observation likelihoods to severe cases of haemophilia B in order to cope with the problem of identical-by-descent mutations.

Amino acid residues which are sequence-conserved in mammalian factor IX proteins and four different human serine proteases are likely to be critical for functions common to all serine proteases. These residues are located predominantly in the interior of factor IX, within α helices or β turns, and substitutions cluster in the Gla domain, the EGF domains and the serine protease domain (around the reactive site and oxyanion hole). All but one of the Cys residues known to be involved in disulphide bonding were affected by mutation as were the reactive site residues, residues involved in carboxylase recognition and activation peptide cleavage site, and residues which contribute to factor X binding. Mutations at factor IX residues which are sequence-conserved in mammals and one or two other serine proteases cluster in the serine protease domain and also at domain boundaries within the protein structure. Presumably, docking of the constituent domains of factor IX and the other homologous serine proteases during protein folding requires amino acid conservation at these boundaries. Finally, residues which are sequence-conserved between mammals but not between other serine proteases appear to be exclusively important for the structure and function of factor IX. When the three-dimensional structure of the protein was considered, spatially clustered groups of mutations at such residues became apparent on the surface of the EGF domains, in regions implicated in the binding of factors Va and VIII, and in the serine protease domain.

Regions of similar functionality in different homologous proteins may not only be sequence-conserved but may also exhibit structural conservation (Greer, 1990). Although sequence-conserved regions of mammalian serine proteases are invariably structurally conserved, structurally conserved regions (SCRs) may differ markedly with respect to their amino acid sequences. SCRs were defined by Greer (1990) as those portions of known protein structures that overlap very well when superimposed. In serine proteases, SCRs usually comprise secondary structure elements, the active site and other essential structural framework residues of the molecule.

Structural conservation in human factor IXa was determined by Wacey *et al.* (1994) utilizing their computer-derived protein model. No clear relationship was noted between the severity of the haemophilia B phenotype and the level of structural conservation of a mutated factor IX amino acid residue, although substitutions at structurally conserved residues were estimated to have an approximately two-fold higher likelihood of resulting in a disease state than mutations at non-conserved residues. Furthermore, mutated sites which were not sequence-conserved were nearly all structurally conserved. The only exception involved two missense mutations at Gly59. However, Gly59 lies immediately adjacent to a type β hairpin SCR and would be predicted to be critical in defining this structural element (Swindells and Thornton, 1991). Some SCRs, although not sequence-conserved, therefore probably serve as structural supports through their backbone interactions and should be regarded as 'scaffolding' residues rather than 'spacers' (Bottema *et al.*, 1991).

The topological properties of a mutated factor IX amino acid residue are also important for determining clinical severity. Mutations at residues with their side chains pointing away from the solvent ('buried residues') were found to cause severe, rather than mild or moderate, haemophilia B 1.5 times more often than mutations at residues with solvent-accessible side chains (Wacey *et al.*, 1994). A more refined analysis revealed that the likelihood of resulting in a severe disease phenotype was higher for substitutions in hydrophobic as opposed to polar regions, probably because of the critical importance of these residues for correct protein folding. The observations were consistent with the conclusions of other workers that amino acid substitutions occurring in the protein core give rise to a 'continuum of increasingly non-native properties' affecting the stability and/or the folding dynamics of the protein (Alber, 1989; Pakula and Sauer, 1989; Lim *et al.*, 1992).

Evolution of the vitamin K-dependent coagulation factors

3.1 Introduction

The vitamin K-dependent serine proteinases of coagulation (factors VII, IX and X, prothrombin and protein C) exhibit substantial sequence and structural homology (Greer, 1990). They all contain an N-terminal domain of glutamic acid (Gla) residues which becomes carboxylated by a liver microsomal enzyme system in a post-translational modification process, requiring the presence of vitamin K. Other sequence similarities and homologies have also been recognized. Thus, factors VII, IX, X and protein C all contain two epidermal growth factor-like (EGF) domains and a catalytic domain (Blake *et al.*, 1987); prothrombin differs slightly in that it possesses two kringle domains instead of the EGF domains (Gojobori and Ikeo, 1994). The functional correlates of these motifs are beginning to be unravelled although, as yet, little is known of their specific functions.

With the exception of prothrombin, the genes encoding the vitamin K-dependent coagulation factors are of very similar exon/intron organization (Tuddenham and Cooper, 1994), suggesting that they have arisen from a relatively recent common ancestor through a process of gene duplication and divergence (Neurath, 1984; Patthy, 1985, 1990). The organization of these genes also reflects the functional modular assembly of the respective proteins. Therefore, the presence/absence of modules such as the calcium-binding and EGF-like domains or kringles could be used to reconstruct the evolutionary past of the genes. The transfer of modules between genes has presumably arisen via exon shuffling. Patthy (1990) has described at length the principles underlying the assembly of present-day coagulation factor protein genes from their constituent modules. Evolution of these genes has proceeded by repeated insertions, duplications, exchanges and deletions of modules. A diagram illustrating the phylogeny of vitamin K-dependent coagulation factors as deduced by Patthy (1990) is presented in *Figure 3.1*.

How has it been possible to produce a plethora of different proteins/genes by exon shuffling in a comparatively short period of evolutionary time? The answer appears to lie with the close correspondence between exon boundaries and modular domains and with the fact that the ancestors of the important modules all had so-called 'phase 1' introns (codons not interrupted by splice site) at both module boundaries. Phase

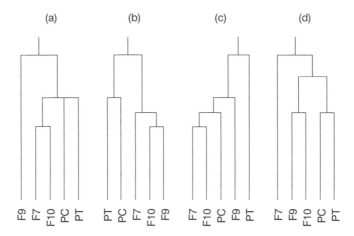

Figure 3.1. Proposed phylogenies for the vitamin K-dependent factors of coagulation. (a) Krawczak *et al.* (1996), (b) Doolittle (1993), (c) Doolittle and Feng (1987), (d) Patthy (1990).

compatibility between modules, on the other hand, is essential to maintain the reading frame and thus the functionality of a novel protein product.

Doolittle (1993) has proposed the following tentative scheme for the evolution of the vitamin K-dependent coagulation factors (*Figure 3.1*). First, an ancestral prothrombin emerged by serine proteinase duplication and the acquisition of Gla and EGF domains. Prothrombin's EGF domain(s) served as a site for the binding of tissue factor which at this time possessed the ability to activate it to yield thrombin. After the emergence of fibrinogen, factor X appeared as a result of a prothrombin duplication. The ability of factor X to activate prothrombin released the latter from its dependence on tissue factor. Factor VII, duplicated from factor X, was able to bind tissue factor and to activate factor X. Factor IX emerged last, again duplicated from factor X. Prothrombin then acquired kringle domains via exon shuffling (Rogers, 1985) allowing it to bind fibrin. The concomitant loss of prothrombin's EGF domains abolished its now redundant interaction with tissue factor.

3.2 Reconstruction of mammalian ancestral cDNAs

Proteins from extinct organisms can be studied by the analysis of DNA recovered from preserved organic specimens. This approach, however, requires biological material that has been completely protected from the oxidative effects of oxygen and water (DeSalle *et al.*, 1992; Woodward *et al.*, 1994). Alternatively, the amino acid sequences of an ancient protein can be reconstructed from the sequences of its extant descendants. This methodology relies upon the principle of maximum parsimony (i.e. the assumption that extant proteins have evolved from that of an extinct ancestor by the smallest number of mutational changes). Use of this strategy has permitted, for example, the investigation of ribosomal DNA phylogeny (Friedrich and Tautz, 1995). Adopting the maximum parsimony principle and employing a novel cDNA-based strategy, Krawczak *et al.* (1996) reconstructed the catalytic domains of the early mammalian ancestors of the vitamin K-dependent factors and, based upon their results, carried on to reconstruct the common ancestor of all five proteins from an earlier stage of vertebrate evolution.

The study of Krawczak *et al.* (1996) was based upon mammalian phylogenies from different sources (Vogel and Motulsky, 1986; Nei, 1987; Novacek, 1992) and included cDNA sequences from human, macaque, sheep, pig, rabbit, rat, mouse, dog and cow. Conserved cDNA blocks were used as anchors to align intervening non-conserved cDNA blocks for a particular protein and, since protein function must have been retained during evolution, frameshift mutations were precluded in all alignments. For each protein, the authors then deduced from the alignments the most likely cDNA sequence at each node of the employed mammalian phylogenies, including the roots representing the respective mammalian ancestors. A small number of ambiguities remained as to the ancestral mammalian cDNA sequences which were resolved allowing for relative nearest neighbour-dependent mutation rates in humans. Use of human-derived parameters in other mammals was justified by reference to a study demonstrating the long-term evolutionary stability of relative single base-pair substitution rates (Krawczak and Cooper, 1996).

The evolution of a gene family whose members acquire different functions is invariably accompanied by rapid amino acid sequence divergence in functionally important regions (Ohta, 1991b, 1994). Indeed, specific examples of this phenomenon include the 'accelerated evolution' (hypervariability) of the reactive centre regions of serine proteinase inhibitors (Hill and Hastie, 1987) and the active site regions of some serine proteinases (Creighton and Darby, 1989). For humans, the mutation rates arrived at in the reconstruction process were found by Krawczak *et al.* (1996) to be significantly higher in the factor VII, factor X and protein C lineages than in the factor IX and prothrombin lineages (*Table 3.1*). Similarly in the dog, the factor VII lineage exhibited a higher mutation rate than that of factor IX. That factor IX exhibited a lower mutation rate than the other proteins was indicative of its early emergence (see Section 3.3): once adapted to its functions, factor IX would have had to change rather less over evolutionary time than the other proteins of more recent origin which still had to adapt to their new-found roles.

For factor IX, protein C and prothrombin, mutation rate differences were also apparent between species and exhibited an inverse correlation with generation time (*Table 3.1*). Consistent with previous results (Britten, 1986; Collins and Jukes, 1994), the mutation rate in humans was thus much less than that found in rodents.

Table 3.1. Significant mutation rate differences (Krawczak *et al.*, 1996)

Species	Clusters	Chi square	α (1 df)
Between proteins within a species (seven comparisons)			
Hum	F9, PT < F10, F7, PC	12.5	4.1×10^{-4}
Dog	F9 < F7	8.3	3.9×10^{-3}
Between species within a protein (five comparisons)			
F9	Hum, Rbt, Dog < Shp, Pig, Rat, Mus	25.3	6.0×10^{-7}
PC	Hum, Cow < Mus, Rat	15.3	9.2×10^{-5}
PT	Hum < Cow, Mus, Rat	12.0	5.3×10^{-4}

F9, factor IX; PC, protein C; PT, prothrombin; Hum, human; Mus, mouse; Rbt, rabbit; Shp, sheep; α, error probability.

3.3 Evolutionary divergence

Krawczak et al. (1996) identified a number of highly conserved regions in their recon-structed ancestral mammalian cDNA sequences. Mismatches in these regions were classified on the basis of whether they corresponded to either a silent or a missense mutation during the process of evolutionary divergence (Table 3.2). Interestingly, the numbers of silent and missense mutations did not correlate with each other. This finding was interpreted in terms of the existence of two distinct molecular clocks: one would be based upon silent mutations and would run constantly after the divergence of any two sequences. The other would be based upon missense mutations and would continue to run immediately after gene duplication but would stop, or at least be dra-matically slowed, when the new protein product had first acquired and then become adapted to its new biological function.

Since by far the smallest number of silent mutations in conserved codons had occurred since the divergence of factors VII and X, Krawczak et al. (1996) concluded that these two proteins must have their most recent root in common (Figure 3.1). This assertion was further supported by the fact that the human factor VII (F7) and factor X (F10) genes are not only syntenic but are also very closely linked on chromosome 13q34 (Miao et al., 1992), consistent with their recent emergence through a process of duplication and divergence. Since the conserved cDNA blocks of factor IX exhibited the largest number of neutral differences to other proteins, this was held to imply that factor IX was the first to diverge from the other proteins. However, a potential pitfall with this conclusion could have been a faster rate of substitution at the X-linked fac-tor IX (F9) locus compared with the other, autosomal genes. However, at least for human, substitution rates in the F9 gene were significantly lower than in the factor F7, F10 and protein C (PROC) genes (Table 3.1), which also appeared to be the case in dog. Thus, divergence time rather than a higher propensity to mutate was held to be responsible for the large number of silent substitutions which separate the F9 cDNA sequence from the other cDNAs. The precise order of divergence events for protein C, prothrombin and the common ancestor of factors VII and X could not be clarified by Krawczak et al. (1996) on the basis of their data.

The cDNA-based phylogeny of the five vitamin K-dependent factors presented in Figure 3.1 emphasizes the previously recognized relatedness of: (i) protein C and pro-thrombin, and (ii) factors VII and X. However, it is markedly different from other proposed phylogenies (Patthy, 1985, 1990; Doolittle and Feng, 1987; Doolittle, 1993; Figure 3.1). This may have been because earlier attempts employed alignments of

Table 3.2. Mismatches noted in 50 codons which are highly conserved between the mammalian ancestral sequences of the five vitamin K-dependent factors

	Factor VII	Factor IX	Factor X	Protein C	Prothrombin
Factor VII	—	8	9	9	8
Factor IX	25	—	10	8	9
Factor X	8	19	—	10	13
Protein C	17	26	16	—	6
Prothrombin	17	17	14	18	—

Upper right half, minimum number of missense mutations per pair; lower left half, minimum number of silent mutations per pair.

amino acid rather than cDNA sequences. The phylogeny of Krawczak *et al.* (1996) was closest to that presented by Doolittle and Feng (1987): both phylogenies claimed a comparatively recent common root for the catalytic domains of factors VII and X. However, the major difference lies in the much earlier branching out of the catalytic domain of factor IX, postulated by Krawczak *et al.* (1996).

Gene duplication has played a very important role in the evolution of the genomes of higher organisms (Ohno, 1970). Indeed, there is now considerable evidence for saltatory increases in the number of genes around the time of the emergence of the vertebrates 500 million years ago (Bird, 1995). These increases appear to have been caused by the duplication and subsequent divergence of many different gene sequences including for example those of the *Hox* gene family. It is still impossible to say for certain, however, when the duplication and divergence of the vitamin K-dependent serine proteinases of coagulation occurred during vertebrate evolution. Prothrombin is present in bony fish (trout), cartilaginous fish (dogfish) and in the hagfish, one of the modern representatives of the jawless *Agnatha* (Banfield and MacGillivray, 1992; Doolittle, 1993). Although thrombin is thus found in the most primitive of modern vertebrates, there is no evidence for its existence in either proto-chordates or echinoderms. Whether the other four vitamin K-dependent factors of coagulation are present in these types of fish is as yet unclear (Doolittle, 1993). If they are, the adaptive radiation of the vitamin K-dependent factors of haemostasis must have occurred during the space of some 50 million years between the divergence of the protochordates and the appearance of the *Agnatha*, some 450 million years ago (Doolittle, 1993). The processes of gene duplication and divergence that have led to the emergence of the five present day vitamin K-dependent factors of coagulation pro-vide further evidence of what must have been a very active phase in the evolution of the modern vertebrate genome.

3.4 Reconstruction of a common ancestor of the vitamin K-dependent coagulation factors

The evolutionary scenario depicted in *Figure 3.1* is based solely upon the highly con-served regions of the reconstructed ancestral mammalian cDNA sequences. However, this phylogeny was used by Krawczak *et al.* (1996) to attempt to reconstruct the com-mon ancestor of all five vitamin K-dependent factors by means of molecular model-ling. To this end, the less well conserved regions of the mammalian ancestral sequences were first aligned in the order stipulated by the phylogeny. The alignment was then used to determine the nucleotides present at each node of the phylogeny, moving upwards through the tree to its root. The resulting cDNA sequence, repre-senting the common ancestor of all five vitamin K-dependent factors, contained sev-eral gaps which yielded ambiguities in the sequence of the ancestral protein. However, since the protein core had to contain a full complement of residues for meaningful molecular modelling to be possible, all such amino acids were replaced with the anal-ogous residue found in extant human thrombin as a 'best guess'.

One way to examine the plausibility of the deduced amino acid sequence of the vit-amin K-dependent factor ancestral protein would have been to express it *in vitro* and to characterize it biochemically (Malcolm *et al.*, 1990; Shih *et al.*, 1993). An alternative strategy was to construct a model of the tertiary structure of the protein by compara-tive methods and examine its topology and biophysical properties. Such a model was created by Krawczak *et al.* (1996) using the X-ray crystallographic coordinates of the heavy chain of α-thrombin as a template. This template was aligned against the high-resolution structure of seven other serine proteinases and the amino acid sequence of the reconstructed ancestral protein (Greer, 1990). Sequence modifications became necessary in this process at two variable regions, which were longer than the analo-gous loops present in extant serine proteinases, and a single unpaired Cys residue (Cys22), which was replaced by Ala as in extant human thrombin. Extant human thrombin was chosen for comparison in this and other instances since, of all of the modern haemostatic serine proteinases, it bore the strongest sequence homology to the ancestral protein. Moreover, it was assumed (Doolittle, 1993) that the vitamin K-dependent factor ancestral protein would have been capable of performing the end effector function of thrombin in the haemostatic cascade (i.e. the cleavage of soluble fibrinogen to generate insoluble fibrin clot).

The ancestral protein was found to contain 86 charged groups at pH 7.0, very sim-ilar to the number (89) found in extant human thrombin. In both cases, the large majority of these charges were accessible to water. The global electrostatic distribu-tion across the ancestral protein's surface was relatively uniform in comparison with the dipolar distribution evident in extant human thrombin. Both structures still pos-sessed an unbroken equatorial belt of negative charge but the extent of the electrosta-tic field strength in the ancient protein was not as strong as that of extant thrombin. The increase in charge intensity and dispersal over evolutionary time was speculated to reflect a trend toward increasing protein binding specificity.

The fibrinogen-binding exosite (P1) was found to have greatly increased in size during the evolution of thrombin. From the electostatic contour map, only five Arg (126, 165, 233) and three Lys residues (230, 243, 245) of the ancestral protein appeared to contribute to a small patch of positive charge (*Figure 3.2*). Moreover, sev-eral thrombin residues known to be important in the binding of fibrinogen (Tsiang *et al.*, 1995) were not present in the ancient protein, suggesting that the ancestral pro-teinase bound fibrinogen only weakly. A number of residues in thrombin have been shown to be involved in the binding of protein C (Tsiang *et al.*, 1995) and these are also components of the fibrinogen-binding site (Lys36, Trp60D, Lys70, His71, Arg73, Tyr76, Arg77A, Lys81, Lys109, Lys110, Glu217, Arg221A). However, only a fraction of these residues were present in the ancestral protein (Trp60D, His71, Arg77A, Lys110, Glu217, Arg221A). This was not surprising since, at the dawn of vertebrate evolution, protein C had yet to evolve from the vitamin K-dependent fac-tor ancestral protein (Doolittle, 1993). By contrast, both thrombin residues impli-cated in binding thrombomodulin (Gln38, Arg75; Tsiang *et al.*, 1995), the endothe-lial cell-surface thrombin receptor, were present in the ancestral protein. Thus an ancient thrombomodulin-like molecule might have been able to interact with the vitamin K-dependent factor ancestral protein.

When the residues responsible for binding thrombomodulin and protein C were excluded from the fibrinogen-binding patch, five of the remaining seven residues were found to be conserved between the vitamin K-dependent factor ancestral protein and extant human thrombin (Lys60F, Asn60G, Asp186A, Lys186D, Glu192). Both of the altered residues (Thr60I, Asp222) exhibit conservative changes (to Ser and Glu, respectively). These seven residues may have constituted the original fibrinogen-binding patch that was to increase in size and binding affinity as well as diversifying functionally over evolutionary time.

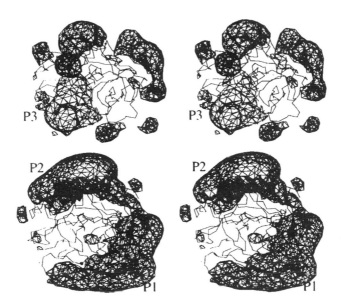

Figure 3.2. Stereo views of the electrostatic profiles of vitamin K-dependent factor ancestral protein (above) and extant human thrombin (below). The view is towards the active site canyon. The positions of the fibrinogen-binding site (P1) and heparin-binding site (P2) are indicated. A large positively charged patch (P3) is also present on the vitamin K-dependent factor ancestral protein. Derived from Krawczak *et al.* (1996).

The extensive positive electrostatic charge associated with the heparin-binding site in extant thrombin was much smaller in size and field strength in the model of the ancient protein (see *Figures 3.2* and *3.3*). In prothrombin, the second kringle domain interacts with the heparin-binding site to slow down antithrombin III/heparin-mediated inhibition (Arni *et al.*, 1993) prior to proteolytic prothrombin activation. The primitive heparin-binding site evident in the ancient proteinase would have been unable to bind the kringle 2 domain as strongly as it does today. This is consistent with the view that the ancient protein contained a light chain of Gla and EGF domains, the latter only being replaced by kringles at a later stage in the evolution of the proteinase (Patthy, 1985). The heparin-binding site then evolved in such a way as to balance the dual requirements of antithrombin III/heparin-mediated inhibition of thrombin and the kringle 2-mediated protection of prothrombin from premature inhibition.

In summary, Krawczak *et al.* (1996) were able to generate a tertiary model of the ancestral vitamin K-dependent serine proteinase that was both energetically satisfactory and possessed a credible fold. Therefore, the authors concluded that their derived cDNA sequence encodes a biophysically plausible protein.

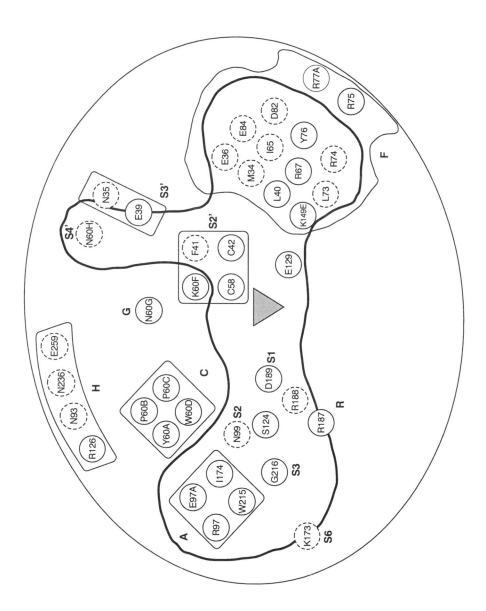

Figure 3.3. Schematic view of the active site canyon (bold contour) of the vitamin K-dependent factor ancestral protein. The active site triad (His57, Asp102, Ser195, chymotrypsin numbering) is denoted by a triangle. H, Heparin-binding site; G, glycosylation site; C, chemotactic region; R, RGD sequence; A, aryl-binding site; F, fibrinogen-binding exosite; S, specificity sites N-terminal to cleavage; S', specificity sites C-terminal to cleavage. Residues that are conserved between the vitamin K-dependent factor ancestral protein and extant human thrombin are circled. Those residues that have not been conserved are circled with a broken line.

Antithrombin III and antithrombin III deficiency

4.1 Introduction

The existence of a plasma cofactor for the anticoagulant effect of heparin was inferred by the discoverers of the polysaccharide (Howell and Holt, 1918). The plasma protein which interacts with heparin and neutralizes thrombin was first isolated in a homogeneous preparation by Abildgaard (1968, 1969). This protein was shown to inactivate thrombin although only slowly or progressively as compared with its effect in the presence of heparin. The protein, termed antithrombin III (ATIII), was purified and shown to bind heparin directly thus accelerating thrombin inhibition (Rosenberg and Damus, 1973; Damus *et al.*, 1973).

It is now clear that ATIII is the most important circulating inhibitor of thrombin. Deficiency of ATIII is associated with a thromboembolic tendency in humans (Cosgriff *et al.*, 1983). Indeed, ATIII deficiency was the first described cause of hereditary thrombophilia (Egeberg, 1965). Intensive investigation of ATIII, both on the basis of its homology to α1-proteinase inhibitor, and through the study of naturally occurring mutations associated with dysfunctional variants, has led to many insights into its structure and function.

4.2 Structure and function of ATIII

4.2.1 Primary sequence, secondary structure and post-translational modifications

ATIII is a member of the family of serine proteinase inhibitors or serpins (reviewed by: Bode and Huber, 1992; Carrell and Evans, 1992; Harper and Carrell, 1994; Potempa *et al.*, 1994; Schulze *et al.*, 1994; Olds *et al.*, 1995). This functionally diverse class of proteins includes the plasma proteins, heparin cofactor 2, C1-inhibitor and plasminogen activator inhibitors (PAI)-1 and -2 as well as the non-inhibitory egg-white protein, ovalbumin (Huber and Carrell, 1989). Amino acid sequence homologies between the most functionally divergent large serpins may be as low as 30%, yet structural homologies between the same proteins are always greater than 70%. Various conformations are evident and these conformations are relevant to the nature of the inhibitory mechanism (Schulze *et al.*, 1994).

The complete amino acid sequence of ATIII was established by Edman degradation (Petersen *et al.*, 1979). It circulates in plasma as a single-chain glycoprotein of M_r 58 000. The 432 residue sequence of the mature protein (after cleavage of the 32-amino-acid residue leader sequence) is approximately 30% homologous to other members of the serpin family (Carrell *et al.*, 1987; see Section 4.2.3). ATIII contains nine α helices (A–I) and three β-sheets (A–C). There are six cysteinyl residues at positions 8, 21, 95, 128, 247 and 430 which participate in disulphide bonding (8–128, 21–95, 247–430). Two of these disulphide bonds serve to link the N terminus to the main body of the protein (Cys8 to Cys128 in the D helix and Cys21 to Cys95 in the C helix). Cys247–Cys430 is essential for heparin binding (Ferguson and Finlay, 1983).

There are four N-linked biantennary oligosaccharide side chains (Franzen *et al.*, 1980) located at Asn residues 96, 135, 155 and 192. Two of these link the 45 residue N-terminal region to the C and D helices of the molecule. No other modified residues have been reported. Asn135 is not glycosylated in a fraction (5%) of human plasma ATIII with increased heparin affinity (Petersen and Blackburn 1985; Brennan *et al.*, 1987). This variant has been termed ATIII β. Site-directed mutagenesis of Asn135 to Gln generated an ATIII variant with higher affinity for heparin than the normal form (ATIII α) but normal inhibitory activity toward factor Xa (Turko *et al.*, 1993). Glycosylation of Asn155 has been shown to be associated with a reduced affinity for heparin (Garone *et al.*, 1996).

4.2.2 Tertiary structure

The structure of ATIII was initially deduced from the crystallographically determined structure of α1-proteinase inhibitor [Loebermann *et al.*, 1984; Carrell *et al.*, 1987; Engh *et al.*, 1990; *Figure 4.1* (see Frontispiece)]. One limitation was that the α1-proteinase inhibitor was crystallized after cleavage of its reactive centre bond, Met358–Ser359 (ATIII homology, Arg393–Ser394), so that the reactive centre 'bait loop' presented to the proteinase was lost. Once the crystal structure of ovalbumin, solved to 1.95 Å, became available (Stein *et al.*, 1990), this structure was used to propose a general model for the reactive centre of serpins (Carrell *et al.*, 1991; *Figure 4.1*). The bait loop of ovalbumin is in the intact conformation (i.e. it is completely exposed and in the form of an α-helix). Although ovalbumin is not itself a proteinase inhibitor, its homology was strong enough to carry conviction. The relevant region of ovalbumin (Ser350–Ser359) corresponding to residues P9–P1′ of α1-proteinase inhibitor and ATIII forms a helical loop with limited contact to the core of the molecule. After cleavage at the reactive site, P1–P1′, this loop relaxes into a position alongside the prominent β sheets spanning the length of the molecule (Evans *et al.*, 1992).

The crystallographic structure of the cleaved form of bovine ATIII was then determined to a resolution of 3.2 Å (Mourey *et al.*, 1993) whilst the crystal structures of the intact and cleaved forms of human ATIII have now been resolved to 3.2 Å (Schreuder *et al.*, 1994) and 3.0 Å (Wardell *et al.*, 1993; Carrell *et al.*, 1994) respectively (*Figure 4.1*). Both structures of the human protein are of a complex between the intact and cleaved forms of ATIII. In the intact conformation, the bait loop is partially inserted into the central sheet A of the molecule whilst another portion of the bait loop is inserted into sheet C of the cleaved molecule, replacing strand s1C [*Figure 4.2* (see Frontispiece)]. The loop makes a β turn at residues 377 and 378 to turn back into the central sheet whilst residue 380 (P14) is rotated such that the loop leaves the sheet at this point. After leaving the central sheet, the intact bait loop makes a β turn at Glu381 to continue parallel to the central sheet. Residues 382–385 (AAAS) of the hinge region are small and flexible, suggestive of high mobility, a requirement for complex formation. Thr386 makes an H bond with Glu378, thereby linking the bait loop to the main body of the intact protein. Docking of the bait loop of ATIII with the active site cleft of thrombin was achieved with minimal alteration (Schreuder *et al.*, 1994) consistent with the accuracy of the ATIII bait loop structure.

The flexibility of the bait loop of ATIII allows it to adopt a series of different conformations. An initially quiescent conformation, reminiscent of ovalbumin, is characterized by a fully exposed bait loop (Carrell *et al.*, 1991; Mourey *et al.*, 1993). An intermediate 'active' conformation, which appears to correspond to the heparin-activated conformation of the protein (Carrell *et al.*, 1991), exhibits a bait loop which is partially inserted into the A sheet (to P14) allowing ATIII to be ready to react with its target proteinase. This transient docking conformation is then finally converted to the 'locked' conformation by insertion to P10 that stabilizes the complex with thrombin. Loop insertion appears to be necessary for the locked conformation that stabilizes the docked serpin–proteinase complex (Crowther *et al.*, 1992; Bjork *et al.*, 1992, 1993; Hopkins *et al.*, 1993). Indeed, mutations in the residues of the hinge region of the bait loop (P9–P14) serve to convert ATIII from an inhibitor to a substrate owing to the inability of these mutants to form stable complexes with the proteinase (Hopkins *et al.*, 1993; see Section 4.4.2, 'Single base-pair substitutions in the *AT3* gene'). A latent form of ATIII also exists with the bait loop fully incorporated into the A sheet as in the latent form of PAI-1 (Mottonen *et al.*, 1992; *Figure 4.2*). This form of ATIII possesses decreased heparin affinity and is neither an inhibitor nor a substrate of thrombin.

4.2.3 Homologies with other proteins

Carrell *et al.* (1979) first demonstrated that ATIII shared considerable sequence homology at the amino acid sequence level with α1-proteinase inhibitor. These serpins are members of a large family of related proteins (including chicken ovalbumin, heparin cofactor II, angiotensinogen and C1 inhibitor; reviewed by Lane and Caso, 1989) which appear to share the same mechanism of action and which are thought to have evolved from a common ancestor over the last 500 million years (Hunt and Dayhoff, 1980). Thus α1-proteinase inhibitor and ATIII exhibit 33% amino acid sequence homology and 46% nucleotide sequence homology (Chandra *et al.*, 1983a). Similarly, α1-antichymotrypsin and ATIII are 33% and 46% homologous at the amino acid and nucleotide sequence levels, respectively (Chandra *et al.*, 1983b). However, amino acid sequence homology is more limited around the C-terminal reactive sites of these proteins, Arg393–Ser394 (ATIII), Met358–Ser359 (α1-proteinase inhibitor) and Leu358–Ser359 (α1-antichymotrypsin). This sequence divergence is clearly the means by which target selectivity is achieved. The importance of these residues in determining specificity is illustrated by a variant α1-proteinase inhibitor (Pittsburgh Met359→Arg) which had antithrombin selectivity and caused a fatal bleeding tendency (Carrell *et al.*, 1989).

4.2.4 Mechanism of proteinase inhibition and preferred targets

ATIII inhibits all the active factors of the intrinsic pathway (reviewed by Olson and Björk, 1992). From *in vitro* studies, it appears that ATIII is the main plasma inhibitor of factor IXa, factor Xa and thrombin; factor XIa is more readily inhibited by α1-proteinase inhibitor and factor XIIa by C1-inhibitor (Scott *et al.*, 1982; De Agostini *et al.*, 1984). ATIII is therefore predominantly involved in modulating the later stages in coagulation. The rate of inhibition of factor VIIa by ATIII is much lower than that of thrombin although comparable to that of factors XIa and XIIa (Lawson *et al.*, 1993). This has suggested that ATIII–heparin inhibition of factor VII/tissue factor might be as important as the tissue factor pathway inhibitor (TFPI) mechanism. Whereas factor VIIa is relatively resistant to ATIII-heparin inhibition, the factor VIIa–tissue factor complex is some 33-fold more susceptible (Broze *et al.*, 1993; Lawson *et al.*, 1993; Rao *et al.*, 1993; Jesty *et al.*, 1996). It is thought that the binding of ATIII to the factor VIIa–tissue factor complex induces an alteration in factor VIIa that impairs its binding to tissue factor (Rao *et al.*, 1995). This effect is reversible by addition of high concentrations of factor VIIa whilst TFPI–factor Xa remains capable of inhibiting the factor VIIa–tissue factor complex (Rao *et al.*, 1995).

ATIII inhibits its target proteinases by acting as a pseudo-substrate, forming a stable stoichiometric one-to-one complex with the proteinase through a stable acyl intermediate (Lawrence *et al.*, 1995; reviewed by Gettins *et al.*, 1993a). Formation of this complex is accompanied by cleavage of a specific bond, Arg393–Ser394 (P1–P1'), located on the S4 or 'bait loop' and termed the reactive centre of the inhibitor (Björk *et al.*, 1982; Wilczynska *et al.*, 1995). The complex is stable to denaturing reagents but can be released by nucleophilic attack (Fish and Björk, 1979) suggesting that a covalent link has been formed. In the cleaved structure of ATIII, the P1 residue is separated from P1' by approximately 70 Å as in α1-proteinase inhibitor (Loebermann *et al.*, 1984).

In solution, the bait loop comprises a protruding element which appears to be highly mobile and capable of moving in and out of the main body of the protein (reviewed by Lane *et al.*, 1992a). In the X-ray crystallographic structure of ovalbumin (Stein *et al.*, 1990), this bait loop adopts an α-helical conformation. In the crystal structure of the latent form of PAI-1 (Mottonen *et al.*, 1992; *Figure 4.1*), the bait loop partially inserts itself into the protein as a β-pleated sheet. Molecular dynamic simulations of the uncleaved structure of α1-proteinase inhibitor (Engh *et al.*, 1990), structural homologies between serpins, binding studies on peptidomimetics of the insertion loop (Björk *et al.*, 1992), peptide-binding assays (Carrell *et al.*, 1991) and epitope mapping data (Björk *et al.*, 1993) are, however, inconsistent with these crystal structures. Rather, these studies suggest that the bait loop of ATIII adopts a 'canonical' conformation, similar to the reactive site loop of the Kunitz inhibitors (Carrell *et al.*, 1991), in which the N-terminal region (P17–P6) of the uncleaved bait loop folds back into the body of the ATIII molecule to be flanked by β sheets 3A and 5A.

Two similar models have been proposed for the mechanism of action of serpins. The first of these, the 'induced conformational change model' (Skriver *et al.*, 1991; Björk *et al.*, 1992), proposes that the target proteinase is trapped by the N-terminal portion of the S4 loop as it inserts between sheets 3A and 5A. The alternative 'pre-equilibrium conformational change model' (Carrell *et al.*, 1991), suggests instead that the inhibitor exists in an equilibrium between two conformational states, one active with the N-terminal portion (P18–P10) of the bait loop folded parallel to the A sheets, and one inactive with the bait loop projecting fully into the solvent. The second model proposes that the serpin alternates between the two conformational states and that the proteinase binds only to the active conformation (Carrell *et al.*, 1991). Both models assume that the partial insertion of the bait loop into the A sheets is essential for serpin–proteinase complex formation; biochemical methods have, however, so far failed to distinguish between the two models (Björk *et al.*, 1993).

The inhibitory activities of ATIII toward thrombin and factor Xa can be dissociated by site-directed mutagenesis. A factor Xa-specific ATIII molecule can be created by replacing the P3–P3' region of ATIII by the sequence around the Arg323–Ile324 factor Xa cleavage site in prothrombin (Theunissen *et al.*, 1993). A factor Xa-specific ATIII molecule also results from substitution of the P1' Ser residue with virtually any other amino acid (Theunissen *et al.*, 1993). These experiments have demonstrated the strict structural requirements of the ATIII reactive site for inhibition of thrombin. Indeed, the S1' pocket of factor Xa possesses more room for the P1' residue of ATIII than does the S1' pocket of thrombin. Le Bonniec *et al.* (1995) have shown that residue 192 and the '60' insertion loop of thrombin are important for its interaction with ATIII.

The concentration of ATIII in plasma (2 μM) is far greater than the concentration of its target proteinases generated during coagulation. However, the rates of inhibition of thrombin and factor Xa in plasma (c. 1 min) and of factor IXa (c. 10 min) are far too slow to prevent coagulation. The rates of inhibition are dramatically enhanced by heparin, becoming virtually instantaneous, which accounts for both the *in vivo* and *in vitro* effect of heparin.

4.2.5 ATIII and heparin

The relatively slow rate of inhibition of thrombin and factor Xa is increased by up to 10 000-fold when, acting as a catalyst, the multimeric anionic glycosaminoglycan, heparin, is present (Jordan *et al.*, 1979; Hoylaerts *et al.*, 1984; reviewed by Olson and Björk, 1992). The influence of heparin in this reaction is due, at least in part, to binding ATIII which has been thought to occur as a two-step process; a high-affinity step is accompanied by an approximately 10-fold increase in the affinity of heparin for ATIII (Evans *et al.*, 1992) and leads to a conformational change in the serpin which involves the P1 residue (Evans *et al.*, 1992; Olson *et al.*, 1992; Gettins *et al.*, 1993a). This change results in turn in a further approximately 100- to 300-fold increase in the affinity of heparin for ATIII (Evans *et al.*, 1992; Olson *et al.*, 1992). Although the heparin-induced conformational change in ATIII enhances the rate of factor Xa inhibition, it probably does not significantly alter the inhibitory activity of the ATIII–heparin complex towards thrombin (Olson and Björk, 1991; Evans *et al.*, 1992; Gettins *et al.*, 1993b). Heparin thus promotes the interaction of thrombin and ATIII primarily by approximating the proteinase and the serpin on its surface. Heparin then promotes the formation of the initial thrombin–ATIII complex, thereby increasing the affinity of the proteinase for the serpin 1000-fold (Olson and Shore, 1982). The serpin-proteinase complex is then converted into a stable covalent complex (reviewed by Pratt and Church, 1993).

Heparin binds to, and modulates the activity of, a variety of different proteins (Margalit *et al.*, 1993). Commercially available heparin is an alternating chain of highly sulphated 1,4-linked hexuronic acid and D-glucosamine residues. Heparin with high affinity for ATIII contains a unique pentasaccharide, not found in low-affinity preparations, which contributes more than 90% of the ATIII-binding affinity of the oligosaccharide (Thunberg *et al.*, 1982; Atha *et al.*, 1985). However, this pentasaccharide only enhances the ATIII–thrombin reaction rate some 1.7-fold as compared with 4300-fold for a full-length heparin of approximately 26 saccharide units (Olson *et al.*, 1992). The ATIII–thrombin reaction exhibits a sharp threshold for rate enhancement at 14–16 saccharide units (Pratt et al., 1992; reviewed by Lane et al., 1992a; Olson and Björk, 1992). This threshold effect is probably explicable in terms of a requirement for a certain number of saccharide units in promoting the approximation of ATIII and thrombin on the surface of heparin in order to form the ternary complex. Finally, heparin has been fractionated into at least three populations of molecules with affinity for ATIII (Edens et al., 1995); the affinity of these heparin species for ATIII appears to vary inversely with their length.

It is clear from kinetic analysis that heparin induces a conformational change in ATIII which locks it on to the surface of the inhibitor (Olson et al., 1981). Binding of heparin to ATIII alters ultraviolet absorption, circular dichroism and tryptophan fluorescence (Olson and Shore, 1981; Evans et al., 1992; reviewed by Lane et al., 1992a). The induced conformational change in ATIII structure then exposes Lys236 (Chang, 1989). The negatively charged sulphate and carboxylate groups of heparin are proposed to interact with the positively charged residues of ATIII (Huber and Carrell, 1989). These residues have been identified largely through chemical modification experiments (reviewed by Lane et al., 1992a) and by studies of naturally occurring mutant ATIII molecules deficient in their capacity to bind heparin (reviewed by Lane et al., 1993). Chemical modification of various residues in ATIII prevents heparin binding (e.g. Trp49, Lys107, Lys114, Lys125, Arg129, Lys136 and Arg145; Blackburn et al., 1984; Liu and Chang, 1987; Petersen et al., 1987; Chang, 1989; Sun and Chang, 1990). A reduction in heparin binding, consequent to mutation at Trp49, has also been demonstrated by in vitro mutagenesis/expression experiments (Gettins et al., 1992); however, although Trp49 may contact heparin, its substitution by Lys does not affect the heparin-induced conformational change or the enhancement of the rate of factor Xa inhibition. Identification of amino acid residues involved in the binding of heparin (e.g. His120) has also come from the use of nuclear magnetic resonance (NMR) spectroscopy as a tool for examining the structural consequences of site-directed mutagensis (Fan et al., 1994).

A synthetic peptide corresponding to amino acid residues Ala124 – Arg145 has been used to raise an antibody that blocked heparin binding to ATIII (Smith et al., 1990). V8 proteinase digestion experiments have demonstrated that negatively charged residues within the region Glu34–Leu51 of the A helix of ATIII bind heparin (reviewed by Lane et al., 1992a). Finally, structural analysis of naturally occurring ATIII variants manifesting reduced binding to heparin suggest that residues Ile7, Met20, Arg24, Pro41, Arg47, Leu99, Ser116, Gln118 and Arg129 are within or near the path of heparin binding (Lane et al., 1993; Okajima et al., 1993; Carrell et al., 1994; Chowdhury et al., 1995; see Section 4.4.2, 'Single base-pair substitutions in the AT3 gene'). The first five of these residues are located on the A helix, the latter three residues are located on the D helix, whilst residue 99 is located on the C helix. Evans et al. (1992) identified homologous positively charged residues present in ATIII and C1 inhibitor but absent from α1-proteinase inhibitor. They proposed that a ridge of positively charged residues extended from a high-affinity heparin binding site (involving residues 47, 121, 125, 129, 132 and 133) to a second partially contiguous low-affinity heparin-binding site involving residues Asn135, Lys136, Lys257, Arg235, Lys236 and Lys275. Binding of heparin to the high-affinity site is accompanied by the conformational change whilst binding to the low-affinity site allows charge neutralization of the the reactive centre pole of the molecule as well as providing a bridge between the serpin and the proteinase to yield the stable ternary complex.

Carrell *et al.* (1994) proposed that the binding of the heparin pentasaccharide induces a shift of helix D together with an opening of the A sheet. This is accompanied by entry of the bait loop to P14 with disruption of the salt bridge between Arg393 and Glu255 to allow the arginine to rotate to the exterior of the molecule. Further insertion of the loop, either upon cleavage or upon formation of the complex with thrombin, would lead to a shift in the loop between helix D and the A sheet accompanied by loss of affinity for heparin. Full activation of ATIII would, however, require the bridging of heparin across ATIII to a second heparin-binding site on thrombin to yield the ternary complex. van Boeckel *et al.* (1994) have suggested that heparin binding induces a change in the conformation of the peptide linking helix D and sheet A, resulting in expulsion of the loop from sheet A.

After binding heparin, a secondary conformational change is induced in ATIII that makes the reactive centre bind more rapidly to its target proteinases. Charge neutralization on the positively charged surface of ATIII to allow it to interact with thrombin has also been held to be an important function of heparin (Beresford and Owen, 1990). However, Olson and Björk (1991) have demonstrated that the interaction between thrombin and ATIII does not appear to be influenced by the surface charge of these proteins either in the presence or absence of heparin. Heparin's major role, therefore, appears to be to promote the interaction of thrombin and ATIII by approximating these proteins on the surface of the polysaccharide, thereby facilitating the formation of the ternary complex (Griffith, 1982; Pletcher and Nelsestuen, 1983).

To accelerate the rate of the ATIII–thrombin reaction, heparin must also bind to thrombin. This it does much less specifically and with much weaker affinity than it binds ATIII (Olson, 1988; reviewed by Olson and Björk, 1992). Olson *et al.* (1991a) proposed that thrombin interacts with five or six ionic groups in a three-disaccharide binding unit of heparin. The heparin-binding domain of thrombin has been identified by chemical modification experiments (reviewed by Stubbs and Bode, 1993a, b); centred around four residues (Arg93, Arg126, Lys236 and Lys240; chymotrypsin numbering) on the 'north face' of the molecule, this region extends over the surface of thrombin to its 'west face', encompassing residues Arg97, Arg101, Arg165, Lys169, Arg173, Arg175, Arg233 and Lys235 ('anion-binding exosite II') (Sheehan and Sadler, 1994; Gan *et al.*, 1994; reviewed by Stubbs and Bode, 1993a). Docking ATIII with thrombin (Schreuder *et al.*, 1994) confirmed the probable involvement of these residues as well as Arg60. Thrombin, binding heparin, may then move toward the bait loop of ATIII, possibly by rolling (Hoylaerts *et al.*, 1984) to form the ternary complex. The relative importance of the heparin–thrombin interaction as compared to the heparin/ATIII conformational change remains controversial (see Rosenberg, 1987).

4.3 The antithrombin III (*AT3*) gene

4.3.1 cDNA cloning

A number of cDNA clones for human ATIII have been isolated from liver cDNA libraries (Bock *et al.*, 1982; Chandra *et al.*, 1983a; Prochownik *et al.*, 1983a; Wasley *et al.*, 1987). A full-length cDNA encoding ATIII (432 amino acids and a 32-amino-acid signal sequence) includes an 84 nucleotide 3′ untranslated region.

Sheffield *et al.* (1992) and Niessen *et al.* (1992) have cloned rabbit and ovine *AT3* cDNAs respectively; the rabbit and ovine amino acid sequences of the mature proteins are 84% and 89% identical to their human counterpart. Interestingly, portions of the rabbit and human ATIII molecules are not interchangeable despite their high level of homology. Replacement of residues in the C-terminal half of the molecule by those of the other species resulted in fusion proteins with lowered ability to bind thrombin (Sheffield *et al.*, 1992).

A chicken *AT3* cDNA has also been isolated which is 64% homologous at the nucleotide sequence level to that of human (Tejada and Deeley, 1995). A partial baboon *AT3* cDNA clone (Stackhouse *et al.*, 1983) is also available for sequence comparison.

4.3.2 Structure and evolution

The cloning and characterization of the human *AT3* gene was first accomplished by Prochownik *et al.* (1985); the gene comprises seven exons spaced out over about 14 kb of genomic DNA (*Figure 4.3*). Initially only six exons were found in the gene; later analysis located a further intron within what had been thought to be exon 3. The two 'new' exons were thus termed 3a and 3b.

The DNA sequence of about 1.5 kb of intronic DNA flanking the exon/intron boundaries and over 300 bp of 5' flanking region was first established by Bock *et al.* (1988). The complete sequence of the human *AT3* gene has since been determined (Olds *et al.*, 1993a); the gene spans some 13 480 bp from transcriptional initiation site to poly (A) addition signal. Nine complete and one partial *Alu* repetitive sequence elements have been identified in introns 1, 2, 3b, 4 and 5; all but one are oriented in the reverse direction. Together, these repetitive elements comprise some 22% of the intronic sequence.

Both Prochownik *et al.* (1985) and Jagd *et al.* (1985) compared the exon/intron distribution of the human *AT3* gene with that of other members of the serpin multigene family: most of the six *AT3* introns occur at non-homologous positions when compared with the four introns of the human α1-proteinase inhibitor (*PI*) and rat angiotensinogen (AGT) genes and the seven introns of the chicken ovalbumin gene. However, intron 2 of the human *AT3* gene and intron b of the chicken ovalbumin

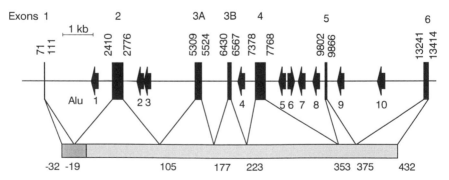

Figure 4.3. Organization of the human *AT3* gene (upper portion) and its encoded protein (lower portion). Exons are denoted by solid boxes. The positions of the first and last nucleotides of each exon are shown. Arrows denote position and orientation of *Alu* repetitive seuence elements. The ATIII protein is shown with dark hatching representing the signal peptide and light hatching representing the secreted protein. Codons initiating each exon are indicated. Reproduced from Lane *et al.* (1993) with permission from F.K. Schattauer Verlagsgesellschaft mbH.

gene occur at identical nucleotide positions. Intron 3 is misaligned by only one nucleotide when compared with the B introns of the *PI* and *AGT* genes. These genes have therefore diverged from a common ancestral gene sequence.

The 5' end of the human *AT3* gene has been sequenced but few structural or functional studies have been attempted. Primer extension analysis mapped the *AT3* transcriptional initiation site: (i) in liver cells to a position 72 bp upstream of the ATG translation initiation codon (Prochownik and Orkin, 1984; Bock *et al.*, 1988). No TATA box is evident immediately upstream of i.

Prochownik (1985) located probable enhancer elements in the 5' flanking region of the human *AT3* gene, by virtue of their homology with enhancer elements present in the J_κ–C_κ intron of both human and murine immunoglobulin κ-chain genes. Two blocks of homology occur, between −52 and −40 and between −13 and +7 (relative to i). Both regions are also similar to known viral enhancer elements. Sequence similarities between the *AT3* gene promoter and the promoters of other liver-specific genes are evident (e.g. Ochoa *et al.*, 1989). Ochoa *et al.* (1989) have provided evidence for the binding of a liver-expressed nuclear protein, *LF1*, between −89 and −68 (relative to i).

The function of DNA sequence in the 3' untranslated region of the human *AT3* gene was explored by Prochownik *et al.* (1987). A series of deletions were introduced into the 3' end of the *AT3* gene and these were inserted downstream of a reporter gene, *neo*, whose level of transcription was assayed in transfected COS-1 cells. Removal of the normal poly (A) cleavage site did not prevent cleavage but did alter its accuracy. Deletion of the AATAAA motif and downstream sequences abolished cleavage but did not affect mRNA transcript levels. Analysis of shorter deletion constructs pinpointed the location of two regions downstream of the endonucleolytic cleavage site which appeared to be important for the cleavage process. One appears to be involved in determining the distance of the cleavage site from the AATAAA signal. Duplication of DNA sequence containing the AATAAA motif, with or without downstream sequences, resulted in cleavage of at least two sites, but duplication of sequence either upstream or downstream of AATAAA produced no observable effect on cleavage efficiency or site selection. These results suggested that the distance between the AATAAA motif and downstream sequences may be critical in determining the precise site of cleavage.

4.3.3 Polymorphisms

A number of DNA sequence polymorphisms have been identified within the coding region of the human *AT3* gene (Lane *et al.*, 1993). The CAA/CAG polymorphism at codon 305 is detectable by virtue of the presence/absence of a *Pst*I restriction site (Prochownik *et al.*, 1983a) and has been formatted for polymerase chain reaction (PCR) by Olds *et al.* (1990b). Bock and Radziejewska (1991) have reported an *Nhe*I restriction fragment length polymorphism (RFLP) in intron 4 of the gene; alternative alleles of 20 kb and 9.0/11.0 kb occur at approximate frequencies of 0.64 and 0.36 respectively. A *Dde*I RFLP has been detected in intron 5 of the human *AT3* gene (alternative alleles, 127 bp or 53 + 74 bp occur with respective frequencies of 0.83 and 0.17; Daly and Perry, 1990).

A further RFLP, detected 5' to the *AT3* gene with *Bam*HI, exhibits alternative alleles of approximately 1.45 and 1.55 kb which occur at respective frequencies of 0.75 and 0.25 in the US population (Bock and Levitan, 1983) and 0.79 and 0.21 in the

Northern Irish population (Winter *et al.*, 1995). The molecular basis of this length RFLP is not however simply the insertion of 100 bp into the shorter allele but rather is the alternative presence of two quite distinct sequences at the same position, 273 bp upstream of the transcriptional initiation site. Sequence analysis (Bock and Levitan, 1983) determined the length of these alternative sequences to be 108 and 32 bp respectively, giving rise to a 76 bp length polymorphism. Wu *et al.* (1989) reported discrimination of the long and short alleles by PCR. These alleles possess sequences, nine or 10 nucleotides in length, at their 5′ and 3′ ends which are capable of forming perfect intra-strand stem–loop structures with sequences immediately upstream or downstream. Winter *et al.* (1995) also noted residual homology between the alternative sequences, suggesting that the short allele might have been derived from the long allele during evolution by partial deletion followed by sequence divergence.

ATIII Dublin (Val3→Glu; Daly *et al.*, 1990) was first described as an electrophoretic charge variant and results from a mutation which gives rise to aberrant signal peptidase cleavage at a site two amino acids downstream of the normal N-terminal histidine. However, since different families with this mutation appear to differ markedly with respect to their risk of thrombosis, the relevance of this variant is unclear. Moreover, additional mutations have been found in at least two patients bearing the −3 variant which may be sufficient to account for the observed clinical phenotype. It occurs in the German population at a frequency of 0.007 (Dürr *et al.*, 1992) and Lane *et al.* (1993) have listed it as a 'polymorphism'. A novel TAC/TGC polymorphism (Tyr/Cys) at residue 158 has also been noted in individuals of Scandinavian origin (Millar *et al.*, 1994a).

Finally, two $(ATT)_n$ copy number polymorphisms have been found in two of the intron 5 *Alu* repeat sequences (*Alu* 5 and *Alu* 8) (Lane *et al.*, 1993; Olds *et al.*, 1993a, 1994b; Perry 1993; Ni *et al.*, 1994). The *Alu* 5 polymorphism has two alleles whilst the *Alu* 8 variable number tandem repeat polymorphism, with in excess of 10 alleles, promises to be extremely informative in linkage analyses. Indeed, Olds *et al.* (1994b) used the *Alu* 8 polymorphism to distinguish between identity by descent and recurrent mutation for multiple recurrences of two *AT3* gene lesions.

4.3.4 Expression

The liver is the major site of synthesis of ATIII (reviewed by Thaler and Lechner, 1981; Fair and Bahnak, 1984). From the representation of *AT3* cDNA clones in human liver cDNA libraries, we may estimate that between 0.075% (Chandra *et al.*, 1983a) and 0.02% (Prochownik *et al.*, 1983a) total mRNA encodes ATIII. Hassan *et al.* (1990) have shown that the level of *AT3* mRNA increases in the human fetus from 30% to 50% of the adult level between 5 and 10 weeks' gestation. Liver is also the major site of *AT3* mRNA synthesis in the rat with kidney and spleen showing lower levels (Kourteva *et al.*, 1995); no *AT3* mRNA was detectable in brain, lung, heart or skeletal muscle.

The transcription of the human *AT3* gene was first studied using an artificial *AT3* 'minigene' containing 300 bp 5′ flanking region (Prochownik and Orkin, 1984). Primer extension experiments using liver mRNA placed the transcriptional initiation site to a unique location 72 bp upstream of the Met initiation codon (Prochownik and Orkin, 1984; Bock *et al.*, 1988). However, in HeLa and African green monkey kidney

(COS-1) cells, *AT3* gene transcription initiated downstream but within 9 bp of the liver start site.

More recently, Tremp *et al.* (1995) demonstrated that a 700 bp fragment of the promoter was sufficient to confer tissue-specific expression upon a downstream gene in the liver and kidney of transgenic mice. An initial characterization of the human *AT3* gene promoter region was performed by DNase I footprinting and four protected regions were identified using nuclear extracts from liver: region I (–138 to –123 from transcriptional initiation site), region II (–96 to –114), region III (–90 to –67) and region IV (–17 to –50). By means of gel retardation analysis (gel shift assays), region III was shown to interact with hepatocyte nuclear factor 4 (HNF4) whilst region I was shown to interact with hepatocyte nuclear factor 3 (HNF3).

Using transient transfection of *AT3* promoter/luciferase reporter gene constructs in hepatoma cells, Winter *et al.* (1995) showed that there was little if any significant difference between the long and short alleles of the promoter length polymorphism (Section 4.3.3). Moreover, mean plasma ATIII activity levels from individuals bearing the different, possible genotypes (short/short, long/long, short/long) were not significantly different, suggesting that this unusual polymorphism does not contribute to the variation in plasma ATIII activity that occurs in the general population. Similar findings were reported by Niessen *et al.* (1996).

Phenotypically, glucocorticoids are known to stimulate *AT3* gene expression whereas oestrogen reduces it (Thaler and Lechner, 1981). Oestrogen down-regulates the expression of the chicken *AT3* gene in the liver by about three-fold (Tejada and Deelcy, 1995). An understanding of the mechanisms of *AT3* gene induction and repression, however, awaits the functional analysis of the 5′ flanking region of the gene.

4.3.5 Chromosomal localization

Kao *et al.* (1984a) localized the *AT3* locus to chromosome 1 by Southern blotting of somatic cell hybrids, and refined this localization to 1p31.3–qter (Kao *et al.*, 1984b). Bock *et al.* (1985a) using *in situ* hybridization, further localized the *AT3* gene to the region 1q21.1–q25. In addition, these authors carried out gene dosage experiments on DNA isolated from cultured fibroblasts derived from two patients with deletions of the long arm of chromosome 1. The dosage of the *AT3* gene was found to be lower than in DNA from the cell line GM2025 (del 1q21–q25) but not in the DNA from cell line GM803 (del 1q21.3–q23). Taken together, these results suggest that the *AT3* gene is located between 1q23 and 1q25.

4.4 ATIII deficiency

4.4.1 Clinical studies

ATIII deficiency and its prevalence. The normal circulating level of ATIII in human is about 200 mg l^{-1} (4 μM) and about half of this is intravascular (Collen *et al.*, 1977). The half-life of radiolabelled ATIII in plasma is just under 3 days (Collen *et al.*, 1977). Laboratory assays are either immunological or functional. Functional assays either measure thrombin or factor Xa inactivation in the presence of heparin (heparin cofactor activity) or the proteinase inhibitory activity of ATIII in the absence of heparin (progessive antithrombin activity).

In clinical studies, the level is usually expressed as a percentage of a standard pool of normal plasmas with a range of 70–130 U dl⁻¹. A study of nearly 10 000 healthy blood donors revealed a mean ATIII level of 105.6 U dl⁻¹, SD 11.2 (Tait et al., 1993a). A broadly similar range has been reported by Rodeghiero and Tosetto (1996). Pre-menopausal women exhibited lower mean ATIII levels than men of the same age whereas post-menopausal woman had higher mean ATIII levels than both men of the same age and pre-menopausal women. Similar results have been reported by Meade et al. (1990) and Dolan et al. (1994).

The importance of maintaining a sufficiently high level of ATIII in the circulation can be gauged from the fact that individuals with heterozygous ATIII deficiency often possess plasma ATIII levels as high as 65 U dl⁻¹ and yet still suffer from repeated thromboembolism. The first case of ATIII deficiency in a Norwegian family with recurrent venous thrombosis and half the normal levels of progressive antithrombin (assayed in the absence of heparin), heparin cofactor activity (assayed in the presence of heparin), and ATIII antigen was reported by Egeberg (1965). ATIII deficiency is now recognized as an important cause of venous thrombosis (reviewed by Blajchman et al., 1992b). It may also be a risk factor in arterial disease (Johnson et al., 1990; Arima et al., 1992) although both high and low levels of plasma ATIII appear to be associated with a high risk of arterial disease (Meade et al., 1991). Finally, ATIII activity has been found to be increased in men with moderate peripheral atherosclerosis but decreased in men with more severe atherosclerosis (van der Bom et al., 1996).

Clinically symptomatic familial ATIII deficiency is not uncommon, occurring at a frequency of between 1/5000 (Abildgaard, 1981; Tait et al., 1991b) and 1/2000 (Rosenberg, 1975) in the general population (reviewed by Beresford, 1988; Blajchman et al., 1992b; Lane et al., 1992a). The asymptomatic deficiency state could in principle occur much more frequently. Indeed, Tait et al. (1991b) have estimated that the asymptomatic type IIc deficiency state (heparin binding site defects) occurs with a frequency of 1/350. Further, Tait et al. (1994) identified 16 cases of congenital ATIII deficiency (0.17%) by family studies and genetic analysis in a sample of 9669 blood donors; of these, 14 exhibited a type II deficiency state and two a type I deficiency state. A very similar value for the prevalence of congenital asymptomatic ATIII deficiency (0.20%) was arrived at by Wells et al. (1994).

ATIII deficiency (reviewed by Beresford, 1988; Hathaway, 1991; Lane et al., 1992a; Blajchman et al., 1992b; Olds et al., 1995) normally exhibits an autosomal dominant mode of inheritance, with most affected individuals being heterozygous for the defect and presenting with plasma ATIII levels of between 40% and 70% of normal. The risk of thrombosis increases with age; 85% of these patients over the age of 50 will have experienced at least one thrombotic episode (Thaler and Lechner, 1981; Cosgriff et al., 1983; Pabinger and Schneider, 1996) and such individuals have a recurrence risk of 63% (Pabinger and Schneider, 1996). The clinical features of thromboembolism are similar in ATIII, protein C and protein S deficiencies although thrombotic symptoms tend to develop earlier in women with ATIII deficiency and ATIII-deficient individuals appear to have a lower frequency of superficial thrombophlebitis (Pabinger and Schneider, 1996).

Classification. Classically, at least two main types of ATIII deficiency have been recognized at the clinical level. In type I deficiency, both ATIII activity and antigen are reduced to the same extent. Type II patients exhibit a higher level of ATIII antigen than activity.

A number of different systems of classification of ATIII deficiency have been devised (e.g. Girolami *et al.*, 1988; Hultin *et al.*, 1988) on the basis of the plasma activity/antigen ratio and the functional properties of the variant protein (e.g. heparin binding). The following classification system has been widely used:

(i)   Type I: low ATIII activity and antigen levels.
      Subtype Ia: reduced synthesis/increased turnover of a normal ATIII protein.
      Subtype Ib: reduced synthesis/increased turnover of ATIII exhibiting abnormal heparin binding.
(ii) Type II: low ATIII activity but normal antigen level.
      Subtype IIa: abnormality of reactive site and heparin-binding site.
      Subtype IIb: abnormality of reactive site only.
      Subtype IIc: abnormality of heparin-binding site only.

Lane *et al.* (1993) proposed that type II deficiency may be subdivided into three distinct subtypes: RS (reactive site defects), HBS (heparin binding site defects) and PE (describing those mutations which exert pleiotropic effects). Although the classical classification system is adhered to here, the terminology of Lane *et al.* (1993) may ultimately replace it if it receives formal approval from the scientific community.

ATIII deficiency and thrombotic disease. Between 1 and 6% of all patients presenting with a history of venous thrombotic disease have an inherited deficiency of ATIII, the exact frequency depending upon patient selection criteria (Gladson *et al.*, 1988; Scharrer *et al.*, 1988b; Butler and ten Cate, 1989; Hirsh *et al.*, 1989; Heijboer *et al.*, 1990; Harper *et al.*, 1991; Melissari *et al.*, 1992; Pabinger *et al.*, 1992b; Awidi *et al.*, 1993; Koster *et al.*, 1995b). Extrapolation from the incidence figures presented in the study of Gladson *et al.* (1988) yields a prevalence of clinically overt ATIII deficiency for the general population of 1 in 48 000 (Tait *et al.*, 1994). The best direct estimate of prevalence of ATIII deficiency in the general population comes from a large study of blood donors in the west of Scotland (Tait *et al.*, 1994); 16/9669 (0.18%) individuals exhibited a congenital deficiency of ATIII. Interestingly, 13 of the ATIII variants detected in this way were type II and most of these (11) were the Cambridge II variant (Ala384→Ser). Two patients with a type I deficiency had relatives who had experienced thrombotic episodes but only two of the 11 individuals with the Ser384 variant were symptomatic. This implies that 'covert' asymptomatic ATIII deficiency is much more common in the general population than the clinically overt form of the condition. Thus, as with protein C deficiency, other factors, either genetic (e.g. epistatic effects emanating from other loci) or environmental (e.g. surgery, trauma, oral contraceptive use), must play a role in determining whether or not an individual with ATIII deficiency will come to clinical attention.

The great majority of known cases of ATIII deficiency are inherited as autosomal dominant traits but there is one notable exception: type IIc heparin-binding site defects (see Section 4.4.2). Homozygous type I deficiency is extremely rare. Blajchman *et al.* (1992b) quote the single case of two homozygously affected offspring who died within the first 3 months of life. Lane *et al.* (1996a) reported the case of a child homozygous for a Leu99→Phe substitution who suffered a right-sided hemiparesis at the age of 4 months due to occlusion of the left middle cerebral artery and a large left-sided infarct followed by further thromboembolic events. Thus, for highly deleterious mutations, the homozygous deficiency state may be incompatible with life. Less severe homozygous or compound heterozygous ATIII deficiency can, however, occur. One example is that of two brothers with 13–16% ATIII (heparin cofactor) activity but near-normal progressive ATIII activity (92–110%) and antigen (73–85%) levels (Cucuianu *et al.*, 1994).

In ATIII deficiency, thrombotic risk clearly also varies with the nature of the inherited lesion. Heterozygous type I ATIII deficiency (by far the most common form) has been reported to be associated with an incidence of thrombosis of 52–54% whilst the corresponding figure for type IIa/IIb deficiency was said to be between 58 and 66% (Finazzi *et al.*, 1987). By contrast, type IIc heterozygotes were attributed a risk of 6% whereas all homozygotes appear to be affected. Such figures are, however, likely to be severely biased by the selective reporting of cases with a positive clinical history.

Approximately 50% of ATIII-deficient individuals have their first thrombotic episode in the absence of any obvious predisposing factor (Cosgriff *et al.*, 1983). The risk of thromboembolism in those with ATIII deficiency is however increased in the presence of other predisposing factors such as surgery, immobility, pregnancy and the use of oral contraceptives. Perhaps the best characterized family with inherited ATIII deficiency is that described by Demers *et al.* (1992); 6/31 (19%) of ATIII-deficient individuals had experienced thrombotic events compared with 0/36 non-deficient family members. In a review of the literature involving 62 families with ATIII deficiency (Demers *et al.*, 1992), the mean and median prevalences of venous thrombosis were 51% and 55% respectively (cf. 1.5% in non-deficient individuals). Risk factors for thrombosis were associated with 32% of thrombotic episodes. The risk for thrombosis peaked between the ages of 15 and 35. Thus for instance, the incidence of thrombosis per patient year in ATIII-deficient women taking oral contraceptives was found to be 27.5% (Pabinger *et al.*, 1994a).

In a population-based patient–control study, Koster *et al.* (1995b) estimated the increase in risk of thrombotic disease associated with the manifestation of ATIII deficiency. ATIII deficiency was defined as a measurement of less than 0.80 U ml^{-1}. When only the first measurement of plasma ATIII acivity was used, the odds ratio was found to be 2.2 (95% CI, 1.0–4.7). When the first and second measurements of ATIII were used, the odds ratio was 5.0 (95% CI, 0.7–34).

A minor form of ATIII, cleaved by thrombin at the active site, has been detected *in vivo* and occurs at a level of 1–4% of plasma ATIII (Lindo *et al.*, 1995). This form of ATIII reportedly occurs at a significantly higher level in patients with venous thrombosis than in healthy controls.

ATIII deficiency and mortality. The extent to which congenital ATIII deficiency gives rise to excess mortality is highly contentious. Rosendaal *et al.* (1991) studied 171 individuals from 10 families with hereditary ATIII deficiency and found no excess mortality. On this basis, they recommended against prophylactic anticoagulant therapy. However, this study did not quote causes of death or clinical penetrance of the deficiency state. These workers also studied nine ATIII-deficient Dutch families (in eight of whom the underlying *AT3* gene lesion was known): no excess mortality was observed (van Boven *et al.*, 1994).

De Stefano and Leone (1991) performed a meta-analysis of 552 individuals from 37 families reported in the literature; 43% of individuals had experienced a thrombotic episode and 42% of recorded deaths had been attributed to either venous or arterial thrombosis. These authors claimed that even using very conservative criteria including post-mortem examination, 13% of recorded deaths were due to either pulmonary embolism or mesenteric vein thrombosis. Another consideration is that some families are clearly more affected than others. Thus Leone *et al.* (1987) reported a two-generation ATIII-deficient family with nine deaths from thrombosis out of 35 family members. In such families, long-term prophylaxis would appear to be a reasonable course of action. However, as Rosendaal and Heijboer (1991) have pointed out, anticoagulation itself also carries risks of mortality and morbidity. A cost–benefit analysis would therefore seem to be appropriate on a case-by-case basis.

Intra-uterine mortality was not considered by the above authors. Sanson *et al.* (1996) estimate that the risk of natural abortion and stillbirth is 2.5-fold higher for ATIII-deficient women than non-deficient women.

Treatment. Menache *et al.* (1990) described a study of the administration of human ATIII to 18 patients with an inherited ATIII deficiency. ATIII was deemed efficacious as assessed by the absence of thrombotic manifestations after surgery and/or parturition, and the non-extension and non-recurrence of thrombosis in patients who had exhibited an acute thrombotic episode. No side effects were noted and the half-life of the infused ATIII was 43–77 h.

4.4.2 Molecular genetic analysis

Although the subject of recent reviews (e.g. Aiach *et al.*, 1995; Olds *et al.*, 1995; Perry, 1995; Lane *et al.*, 1996a; Perry and Carrell, 1996), the molecular genetic analysis of ATIII deficiency will now be reviewed in some detail. Although it is clear that the nature of the characterized mutations broadly reflects the clinical classification, this classification will probably require continuing revision in the light of further elucidation of new types of lesion in the *AT3* gene.

Deletions of the AT3 gene. The first deletions of the *AT3* gene to be described predated the cloning of the human gene; Winter *et al.* (1982) reported two patients with deletions of chromosome 1q who were shown to possess only 50% of the normal level of ATIII activity. Subsequently, another gross *AT3* gene deletion was inferred from RFLP segregation data (Prochownik *et al.* 1983b); two affected brothers appeared homo- or hemizygous for the (+) *Pst*I allele while their mother was either homo- or hemizygous for the alternative (–) allele. The brothers were presumed to be (+)/del having inherited the deletion-bearing chromosome from their mother of genotype (–)/del. This interpretation was later confirmed by densitometric gene dosage analysis (Bock and Prochownik, 1987). Perhaps the best characterized of the gross *AT3* gene deletions is that removing 2761 bp of the gene including exon 5 and flanking regions (Olds *et al.*, 1993a); the breakpoints were both located within intronic *Alu* repetitive sequences consistent with a model of homologous recombination.

A 105 bp deletion in exon 4 resulting in the in-frame deletion of 35 amino acids from Tyr240 to Gly276 has been described (Emmerich *et al.*, 1994a). This lesion, which results in type I ATIII deficiency, is presumed to generate an unstable protein since no cross-reacting material was detectable immunologically. This may be because Cys247, deleted in this variant, is involved in disulphide bridging. The deleted sequence is bounded by an imperfect direct repeat-suggesting that the deletion probably arose by slipped mispairing (Krawczak and Cooper, 1991).

Despite these relatively well characterized examples, major deletions of the *AT3* gene are comparatively rare. Blajchman *et al.* (1992b) estimated that less than 10% of cases of type I ATIII deficiency are of this type.

Some 25 different short deletions of the *AT3* gene have now been reported (Daly *et al.*, 1992, 1996a; Fernandez-Rachubinski *et al.*, 1992; Chowdhury *et al.*, 1993; Lane *et al.*, 1993, 1994; Emmerich *et al.*, 1994a; Jochmans *et al.*, 1994b; van Boven *et al.*, 1994). These deletions vary between 1 bp and 30 bp in length. All result in a type I deficiency (Lane *et al.*, 1993, 1994); indeed, the majority of cases of type I ATIII deficiency appear to be caused by short deletions and insertions. Although scattered throughout the *AT3* gene coding region, three 'hotspots' are apparent: non-identical deletions occur at codons 244/245 (Grundy *et al.*, 1991b) and at codon 81 (Lane *et al.*, 1993) whereas identical 6 bp deletions to have occurred at codons 106/107 have been shown by RFLP haplotyping not to be identical by descent (Olds *et al.*, 1993b). All three hotspots are closely associated with sequences matching the deletion hotspot consensus sequence (T G A/G A/G G/T A/C) of Cooper and Krawczak (1993).

Van Boven *et al.* (1994) reported the case of a 7768delG deletion which, since it involves the last nucleotide of exon 4, is presumed to abolish correct splicing. However, no ectopic transcript analysis was attempted to confirm this expectation.

Insertions in the AT3 *gene.* A number of single base insertions have been noted as a cause of type I ATIII deficiency (Lane *et al.*, 1993, 1994; Daly *et al.*, 1996a); these include T (codon 48), T (codon 102), A (codon 169), A (codon 208), A (codon 228), A (codon 408), A (codon 421), G (codon 423). All result in premature translational termination.

Single base-pair substitutions in the AT3 *gene.* Over 100 mutant *AT3* alleles have been characterized and shown to contain point mutations (single base-pair substitutions) (Appendix; Lane *et al.* 1991, 1993, 1994, 1997); 42 of these are different. The search for mutations in the *AT3* gene has been performed by sequencing all seven exons (Grundy *et al.*, 1991b) or by sequencing lymphocyte-derived ectopic transcripts (Perry, 1995). Mutation screening procedure such as single-strand conformation polymorphism analysis (Millar *et al.*, 1993c) or the Hydrolink heteroduplex detection procedure (Chowdhury *et al.*, 1993) have also been employed to pinpoint lesions.

Nearly 50 different missense mutations in the *AT3* gene responsible for ATIII deficiency and recurrent thrombosis have now been characterized (Lane *et al.*, 1991, 1993, 1994). Known mutations fall naturally into different groups which reflect the categorization employed at the clinical level. When the known missense mutations are plotted on to the three-dimensional structure of the ATIII molecule, they may be seen to form clusters which reflect the domain structure of the protein. The structural perturbations associated with the different known dysfunctional ATIII molecules have been comprehensively reviewed by Stein and Carrell (1995).

Type I deficiency. Eleven different single base-pair substitutions have been reported as a cause of type I ATIII deficiency (Lane *et al.*, 1993, 1994). Four of these are nonsense mutations; the Arg129→Term substitution, caused by a CGA→TGA transition, has been noted nine times independently and is responsible for the majority of repeat mutations in this type of ATIII deficiency. The remaining missense mutations presumably exert their deleterious effects on the structure and hence stability of the protein rather than on its function. One example of this is provided by the Cys430→Phe substitution which must abolish the normal disulphide linkage between Cys247 and Cys430 (van Boven *et al.*, 1994).

A single splice site defect has also been noted to cause type I deficiency: the underlying lesion (which had arisen *de novo*) was a AAG→AAA transition in the last base of exon 3a at Lys176 which did not alter the encoded amino acid (Berg *et al.*, 1992). Ectopic transcript analysis was used to demonstrate that this mutation resulted in the skipping of exon 3a in blood lymphocytes of the patient (Berg *et al.*, 1992). Oligonucleotide discriminant hybridization of the wild-type sized *AT3* mRNA transcript derived from the patient further demonstrated that correct splicing from the mutant allele had been completely abolished (Berg *et al.*, 1992). Ectopic transcript analysis has also been used to study the phenoypic effects of another splice site mutation in the *AT3* gene: a G→A transition at nucleotide 9787, 14 bp 5′ to exon 5 was shown to activate a cryptic acceptor splice site resulting in the inclusion of an extra 12 bp of intron 4 into the *AT3* cDNA (Jochmans *et al.*, 1994b).

Type II deficiency. Most single base-pair substitutions in the *AT3* gene result in a type II deficiency state. Of the different mutations known to cause a reactive site (RS) defect, resulting in the reduction or abolition of thrombin binding, the vast majority occur in two distinct clusters: (i) at residues Ala382 (P12) and Ala384 (P10) in the 'hinge region' where sheet 4A of the bait loop turns and joins the A sheet as strand 5A, and (ii) around the reactive site at residues 392, 393 and 394 (P2, P1 and P1′) (Lane *et al.*, 1993).

Type II deficiency: hinge region mutations. Hinge region variants Ala382→Thr, Ala384→Ser, Ala384→Pro exhibit a normal association rate but form less stable complexes with thrombin (Hopkins *et al.*, 1993). Although unable to inhibit thrombin, these variants are nevertheless capable of serving as substrates for the proteinase (Devraj-Kizuk *et al.*, 1988; Molho-Sabatier *et al.*, 1989; Perry *et al.*, 1991; Austin *et al.*, 1991a; Caso *et al.*, 1991; Ireland *et al.*, 1991a).

The hinge region of all inhibitory serpins is characterized by the presence of amino acids with small aliphatic side chains (Hopkins *et al.*, 1993). In inhibitory serpins, this portion of the bait loop (strand 4A) is partially inserted into the A sheet. This allows the bait loop to take up a taut 'canonical' conformation thereby promoting the formation of a stable complex between ATIII and its target proteinase (Carrell *et al.*, 1991). The naturally occurring mutations within this region introduce bulky or polar side chains that would be predicted to interfere with insertion of the bait loop into sheet A required for the adoption of the inhibitory conformation. Mutation of Ala382 leads to a major loss of inhibitory activity whereas the Ala384→Ser substitution results in more minor changes in inhibitory kinetics, suggesting that adoption of the inhibitory conformation probably requires insertion of the bait loop to P12 and possibly as far as P10 (Carrell and Evans, 1992). However, studies of the stability to denaturation of ATIII Hamilton (Ala382→Thr) have shown that this lesion does not affect the transformation to the more stable form or insertion of strand 4A into the A sheet (Wright and Blajchman, 1994).

Both mutations at Ala382 and Ala384 do, however, convert ATIII from an inhibitor of thrombin to a substrate (analogous to ovalbumin). These substitutions appear to alter the conformation of the the bait loop in such a way that it becomes exposed and available for proteolytic cleavage. This postulate has received support generally from studies of mutants in the P10–P12 region of other serpins (reviewed by Olds *et al.*, 1994a) and specifically by the study of the binding to ATIII of a synthetic tetadecapeptide corresponding to the P1–P14 region of the bait loop (Björk *et al.*, 1992b); insertion of the bait loop in such a complex is blocked and the ATIII molecule is not only unable to inhibit thrombin but is also cleaved as a substrate.

Austin *et al.* (1991b) performed site-directed mutagenesis on residue Ala382; mutations exerted a variable effect on ATIII–thrombin complex formation; substitutions of Ala by Gly, Ile, Leu and Val had a negligible effect whereas substitution by Thr and Gln exhibited low inhibitory activity and substitution by Lys abolished inhibitory activity. These different effects are determined either by the relative effect of the substitution on the stability of the hinge region or by the inability of the substituting residue at position 382 to interact with the hydrophobic pocket comprising residues Phe77, Phe221, Ile412 and Phe422. The naturally occurring lesions at residues 382 and 384 are detectable by restriction enzyme analysis (Fernandez-Rachubinski and Blajchman, 1992).

Type II deficiency: reactive site mutations. The variant antithrombins with mutations around the reactive site (Arg393–Ser394) are defective in their interaction with thrombin whilst exhibiting normal or even slightly increased heparin affinity. Substitution of Gly392 by Asp (ATIII Stockholm; Blajchman *et al.*, 1992a) yields an ATIII molecule that is unable to form an inhibitory complex with thrombin efficiently; thrombin binding to the variant molecule was, however, acelerated by heparin but still to subnormal levels. Sheffield and Blajchman (1994) used site-directed mutagenesis to generate seven variants at this position (Pro, Met, Gln, Val, Lys, Glu, Asp). When tested for their ability to form complexes with thrombin and factor Xa, Pro and Gln exhibited minimal impairment, Lys (positively charged) and Met (hydrophobic) moderate impairment and Asp, Glu (negatively charged) and Val (hydrophobic) severe impairment. Interestingly, the reactivity of the different mutants did not correlate with the size or hydrophobicity of their side chains.

Substitutions of residue Arg393 by Cys, His and Pro completely block thrombin inhibition (Erdjument *et al.*, 1988a, b, 1989; Owen *et al.*, 1988; Thein and Lane, 1988; Lane *et al.*, 1989a, b; Molho-Sabatier *et al.*, 1989; Ireland *et al.*, 1991b). Both ATIII Glasgow (Arg393→His) and ATIII Pescara (Arg393→Pro) exhibit increased heparin affinity and are resistant to cleavage by neutrophil elastase (Owen *et al.*, 1991). Lacking the P1 residue, these variants are unable to form the salt bridge normally present between Arg393 and Glu255 which may thus lead to their adopting a high-affinity conformation resembling the normal heparin-induced conformation. The Arg393→Cys variant is found in high-molecular-weight complexes due to the formation of a disulphide bond with albumin (Erdjument *et al.*, 1987).

Several examples of the Ser394→Leu substitution at position P1′ are known (ATIII Denver; Stephens *et al.*, 1987. ATIII Milano 2; Olds *et al.*, 1989). Site-directed mutagenesis of Ser→Leu at this position resulted in an ATIII variant with the same physical and functional properties as ATIII Denver (Stephens *et al.*, 1988). This variant is associated with a 430-fold reduction in the second order rate constant for the inhibition of thrombin but has less effect upon inhibitory function against factor Xa (Olson *et al.*, 1991). The P1′ residue is thought to play a critical role in determining the rate of proteinase inhibition (Theunissen *et al.*, 1993).

Type II deficiency: heparin-binding site mutations. A number of ATIII variants exhibit reduced affinity for heparin. These are of two types: mutations affecting residues that make direct contact with heparin and mutations which contribute to the overall conformation of the heparin-binding domain of ATIII. Although the 11 known 'heparin-binding site' (HBS) mutations are generally associated with a low risk of venous thrombosis, they have proven very useful in defining the heparin-binding domain of ATIII. These variants lead to loss of heparin binding but they retain normal antithrombin activity as assayed in the absence of heparin. The majority of reported HBS mutations appear to be due to missense mutations in residues 41, 47, 99 and 129 although isolated examples of mutations affecting heparin binding have also been found in residues 7, 20, 24, 99, 116, 118, 166, 237 and 368 (Borg *et al.*, 1988; Brennan *et al.*, 1988; Lane *et al.*, 1993; Okajima *et al.*, 1993; Park *et al.*, 1993; Chowdhury *et al.*, 1995). The locations of these lesions are consistent with the results of chemical modification and peptide/antibody-binding experiments (reviewed by Blajchman *et al.*, 1992b; Olds *et al.*, 1994a). Residues 7, 20, 24, 41 and 47 are located on the A α helix, residue 99 on the C helix and residues 116 and 129 on the D helix, respectively. The best characterized of these variants will now be described.

The Pro41→Leu ATIII variant was first reported as ATIII Basel (Chang and Tran, 1986). Chang and Tran (1986) considered that this variant was not involved directly in binding heparin but rather that the replacement of Pro at this position was likely to induce a structural change in the vicinity of the actual binding site through increasing helical structure and hydrophobicity. This conclusion was supported by de Roux *et al.* (1990) who examined the rate of thrombin inhibition by the Leu41-containing variant by comparison with the wild-type in the presence of an increasing concentration of heparin; kinetic analysis suggested that Pro41 was more likely to be involved in conformational changes induced by heparin than in the binding of the polysaccharide.

Heparin binds to ATIII by ionic bonding ('salt bridges') between the negatively charged sulphate groups of the polysaccharide and the positively charged basic amino acids (Arg, Lys) of the serpin. The observation that the Arg47→Cys substitution (ATIII Toyama; Sukuragawa *et al.*, 1983; Koide *et al.*, 1984. Padua 2; Olds *et al.*, 1990a) completely abolished heparin binding suggested that Arg47 might lie on the binding path of heparin. Substitutions of the same residue by His (ATIII Padua 1; Caso *et al.*, 1990. ATIII Rouen I; Owen *et al.*, 1987) and by Ser (ATIII Rouen II; Borg *et al.*, 1988) also impaired heparin binding, consistent with this postulate. Substitutions of Arg47 by His and Cys are associated with reductions in heparin affinity of 11- and 924-fold respectively (Watton *et al.*, 1993). Cys is neutral but His is capable of carrying a positive charge at low pH. Thus, at pH 6.0, ATIII Rouen I exhibits normal heparin affinity but at pH 8.0, the variant protein's affinity for heparin is greatly reduced (Evans *et al.*, 1992). Positive charge is therefore very important for the interactions between ATIII and heparin and may account for the dramatic difference between these two variants at the same residue.

The Leu99→Phe substitution (ATIII Budapest 3; Olds *et al.*, 1992b) is associated with a 14-fold reduction in heparin affinity (Watton *et al.*, 1993). Leu99 is located in helix C, which lies under helix D which is implicated in heparin binding. The introduction of a bulky Phe residue at this position may therefore distort the heparin-binding domain or alternatively may affect the glycosylation site at Asn96 (Olds *et al.*, 1992b).

Substitution of Gln for Arg129 (ATIII Geneva; Gandrille *et al.*, 1990; Evans *et al.*, 1992) results in diminished heparin binding but some residual heparin cofactor activity is retained. This variant is associated with a 136-fold reduction in heparin affinity (Watton *et al.*, 1993). Substitution by an uncharged Gln residue would be predicted to interfere with interaction with the negatively charged sulphate groups of heparin. The Arg129 residue is thought to play an important role both in establishing the conformation of the ATIII molecule and in enhancing the inhibition of factor Xa; the Gln variant, purified from patient plasma, can be partially reactivated by heparin for inhibition of thrombin but not factor Xa (Najjam *et al.*, 1994).

The substitution of Ile7 by Asn (Brennan *et al.*, 1988) creates a novel glycosylation site. The introduced carbohydrate side chain is predicted to result in steric hindrance of the interaction between heparin and ATIII.

It has been suggested that the reactive site variant, ATIII Kumamoto II (Arg393→His) has an increased affinity for binding heparin (Okajima *et al.*, 1995). Thus amino acid substitutions do not invariably decrease heparin binding but may also increase it.

Type II deficiency: pleiotropic mutations. A number of different mutations with pleiotropic effects have been reported in residues 402, 404, 405, 406, 407 and 429 (Lane *et al.*, 1993). Six of these PE variants (Phe402→Cys, Phe402→Ser, Phe402→Leu, Ala404→Thr, Asn405→Lys and Pro407→Thr), located in strand 1C and the polypeptide leading into strand 4B close to the C-terminal of the protein, were studied in detail by Lane *et al.* (1992b); all exhibited a heparin-binding abnormality as detected by crossed immunoelectrophoresis, five manifested a reduced plasma ATIII antigen concentration and two were shown to have greatly reduced heparin cofactor and progressive inhibitor activities *in vitro*. Thus mutations within strand 1C not only impair the function of the reactive site but may also result in a reduction in the plasma ATIII antigen level, presumably because this strand is essential for the structural integrity and stability of ATIII. Lane *et al.* (1992b) proposed that the defective interaction of these mutant proteins with thrombin resulted from the close proximity of strand 1C to the reactive site. They also speculated that the altered affinity for heparin resulted from conformational linkage between the reactive site and heparin-binding site regions of the molecule. Thus mutations in strand 1C would relay structural changes to the distal heparin-binding site by perturbing the B sheet and the core of the molecule.

ATIII Budapest (Pro429→Leu; Olds *et al.*, 1992a) and ATIII Kyoto (Arg406→Met; Nakagawa *et al.*, 1991) also manifest impaired thrombin-inhibitory and heparin-binding activities and are associated with a reduced plasma level of ATIII antigen. Pro429 lies adjacent to strand 1C in the three-dimensional structure of the protein and its replacement would disrupt a β turn. ATIII Utah (Pro407→Leu; Bock *et al.*, 1988) exhibits a reduced plasma antigen level, defective interaction with thrombin but normal affinity for heparin.

Mutations affecting residues Phe402, Ala404, Asn405, Pro407 and Pro429 are associated with reduced levels of ATIII antigen with reductions in heparin affinity of between 194- and 1241-fold (Watton *et al.*, 1993). Watton *et al.* (1993) suggested that these substitutions may not prevent heparin from forming its initial loose complex with ATIII but might interfere with high-affinity binding associated with the heparin-induced conformational change.

The reduced plasma level of ATIII could also be a consequence of another mechanism. Pleiotropic mutations could serve to expose a cryptic site, corresponding to residues 408–412, which in other serpins acts as an hepatocyte receptor recognition site (Joslin et al., 1991; Mast et al., 1991). These substitutions could thus enhance the removal of these ATIII variants from the plasma.

Type II deficiency: a temperature-sensitive variant. A highly unusual ATIII variant (ATIII Rouen VI) has been described which is caused by the substitution of Asn187 by Asp (Bruce et al., 1994). The proposita and her father had both experienced deep vein thrombosis during pyrexial infections. This ATIII variant exhibited an increased heparin affinity but appeared to possess normal inhibitory activity. However, the ATIII activity level declined rapidly after storage of plasma at 4°C. This loss of activity was found to occur only slowly at 37°C but was greatly accelerated by incubation at fever temperature, 41°C.

Biochemical studies showed that this loss of activity was accompanied by the formation of a new stable inactive and uncleaved (L) form of ATIII (latent antithrombin) in which the bait loop becomes fully inserted into the A sheet. Polymerization also occurred with the bait loop of one variant ATIII molecule becoming inserted into the A sheet of another. Elevated temperature was shown to accelerate the formation of the latent form of ATIII and bait loop–A sheet polymers resulting in a loss of ATIII activity.

Insertion of a synthetic 11-mer peptide into ATIII Rouen VI was faster than for wild-type ATIII, indicating that the variant is the better acceptor of the bait loop, consistent with an increased propensity to form latent antithrombin. Cleavage of the P4–P5 peptide bond in ATIII Rouen VI by neutrophil elastase was eight-fold slower than for wild-type ATIII, indicating that this bond is less accessible in the variant ATIII molecule. Thus the mutation increases accessibility of the A sheet for loop incorporation, resulting in increased formation of latent ATIII and ATIII polymerization. The thermolability of ATIII Rouen VI provides an explanation for why thrombotic episodes in this family occurred in association with fever.

Bruce et al. (1994) proposed that the increased accessability of the A sheet was due to the fact that Asp187 forms an H bond to a carbonyl group in the peptide loop that connects the F helix to strand 3A. The F helix is involved in the sliding movement that opens the A sheet to allow insertion of strand 4A and interacts with strand 5A of the A sheet to promote stabilization of the A sheet. Substitution of Asn187 by Asp breaks the link between the F helix and strand 3A allowing movement of strands 1A, 2A and 3A and the opening of the A sheet between strands 3A and 5A.

Mutations in Asp187. Two different single base-pair substitutions have been reported in the highly conserved Asn187 residue converting it to Asp and Lys respectively (Perry et al., 1995a). The Asn187→Asp substitution is associated with a type II deficiency state whilst the Asn187→Lys substitution is associated with a type I deficiency. Asn187 is located in the F helix and is linked to Ile202 of strand 3 of the A sheet by a hydrogen bond. The F sheet overlies the A sheet and moves with strands 2 and 3 of this sheet as they open to allow entry of the reactive site loop to form strand

4. As we have seen, this movement is essential for the ATIII molecule to adopt its inhibitory conformation. Substitution of Asn187 probably disrupts this sliding movement, leading to a loss of inhibitory activity.

Mutations in CpG dinucleotides. As found in many other human genes, CpG mutations appear to be very frequent in the *AT3* gene (Perry and Carrell, 1989). In terms of individual patients in whom single base-pair substitutions have been characterized, over 50% are C→T or G→A transitions within CpG dinucleotides compatible with a model of methylation-mediated deamination (Cooper and Krawczak, 1993). However, if only independent mutations are considered, this proportion falls to 31%, close to the average for human disease genes (Cooper and Krawczak, 1993). CpG mutations contribute disproportionately to heparin-binding defects through recurring mutations in residues 41 and 47. CpG mutations are also evident in the reactive site residues 393 and 394. In cases of recurring mutations at CpG sites, it is often unclear whether they are truly recurrent (i.e. of independent origin) or are instead identical by descent. These types of lesion may be distinguished by determining the haplotypes of the *AT3* alleles bearing the lesions. Thus, using the *Alu* 8 VNTR, Olds *et al.* (1994b) demonstrated that a Leu99→Phe lesion was associated with the same haplotype in five apparently unrelated families and was therefore likely to be identical by descent. However, these authors also demonstrated that a CGA→TGA transition converting Arg129→Term was associated with two distinct haplotypes, consistent with the hypothesis of recurrent mutation at this site.

Mode of inheritance. The vast majority of known cases of ATIII deficiency are heterozygous for an *AT3* gene defect. The deficiency state is therefore normally inherited as an autosomal dominant trait. There are, however, several notable exceptions to this rule. ATIII Tours is an asymptomatic heterozygous type II (HBS) ATIII deficiency caused by an Arg47→Cys substitution (Duchange *et al.*, 1987). However, when the same mutation occurs in the homozygous state as in ATIII Toyama (Koide *et al.*, 1984), ATIII Alger (Brunel *et al.*, 1987) and ATIII Kumamoto (Ueyama *et al.*, 1989, 1990), recurrent episodes of both venous and arterial thrombosis occur.

Several examples of individuals homozygous for the heparin-binding site mutation, Leu99→Phe, have also been reported (Olds *et al.*, 1992b; Chowdhury *et al.*, 1994); again the individuals concerned suffered episodes of both venous and arterial thrombosis. As with the Arg47→Cys substitution, heterozygous relatives of the patients homozygous for the Leu99→Phe substitution were asymptomatic. A further patient (ATIII Fontainebleau), the product of a consanguineous union, has been described with a hitherto uncharacterized homozygous mutation (Boyer *et al.*, 1986).

A single report of a family with homozygous type I deficiency has been reported (quoted in Blajchman *et al.*, 1992b); two homozygously affected offspring died within the first 3 months of life. Thus for highly deleterious mutations, the homozygous deficiency state may be lethal.

Co-inheritance of APC resistance and ATIII deficiency. There are several examples of a particularly severe form of ATIII deficiency segregating in some families. The recognition of the factor V Leiden variant as a risk factor in venous thrombosis has provided a potential explanation for this phenomenon. This is because the *AT3* (chromosome 1q23–q25.1) and the *F5* (1q21–q25) genes are both present in the same region of the long arm of chromosome 1 and therefore have the potential to co-segregate through family pedigrees. van Boven *et al.* (1996) studied the co-inheritance of factor V Leiden and inherited ATIII deficiency in a total of 128 families with ATIII deficiency. Altogether, 127 probands and 188 of their relatives were screened for the factor V Leiden variant: the variant was found in 18 families. Nine families were available for assessment of the mode of inheritance and clinical significance of the combined deficiency state. Co-segregation of both defects was observed in four families.

A total of 11/12 individuals possessing both a reactive site or pleiotropic effect ATIII mutation and the factor V Leiden variant had experienced thrombotic symptoms. By contrast, for individuals bearing both a heparin-binding site defect of ATIII and the factor V Leiden variant, only 2/6 manifested thrombotic symptoms. The severity of the ATIII defect may therefore have a bearing on the likelihood of the combined deficiency state coming to clinical attention. Lane *et al.* (1996b) reported the first prenatal diagnosis in a case of combined ATIII deficiency and factor V Leiden.

ATIII deficiency: other influences on the clinical phenotype. ATIII levels are known to be significantly higher in women than men and higher in Blacks than in Whites (Conlan *et al.*, 1994). However, ATIII levels decreased with age in men whereas they increased with age in women (Conlan *et al.*, 1994). In age- and race-adjusted analyses, ATIII levels were positively correlated with smoking, high density lipoprotein–cholesterol level, triglyceride level (men only) and, in women, with diabetes and lipoprotein (a) level (Conlan *et al.*, 1994). ATIII levels were negatively correlated with educational level, body mass index in men and hormone treatment in females (Conlan *et al.*, 1994). Finally, glucocorticoids are known to stimulate ATIII expression whereas oestrogen reduces it (Thaler and Lechner, 1981).

One genetic factor influencing the probability of an individual with ATIII deficiency coming to clinical attention is *epistasis*, the contribution from another distinct locus. The obvious candidate here is factor V since the factor V Leiden variant is an important independent risk factor for venous thrombosis and the *AT3* and *F5* genes both map to the same region (q21–q25) of chromosome 1. van Boven *et al.* (1995) studied 111 families with ATIII deficiency and found 13 (11.7%) to have the factor V Leiden variant. The two defects were found to co-segregate in some families but not in others, presumably due to recombination between the *AT3* and *F5* loci. The mean age of onset of thrombosis in individuals with both traits was 16 ($n=11$), compared with 26 ($n=15$) for those with only the ATIII defect and 44 for those with only the factor V Leiden variant. The presence of the factor V Leiden variant therefore appears to be an additive risk factor for those already predisposed to thrombosis by virtue of an inherited ATIII defect.

Polymorphic variation in the promoter region of the *AT3* gene could in principle have some clinical influence by affecting the expression of the downstream gene. However, by means of both *in vitro* and *in vivo* assays, Winter *et al.* (1995) have shown that there is little or no difference between the long and short alleles of the promoter length polymorphism (Sections 4.3.3 and 4.3.4).

Protein C and protein C deficiency

5.1 Introduction

Protein C is a vitamin K-dependent plasma protein which, although highly homologous to factors X, IX and VII, functions as a natural anticoagulant through the inactivation of factors Va and VIIIa. Protein C, like other coagulation proteins, circulates in an inactive form which is activated by limited cleavage with thrombin. This step is greatly accelerated by thrombomodulin, an endothelial surface protein cofactor. Activated protein C (APC) itself requires a cofactor, protein S (see Chapter 6), for efficient inactivation of factors Va and VIIIa. Protein C circulates in human plasma at a concentration of 4 mg l^{-1} with a half-life of approximately 10 h (Riess *et al.*, 1985).

Deficiency of protein C has been shown to be one of the most common causes of an inherited thrombotic tendency. Although many mutations underlying protein C deficiency have been identified, it is not yet understood why protein C deficiency should be so common in the general population (approx. 1 in 200) but associated with thrombosis in only a minority of affected families. For general reviews, see Clause and Comp (1986), Bertina (1988a), Dolan *et al.* (1989), Esmon (1989b) and Dahlbäck (1995a).

5.2 Structure and function of protein C

5.2.1 Primary sequence and proteolytic processing

Protein C circulates predominantly as a covalently linked dimer (Stenflo, 1976) although 10–15% of protein C circulates in a single-chain form, lacking the removal of the internal Arg–Lys dipeptide (Miletich *et al.*, 1983). The amino acid sequence of the protein, predicted from cDNA sequence (Foster *et al.*, 1985), indicates the dimer to be derived by proteolytic processing from a single-chain precursor. The amino acid sequence of pre-pro-protein C is presented in *Figure 5.1*. The pre-pro-sequence of 42 residues (–1 to –42) is cleaved by a trypsin-like processing proteinase to reveal the light chain N-terminal sequence ANSFL. The pre-pro-leader sequence functions in two ways. The pre-sequence (–42 to –25) contains a hydrophobic core or signal peptide (–40 to –29) which targets the protein for secretion (Gierasch, 1989; Nothwehr and Gordon, 1990). During translocation through the membrane of the rough endoplasmic reticulum, the signal peptide is cleaved by a signal peptidase. The pro-sequence (–24 to –1) contains the conserved residues and sequence homology common to all Gla proteins which determines vitamin K-dependent γ-carboxylation (Vermeer, 1990). The propeptide is thought to be removed when the precursor protein reaches the Golgi body. Foster *et al.* (1987) have shown that residues Arg1 and Arg4 of the propeptide are important for pro-peptide release. Since the pre-pro-peptides of human protein C, factor VII and factor IX are interchangeable without loss of function, this region of protein C would appear only to possess a general rather than a specific role in γ-carboxylation and protein export (Geng and Castellino, 1996).

A dipeptide, Lys–Arg, which connects Leu155 to Asp158, is removed prior to secretion by a carboxypeptidase B-like enzyme in a calcium-dependent reaction (Foster *et al.*, 1991; McClure *et al.*, 1992; *Figure 5.2*). The wild-type sequence presents a dibasic

Figure 5.1. Amino acid sequence and tentative secondary structure of human pre-pro-protein C. The locations of the seven introns (A – G) are shown in the various regions of the protein. An eighth intron in the 5′ untranslated region is not shown. The pre-pro-leader sequence (–42 to –1) is cleaved during protein biosynthesis to give rise to the mature protein with an N-terminal sequence of ANSF for the light chain. The Lys156–Arg157 dipeptide that connects the light and heavy chains during biosynthesis is not shown. The Arg169–Leu170 peptide bond cleaved by the thrombin–thrombomodulin complex during the activation reaction is shown with an arrow. The three amino acids of the catalytic triad are circled whilst the four potential carbohydrate attachment sites are shown with solid diamonds. Adapted from Reiner and Davie (1995) by kind permission of McGraw-Hill Inc, New York.

target flanked by Ser–His–Leu on the N-terminal side. Foster *et al.* (1990) studied the effects of sequence alterations on the efficiency of cleavage of protein C. In factor X, the corresponding sequence is Ser–Glu–Arg. Replacement of His154 by Arg by site-directed mutagenesis led to a far higher proportion of two-chain processed protein C being released from baby hamster kidney cells *in vitro*, emphasizing the importance of this residue for recognition by the processing proteinase(s). Substitution of the basic amino acids (Lys156, Arg157) also led to loss of processing. Cleavage of the Lys156–Arg157 dipeptide is effected by the paired basic amino acid cleaving enzyme, furin (Drews *et al.*, 1995; Denault *et al.*, 1996).

The pathway of processing and secretion of protein C is not yet clearly understood but the relative timing of the different post-translational modifications has begun to be elucidated (McClure *et al.*, 1992; *Figure 5.3*).

5.2.2 Secondary structure and domains

The putative secondary structure of protein C (*Figure 5.1*) is very similar to those of factors IX, X and VII. The 37 N-terminal residues of the mature protein (after signal peptide cleavage) comprise the Gla domain, connected by an amphipathic helix or 'aromatic stack' (residues 38–45) to two epidermal growth factor (EGF) homologous domains (46–91 and 92–136). The pattern of disulphide linkage in the first growth factor domain is highly unusual, with four bonds rather than the three common to other EGF domains. The additional bond (Cys59–Cys64) is within the first loop and may distort the three-dimensional fold of this domain. The Gla and EGF domains comprise the light

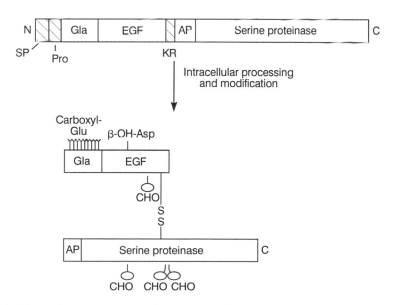

Figure 5.2. Schematic representation of nascent and fully processed human protein C. The relative locations of the functional domains of the protein, modified amino acids β-hydroxyaspartic acid (β-OH-Asp) and γ-carboxyglutamic acid (Y), and N-linked glycosylation sites (CHO) are indicated. SP, signal peptide; Pro, propeptide; Gla, γ-carboxyglutamic acid domain; EGF, epidermal growth factor-like domains; KR, Lys–Arg dipeptide; AP, 12-amino-acid activation peptide. Derived from McClure *et al.* (1992) with permission from the American Society for Biological Chemistry.

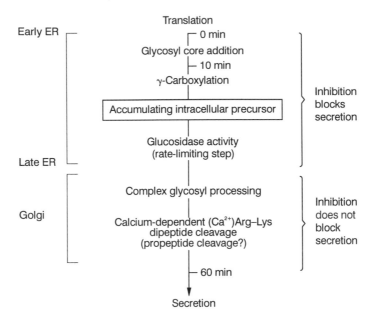

Figure 5.3. Time course for the processing and secretion of recombinant protein C in the endoplasmic reticulum (ER) and golgi of human cells. Derived from McClure *et al.* (1992) with permission from the American Society for Biological Chemistry.

chain after activation. The activation peptide region contains the dipeptide (Lys156–Arg157) released during late processing of the protein, and the scissile bond (Arg169–Leu170) that is cleaved on activation of protein C by thrombin. The heavy chain of the mature protein therefore comprises the activation peptide and the catalytic domain. It is homologous to other serine proteinases with fairly close similarities to factors X, IX and VII, although detailed analysis reveals regions of highly conserved sequence separated by variable regions (Furie *et al.*, 1982; Greer, 1990).

Protein C, in common with factors VII, IX, X and prothrombin, is a member of the family of trypsin-like serine proteinases which all possess His, Asp and Ser residues closely juxtaposed in a catalytic triad. The catalytic triad residues of protein C are located at His211, Asp257 and Ser360: cleavage of substrate peptide bonds proceeds using a 'charge relay' mechanism which potentiates the partial ionization of the side chain of Ser360. Complex formation between enzyme and substrate is followed by an attack on the carbonyl carbon atom of the scissile bond by the serine whose nucleophilicity is enhanced by the presence of the adjacent histidine which functions as a base catalyst. Proton donation by the histidine to the newly formed alcohol or amine group then results in the dissociation of the first product and formation of an acyl–enzyme intermediate. Deacylation then occurs with the attacking nucleophile provided by a water molecule. Each step proceeds through a tetrahydral intermediate. The Asp residue appears to be important but may not be absolutely required for catalysis to occur. This charge relay mechanism is capable of accelerating the rate of hydrolysis of the scissile peptide bond by a factor of more than 10^9 relative to the uncatalysed reaction.

The heavy and light chains of protein C are covalently linked by a single disulphide bond (Cys141–Cys277). The remaining six Cys residues form intra-chain disulphide linkages (Cys196–Cys212, Cys331–Cys345, Cys356–Cys 384).

5.2.3 The γ-carboxyglutamic acid (Gla) domain

The γ-carboxyglutamic acid (Gla) domain of protein C functions by providing a number of calcium-binding sites that allow protein C to undergo a conformational change to a calcium-dependent form necessary for binding to phospholipid membranes. The Gla domain appears to perform the same function(s) in the other Gla-containing proteins of coagulation since it has been shown that the Gla regions of the different proteins may be interchanged with little alteration in physiological function (Christiansen and Castellino, 1994; Freedman *et al.*, 1996).

In protein C, nine glutamic acid residues at positions 6, 7, 14, 16, 19, 20, 25, 26 and 29 are modified to Gla. The importance of this modification is evidenced by the loss of protein C anticoagulant activity after blockage of γ-carboxylation through administration of the vitamin K antagonist, warfarin.

The pro-sequence of protein C (residues −24 to −1) is required for vitamin K-dependent carboxylase recognition (Suttie *et al.*, 1987; reviewed by Suttie, 1993). The recognition process has been analysed by site-directed mutagenesis and *in vitro* expression (Foster *et al.*, 1987); thus residues −12 to −17 are known to be important for γ-carboxylation since a protein C mutant lacking these residues was unable to bind barium citrate, had no anticoagulant activity, and on sequencing yielded glutamic acid at the first two potential Gla residues (6 and 7). Residue −16 (Phe) has been implicated in carboxylase recognition by other work in which this residue was mutated, leading to abolition of γ-carboxylation (Jorgensen *et al.*, 1987). Residues −1 to −10 can support partial γ-carboxylation *in vitro* in a cell-free assay system (Suttie *et al.*, 1987). Probably the entire region from −1 to −17 contains recognition motifs for the carboxylase enzyme, since it is highly conserved between Gla proteins. Zhang and Castellino (1991) have suggested that residues 17–22 of the hexapeptide disulphide loop may also be important for substrate recognition by the γ-carboxylase.

Mutagenesis of the highly conserved Gla residues at positions 6 and 7 were found to greatly reduce functional protein C activity although this did not alter γ-carboxylation at the other Gla residues (Zhang and Castellino, 1990). Similarly, *in vitro* mutagenesis of protein C residues Gla19, Gla20 and Gla22 resulted in virtually complete inactivation of its calcium-dependent activity in a clotting assay (Zhang and Castellino, 1991). This result suggests that the hexapeptide disulphide loop of protein C is important for the calcium effect required for inhibition of protein C activation by thrombin as well as for the parallel stimulation of activation by the thrombin–thrombomodulin complex. Mutation of each Gla residue individually has shown that residues 7, 16, 20, 25 and 26 are critical for calcium-dependent anticoagulant activity whilst mutational substitution of Gla residues 6, 14 and 19 and Arg15 result in protein C molecules with substantial residual anticoagulant activity relative to wild-type activated protein C (Zhang *et al.*, 1992; Zhang and Castellino, 1993; Jhingan *et al.*, 1994). Mutation of Gla residue 29 resulted in an intermediate level of anticoagulant activity.

5.2.4 Interaction with phospholipid

Protein C interacts with phospholipid vesicles through its Gla domain (reviewed by Mann *et al.*, 1990) via a two-stage process. First, a conformational change occurs as a consequence of the occupancy of two or three cation binding sites in the Gla domain. These sites are relatively non-specific for the type of multivalent cation present. Then a second group of four or five sites, with much greater specificity for Ca^{2+}, bind the cation in an interaction which is required for binding to phospholipid. Zhang and Castellino (1992, 1993) have shown that Gla16 and Gla26 (which coordinate calcium ions Ca-3, Ca-4 and Ca-5 not exposed on the surface) are critical for the maintenance of the Ca^{2+}-dependent structure of the Gla domain, whereas residues Gla20 and Arg15 (which coordinate the surface exposed Ca-6) are critical for the direct binding of protein C to the phospholipid surface. Gla25 and Gla29 coordinate Ca-1 which is important for the orientation of protein C on the phospholipid surface. Gla14 and Gla19 (which coordinate the surface-exposed Ca-7) are not critical to the function of protein C (Zhang and Castellino, 1992, 1993; Christiansen et al., 1994). Zhang and Castellino (1994) have also demonstrated the importance of residue Leu5 in the binding of protein C to phospholipid vesicles; mutation of this residue reduced protein C anticoagulant activity in a phospholipid-dependent assay despite retention of the calcium-dependent conformation. Christiansen et al. (1994) have shown that Gla6 (which coordinates Ca-4 and Ca-5) is essential for calcium binding.

5.2.5 Other post-translational modifications

A β-hydroxyaspartic acid residue is located at residue 71 (Fernlund and Stenflo, 1983). This residue appears to be directly involved in calcium binding since replacement of β-hydroxyaspartic acid by glutamic acid removed the recombinant protein's calcium-dependent epitope and reduced its biological activity to approximately 10% of normal (Öhlin et al., 1988b). Similarly, mutagenesis of this residue to Ala reduced Ca^{2+} binding to the EGF1 domain (Geng et al., 1996).

Based on consensus sequence and 23% carbohydrate content by weight (Kisiel, 1979), human protein C probably has N-linked carbohydrate attached to Asn97, Asn248, Asn313 and Asn329. Glycosylation occurs before γ-carboxylation and the two processes are not coupled (McClure et al., 1992; Figure 5.2). The extent of glycosylation of protein C is variable both in vivo and in vitro (Greffe et al., 1989; Yan et al., 1990) but is required for efficient secretion (Grinnell et al., 1991; McClure et al., 1992). Indeed, approximately 30% of protein C in human plasma is not glycosylated at Asn329; however, this 'β-form' of protein C retains full biological activity (Miletich and Broze, 1990). The carbohydrate groups may play a role in the inhibition of E-selectin-mediated cell adhesion to cultured endothelial cells, an important event in inflammation (Grinnell et al., 1994).

5.2.6 Activation of protein C

Role of thrombomodulin. An endothelial cell cofactor which binds thrombin with high affinity and enhances the rate of activation of protein C by thrombin 20 000-fold was identified by Esmon and Owen (1980, 1981). This cofactor, termed thrombomodulin, has been isolated and characterized (see Chapter 8; for reviews, see Esmon 1989a, b; Dittman and Majerus, 1990). Thrombomodulin is anchored to the endothelial cell surface by a transmembrane domain to which is attached an extracellular portion with six EGF-like domains. The two EGF domains nearest the cell surface (EGF5 and EGF6) contain the thrombin-binding region. EGF4 is required for protein C activation (Zushi et al., 1991). The thrombomodulin–thrombin complex appears to interact with both the Gla and EGF domains of protein C (Hogg et al., 1992; Olsen et al., 1992). Residues Lys191, Lys192 and Lys 193 appear necessary for efficient activation by the thrombin–thrombomodulin complex although not for activation by free thrombin (Gerlitz and Grinnell, 1996).

Thrombomodulin provides about 50–60% of thrombin-binding sites on the endothelial cell surface, approximately 10^5 thrombomodulin molecules per endothelial cell (Maruyama and Majerus, 1985). Thrombomodulin is located predominantly in the microvasculature which accounts for 99% of the total endothelial surface presented to blood. The concentration of membrane-bound thrombomodulin is more than 1000-fold higher in the microvasculature than in the major vessels. Thus whilst thrombin is free in the major vessels, it is bound by thrombomodulin as soon as it enters the microvasculature.

Upon binding thrombomodulin, thrombin loses its ability to activate factors V and VIII and to clot fibrinogen whilst its activity toward protein C is enhanced (reviewed by Walker and Fay, 1992). The half-life of the thrombin–thrombomodulin complex on the cell membrane is less than 15 sec (Esmon, 1987), the complex being internalized by endocytosis. Thrombomodulin also stimulates the rate of inactivation of thrombin by ATIII (Hofsteenge et al., 1986; Preissner et al., 1987), blocks the ability of thrombin to activate platelets and inhibits the factor Xa-mediated activation of prothrombin (Thompson and Salem, 1986). Recombinant human thrombomodulin exhibits an antithrombotic effect on thrombin-induced thromboembolism in rats and mice (Kumada et al., 1988; Gomi et al., 1990; Solis et al., 1991).

Recent results have indicated that prothrombin activation products are also capable of activating protein C (Hackeng et al., 1996). Specifically, meizothrombin, formed during the initial phase of prothrombin activation, activates protein C on the endothelial cell surface with a six-fold higher efficiency than α-thrombin. In so doing, it down-regulates its own formation as well as that of thrombin.

Cleavage by thrombin and factor X. Protein C is converted to its activated form (APC) by a single cleavage of the Arg169–Leu170 peptide bond, releasing a dodecapeptide from the heavy chain. This activation peptide is present in the blood of healthy individuals although it is rapidly removed from the circulation (Bauer et al., 1984). APC circulates in blood at a very low level owing to its very short half-life of 20–30 min. About 10% of protein C is not cleaved within the activation peptide and circulates as a single-chain molecule (Miletich et al., 1983).

The sequence of the thrombin cleavage site in protein C is Asp–Pro–Arg–Leu–Ile–Asp (residues 167–172). Replacement of either of the Asp residues at the P3 or P3′ positions with Gly increased the rate of thrombin cleavage approximately 30-fold (Ehrlich et al., 1990; Richardson et al., 1992). These two Asp residues are thus inhibitory to thrombin proteolysis and serve to make protein C a poor substrate for thrombin. This they might bring about by binding near acidic residues in thrombin, resulting in charge repulsion. Residues Glu39 and Glu192 (chymotrypsin numbering) of thrombin have been proposed to interact with the P3 and P3′ residues of protein C (Le Bonniec et al., 1991a, b). Mutation of Glu192 to Gln yields a thrombin molecule which cleaves the activation peptide from protein C some 20 times faster than wild-type thrombin (Le Bonniec et al., 1991b). However, substitution of both the P3 and P3′ Asp residues by Gly decreases the difference in hydrolysis rate between the mutant and the wild-type thrombin to less than four-fold. Thus the thrombin mutant facilitates the cleavage of substrates with acidic residues in the P3 and P3′ positions. Activation by the thrombin mutant in the presence of thrombomodulin was only two-fold faster than that obtained with the wild-type thrombin–thrombomodulin complex. The thrombin mutant therefore only results in greatly enhanced cleavage of protein C in the absence of thrombomodulin; the presence of thrombomodulin essentially abolishes the distinction between the wild-type and mutant thrombins.

Esmon (1993b) proposed a model to explain these findings: the Glu192 residue of thrombin is involved in slowing protein C cleavage whilst thrombomodulin binding to thrombin causes the side chain of Glu192 to move, thereby alleviating its inhibitory influence. In effect, the Glu192 to Gln substitution mimics the catalytic switch normally induced by thrombomodulin, thus promoting protein C activation.

Replacement of Pro at the P2 position by either Val or Leu influences the calcium-dependent conformational changes in protein C that control activation (Rezaie and Esmon, 1994b). These mutations lead to greatly reduced secretion since the protein is retained within the cell. Replacement of the P3 and P3' Asp residues in human protein C by Asn did not affect thrombin activation or the requirement for calcium by the thrombin–thrombomodulin activation complex (Rezaie and Esmon, 1994a).

A small number of residues in the extended active site pocket, or even remote from the active site pocket itself, are important in conferring substrate specificity upon protein C. Studies such as that of Rezaie (1996), who has implicated residue Thr268 in conferring specificity at the S2 subsite, are important in defining the determinants of differential specificity between the various serine proteinases of coagulation.

The residues of thrombin which interact with protein C have been studied by Tsiang et al. (1995) by means of site-directed mutagenesis/alanine scanning: Lys21, Trp50, Lys65, His66, Arg68, Tyr71, Arg73, Lys77, Lys106, Lys107, Glu229 and Arg233 of thrombin all appear to be involved. The substitution of Ala for Glu at residue 229 abolishes thrombin's procoagulant function but does not adversely affect its anticoagulant function, that is its ability to activate protein C in the presence of thrombomodulin (Gibbs et al., 1995).

Berg et al. (1996a) have shown that a modulator of the enzymatic activity of human thrombin, LY254603, was capable of both enhancing the thrombin-catalysed generation of protein C and inhibiting thrombin-dependent clotting of fibrinogen. This was achieved through altering thrombin's S3 substrate recognition site. This type of approach could be used in the development of a new generation of antithrombotic drugs.

Factor Xa is also capable of activating protein C through cleavage of the heavy chain which is indistinguishable from that produced by thrombin (Haley et al., 1989; Freyssinet et al., 1989). This reaction requires calcium ions and phospholipid and proceeds at a rate which is 10-fold greater than that found with α-thrombin. Thrombomodulin also serves as a cofactor for this reaction the rate of protein C activation is stimulated to the same extent as with thrombin.

The role of factor V in protein C activation. Factor Va, and more specifically its light chain, serves as a cofactor in the thrombin-mediated activation of protein C (Salem et al., 1983a, b). However, the activity of factor Va as a cofactor is only 5% of that of thrombomodulin (Salem et al., 1984) and its functional significance is unclear.

The role of calcium in protein C activation. Calcium plays two roles in protein C activation: it inhibits activation by α-thrombin (Amphlett et al., 1981; Rezaie and Esmon, 1992) but is required for activation by the thrombomodulin–thrombin complex (Rezaie and Esmon, 1992). In both reactions, calcium brings about a 10- to 100-fold change in rate. These effects of calcium are mediated through calcium-induced conformational changes in protein C. The Gla domain binds calcium, inducing a conformational change which is required for interaction of protein C with phospholipid. A high-affinity calcium-binding site is also present in the EGF domains of the light chain (Öhlin and Stenflo, 1987; Öhlin et al., 1988a, 1990; reviewed by Esmon, 1989) and binding of calcium to this site is required for protein C activation. Evidence that this calcium-binding site involves the β-hydroxyaspartic acid at residue 71 in the first EGF domain was presented by Öhlin et al. (1988b): calcium-binding to this region induced a conformational change affecting not only the EGF domains but also the heavy chain activation site (Stearns et al., 1988; Orthner et al., 1989). However, protein C deletion mutants lacking the Gla domain and either with or without the first EGF domain were indistinguishable with respect to calcium response (Rezaie et al., 1992). The high-affinity binding site of protein C does not simply reside within the first EGF domain. Esmon (1993b) proposed that the first EGF domain of protein C binds calcium with high-affinity only in the presence of the Gla domain and that the Gla domain serves to convert the calcium-binding site in the the first EGF domain from a low- to a high-affinity site.

A Glu80→Lys mutation results in a protein C molecule which no longer requires calcium for activation by the thrombin–thrombomodulin complex and whose activation by thrombin alone is not inhibited by calcium (Rezaie et al., 1994). It is thought that this substitution may serve to introduce a salt bridge with Glu70 which then stabilizes the protein C zymogen in a conformation similar or identical to that required for rapid activation which is normally stabilized by calcium.

5.2.7 Function of APC and requirement for protein S

APC generated by the thrombin–thrombomodulin complex is a serine proteinase with a relatively long half-life (20–30 min; Okajima et al., 1990) and a very restricted substrate specificity. APC cleaves factors V and VIII at specific sites in each cofactor, rendering them inactive. In the case of factor VIII, the scissile bonds are at Arg336 –Met337, Arg562 –Gly563 and probably Arg740–Ser741 within the heavy chain (Fay et al., 1991a). APC cleaves membrane-bound human factor V at four sites: Arg306, Arg506, Arg679 and Lys994 but, in the absence of membrane, no cleavage occurs (Kalafatis et al., 1994). By contrast APC cleaves membrane-bound human factor Va at three sites: Arg306, Arg506 and Arg679 (Kalafatis et al., 1994). Inactivation of both factors V and Va requires the presence of membrane and cleavage of Arg306 (Kalafatis et al., 1994). Cleavage at Arg506 precedes, and is required for, subsequent cleavage at Arg306 and Arg679 (Kalafatis et al., 1994). In the absence of membrane, APC cleaves the heavy chain of factor V only at Arg506 and Arg679 (Kalafatis et al., 1994). Both factors V and VIII are much more susceptible to attack by protein C after they have been activated by thrombin. The essential role of protein C in the inactivation of factor Va is evidenced by the significant inhibition of the inactivation of factor Va in vitro by addition of anti-protein C antibodies (Jackson et al., 1994).

Three regions of APC are essential for anticoagulant activity and recognition of factor Va: residues 142–155 within the connecting region (Mesters et al., 1993a) and residues 311–330 and 390–404 within the heavy chain (Mesters et al., 1991, 1993b). The APC-binding sites on factor Va and factor VIIIa are located at residues 1865–1874 and 2009–2018 respectively (Walker *et al.*, 1990). Inactivation of factor VIIIa by APC is reduced when the cofactor is complexed with either von Willebrand factor (Fay *et al.*, 1991b) or factor IXa (Bertina *et al.*, 1984). In an analogous fashion, factor Xa reduces the inactivation of factor Va by APC (Nesheim *et al.*, 1982).

Walker (1980) showed that protein S accelerated the rate of inactivation of factor V by APC. This effect requires the presence of phospholipid and Ca^{2+}. Protein C appears to interact with protein S through its EGF domains (Öhlin *et al.*, 1988b, 1990). In a more detailed kinetic study, protein S was shown to have only a modest effect on the rate of factor Va inactivation (Solymoss *et al.*, 1988). However, protein S abrogated the protective effect of factor Xa, suggesting that the physiological effect of protein S is to allow APC inactivation of factor Va in the presence of factor Xa. Whatever its precise mode of action, protein S is clearly important physiologically since its deficiency is associated with clinical thrombophilia.

Mesters *et al.* (1991) showed that a peptide corresponding to residues 390–404 of protein C inhibited APC anticoagulant activity in coagulometric assays. So did polyclonal antibodies raised against this peptide. These authors suggested that residues 390–404 represent an exosite on APC which is necessary for its anticoagulant activity and for the recognition of its substrates factors Va and VIIIa. Since neither the peptide nor the antibodies inhibited the amidolytic activity of APC, this exosite is distinct from that of the active site.

A recombinant protein C molecule whose EGF-like domains had been replaced by those of factor IX exhibited 30% normal activity toward factor Va and less than 10% normal activity toward factor VIIIa (Yu *et al.*, 1994). Thus the EGF domains may contribute to the anticoagulant activity of APC, perhaps by helping to align its substrates on the phospholipid surface.

APC can also ameliorate the effects of endotoxaemia in mammals: for example, it prevents the lethal effects of lipopolysaccharide-induced septic shock in baboons (Taylor *et al.*, 1987). It is thought that this effect is brought about by the suppression of cytokine production, specifically tumour necrosis factor α (TNFα; Grey *et al.*, 1994). Protein C zymogen and active site-blocked APC were, however, ineffective in inhibiting TNFα (Grey *et al.*, 1994). A role for protein C in the inflammatory response is also suggested by the observation that APC decreases lipopolysaccharide-induced cytokine production by cultured human monocytes, inhibits CD14 down-regulation induced by interferon and inhibits CD11 down-regulation induced by lipopolysaccharides (Grey *et al.*, 1993).

5.2.8 Protein C and fibrinolysis

Using a clot lysis test, Comp and Esmon (1981) demonstrated that infusion of APC into dogs enhanced fibrinolysis. The precise mechanism of this fibrinolytic effect in the whole animal was initially unclear (Bajzar *et al.*, 1990) but *in vitro* studies suggested that it might be attributed specifically to the attenuation of prothrombin activation (Bajzar and Nesheim, 1993; de Fouw *et al.*, 1993). This would mean that prothrombin activation and plasminogen activation are coupled and that the profibrinolytic effect of APC is expressed through the inhibition of prothrombin activation. Gruber *et al.* (1994) proposed that APC might enhance the efficacy of fibrinolysis by reducing the relative mass of fibrin within thrombi.

APC added to cultured endothelial cells forms complexes with plasminogen activator inhibitor-1 (PAI-1), albeit slowly, thus neutralizing PAI-1 (Sakata *et al.*, 1986; de Fouw *et al.*, 1987, 1988). Although this would be predicted to shift the balance towards fibrinolysis by allowing unopposed action of tissue plasminogen activator, the interaction between APC and PAI-1 does not automatically exert an effect on fibrinolysis (Weinstein and Walker, 1991). The latent form of PAI-1 is not susceptible to formation of complexes with APC (Sakata *et al.*, 1988) so that only partial inactivation of PAI-1 is achieved by APC when added to culture medium.

The mechanism for the thrombin-dependent profibrinolytic action of protein C has now been investigated in some detail and involves a thrombin-activatable fibrinolysis inhibitor (TAFI), also known as plasma procarboxypeptidase B (Bajzar *et al.*, 1995). TAFI, once activated by thrombin, attenuates fibrinolysis by catalytic removal of C-terminal lysines from partially degraded fibrin (Redlitz *et al.*, 1995; Bajzar *et al.*, 1996a), thereby inhibiting fibrinolysis. Thrombin activation of TAFI is accelerated 1000-fold by thrombomodulin (Bajzar *et al.*, 1996a) thereby providing the link between coagulation and fibrinolysis and maintaining a balance between these pathways. Bajzar *et al.* (1996b) have now shown that APC promotes fibrinolysis by preventing the formation of thrombin, thereby preventing activation of TAFI and subsequent inhibition of fibrinolysis.

Protein S appears to play a role in promoting the fibrinolytic activity of APC (de Fouw *et al.*, 1990; Weinstein and Walker, 1991). Studies on patients with either heterozygous (Conard *et al.*, 1984) or homozygous (Aznar *et al.*, 1986) protein C deficiency have however failed to demonstrate any reduction in fibrinolytic potential.

Finally, plasmin has been shown to both activate and inactivate protein C (Varadi *et al.*, 1994) which may account for the procoagulant response sometimes noted in fibriolytic therapy.

5.2.9 APC resistance

The addition of APC to plasma normally prolongs the activated partial thromboplastin time (APTT) because APC cleaves factors Va and VIIIa. By contrast, plasma from some individuals may be partially or almost completely resistant to the anticoagulant activity of APC (reviewed by Dahlbäck, 1994). This phenomenon is reviewed extensively in Chapter 6.

5.2.10 Inhibitors of protein C

There are two major inhibitors of APC in plasma: protein C inhibitor (PCI) and α1-proteinase inhibitor. PCI was first described by Marlar and Griffin (1980) and purified by Suzuki *et al.* (1983b). PCI is synthesised in the liver and is secreted into the blood (plasma concentration, 5 μg ml^{-1}) as a single-chain glycoprotein (M_r 57 kDa). The physiology and biochemistry of this inhibitor has been reviewed by Suzuki *et al.* (1989). PCI inhibits APC with a K_i of 5.8×10^{-8} M. It forms a 1:1 molar complex with APC and in the process is cleaved at the reactive site (Arg354/Ser355) to release a C-terminal peptide. The rate of this reaction *in vitro* is enhanced by both dextran sulphate and heparin. PCI also acts as a potent inhibitor of the thrombin–thrombomodulin complex (Rezaie *et al.*, 1995). Based on an energy-minimized three-dimensional model of PCI derived from the coordinates of α1-proteinase inhibitor, Kuhn *et al.* (1990) predicted a strongly charged amphipathic helix at the N terminus of the protein which they hypothesized was involved in heparin binding. Considerable amino acid sequence homology with the other serine proteinase inhibitors, α1-antichymotrypsin, α1-proteinase inhibitor, ATIII and angiotensinogen has been noted (Suzuki *et al.*, 1987a). The cloning and characterization of the PCI gene (*PCI*) is described in Section 5.5.

APC is also inhibited by α1-proteinase inhibitor (Meijers *et al.*, 1988; Heeb *et al.*, 1989; van der Meer *et al.*, 1989), α2-macroglobulin and α2-antiplasmin (Heeb *et al.*, 1991; Hoogendoorn *et al.*, 1991).

5.2.11 Protein C receptor

A receptor has been described on human umbilical vein endothelial cells which binds specifically in a Ca^{2+}-dependent fashion to both protein C and APC (Fukudome and Esmon, 1994). Antibody-binding studies indicated that the endothelial cell protein C receptor (EPCR) was distinct from thrombomodulin and that binding did not involve protein S. APC/cell receptor binding studies yielded a K_d of 30 nM and an estimate for the number of sites per cell of 7000. Binding required the Gla domain of protein C but this was a specific interaction since the Gla domains of protein S and factor X failed to compete with APC for binding. That EPCR is necessary for optimal activation of protein C by the thrombin–thrombomodulin complex is evidenced by the observation that antibodies that block protein C binding to EPCR reduce the rate of protein C activation (Stearns-Kurosawa *et al.*, 1996). However, antibodies that bind to EPCR without blocking protein C binding have no effect on protein C activation (Stearns-Kurosawa *et al.*, 1996). This suggests that EPCR forms part of a tetra-molecular complex on the endothelial cell surface with protein C, thrombin and thrombomodulin.

A cDNA encoding the human EPCR was isolated by expression cloning (Fukudome and Esmon, 1994). This cDNA contained 1302 bp and predicted a protein of 238 amino acids including a 15-amino-acid signal sequence. The protein contains a potential transmembrane domain of 25 amino acids at the C-terminal end and appears to be related to the cell cycle-dependent murine protein CCD41 (also termed centrocyclin). The 1.3 kb *EPCR* mRNA was only detectable at high levels in endothelial cells and is down-regulated by TNF. The bovine EPCR has also been isolated and exhibits fairly strong homology to its human counterpart (Fukudome and Esmon, 1995).

5.3 The protein C (*PROC*) gene

5.3.1 cDNA cloning

Human protein C cDNA clones have been isolated by Foster and Davie (1984) and Beckmann *et al.* (1985). The 461-amino-acid protein encoded includes a 42-amino-acid leader sequence which probably consists of a 33-amino-acid signal peptide and a nine-amino-acid linking propeptide, analogous to the structure of the other vitamin K-dependent clotting factors. The 1795 bp cDNA sequence comprises a 75 bp 5′ non-coding region, 1383 bp encoding the mature protein, 294 bp 3′ non-coding region and a 38 bp poly(A) tail.

Two alternative polyadenylation sites (ATTAAA) are present at the 3′ end of the *PROC* gene, giving rise to either 68 bp or 294 bp 3′ non-coding sequences in the cDNA. The two ATTAAA sequences are 16 bp and 9 bp upstream, respectively, of the site of poly(A) addition. Consistent with dual usage of these alternative sites, two different *PROC* mRNA species are observable on a Northern blot, 1850 and 1650 nucleotides in length (Beckmann *et al.*, 1985). This dual usage is also reflected in the roughly equal number of cDNA clones isolated corresponding to each of these size classes (Beckmann *et al.*, 1985; Foster *et al.*, 1985).

A bovine *PROC* cDNA has also been cloned (Long *et al.*, 1984) and sequencing has permitted the comparison of the human and bovine protein sequences; the human and bovine heavy chains are approximately 72% homologous at the amino acid sequence level (Foster and Davie, 1984). A cDNA encoding rat protein C has also been characterized which exhibits 75% nucleotide sequence similarity to the human sequence and predicts 68% homology to the human sequence at the amino acid sequence level (Okafuji *et al.*, 1992). Tada *et al.* (1992) have reported the sequence of a murine protein C cDNA.

Murakawa *et al.* (1994) determined the nucleotide sequences from the portion of the *PROC* genes encoding the catalytic domain in six mammals (rhesus monkey, dog, cat, goat, horse and mouse). These sequences exhibited between 69% (dog vs. mouse) and 96% (monkey vs. human) amino acid sequence homology with each other. Moreover, 56% of the amino acids were identical in all nine species from which sequence data were available, including five Cys residues thought to be involved in disulphide bonding, the Asn residues involved in carbohydrate attachment, and the catalytic site residues. Protein C-specific segments were noted in regions VR7 and the N-terminal region of CR5, suggesting that these regions might play a role in the interactions of protein C with other proteins.

5.3.2 Structure and evolution

Genomic clones spanning the 11 kb of the human *PROC* gene have been isolated and characterized (Foster *et al.*, 1985; Plutzky *et al.*, 1986). The distribution of the nine exons making up the gene and a partial restriction map of the region are shown in *Figure 5.4*. The sizes of these exons (*Table 5.1*) range from as little as 25 bp (exon 4) to 587 bp (exon 9). Initially, Foster *et al.* (1985) detected only eight exons; the discovery of a ninth exon containing 53 bp of 5' non-coding region soon followed (Plutzky *et al.*, 1986). Intron boundaries occur in homologous positions and exhibit the same splice junction type as the introns in the corresponding locations in the genes for factors VII, IX and X. By contrast, the *PROC* and prothrombin (*F2*) genes possess only

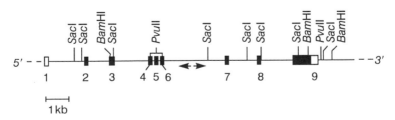

Figure 5.4. Exon/intron distribution in the human *PROC* gene (after Romeo *et al.*, 1987; Foster *et al.*, 1985; Plutzky *et al.*, 1986). Coding exons are denoted by solid bars while the non-coding exon 1 is denoted by an open bar. Arrows denotes the position of two *Alu* repetitive elements found in intron 6 of the *PROC* gene (Foster *et al.*, 1985). The direction of the arrow is 5'→3'.

Table 5.1. Exon structure in the human *PROC* gene

Exon	Length of nucleotide sequence (bp)	Amino acids	Protein domain
1	65	—	(5' untranslated region)
2	70	-42 to -20	Leader peptide
3	167	-19 to 37	Leader peptide and Gla domain
4	25	38–45	Connecting region ('aromatic stack')
5	138	46–91	EGF
6	135	92–136	EGF
7	143	137–184	Connecting region/ activation peptide
8	115	185–222	Catalytic domain
9	587	223–419	Catalytic domain and 3' untranslated region

Data from Foster *et al.* (1985), Plutzky *et al.* (1986), Miao *et al.* (1996).

exons 2, 3 and 4 in homologous positions (reviewed by Furie and Furie, 1988) despite their overall slightly closer evolutionary relationship (reviewed by Patthy, 1985).

Primer extension analysis initially localized the transcriptional initiation site (i) to –1501 (Plutzky *et al.*, 1986). However, Miao *et al.* (1996) recently located the major i site to –1513 thus yielding a length for exon 1 of 65 bp. A TATA box-like sequence (AATATTT) is present between 22 and 14 bp upstream of i.

Two truncated *Alu* sequences are present in intron 6 in the *PROC* gene (Foster *et al.*, 1985). The first is 300 bp in length and is present in reverse (3'→5') orientation. The second, partially homologous to the first, is approximately 146 bp long and is present in a 5'→3' orientation.

An alternative splicing event, involving the skipping of exons 7 and 8, has been noted in healthy controls by means of ectopic transcript analysis (Lind *et al.*, 1993). This would be predicted to result in an in-frame splicing of exons 6–9 and the translation of a truncated protein C molecule lacking 87 internal amino acid residues including three Cys residues involved in disulphide bridging. It is presumed that this truncated protein product would be retained and proteolytically degraded in the endoplasmic reticulum, since no excess of protein C antigen over activity was noted in these individuals. The relative amount of this alternative splicing product appears to be much lower in hepatocyte mRNA as compared to lymphocyte mRNA, suggesting that it is unlikely to be phenotypically significant.

5.3.3 DNA polymorphisms

A number of intragenic DNA polymorphisms are known within the *PROC* gene (reviewed by Reitsma *et al.*, 1993). One is located in exon 1 within the 5' non-coding region; this A/T polymorphism, at nucleotide –1476 (nucleotide numbering of Foster *et al.*, 1985; i.e. nucleotide +37 relative to i) occurred at frequencies of 0.59 and 0.41 in a Caucasian population (Spek *et al.*, 1994). Two further polymorphisms occur within the promoter region [C/T at –1654 (141 bp upstream of i) and A/G at –1641 (128 bp upstream of i)] with respective frequencies of 0.64/0.36 and 0.59/0.41 in the Dutch population (Spek *et al.*, 1994). In the British population, allele frequencies were

0.68/0.32 (C/T at −1654) and 0.61/0.39 (A/G at −1641) (Scopes *et al.*, 1995). High linkage disequilibrium was noted between these polymorphisms; in the British and Dutch population, 25–33% of control individuals possess the C......A haplotype, 27–29% the T......A haplotype, 32–36% the C......G haplotype and 4% the T......G haplotype (Scopes *et al.*, 1995; Spek *et al.*, 1995a). Soria *et al.* (1995a) also described a G/T polymorphism at nucleotide 6376 in intron 7 of the *PROC* gene with allele frequencies of 0.83 and 0.17 respectively.

Silent substitutions within codons Ser99 (TCG/TCT), Asp214 (GAT/GAC) and Asp255 (GAT/GAC) have also been reported (Reitsma *et al.*, 1993; Tsay *et al.*, 1994). The Asp214 polymorphism is detectable by denaturing gradient gel electrophoresis (Gandrille and Aiach, 1991) whilst the Ser99 polymorphism may be detected by PCR analysis using 'mutagenic primers' (Yamamoto *et al.*, 1991a). Yamamoto *et al.* (1991b) have reported two PCR-formatted DNA sequence polymorphisms: These are silent transitions at residues Arg87 and Lys156 which occur with respective allele frequencies of 0.875/0.125 and 0.64/0.36 in the Japanese population. Tsay *et al.* (1994) have calculated the frequencies of some of the above polymorphisms in various ethnic groups.

Several extragenic polymorphisms have also been reported. *Msp*I and *Apa*I RFLPs, located approximately 7 kb 5′ to the *PROC* gene (Lintel-Hekkert *et al.*, 1988; Reitsma *et al.*, 1990), occur with frequencies of 0.69 and 0.31 in a Caucasian population. Koenhen *et al.* (1989) have reported a second *Msp*I RFLP which maps to intron 8 of the *PROC* gene; alternative alleles occur with relative frequencies of 0.99 and 0.01 in European Caucasians.

5.3.4 Gene expression

Protein C expression has been detected by radio-immunoassay in 8-day cultures of HepG2 cells (Fair and Marlar, 1986) and in human liver (Wion *et al.*, 1985). On the basis of representation of cDNA clones in human liver cDNA libraries, *PROC* mRNA comprises approximately 0.02% of total liver mRNA (Beckmann *et al.*, 1985).

Northern blotting reveals two *PROC* mRNA species of 1850 and 1650 nucleotides in human liver poly(A)⁺ mRNA (Beckmann *et al.*, 1985). These species probably result from the use of alternative polyadenylation sites. *PROC* mRNA has also detected in human HepG2 cells (Beckmann *et al.*, 1985) and umbilical vein endothelial cells (Tanabe *et al.*, 1991) but not in cultured human fibroblasts. In the rat, *PROC* mRNA is detectable in liver (major site), kidney, diaphragm, stomach, intestine, uterus and placenta but not in brain, heart, lung, spleen, ovary or bladder (Jamison *et al.*, 1995).

Hassan *et al.* (1990) have shown that the level of *PROC* mRNA increases in the human fetus from 30 to 50% of the adult level between the age of 5 and 10 weeks gestation. In the rat, *PROC* mRNA is expressed at very low levels at prenatal day 18 and these levels increase to a maximum by postnatal day 13 (Jamison *et al.*, 1995).

Transfection experiments have demonstrated that a CAT reporter gene construct containing 626 bp of the putative *PROC* gene promoter was capable of driving CAT expression in HepG2 hepatoma cells (Berg *et al.*, 1994). Reporter gene expression and gel retardation experiments with a *PROC* gene promoter bearing a point mutation within a sequence strongly homologous to the consensus binding site for hepatocyte nuclear factor 1 (IINF1) further demonstrated that HNF1 is important for the expression of the *PROC* gene *in vivo* (Berg *et al.*, 1994). Taken together with the identification of a human hepatoma cell line which contains HNF1 but which does not express protein C, Berg *et al.* (1994) proposed that HNF1 is necessary, although not sufficient, for *PROC* gene expression in the liver.

Miao *et al.* (1996) have performed a thorough characterization of the *PROC* gene promoter by *in vitro* expression experiments, deoxyribonuclease I (DNase I) footprinting analysis and gel retardation analysis. Among promoter elements identified were two overlapping and oppositely oriented HNF1-responsive elements (−10 to +9), an HNF3-responsive element (−25 to −11), PCE1 (a unique liver-specific regulatory element in exon 1 from +12 to +30), an Sp1 binding site in exon 1 (+58 to +65; recognized in co-transfection experiments with an Sp1 expression plasmid) and a silencer element (PCS1) between −162 and −82.

5.3.5 Chromosomal localization

The human protein C gene (*PROC*) was first allocated to chromosome 2 by hybridization of a cDNA clone to a human–hamster somatic cell hybrid panel (Rocchi *et al.*, 1986). This localization was confirmed by Long *et al.* (1988) using a mouse–human cell hybrid panel. *In situ* hybridization permitted the regional localization of the *PROC* gene to 2q14→q21 (Kato *et al.*, 1988). Patracchini *et al.* (1989) reported further sublocalization of the human *PROC* gene to chromosome 2q13→q14 also by *in situ* hybridization.

5.3.6 In vitro *expression*

The successful *in vitro* expression of protein C in mammalian cells has also been reported (Foster *et al.*, 1987, 1990; Yan *et al.*, 1990). However, some variation was noted in the extent to which the different transfected cell lines were able to carry out the processing of recombinant protein C precursor to its mature two-chain form. Foster *et al.* (1991) have described the complete conversion of the precursor to its two-chain form through co-expression of the *Kex2* endopeptidase from yeast.

Large amounts of recombinant protein C displaying complete γ-carboxylation have been produced using the human kidney cell line 293 using an expression vector under the control of a hybrid adenovirus major late promoter/BK virus enhancer (Grinnell *et al.*, 1987). In this study, expression levels as high as 12 mg/10^6 cells of fully active protein were achieved. After purification, 90% of the molecule was found to be in the two-chain form, the cleavage presumably occurring inside the cells. The biological activity of the recombinant molecule has been tested either by assessing its anticoagulant activity in a clotting assay or its amidolytic activity as determined by the hydrolysis of a tripeptide substrate. Further analysis of this recombinant product will be necessary to assess its efficacy in the prevention and treatment of thrombotic disease and its role in the prevention of septic shock.

Ehrlich *et al.* (1989) described a successful strategy for the *in vitro* expression of human protein C in mammalian cells and its activation during post-translational processing. This was accomplished by replacing the 12-residue activation peptide of wild-type protein C with an eight-residue sequence involved in the proteolytic processing of the human insulin receptor.

The high level expression of human protein C in the milk of transgenic pigs has been achieved (Velander *et al.*, 1992); a yield of 1 mg ml^{-1} h^{-1} was reported using a murine whey acid protein gene promoter.

5.4 Protein C deficiency

5.4.1 Introduction and clinical studies

Plasma protein C levels are quite variable in the general population (Dolan *et al.*, 1994); amongst a group of healthy blood donors, protein C concentrations were found to be higher in males (mean 1.07 μM l^{-1}, range 0.37–2.11) than in females (mean 1.01 μM l^{-1}, range 0.59–1.61) and levels increased with age in both sexes. Henkens *et al.* (1995b) and Rodeghiero and Tosetto (1996) have presented similar data.

The first case of inherited protein C deficiency presenting with recurrent thrombosis was reported by Griffin *et al.* (1981). Since then, numerous studies have confirmed the association of protein C deficiency and venous thrombosis (reviewed by Broekmans and Conard, 1988). Protein C deficiency is usually but not always inherited as an autosomal dominant trait, heterozygotes being at risk for superficial thrombophlebitis, deep vein thrombosis and pulmonary embolism. Heterozygotes may, however, exhibit quite a wide range of protein C activity (19–82%) and antigen (22–88%) values and there is some overlap with the normal range (Pabinger *et al.*, 1992a; Allaart *et al.*, 1993; Dolan *et al.*, 1994). Indeed, Allaart *et al.* (1993) have pointed out that using 65% plasma protein C activity as the lower limit of the normal range, 15% of heterozygotes would not be identified and 5% of normal healthy individuals would be labelled as deficient. Unequivocal determination of heterozygosity therefore requires genetic analysis.

Thrombotic disease is usually much more severe in individuals bearing two defective *PROC* alleles. Very low levels of plasma protein C (less than 2%) are associated with a neonatal coagulopathy, purpura fulminans, which is usually lethal if left untreated (Seligsohn *et al.*, 1984; Sills *et al.*, 1984; Marciniak *et al.*, 1985; Manco-Johnson *et al.*, 1988; Marlar *et al.*, 1989). If affected individuals have not already suffered irreversible brain or eye damage, treatment with fresh-frozen plasma, protein C concentrate or APC can provide an effective treatment (Dreyfus *et al.*, 1991, 1995; Cassels-Brown *et al.*, 1993; Wada *et al.*, 1993). Protein C concentrate has also been used prophylactically together with oral anticoagulation in several thrombotic patients with inherited protein C deficiency (De Stefano *et al.*, 1993).

Two types of hereditary protein C deficiency have been recognized as a result of laboratory analysis: type 1 deficiency, the most common, is characterized by parallel reduction of functional and immunological protein C levels due to the reduced synthesis of normally functioning molecules. In type II deficiency, the functional protein C level is reduced to a greater extent than the immunological level. This is presumably due to the synthesis of an abnormal protein C molecule with reduced specific activity. The two types of protein C deficiency are indistinguishable in terms of clinical symptoms, frequency and age of onset. In both types, the plasma concentration of the other vitamin K-dependent factors is within the normal range. In case of treatment with vitamin K antagonists, the level of protein C activity/antigen is sharply reduced but the deficiency state may still be detectable by virtue of a lower level of protein C antigen compared to the other vitamin K-dependent factors.

A number of different functional assays are available to measure protein C activity (Preissner, 1990). Type II deficiencies may often exhibit different laboratory patterns due to the heterogeneity of the underlying defects. Two main categories of dysfunctional protein C deficiency may be distinguished; the first (type IIa) is characterized by a reduction of protein C activity regardless of the functional assay method employed (e.g. defects of the active site itself). The second (type IIb) presents as a discrepancy between chromogenic and coagulometric values; the former is in the normal range and usually in accord with the antigen level whilst the latter is below the lower limit of the normal range. Such a discrepancy may be due to a defect of the protein C molecule which impairs the interaction of protein C with protein S, thrombomodulin, phospholipid or its substrates. Since these dysfunctional protein C molecules can be activated normally and their active sites are unaffected, chromogenic activity is normal whilst coagulometric activity is low. Several examples of type IIb protein C deficiency are now known (Vasse et al., 1989; Girolami et al., 1993; Mimuro et al., 1993).

It is still difficult to obtain reliable data on the incidence/prevalence of heterozygous protein C deficiency either in patients with thrombotic disease or in the general population. This is due to several defects in the studies reported to date; inadequate description of the population under study, incomplete documentation of the presence of the thrombotic disease, or shortcomings of current laboratory tests. The prevalence of protein C deficiency in patients presenting with venous thrombotic disease is related to the age distribution of the patient sample; the younger the patients, the higher the proportion that are protein C deficient (reviewed by Dolan et al., 1989). This proportion is increased still further if only patients with recurrent thromboembolism are considered. Thus, Gladson et al. (1988) reported prevalences of heterozygous type I protein C deficiency of 4% in a group of young (under 45 years old) unrelated patients with venous thrombotic disease, and 12% of patients with recurrent thromboembolism. In a study of 277 patients, Heijboer et al. (1990) reported that 3.2% of patients with deep vein thrombosis manifested a deficiency of protein C whilst only 1.5% of controls did so. Melissari et al. (1992) reported a frequency of 9.2% in their 393 patient study with a history of acute venous thrombosis whilst the corresponding frequency from the 217 patient study of Awidi et al. (1993) was 7.8%. Lensen et al. (1996) found that the median age of the first thrombotic event for protein C deficient members of thrombophilia families was 31.5 years.

The data of Broekmans et al. (1983) and Gladson et al. (1988) may be used to estimate that heterozygous 'overt' protein C deficiency occurs at a frequency of between 1/16 000 and 1/36 000 in the general population. However, since these prevalence figures were derived from patients with venous thrombosis, they cannot provide a measure of the frequency of clinically asymptomatic ('covert') protein C deficiency in the general population.

Miletich et al. (1987) reported that 1/70 of a sample of 5422 healthy adult blood donors exhibited plasma protein C antigen levels consistent either with the lower end of a normal distribution or with heterozygous type I deficiency (55–65% normal). Ten individuals with protein C levels between 33 and 51% were found. Confirmation of the autosomal inheritance pattern of a heterozygous deficiency came from phenotypic studies in four families, and a frequency of 1/200 to 1/300 for the 'covert' deficiency state was inferred. If a plasma protein C level below 65% was by itself a risk factor for thrombosis, then at least a proportion of the 79 putative heterozygotes or 'low normals' identified in this study should have been clinically affected, yet none had a history of thrombosis. The validity of these findings is however supported by the fact that the postulated prevalence of heterozygous protein C deficiency in the general population is consistent with the observed frequency of homozygosity for the disorder.

The finding that heterozygous protein C deficiency occurs in the asymptomatic normal population is intriguing. Although the immunological assay used by Miletich *et al.* (1987) would not have been capable of detecting functional defects in the protein C molecule, it is hard to see how functional defects could correlate with thrombotic risk in the absence of such a correlation with lowered antigen levels. Bertina *et al.* (1988) suggested that the findings of Miletich *et al.* (1987) were merely a reflection of the clinical heterogeneity manifested by protein C deficiency. This interpretation has been borne out by the findings of Bovill *et al.* (1989) who studied 184 members of a New England kindred with protein C deficiency and a high incidence of venous thromboembolism. Positive thrombotic histories were exhibited by 13/46 protein C-deficient family members (cf. 5/138 unaffected relatives) consistent with heterozygous protein C deficiency being an important independent risk factor for thrombotic disease. However, the absence of disease in many protein C-deficient family members indicates that other features must also play a role in the clinical expression of this condition. Bovill *et al.* (1989) were able to exclude deficiencies of antithrombin III or protein S on the basis of clinical assays but suggested that other possible defects could include the reduced production of thrombomodulin or heparan sulphate or defects in the fibrinolytic system.

This issue was largely resolved by the study of Tait *et al.* (1995). The plasma protein C activity of some 9854 healthy blood donors was measured, initially yielding 255 (2.6%) with an activity level below 68% normal. However, only 77 of these remained below this value upon retesting, indicating the importance of multiple testing to avoid the effects of intra-assay and intra-individual variation. After family studies and gene analysis, the observed prevalence of inherited asymptomatic protein C deficiency in the general population was found to be 0.145% although the authors concluded that it might be as high as 0.2%. The frequent occurrence of the 'covert' deficiency state argues for the importance of other gene defects acting in combination with protein C deficiency to confer a high thrombotic risk.

The probability of thromboembolism in a heterozygous protein C-deficient patient increases with age. In a series of Dutch patients, age at the first episode of thrombosis was often between 20 and 30; about 50% of all patients experienced thrombotic symptoms by the age of 30 whilst 80% were symptomatic by the age of 40 (Broekmans, 1985; Pabinger and Schneider, 1996). The recurrence risk in one of these (retrospective) studies was found to be 63% (Pabinger and Schneider, 1996). Allaart *et al.* (1993) reported that about 50% of all such patients experience at least one thrombotic episode before the age of 45 compared with 10% of their normal relatives. However, these patients are very unlikely to experience clinical symptoms before the age of 15 (Broekmans, 1985). The cause of the age effect is unclear. However, Bauer *et al.* (1987) reported increased plasma concentrations of protein C activation peptide with age, indicative of a hyperactive haemostatic system; these authors speculated that a further reduction in protein C functional activity in individuals with protein C deficiency could explain this age effect.

The absence of thrombosis in many putative heterozygotes would seem to argue against plasma protein C concentration being a prime determinant of disease (Bovill *et al.*, 1989; Allaart *et al.*, 1993). Indeed, Henkens *et al.* (1993) reported that there is no relationship between the level of protein C activity or antigen in the plasma and the clinical expression of protein C deficiency. Nevertheless, in many individuals/families, heterozygous protein C deficiency is clearly an important independent risk factor for thrombotic disease (Bovill *et al.*, 1989; Allaart *et al.*, 1993). Thus Henkens *et al.* (1993) found that 30% of individuals with heterozygous protein C deficiency had experienced thromboembolism whereas the comparable figure for those without protein C deficiency was 3%. Further, a prospective cohort study of 20 asymptomatic protein C-deficient individuals (median age 20 years) demonstrated an incidence of thromboembolism of 2.5% per patient year (Pabinger *et al.*, 1994b).

Known risk factors for thrombosis were found in 54% of protein C-deficient patients in a Dutch–French study (Broekmans *et al.*, 1986); pregnancy in 20%, surgery in 15%, immobilization in 9% and oral contraceptive use in 7%. In the other 46% of protein C-deficient patients, the first thrombotic event occurred in the absence of an obvious predisposing factor. The statistical significance of these data are, however, hard to assess.

In their study of 77 individuals with a genetically confirmed heterozygous protein C deficiency, Allaart *et al.* (1993) reported that thrombotic events displayed an association with immobility or surgery. However, 50% of all first thrombotic episodes and 65% of recurring episodes were spontaneous in that no obvious predisposing factor was apparent. The proportions from the study of Henkens *et al.* (1993) were 64% and 36% respectively. Pregnancy was a risk factor in protein C-deficient women with no previous clinical history (2.6% vs. 0.3% in pregnancy and the puerperium) and in protein C-deficient women with a previous history of thrombotic disease (36% of pregnancies affected). This is despite raised levels of protein C antigen during pregnancy and the puerperium (Mannucci *et al.*, 1984). Interestingly, having a protein C-deficient mother is also a risk factor for the fetus: Sanson *et al.* (1996) have estimated that protein C-deficient women have a 2.5-fold higher risk of a natural termination or stillbirth. Oral contraceptive use may also be a risk factor in individuals with protein C deficiency (Allaart *et al.*, 1993; Pabinger *et al.*, 1994a) despite being apparently associated with an increase in protein C antigen (Meade *et al.*, 1985).

Since, however, we are talking about risk factors and not a highly penetrant genetic disorder, other hitherto unknown factors (e.g. separate compensatory mechanisms, epistatic effects, environmental 'triggers' or the specific nature of the *PROC* gene lesion) probably also contribute to the likelihood of a thrombotic event.

In a population-based patient–control study, Koster *et al.* (1995b) estimated the increase in risk of thrombotic disease associated with the manifestation of protein C deficiency. Protein C deficiency was defined as a measurement of less than 0.67 U ml^{-1} or less than 0.33 U ml^{-1} when treated with coumarins. When only the first measurement of plasma protein protein C activity was used, the odds ratio was found to be 3.1 (95% confidence interval, 1.7–7.0). When the first and second measurements of protein C activity were used, the odds ratio increased to 3.8 (95% confidence interval, 0.7–34). When the most stringent criterion of a protein C deficiency was also employed (i.e. a *PROC* gene lesion confirmed by DNA analysis), the odds ratio increased still further to 6.5 (95% confidence interval, 1.8–24).

Heterozygous protein C deficiency is also sometimes associated with arterial thrombosis (e.g. DeStefano *et al.*, 1991; Deguchi *et al.*, 1992; Matsushita *et al.*, 1992; Kazui *et al.*, 1993). Kario *et al.* (1992) have suggested that this might be a consequence of the procoagulant activity of factor VII being enhanced by low protein C levels.

Future therapeutic approaches to severe protein C deficiency will probably include gene transfer/therapy, perhaps involving the transfer of the *PROC* gene into the liver or endothelium. Several lines of evidence now support the view that protein

C augmentation may prove of benefit in the prevention of venous or arterial thrombosis. Treatment of individuals homozygous for protein C deficiency with protein C concentrate results in a decrease in thrombin generation (Conard et al., 1993) and provides effective prophylaxis (Marlar et al., 1992; Minford et al., 1996). Protein C activation is known to occur rapidly in ischaemic tissues and the inhibition of protein C activation exacerbates the damage caused by arterial occlusion in models of ischaemic injury (Snow et al., 1991). APC prevents or reduces thrombus formation in baboon (Gruber et al., 1990), canine (Sakamoto et al., 1994) and rabbit (Arnljots et al., 1994; Arnljots and Dahlbäck, 1995) models of arterial thrombosis. Activated protein C also inhibits thrombus formation on thrombogenic grafts under arterial flow conditions (Hanson et al., 1993; Lozano et al., 1996).

5.4.2 Molecular genetics

A considerable number of mutations in the *PROC* gene have now been reported (Reitsma et al., 1993, 1995; Aiach et al., 1995; Reitsma, 1996). Those that have been fully reported in the literature are logged in the Appendix. The underlying principles of mutation in the *PROC* gene will now be reviewed using some of these lesions as examples.

Single base-pair substitutions. Some 132 different single base-pair substitutions in the *PROC* gene were logged in the last version of the protein C mutation database (Reitsma et al., 1993, 1995; Reitsma, 1996). In contrast with the situation found in ATIII deficiency, the majority of characterized *PROC* gene lesions are associated with a type I deficiency state. Indeed, all known nonsense mutations, promoter mutations, mRNA splicing mutations, insertions and deletions (in-frame or otherwise) in the *PROC* gene are associated with a type I deficiency. Most type I and all type II mutations are of the missense variety. Most mutations found in the Gla domain (exon 3) are associated with a type II deficiency state. Other type II mutations include those at Arg1 (propeptide cleavage site), Arg169 (thrombin cleavage site) and His211 (active site residue). Murakawa et al. (1994) noted that missense substitutions in the catalytic domain occurred invariably within residues which are evolutionarily conserved across nine vertebrate species.

Mutation screening. The search for mutations in the *PROC* gene has been performed either by sequencing all nine exons (Millar et al., 1993b) or by means of a screening procedure such as SSCP (Soria et al., 1992; Zheng et al., 1994; Miyata et al., 1996), denaturing gradient gel electrophoresis (DGGE; Gandrille et al., 1993a, 1994) or temperature gradient gel electrophoresis (Hernandez et al., 1995). However, mutations have not always been detected in patients thought likely to harbour them. In their study of 14 patients with familial protein C deficiency, Poort et al. (1993) found that one type I deficiency patient did not possess any detectable mutation within the coding region, splice junctions or immediate promoter region of the *PROC* gene. It may well be, therefore, that some mutations reside in another location.

The CpG dinucleotide as a mutation hotspot. Of the 132 known *PROC* gene lesions, 42 (32%) occur in CpG dinucleotides and are C→T or G→A transitions compatible with a model of methylation-mediated deamination (Cooper and Krawczak,

1993; Reitsma *et al.*, 1995). Reitsma *et al.* (1993) showed, however, that the intragenic distribution of CpG mutations in the *PROC* gene is strikingly non-random: 9/12 (75%) single base-pair substitutions in exon 7 occurred in CpG dinucleotides whereas none of the 12 point mutations reported in exons 5 and 6 were in CpG dinucleotides. It should be noted that 'CpG suppression', as measured by the ratio of CpG/GpC frequencies, is at a minimum for the region around exons 4–6. This relatively high CpG frequency is therefore both indicative of a lower level of cytosine methylation and consistent with the absence of methylation-mediated deamination events.

In view of the hypermutability of the CpG dinucleotide, it is nevertheless not surprising that some cases of CpG mutation in the *PROC* gene have originated independently. Recurrent mutation has been demonstrated for the Arg169→Trp substitution on the grounds of extreme geographical separation of cases (Matsuda *et al.*, 1988; Grundy *et al.*, 1989) and for the Arg306→Term mutation by means of RFLP haplotyping and likelihood analysis (Grundy *et al.*, 1992d).

Altogether, some 18 different C→T or G→A transitions in a total of 16 different CpG dinucleotides (codons –3, 15, 147, 157, 168, 169, 178, 230, 279, 282, 286, 292, 297, 298, 306 and 391) account for the pathological lesion in 130/320 (41%) of patients. However, in the absence of haplotyping data, it is not possible to distinguish between recurrent and identical-by-descent lesions.

Mutational spectra. The largest study of familial protein C deficiency to date has been that of Reitsma *et al.* (1991) who characterized the mutations in a total of 40 probands with type I protein C deficiency. Fifteen different mutations were found of which six accounted for 75% of the defects observed in the Dutch population. These six lesions involved missense mutations at Phe76, Cys105, Arg230, Ile403 and nonsense mutations at Gln132 and Arg306. Chromosomes bearing the Arg230 mutation possess identical RFLP haplotypes suggesting that they are identical by descent rather than due to recurrent mutation (Reitsma *et al.*, 1991). Reitsma *et al.* (1991) showed that this mutation occurred in three Dutch patients who could be traced back to a single common ancestor 250 years ago. No such founder effect is, however, apparent in the Austrian population; no repeat mutations were found in a total of 14 cases studied and only two of these lesions have been previously reported in another population (Poort *et al.*, 1993). Judging from the considerable variety of mutations already known (Reitsma *et al.*, 1993, 1995), extensive allelic heterogeneity appears likely to be the general rule.

Gandrille and Aiach (1995) have determined the spectrum of mutational lesions in a cohort of 121 consecutive type I deficiency patients of French origin. A total of 55 different mutations were detected in 90 individuals; 16 of these lesions were found in more than one patient and nine of these repeat mutations were located in a CpG dinucleotide. Of the 31 patients (24%) in whom no mutation was found, half exhibited plasma protein C levels of between 60% and 70% and may have corresponded to the lower end of the normal range. The reason for the absence of a mutation in the remaining patients is currently unclear.

The use of molecular models to examine the relationship between structure and function. Wacey *et al.* (1993) descibed a molecular model of the serine proteinase domain of APC, constructed by comparative methods, which was used to study missense mutations causing protein C deficiency [*Figure 5.5* (see Frontispiece)]. Generally, amino acid substitutions causing type I deficiency were found to yield energetically unfavourable proteins since substituting residues displayed adverse interactions with neighbouring amino acids, probably leading to disruption of protein folding. By contrast, substitutions causing type II deficiency were confined to regions of functional significance, were generally solvent-accessible and were thus not predicted to interact adversely with neighbouring amino acids.

The above conclusions were confirmed by Greengard et al. (1994c, d) who used their own homology model of the serine proteinase domain of activated protein C (Fisher et al., 1994) to study over 50 known missense mutations, thereby greatly extending our knowledge of structure and function of protein C (see below).

Type I deficiency. Mutations responsible for type I deficiency are expected to affect adversely the structure, folding or secretion of the protein C molecule. Greengard et al. (1994c, d) described a number of different ways in which an amino acid substitution can destabilize the protein C molecule:

Mutations which disrupt hydrophobic packing. Many missense mutations causing type I deficiency replace internal highly conserved, predominantly hydrophobic residues (e.g. Gly197→Glu, Ile201→Thr, Leu223→Phe, Ala259→Val, Ala267→Thr, Val297→Met, Thr298→Lys, Leu318→Phe, Ile323→Phe, Ala346→Thr, Ala346→Val, Met364→Ile, Gly376→Asp, Thr394→Asn) or nearby interacting residues (e.g. Gln184→His, Thr298→Met, Ile403→Met) thereby sterically disrupting the tight packing of internal hydrophobic side chains in the two β-barrel cores of the protein. Val325→Ala and Gln293→His do not cause adverse steric interactions but rather hinder β-barrel structure formation through the replacement of residues with high propensity for β sheet formation with residues with lower propensity. The Ser270→Pro substitution occurs in the first residue of a β turn and is predicted to twist the backbone of the loop connecting the two β barrels. Conversely, the Glu382→Lys substitution, which occurs in a residue at the end of a short β sheet on the edge of the active site, substitutes a strong β breaking residue for a weak one.

Proline mutations. Proline possesses an almost rigid side chain and fulfils the important structural function of introducing bends into α helices. The Pro279→Leu substitution occurs in a connecting a helical region between two β barrels and may thus alter the register of β sheets. Pro247→Leu may alter the angle of connection between β barrels; Pro247 is necessary for a 90° turn required to go from one β strand to the next.

Mutations affecting ion pairing. Another group of type I mutations replaces hydrophilic residues thereby disrupting ion-pair formation. One example of this type of lesion is the Arg178→Gln substitution within the activation peptide. This mutation probably abolishes the putative ion pair between Arg178 and Glu232 required for stable folding and is likely to lead to an altered conformation (Wacey et al., 1993). The variant protein has been expressed *in vitro*: the level of the variant protein secreted was 38% of that of the wild-type protein whilst the amount of protein present in the cell lysate was not significantly different from normal (Sugahara et al., 1994). It would thus appear that extracellular secretion of the mutant is impaired. Since the variant protein did not accumulate in the cell, it may also be assumed that the abnormal conformation of the protein is recognized and it is rapidly degraded. A similar explanation may be required for the Glu232→Lys substitution which will also abolish the ion pair between Arg178 and Glu232.

The Asn256→Asp substitution introduces a negatively charged residue close to

Arg398 and may interfere with the formation of an ion pair between Arg398 and Glu341.

Mutations which disrupt α helices. Several mutations interfere with helix formation either by replacing residues of high helix-forming propensity by residues of low propensity (Met335→Thr), steric disruption (Ile403→Leu) or removal of a bend in the helix (Pro327→Leu).

Mutations affecting the putative EGF-interacting region. Fisher *et al.* (1994) proposed that the second EGF domain interacts with the serine proteinase domain through a hydrophobic patch surrounding residue Cys277 which is disulphide-linked to Cys133. Gly282 lies on the edge of this patch and substitutions of this residue (Gly282→Arg and Gly282→Ser) are predicted to interfere with the interaction of the two domains and probably lead to the misfolding of the global structure of this multidomain protein.

Mutations in the activation peptide. Mutations at the thrombin activation site (Pro168→Leu, Arg169→Trp) destabilize the region containing the activation peptide. Pro168→Leu introduces a hydrophobic side chain into a very hydrophilic environment and influences the calcium-dependent conformational change in protein C necessary for thrombin activation (Rezaie and Esmon, 1994b); *in vitro* expression experiments have demonstrated that the mutant protein is retained intracellularly and secreted at greatly reduced levels (Rezaie and Esmon, 1994b). The Arg169→Trp substitution introduces a hydrophobic side chain to the surface of the molecule and probably results in the loss of ion pairing between Arg169 and Glu163 which normally stabilizes the activation peptide.

Variants exhibiting abnormal disulphide linkages. The Cys331→Arg mutation serves to abolish the disulphide linkage between Cys331 and Cys345. The variant protein has been expressed *in vitro*: the level of the variant protein secreted was only 0.3% of that of the wild-type protein whilst the amount of protein present in the cell lysate was not significantly different from normal (Sugahara *et al.*, 1994). The extracellular secretion of the mutant is thus abolished. The removal of the disulphide bond presumably leads to an altered conformation of the protein which leads to its rapid degradation.

Mutations which impair secretion. *In vitro* expression studies have been employed to demonstrate that three mutations causing type I protein C deficiency [Gly376→Asp (Sugahara *et al.*, 1992), Arg178→Gln, Cys331→Arg (Sugahara *et al.*, 1994)] impair secretion of the protein. Presumably, these lesions prevent the protein from folding into its native conformation. Miura *et al.* (1993) have suggested that such abnormal proteins may not be transported from the rough endoplasmic reticulum (where they are synthesised and glycosylated) to the Golgi apparatus (Pfeffer and Rothman, 1987; Lodish, 1988).

The most detailed study of impaired secretion has been performed on five site-directed mutants at Arg15 (mutated to Gly, Trp, Gln, Leu, Pro); secretion was impaired by varying extents (13–42% normal) with an order that correlated well with relative hydrophobicity of the substituting amino acid (Tokunaga *et al.*, 1996).

Type II deficiency. About 10% of protein C-deficient patients possess a dysfunctional protein C molecule characteristic of type II deficiency. Most of these lesions affect either the Gla or serine proteinase domain of the protein (see below). By definition, these substitutions do not affect the folding, secretion or stability of the resulting proteins significantly but rather alter or abolish its function. Thus from the standpoint of examining the relationship between structure and function, these are the most interesting lesions and so we shall devote some considerable space to them. Greengard *et al.* (1994c, d) demonstrated that missense mutations causing type II deficiency invariably involve solvent-exposed residues and cluster either in a positively charged cluster originally identified by Wacey *et al.* (1993) or are located in or near the active site region.

As mentioned in Section 5.4.1, two forms of type II protein C deficiency have been distinguished: type IIa, no difference between coagulometric and chromogenic assays; type IIb, where a discrepancy between these assays is evident. The majority of cases of type II deficiency fall into the latter category but several examples of type IIa deficiency are known [e.g. the His211→Gln active site mutation, Met343→Ile (Poort *et al.*, 1993), Gly381→Ser (Marchetti *et al.*, 1993) and Gly385→Arg (Miyata *et al.*, 1994]. An example of a type IIb protein C deficiency is described under propeptide cleavage site mutations below.

Propeptide cleavage site mutations. An example of type IIb protein C deficiency is that of Girolami *et al.* (1993) who studied a patient (protein C Padua 2) with a normal level of protein C antigen/chromogenic activity but a reduced coagulometric activity. The dysfunctional molecule responsible exhibited reduced affinity for a calcium-dependent anti-protein C monoclonal antibody. Protein C Padua 2 had an increased molecular weight (95 kDa instead of 65 kDa). This was found to be due to a Arg−1→Cys substitution in the propeptide cleavage site leading to failure to remove the propeptide. Bristol *et al.* (1994) have shown that the highly homologous profactor IX, even when fully γ-carboxylated, is biologically inactive due to the presence of the propeptide sequence.

By contrast, another mutation in the propeptide cleavage site (Arg−1→Ser) did not lead to an increase in molecular weight (Miyata *et al.*, 1995). Instead, it resulted in normal propeptide cleavage with the addition of a Ser residue at the amino-terminal end of the protein. γ-Carboxylation of the protein was normal and the loss of anticoagulant activity was presumed to be due to an alteration in the conformation of the Gla domain.

Gla domain mutations. One of the best characterized of the Gla domain mutations is Glu20→Ala (Lu *et al.*, 1994a, b). This mutant was found to be defective in coagulometric assays, in factor Va inactivation and in fibrinolysis resulting in impaired membrane binding of APC. Activation of the Ala20 variant by thrombin–thrombomodulin was not enhanced by phospholipid and exhibited a different calcium dependence as compared with Glu20. Further, antibody-binding experiments suggested that the Gla domain bearing the Ala20 variant had lost the ability to undergo a calcium-induced phospholipid-independent conformational change. These results were consistent with those of Zhang and Castellino (1991, 1993) who constructed a protein C variant in which Glu20 was replaced by site-directed mutagenesis.

Protein C Mie (Glu26→Lys) is characterized by reduced anticoagulant activity, a reduced thrombin activation time and a reduced rate of inactivation of factor Va in the presence of phospholipids (Nishioka *et al.*, 1996). The mutant protein was found not to bind to phospholipid or endothelial cells but interacted with protein S and factor Va normally. The authors concluded that Gla26 is critical for acquiring the conformation required for binding to phospholipid, thrombomodulin and the endothelial cell protein C receptor.

Basic dipeptide cleavage site mutation. Arg157 forms part of the basic dipeptide (Lys156–Arg157) removed during intracellular processing of the protein C zymogen to its active two-chain form. The Arg157→Gln substitution may prevent normal prote-olytic processing and the single-chain zymogen may not be functionally equivalent to the two-chain zymogen.

Thrombin cleavage site mutation. The Arg169→Gln substitution abolishes the ionic interaction between Arg169 and Glu163, thereby altering the conformation of the activation peptide. However, this lesion probably exerts its pathological effect by preventing thrombin cleavage at Arg169–Leu170.

Catalytic domain mutations. Dysfunctional mutations in the catalytic domain are essentially of two types: those within an area of positive surface charge proposed to play a role in the binding of thrombin–thrombomodulin in the presence of calcium and those which directly or indirectly affect the function of the active site (Wacey *et al.*, 1993; Greengard *et al.*, 1994a).

The Arg229→Gln and Trp mutations occur within a residue which is in solvent contact with, and is located adjacent to, both the active site and the activation peptide. It is also surrounded by a positively charged cluster of many Arg and Lys residues and is flanked by an RGDS (Arg–Gly–Asp–Ser) protein-binding motif. These findings, together with the observation that the mutant proteins possess normal activity in chromogenic assays but lowered activity in anticoagulant assays, led Wacey *et al.* (1993) to propose that the positively charged cluster could be an exosite involved in the binding of thrombomodulin and/or protein S. Other mutations within the posi-tively charged cluster include Arg147→Trp, Arg157→Gln, Arg314→Cys and Arg352→Trp. The Arg147→Trp and Arg352→Trp substitutions result in the place-ment of hydrophobic amino acids on the surface of the molecule. The second of these should prevent the ion pairing of Arg352 with Asp172 and may therefore affect the activation of protein C. Alternatively, since Arg352 is found at the bottom of the specificity pocket in APC, this lesion may disrupt the formation of the active site.

A considerable number of other dysfunctional variants are thought to alter the function of the active site. His211→Gln actually removes one of the catalytic site residues. The Gly381→Ser, Gly383→Cys, Gly385→Arg and Gly391→Ser substitu-tions occur in residues which control access to the substrate specificity pocket. Larger side chains may deny access to the active site to incoming substrate molecules.

Asp359→Asn involves a residue which normally interacts with the *neo* N-terminal Leu170, thereby preventing the induced conformational change required for activa-tion. Met343→Ile may prevent interaction of incoming substrate molecules since Met343 lies within the active site cleft. The Ser252→Asn substitution serves to intro-duce a novel glycosylation site; were a carbohydrate group to be attached at this site, it might well block access to the active site of protein C.

The Gly301→Asp substitution may disrupt ion pair formation between Asp359 and the *neo* N-terminal Leu170. This is because Asp301 is located close to Leu170 which is inserted into the protein core after activation. Normal formation of the oxyanion hole could then be perturbed.

Mutations which create novel disulphide linkages. Wojcik *et al.* (1996) have shown that the free cysteine residues generated by Arg-1→Cys, Arg9→Cys and Ser12→Cys mutations lead to complex formation between protein C and α1-microglobulin. The substitutions Gly383→Cys and Arg314→Cys also generate free cysteine residues which could participate in novel disulphide linkages. This is per-haps particularly likely in the case of the Gly383 substitution on account of its prox-imity to the Cys384 residue which normally forms a disulphide bridge with Cys356.

Mutation affecting glycosylation. Simioni *et al.* (1996c) have described an Asn329→Thr substitution occurring in the residue where lack of glycosylation in approximately 30% of plasma protein C gives rise to the β form of protein C. This substitution resulted in the reduced synthesis of a β-protein C variant with decreased functional activity. No differences were observed between the activation rates of wild-type protein C and the Thr329 variant. However, activated protein C-Thr329 inactivated factor Va at a slower rate than its wild-type counterpart.

Splicing mutations. At least 12 different single base-pair substitutions putatively affecting mRNA splicing have been reported (Reitsma *et al.*, 1995; Miyata *et al.*, 1996). However, only three occurred in an invariant GT or AG dinucleotide. Three occurred in the conserved −1 position at donor splice sites; one of these created a Stop codon, another altered an encoded amino acid and one was silent. A GGCGAGG→GGTGAGG transition, putatively activating a cryptic exon 7 donor splice site, has been noted 7 bp 3′ to the wild-type site. Reitsma *et al.* (1991) reported three non-identical substitutions of the G residue at position +5 (nucleotide 3222; Foster *et al.*, 1985) of the exon 5 donor splice site. This extreme non-randomness is most intriguing. Perhaps significantly, the mutated base occurs within a TGAGGG sequence which is a perfect match for the deletion hotspot consensus of Krawczak and Cooper (1991). Other splice site mutations include a G→T transition at position +5 in the donor splice site of exon 4 (Lind *et al.*, 1995).

Two further probable examples of the activation of a cryptic splice site have been reported in individuals manifesting a type I deficiency state: a C→T transition at nucleotide 6274 (9 bp into intron 7) creates an alternative donor splice site which would result in a frameshift and premature termination of translation 30 codons into exon 8 (Soria *et al.*, 1992). An A→G transition at nucleotide 3318 within the invariant AG of the acceptor splice site in intron 5 abolishes the use of this site and probably activates a cryptic splice site two nucleotides downstream in exon 6 (Soria *et al.*, 1993); this would lead to a frameshift with premature translational termination at codon 119.

The phenotypic consequences of splice site mutations in the *PROC* gene may be studied in lymphocyte RNA by means of ectopic transcript analysis: using this approach, Lind *et al.* (1993) demonstrated that a G→C transversion in the last nucleotide of exon 7 led to skipping of this exon. Ectopic transcript analysis also allowed Soria *et al.* (1996a) to conclude that a 7054G→A intron 7 variant was probably neutral with respect to clinical phenotype.

Several examples of alternatively spliced products in the *PROC* gene have been noted (albeit in ectopic transcripts Berg *et al.*, 1996b) and it is important to be aware of their existence since they could in principle interfere with the characterization of splicing defects.

Promoter mutations. Four different single base-pair substitutions have been found in the putative promoter region of the *PROC* gene causing type I protein C deficiency. One of these is a T→C transition at nucleotide −1515 (i.e. 2 bp upstream of i) within a GGTTATGGATTAAC motif with strong resemblance to the consensus binding site for the liver-enriched transcription factor, HNF1 (Berg *et al.*, 1994). Transfection experiments demonstrated that a CAT reporter gene construct containing 626 bp of the putative *PROC* gene promoter was capable of driving CAT expression in HepG2 hepatoma cells. Levels of CAT expression from constructs bearing the −1515 mutation were found to be drastically reduced by comparison with the wild-type, consistent with the reduced plasma protein C antigen levels observed in the patient. Gel retardation and co-transfection experiments demonstrated that the mutation abolished both the binding and the transactivating ability of HNF1 observed with the wild-type *PROC* gene promoter. This result has subsequently been confirmed by Miao *et al.* (1996): an 85% reduction in reporter gene activity relative to that of the wild-type.

The ability of the mutation to disrupt HNF1 binding appears to be a function not only of the nature of the nucleotide substitution and its position within the recognition sequence, but also of the relative affinity of the wild-type binding site for HNF1. This analysis is therefore consistent with the view that the −2 mutation disrupts the binding of HNF1 to the *PROC* gene promoter region and indicative of an important role for HNF1 in the expression of the *PROC* gene *in vivo*.

An A→G transition at −1533 (i.e. 20 bp upstream of i) occurs within an HNF3 binding site consensus sequence (5′ A/C/G **A** A/T T R T T G/T R Y T Y 3′; Tsay *et al.*, 1993). A further T→A transversion occurs just downstream at −1528 (i.e. 15 bp upstream of i) within a second overlapping HNF3 binding site which occurs in the reverse orientation (Poort *et al.*, 1993). Spek *et al.* (1995b) employed gel retardation assays and UV cross-linking experiments to demonstrate that HNF3 can bind specifically to both HNF3 binding sites. The −1528 mutation reduced promoter activity four-fold in transient transfection experiments (HepG2 cells) and abolished binding to one HNF3 site and reduced binding to the other. This lesion also reduced the transactivation potential of HNF3 by two-fold in HepG2 cells. The −1533 mutation reduced promoter activity five-fold in transient transfection experiments (HepG2 cells), a result confirmed by Miao *et al.* (1996). It also abolished binding to both HNF3 sites and transactivation by HNF3 in HepG2 cells. Since transactivation of the *PROC* gene was noted in HeLa cells (which do not express HNF1) after transfection with HNF3, Spek *et al.* (1995b) concluded that HNF1 was not essential to *PROC* gene expression. However, it would appear likely that some functional interplay does occur between HNF1 and HNF3 in the transcriptional regulation of the *PROC* gene.

A C→T transition at −1511 (+2 relative to i) has also been reported within the HNF1 binding site (Tsay *et al.*, 1993). This lesion reduces the activity of a reporter gene by 90% (Miao *et al.*, 1996).

Deletions. Only one gross deletion of the *PROC* gene has been reported to date: Matsuda *et al.* (1988) employed densitometric analysis of a Southern blot to show that one *PROC* allele of a compound heterozygous Japanese patient with severe protein C deficiency was deleted in its entirety.

A number of short deletions of between 1 and 18 bp have also been noted within the *PROC* gene coding region in other patients (Reitsma *et al.*, 1995). These include three deletions of more than 15 bp, all of which occur between nucleotides 3156 and 3190 (codons 72–82; Poort *et al.*, 1993; Reitsma *et al.*, 1993), a region containing two pairs of long inverted repeats. Other short deletions include delG Arg147 (Grundy *et al.*, 1991a), delG Trp300 (Bernardi *et al.*, 1992), delG Gly362 (Tsay *et al.*, 1993), delCA Asp257, delC Pro413 and del5 3455–3459 3′ to the exon 6 donor splice site (Gandrille and Aiach, 1995), delGAC8478 and delG8857 (Miyata *et al.*, 1996). Several of the other deletions are flanked by direct repeats of between 4 and 6 bp and three occur within, or are flanked by, perfectly matched deletion hotspot consensus sequences (Krawczak and Cooper, 1991).

Yamamoto *et al.* (1992a) reported the study of a type I deficiency patient with a homozygous delGGly381 deletion (protein C Nagoya). This lesion, which leads to a frameshift and the substitution of the last 39 amino acids of the protein with 81 abnormal amino acids, has also been found in other Japanese patients (Tokunaga *et al.*, 1992; Ido *et al.*, 1993) and may represent a relatively common lesion in this population. The elongated protein was expressed *in vitro* and was found not to be exported from the cell (Yamamoto *et al.*, 1992b). Immunoelectron microscopy has shown that protein C Nagoya is retained in the endoplasmic reticulum whilst wild-type protein C can be found in both the endoplasmic reticulum and the Golgi apparatus (Katsumi *et al.*, 1996). Further cross-linking experiments have shown that protein C Nagoya exists in the endoplasmic reticulum as a complex with 78 kDa glucose-regulated protein (GRP78) and 94 kD glucose-regulated protein (GRP94) (Katsumi *et al.*, 1996). Both of these stress proteins are known to function as 'molecular chaperones'. These proteins serve to promote protein maturation by facilitating the folding and assembly of proteins (Welch, 1992). However, they also appear to be involved in the targeting of aberrantly folded proteins for cellular degradation (Welch, 1992).

A 5 bp deletion (TGAGA) in the donor splice site of intron 6 has been described in a patient with type I deficiency but the lesion was not found in the *PROC* genes of either parent (Gandrille *et al.*, 1994a). Transmission of the parental *PROC* alleles was otherwise normal as judged by the inheritance of intragenic polymorphisms. The *de novo* origin of this lesion was demonstrated by confirmation of paternity.

Insertions. Several short insertions have been described in the *PROC* gene as a cause of symptomatic protein C deficiency (Reitsma *et al.*, 1995). For example, 6139insTT results in a frameshift with translational termination at codon 156 (Soria *et al.*, 1992). Tomczak *et al.* (1994) have reported a 3363insC insertion which results in premature translational termination at codon 119. Another (8432insCTGGAC; Poort *et al.*, 1993) results in the in-frame insertion of two amino acids (Leu, Asp) at codon 239. Insertions of C (3363/4) and G (8796–8001) also alter the reading frame (Gandrille and Aiach, 1995). All of these lesions cause a type I deficiency state.

Allelic exclusion. Nonsense mutations, deletions and splice site mutations are, as we have seen, a common cause of type I protein C deficiency. Either directly or indirectly by altering the reading frame, these lesions generate or may generate premature Stop codons and could therefore be expected to result in premature termination of translation. However, studies of other genes (reviewed by Cooper and Krawczak, 1993) have shown that such mutations can also exert their pathological effects at an earlier stage in the expression pathway through being associated with the absence or dramatic reduction in, cytoplasmic RNA ('allelic exclusion'). Soria *et al.* (1996b) studied ectopic *PROC* mRNA transcripts from seven protein C-deficient patients, heterozygous for two nonsense mutations, a 7 bp deletion, a 2 bp insertion and three splice site mutations, by reverse transcriptase (RT)-PCR and direct sequencing. The nonsense mutations and the deletion were absent from the cDNAs, indicating that only mRNA derived from the normal allele had been expressed. Similarly for the splice site mutations, only normal *PROC* cDNAs were obtained. In one case, exclusion of the mutated allele was confirmed by polymorphism analysis. In contrast to these six mutations, the 2 bp insertion was not associated with the loss of mRNA from the mutated allele. In this case, cDNA analysis revealed the absence of 19 bases from the

PROC mRNA consistent with the generation and utilization of a cryptic splice site 3′ to the site of mutation which would result in a frameshift and premature Stop codon.

The mechanism(s) by which some newly created Stop codons result in a decrease in the steady-state level of cytoplasmic mRNA are not yet understood but are under continuing investigation. The rate of transcription may not be altered and a decrease in the steady-state level of cytoplasmic mRNA may arise through reduced efficiency of intranuclear mRNA processing and metabolism, mRNA transport to the cytoplasm or stability of the mature mRNA (Cooper and Krawczak, 1993). Recent experimental data point to reduced nucleo-cytoplasmic mRNA transport in such cases with a possible mechanism linking translation and nuclear transport that could detect the presence of premature Stop codons at the nuclear membrane (Pulak and Anderson, 1993; Hagan *et al.*, 1995). Whatever the mechanistic explanation, allelic exclusion appears to be a relatively common causative mechanism in those cases of type I protein C deficiency which result from mutations that introduce premature Stop codons.

Several examples of alternatively spliced products in the *PROC* gene have been noted (albeit in ectopic transcripts) (Berg *et al.*, 1996b). This must be borne in mind when employing RT-PCR-based procedures since they could in principle interfere with mutation detection.

Asymptomatic protein C deficiency. Tsuda *et al.* (1991) characterized the molecular lesions in three individuals with asymptomatic heterozygous protein C deficiency originally reported by Miletich *et al.* (1987). In the first patient, a C→T transition was noted 9 bp from the exon 7/intron 7 donor splice site. This lesion appears to activate a cryptic splice site and leads to the insertion of 7 bp, causing a frameshift and premature termination. Six family members were found to carry this mutation but none was symptomatic. The other patients were shown to carry two non-identical 18 bp in-frame deletions (residues Ser77–82 and Cys78–Gly83) at a sequence compatible with the deletion hotspot consensus noted by Krawczak and Cooper (1991). Since no trace of the mutant molecules was detectable, they were presumed to be unstable.

Tait *et al.* (1995) characterized nine other *PROC* gene mutations in asymptomatic individuals with protein C deficiency identified by population screening. Seven (Cys78→Gly, Phe139→Leu, Arg147→Trp, Gly282→Ser, Gly301→Val, Thr371→Ala, Arg398→His) were type I variants, Ser12→Cys was a type II variant and a C→T transition at nucleotide −26 occurred in the acceptor splice site of intron 1. Although these lesions would appear to be highly deleterious to the structure/function of the protein C molecules bearing them and despite the fact that only two of these mutations had been found before in symptomatic patients, it is not thought likely that the distinction between symptomatic and asymptomatic protein C deficiency lies in the nature of the gene lesion. After all, these healthy individuals exhibit levels of protein C activity which are also characteristic of individuals with clinically symptomatic protein C deficiency. If allelic heterogeneity is not the sole determinant of clinical phenotype in protein C deficiency, then we must investigate other factors that act in combination with protein C deficiency to increase thrombotic risk.

Severe protein C deficiency. The majority of patients with a familial deficiency of protein C are heterozygous for a *PROC* gene defect; typical protein C activity values in these patients range between 30% and 65%. However, some patients exhibit much lower values, indicative of a homozygous or compound heterozygous deficiency state.

Homozygous protein C deficiency was first described as a cause of neonatal purpura fulminans resulting in rapid death due to massive venous thrombosis (Seligsohn et al., 1984; reviewed by Marlar et al., 1989). Until comparatively recently, it had been assumed to be invariably lethal. Indeed, many cases have now been described in which undetectable protein C activity and antigen levels are accompanied by the early onset of a severe coagulation disturbance (purpura fulminans and/or disseminated intravascular coagulation). Parental consanguinity is a frequent finding in such cases and the phenotypic data are consistent with homozygosity for a *PROC* gene lesion (Marlar et al., 1989; Tuddenham et al., 1989). Two observations have, however, been somewhat puzzling. Firstly, in some of the families with a homozygous severe patient, neither parent manifests any obvious symptom of thrombotic disease. Secondly, some *homozygous* patients have been reported, with very low or even undetectable plasma protein C levels, who exhibit much milder symptoms and/or a later onset of disease (Tuddenham et al., 1989). Clearly, the terms 'homozygous' and 'severe' protein C deficiency are not always synonymous; severity is likely to be a function of the nature, type and location of the specific mutation involved.

Several cases of severe protein C deficiency causing neonatal purpura fulminans have now been characterized by molecular genetic methods. Homozygosity has been demonstrated for a Gln184→His substitution (Soria et al., 1994b), a Ala136→Pro substitution (Dreyfus et al., 1991; Long et al., 1994), an Arg286→His substitution (Long et al., 1994), a Gly292→Ser substitution (Alessi et al., 1996), a Val325→Ala substitution (Witt et al., 1994) and a complex deletion/insertion mutation (3351del4, 3350insA) which resulted in an Asn102→Lys substitution and the removal of codon Gly103 (Millar et al., 1994b). These analyses have provided a direct demonstration that protein C deficiency may sometimes be inherited as an autosomal recessive trait.

Confirmation of homozygosity at the DNA sequence level has now also been accomplished in several cases of protein C deficiency with a comparatively late onset of clinical symptoms (Grundy et al., 1991c; 1992c; Conard et al., 1992; Yamamoto et al., 1992a; Reitsma et al., 1993, 1995). In these families, heterozygotes are usually asymptomatic despite possessing protein C concentrations characteristic of patients with the dominant form of the disease. These studies have also served to demonstrate that a small residual amount of protein C activity (5–6%) is sufficient to avoid neonatal death but not to prevent recurrent thrombosis in adult life. At least seven other cases of late-onset/moderately severe 'homozygous' protein C deficiency (less than 5–16% activity) have been reported in the literature (Tuddenham et al., 1989; Tripodi et al., 1990) which may result from homozygosity for other missense mutations.

It should be remembered that severe protein C deficiency may also be caused by compound heterozygosity for two non-identical *PROC* gene lesions (e.g. the 5 bp deletion and Gly376→Asp missense mutation reported on different alleles in the same patient; Sugahara et al., 1992). Other examples of individuals with two distinct non-identical mutations in their *PROC* genes have been reported (Gandrille et al., 1993b; Ido et al., 1993; Tsay et al., 1993; Sugahara et al., 1994; Tomczak et al., 1994; Alhenc-Gelas et al., 1995; Gandrille and Aiach, 1995; Soria et al., 1995b; Simioni et al. 1996c; see Appendix). Not all such patients are severely affected (e.g. the proposita with one Leu223→Phe substitution and one Ile403→Met substitution on different chromosomes suffered deep vein thrombosis and phlebitis but only from a relatively late age; Gandrille et al., 1993b).

The recognition of distinct dominant and recessive forms of the disorder is most important for those counselling affected families. Without adequate phenotypic/genotypic analysis, these forms could be indistinguishable, leading to incorrect risk assessment. Direct analysis of the lesion(s) in affected families has potentiated successful antenatal diagnoses in at-risk pregnancies (Millar et al., 1994b; Ido et al., 1996).

Mutational demography. As we have seen above, a number of probable examples of recurrent mutation in CpG dinucleotides in the *PROC* gene are now known. There are, however, also several examples of the multiple occurrence of specific lesions outwith CpG dinucleotides (codons 76, 132, 197, 223, 403; Reitsma et al., 1995). These are almost certainly identical by descent rather than due to recurrent mutation.

In principle, the observed mutational spectrum underlying a given genetic disorder is determined by: (i) the basic mechanism(s) of mutation generation, itself a function of the DNA sequence of the underlying gene, (ii) selection acting for or against particular lesions, and (iii) genetic drift, the chance fluctuations (including founder effects) in the gene pool of one or more populations. The relative importance of these three determinants, however, depends critically upon the clinical nature and inheritance pattern of the disease.

Krawczak et al. (1995) examined the geographical distribution and prevalence of 256 single base-pair substitutions (105 of them different) within the *PROC* gene coding region. A significant positive correlation was observed between mutational likelihood and the geographical dispersal of the *PROC* gene lesions within and between 16 different countries. This relationship could be attributed to the fact that, with very few exceptions, high dispersal was only exhibited by CG→TG and CG→CA transitions (i.e. those substitutions that are known to arise *de novo* at the highest frequency; see Section 5.4.2, 'Single base-pair substitutions'). The statistical distribution of mutational likelihoods was as predicted on the basis of the *PROC* cDNA sequence alone, allowing however for the redundancy of the genetic code. These findings suggest that genetic drift and lesion-specific selection have been of relatively minor importance in determining the mutational spectrum observed in the *PROC* gene. Indeed, most multiple reports of particular substitutions in different geographical locations appear to reflect recurrent mutation rather than identity-by-descent.

These findings are in part explicable by the realization that, whereas protein C deficiency may be a recessive disorder with respect to both mortality and reproductive fitness (Allaart et al., 1995), it is dominant with respect to morbidity (albeit with reduced penetrance). Genetic drift, however, can only exert a marked effect when the physical mobility of carriers of the mutation(s) in question is not significantly reduced. The deep and superficial veins of the lower limbs are the most frequent site of thrombosis in protein C-deficient patients, causing pain and swelling. Thus deep vein thrombosis and its other manifestations (e.g. varicose veins and venous ulcers) will have severely impaired an individual's mobility if left untreated.

One reason for selection being incapable of removing defective *PROC* alleles from populations is that the first occurrence of thrombotic symptoms would, at least until comparatively recently, have occurred quite late on in the reproductive lives of genetically predisposed individuals and would have scarcely affected mortality. In any case, Allaart et al. (1995) have recently shown that individuals heterozygous type I protein C deficiency exhibit normal survival as compared with the general population.

Polymorphic variation in the PROC *gene promoter and thrombotic risk.* As we have seen, heterozygous protein C deficiency is an important independent risk factor for venous thrombosis. However, the absence of any clinical symptoms in many putative heterozygotes argues that other factors must contribute to the likelihood of thrombosis in these already predisposed individuals. In principle, one determinant of whether or not protein C deficiency will come to clinical attention could be sequence variation in the promoter region of the *PROC* gene. Differences in transcription efficiency between *PROC* promoter haplotypes might affect the plasma protein C level in an individual already predisposed to venous thrombosis and alter the likelihood that they will come to clinical attention.

Two DNA sequence polymorphisms with highly frequent alleles are known to occur within the *PROC* promoter: C/T and A/G, respectively 141 bp and 128 bp upstream of the major transcriptional initiation site at –1513 (Spek *et al.*, 1994; Section 5.3.3). (The positions of these polymorphisms correspond to –1654 and –1641 in the numbering system of Foster *et al.*, 1985.) In order to explore the possible clinical relevance of these polymorphisms, Scopes *et al.* (1995) transfected human HepG2 hepatoma cells with *PROC* promoter–luciferase reporter gene constructs *in vitro*: the T......A haplotype exhibited at least a two-fold higher transcription efficiency than the C......G haplotype.

However, this difference between haplotypes observed *in vitro* may not necessarily exist *in vivo*. A difference in transcriptional efficiency might not automatically be reflected in plasma protein levels since regulation of the *PROC* mRNA level and/or plasma protein C concentration could occur post-transcriptionally or post-translationally. This would serve to nullify the differential effects on transcription efficiency conferred by the T......A and C......G haplotypes. This question was addressed directly by Spek *et al.* (1995a) who measured the protein C levels associated with the different polymorphic haplotypes in normal healthy controls. These authors demonstrated that the average plasma protein C activity level manifested by individuals homozygous for the T......A haplotype ($n = 28$, mean protein C activity = 116%) was significantly higher than that manifested by individuals homozygous for the C......G haplotype ($n = 40$, mean protein C activity = 94%). (Individuals heterozygous for the two haplotypes manifested a mean protein C activity of 104%, $n=87$.) Thus individuals homozygous for the T......A haplotype exhibited a 22% higher level of plasma protein C activity than those homozygous for the C......G haplotype. These *in vivo* data are therefore quite consistent with the results of the above-mentioned *in vitro* expression data (Scopes *et al.*, 1995) and suggest that the basis for the difference in expression between haplotypes may lie at the transcriptional level. The disparity between plasma levels of the secreted protein product exhibited by the respective homozygotes *in vivo* is clearly much less than the difference in expression noted at the transcriptional level *in vitro*. This implies a dampening effect of post-transcriptional regulation.

The observed difference in expression between the *PROC* promoter haplotypes could be of clinical relevance. For example, if a protein C-deficient individual with one defective allele (most are heterozygous) were to possess a C......G haplotype on the other wild-type allele, then they might be more likely to come to clinical attention than if they had possessed the higher expresssing T......A haplotype. This is because the lower level of expression associated with the C......G haplotype would serve to reduce still further the already low plasma protein C level manifested by the patient so that the postulated threshold between the clinically asymptomatic and symptomatic states would be more likely to be breached. Conversely, a protein C-deficient individual with a T......A haplotype on the healthy allele might be less likely to come to clinical attention since protein C expression from this allele is maximized. Consistent with this postulate, Scopes *et al.* (1995) found the frequency of the C......G haplotype to be slightly higher in a group of 48 unrelated British protein C-deficient patients (0.43) than in controls (0.35) although it failed to attain statistical significance. However, the phase between the inherited *PROC* gene lesion and the promoter alleles was not established in this study and therefore the frequency of the C......G haplotype on the wild-type *PROC* allele could have been underestimated in the patient group.

Spek *et al.* (1995a) approached the same question from a slightly different angle. They determined the polymorphism haplotypes in 130 Dutch patients with an objectively confirmed episode of deep vein thrombosis but no deficiency of protein C. A significantly higher number of individuals with the C......G haplotype were noted in the patient group as compared with the control group. These authors estimated that the possession of the CG haplotype is a risk factor for venous thrombosis (odds ratio 1.6, 95% confidence interval 1.0–2.5). In other words, individuals with the homozygous C......G haplotype have a 50–100% greater probability of developing venous thrombosis than individuals with the homozygous T......A haplotype.

It should be pointed out that a variety of other factors such as lifestyle, age, sex, race, body mass, low-density lipoprotein cholesterol, triglycerides, oral contraceptive use and possibly smoking all influence plasma protein C levels (Conlan *et al.*, 1993; Tait *et al.*, 1993b). Since an individual's protein C level is determined by the combined influence of these different factors, determination of the promoter polymorphism haplotype is unlikely to be of very much use clinically. It may, however, be important in cases where individuals have borderline normal protein C levels and one wishes to establish whether or not they are a carrier of a defective *PROC* allele.

Co-inheritance of APC resistance and protein C deficiency. Co-inheritance of the common factor V Leiden lesion with a *PROC* gene defect could in principle explain the difference in thrombotic risk noted between families with clinically symptomatic protein C deficiency and those with the much more common clinically asymptomatic deficiency state (Miletich *et al.*, 1987; Tait *et al.*, 1995; see Section 5.2.6).

In the Dutch population, co-inheritance of the factor V Leiden mutation with protein C deficiency has been shown to increase the risk of coming to clinical attention (Koeleman *et al.*, 1994); these workers found that 19% of a sample of 48 symptomatic protein C-deficient probands (with known *PROC* gene lesions) possessed the factor V mutation compared with 2% in the general population. When family studies were performed, a thrombotic episode had been experienced by 73% of family members with both a protein C (*PROC*) gene lesion and the factor V (*F5*) gene lesion. By contrast, 31% of family members bearing only the *PROC* gene lesion and 13% of family members bearing only the *F5* gene lesion had experienced a thrombotic episode. Finally, two-locus linkage analysis provided strong support (Lod = 19.6 at θ = 0.00) for the hypothesis that the *F5* and *PROC* gene loci were the two trait loci involved. Even so, at least in one family, not all cases of severe thrombotic disease could be accounted for by co-inheritance of the *PROC* and *F5* gene lesions suggesting that still further factors may be involved.

The factor V Leiden variant has however been found to be less frequent among British (0.06) and Swedish/Danish (0.15) protein C deficiency patients (Hallam *et al.*,

1995a). Indeed, in the British population, the frequency of the factor V Leiden allele was not significantly elevated as compared to healthy controls. Co-inheritance of the factor V Leiden variant is therefore unlikely to be the sole determinant of whether an individual with protein C deficiency will come to clinical attention. Gandrille *et al.* (1995b) have also reported data on the frequency of association of protein C deficiency and the factor V Leiden variant: some 14% of the 113 protein C-deficient patients studied possessed the factor V variant compared with only 1% of controls.

Hallam *et al.* (1995a) also analysed patient data by type of protein C deficiency. It was noted that the frequency of the factor V Leiden variant was 2.8-fold higher in type II deficiency patients compared with type I patients. Hallam *et al.* (1995a) suggested the following possible explanation for this disparity: together with protein S and factor V, APC forms a 1:1:1 stoichiometric complex on the phospholipid surface that is capable of inactivating factor VIIIa (Shen and Dahlbäck, 1994). The factor V Leiden defect is likely to reduce the level of this inactivating complex. This is because thrombin generation is increased in APC-resistant plasma due to the excess of factor Va (Shen and Dahlbäck, 1994). Factor V activated by thrombin is, however, inefficient as a cofactor in the inactivation of factor VIIIa (Shen and Dahlbäck, 1994). Although the reduced synthesis of protein C, characteristic of a type I protein C deficiency state, would presumably reduce the level of factor VIIIa-inactivating complex still further, a normal level of factor V is still available to wild-type protein C molecules for complex formation. By contrast, a dysfunctional (type II) protein C molecule might still be able to interact with protein S and factor V to generate a non-functional complex. Once formed, this inactive complex could sequester factor V and protein S thereby reducing the amount of factor V accessible to wild-type protein C (it might also compete with its functional counterpart for access to factor VIIIa). This reduction might explain why carriership of the factor V Leiden variant increases the likelihood of clinical detection more dramatically for type II protein C deficiency than for type I deficiency patients. However, these authors did not detect all the *PROC* gene lesions responsible for the cases of protein C deficiency and could conceivably have misclassified APC resistance as type II protein C deficiency (see Ireland *et al.*, 1995).

Kalafatis *et al.* (1995b) re-examined their large protein C Vermont pedigree (type II protein C deficiency caused by a Glu20→Ala substitution) and noted that 73% of family members who possessed both the protein C mutation and the factor V Leiden variant had experienced a thrombotic episode. However, whereas recombinant APC is able to cleave factor V R506Q and factor Va R506Q albeit slowly, minimal cleavage occurs when recombinant APC E20A is employed (Kalafatis *et al.*, 1995b). In combination, therefore, these inherited variants serve to generate a stable procofactor (factor V) and a stable active cofactor (factor Va), thereby greatly increasing the probability of thrombosis.

A similar situation has been reported by Brenner *et al.* (1996): all 5/11 clinically symptomatic carriers of a Thr298→Met protein C mutation also carried the factor V Leiden variant whereas venous thromboembolism was not observed in those individuals who carried the *PROC* lesion but lacked the factor V variant. Co-inheritance of the two deficiency states appears therefore to increase the likelihood of an individual coming to clinical attention.

5.5 The protein C inhibitor (*PCI*) gene

A 2.1 kb cDNA clone encoding human PCI has been isolated that encodes a 387-amino-acid mature protein and a preceding 19-amino-acid signal peptide (Suzuki *et al.*, 1987a). It also includes 839 bp of 3′ non-coding region. The human *PCI* gene has recently been cloned and characterized (Meijers and Chung, 1991). It is 11.5 kb in length and consists of five exons. The structure of the human *PCI* gene is very similar to the genes encoding α1-antitrypsin (*PI*) and α1-antichymotrypsin (*AACT*) in terms of the number of exons, the positions and sizes of the introns and the splice junction type. The *PCI* gene has been allocated to human chromosome 14 (Meijers and Chung, 1991) and is therefore closely linked to the genes encoding α1-proteinase inhibitor and α1-antichymotrypsin.

Protein S and protein S deficiency

6.1 Introduction

Protein S was first detected in human plasma during the fractionation of other vitamin K-dependent proteins (DiScipio *et al.*, 1977) some 3 years before its function as a cofactor to activated protein C (APC) was demonstrated (Walker, 1980). The role of this cofactor in coagulation has now been further elucidated. It circulates as a single-chain glycoprotein of M_r 70 kDa, about half complexed to C4b-binding protein (C4bBP) and half free. Free protein S exerts its anticoagulant effect by: (i) acting as a cofactor for the inactivation of factors Va and VIIIa by APC, (ii) inhibiting the activity of the prothrombinase complex by interacting with factors Va and Xa, and (iii) inhibiting the intrinsic activation of factor X through a specific interaction with factor VIII. The clinical importance of protein S is evident from the association of the deficiency state with hereditary thrombophilia (about 5% of all cases) and as a cause of premature arterial thrombosis.

6.2 Structure, function and physiology

6.2.1 Primary structure and domains

Protein S is the vitamin K-dependent protein with the highest isoelectric point (5–5.5). The N-terminal sequence of protein S was established by Edman degradation (DiScipio and Davie, 1979) as Ala–Asn–Ser–Leu–Leu–Gla–Gla. The complete sequence was established by cDNA cloning (Lundwall *et al.*, 1986). The derived sequence contains a 41-amino-acid residue leader sequence with the consensus signal sequence for a γ-carboxylated protein (*Figure 6.1*). The mature protein spans 635 residues, with a predicted molecular mass of 70.69 kDa. There are 34 cysteine residues which are predicted to be involved in 17 disulphide bonds based on structural homologies. The C-terminal sequence of the human protein is Lys–Lys–Thr–Asn–Ser. Bovine and human protein S are 81.6% homologous at the amino acid sequence level.

Viewed by electron microscopy, protein S comprises two domains one of which binds C4bBP. The putative domain structure of protein S, based upon homology to other proteins, is shown in *Figure 6.1* and the constituent amino acid residues are given in *Table 6.1*. Disulphide bonds are based on those determined for bovine protein S by fragmentation and sequencing of the protein. Residues −41 to −18 encode a hydrophobic signal peptide responsible for transport across the endoplasmic reticulum. The pro-peptide (residues −17 to −1) contains the recognition site for the vitamin K-dependent carboxylase and is removed by proteolytic cleavage prior to secretion. The first 36 residues of the N-terminal of the mature protein are strongly homologous to the γ-carboxyglutamic acid (Gla) domains of the other vitamin K-dependent factors. This region is

Venous Thrombosis: from Genes to Clinical Medicine, D.N. Cooper and M. Krawczak.
© 1997 BIOS Scientific Publishers, Oxford.

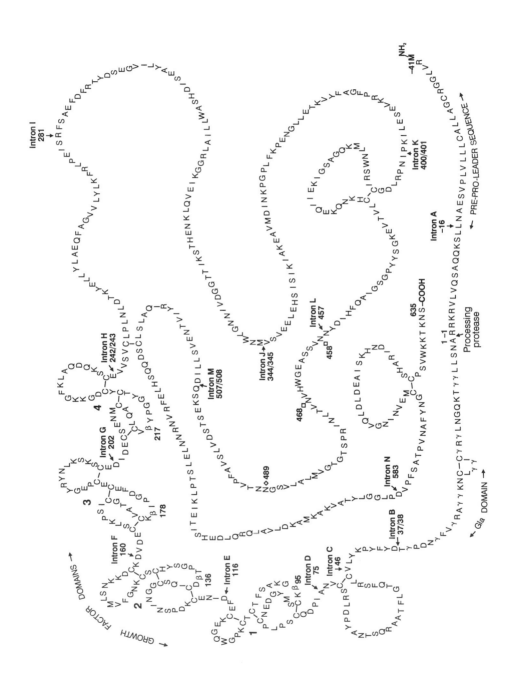

Table 6.1. Exons in the human *PROS* gene

Exon	Length (bp)	Amino acid residues	Protein domain
1	199	-40 to -17	5′ untranslated region/prepeptide
2	158	-16 to 37	Propeptide/Gla domain
3	25	38–45	Hydrophobic region
4	84	46–75	Thrombin-sensitive region
5	123	76–115	EGF1
6	132	116–159	EGF2
7	126	160–201	EGF3
8	122	202–241	EGF4
9	116	242–281	SHBGa
10	190	282–344	SHBGb
11	168	345–400	SHBGc
12	169	401–456	SHBGd
13	152	457–507	SHBGe
14	226	508–582	SHBGf
15	1297	583–635	SHBGg/3′ untranslated region

Data from Ploos van Amstel *et al.* (1990).
EGF, epidermal growth factor-like domain; SHBG, sex hormone-binding globulin-like domain.

followed by a short amphipathic helix (residues 37–46). The next domain, unique to protein S, consists of a disulphide loop with 24 residues interposed between the cysteine bond (Cys47–Cys72). This loop is highly susceptible to thrombin cleavage. Little is known of its structure save for two small loops formed by disulphide bridging (Cys408–Cys434, Cys597–Cys625). Then follow four epidermal growth factor (EGF)-like domains, two more than found in protein C and factors IX and X. The remaining 389 residues (from Cys247) comprise a domain which is unique to protein S. It bears no homology to the serine proteinase domains of the other vitamin K-dependent factors but rather is homologous to sex hormone-binding globulin (SHBG; Gershagen *et al.*, 1987; Ploos van Amstel *et al.*, 1987a) and two basement membrane proteins, laminin A and merosin (Joseph and Baker, 1992). Residues 509–526 correspond to the consensus steroid-binding site in androgen-binding protein (Picado-Leonard and Miller, 1988). The SHBG domain does not appear to be important for protein S cofactor activity (Chang *et al.*, 1994a) but may be involved in binding C4bBP (see Section 6.2.4).

6.2.2 Post-translational modification

There are 11 glutamic acid residues in the N-terminal Gla domain; all are γ-carboxylated in bovine and human protein S (Dahlbäck *et al.*, 1986a; Lundwall *et al.*, 1986). In bovine and human protein S, the EGF domains contain β-hydroxyaspartic acid and

Figure 6.1. Amino acid sequence and tentative secondary structure of human pre-pro-protein S. The locations of the, 14 introns (A–N) are shown. The pre-pro-leader sequence (–41 to –1) is cleaved during protein biosynthesis to give rise to the mature protein with an N-terminal sequence of ANSL. The three potential N-linked carbohydrate attachment sites are shown with open diamonds. Disulphide bonds are positioned by homology with those established for bovine protein S. The amino acids are numbered as follows: –41 to –1, pre-pro-leader; +1 to 41, Gla domain; 42–75, thrombin-sensitive region; 76–242, four epidermal growth factor-like domains; 243 – 635, SHBG homologous region. Reproduced from Reiner and Davie (1995) by kind permission of McGraw-Hill Inc., New York.

β-hydroxyasparagine; the former occurs in the first EGF domain (Asp95), the latter in the three C-terminal EGF domains (Asn135, Asn178, Asn217). The modified residues are all in the erythro form in the protein (Stenflo *et al.*, 1987). The β-hydroxy-asparagine-containing EGF domains contain very high affinity calcium-binding sites (Dahlbäck *et al.*, 1990c). The function of these residues is unclear but recombinant protein S synthesised without β-hydroxylated Asp and Asn residues still retains full cofactor activity (Nelson *et al.*, 1991).

Human protein S contains 7.8% carbohydrate. This is probably N-linked to all or some of three consensus glycosylation sites (Asn458, Asn468, Asn489) in the unique C-terminal domain.

6.2.3 Proteolytic cleavage

Both human and bovine protein S are very sensitive to proteolytic cleavage by throm-bin (Dahlbäck, 1983; Dahlbäck *et al.*, 1986b; Mitchell *et al.*, 1986; Morita *et al.*, 1986). Thrombin readily cleaves protein S in the cysteine loop ('thrombin-sensitive') domain (Cys47–Cys72) After cleavage, the protein comprises two disulphide-linked chains but has lost all its APC cofactor activity (Suzuki *et al.*, 1983a; Walker, 1984b). Human protein S contains three thrombin cleavage sites at Arg49, Arg60 and Arg70 and cleavage at these sites results in the loss of APC cofactor activity (Chang *et al.*, 1994b). The same cleavage can be achieved by APC itself, kallikrein and α-chymo-trypsin at high concentration (Morita *et al.*, 1986).

6.2.4 Interaction with C4b binding protein (C4bBP)

Protein S is present in plasma at a concentration of 20 μg ml^{-1} (0.3 μM) (Schwarz *et al.*, 1986) although values are slightly lower in women (Henkens *et al.*, 1995b). About half of this is bound to C4b-binding protein (C4bBP), in which form protein S is inactive (reviewed by Dahlbäck, 1984). The plasma concentration of C4bBP, a 570 kDa protein, is 200 μg ml^{-1}. The equilibrium rate constant, K_d, for the interaction is about, 10^{-7} M. The interaction between protein S and C4bBP is at least 100-fold tighter in the presence of calcium than without (Dahlbäck *et al.*, 1990a) suggesting that virtually all protein S in plasma should be complexed with C4bBP. The fact that it is not, led Dahlbäck *et al.* (1990a) to postulate the existence of a third component in blood which would regulate the protein S–C4bBP interaction; it was speculated that deficiency of this factor might result in very low levels of free protein S.

The ability of a number of synthetic protein S peptides to inhibit the protein S–C4bBP interaction was studied by Fernandez *et al.* (1993); the region of protein S spanning residues 413–433 appears to be essential for C4bBP binding. The studies of Walker (1989) and Nelson and Long (1992) also implicated residues Gly605–Ile614 of protein S in the binding of C4bBP. However, *in vitro* mutagenesis of protein S residues 608, 609, 611 and 612 did not alter the binding of protein S to C4bBP (Chang *et al.*, 1992). Using deletion variants of recombinant human protein S, Chang *et al.* (1994a) demonstrated that the C-terminal loop of the SHBG-like domain of protein S is involved in the interaction with C4bBP but it remains possible that deletion of this region is responsible for inducing a conformational change that results in a loss of binding affinity for C4bBP elsewhere in the protein S molecule. The site of interac-tion with protein S on the β-chain of C4bBP is discussed in Section 6.5.3.

Residues 413–433 are highly conserved in human, rhesus monkey and bovine pro-tein S each of which can interact successfully with human C4bBP (Greengard *et al.*,

1995a). Porcine protein S, however, has a 10-fold higher K_d for human C4bBP than has human protein S. Greengard *et al.* (1995a) speculated that the non-conservative substitutions Lys429→Ile and Gln607→Pro could result in reduced affinity of porcine protein S for human C4bBP.

Nishioka and Suzuki (1990) have suggested that the protein S–C4bBP complex may competitively inhibit the APC cofactor function of free protein S. Interestingly, rabbit protein S appears not to bind to rabbit C4bBP although it does bind to human C4bBP (He and Dahlbäck, 1993, 1994).

6.2.5 Physiological function and regulation

Free protein S functions as a cofactor for the inactivation of factors Va and VIIIa by APC (Gardiner *et al.*, 1994; Walker, 1984a; reviewed by Heeb and Griffin, 1988). Optimal enhancement of the rate of inactivation of factor Va by APC was achieved when the ratio of protein S to protein C was stoichiometric 1:1 (Walker, 1981). A protein S-binding protein (distinct from C4bBP) that enhances the activity of protein S in a purified system has been described, but not characterized beyond molecular weight (M_r 138 000) and the fact that it contains two chains (M_r 94 000 and 46 000).

Monoclonal antibodies against the Gla domain of protein S inhibit the binding of protein S to phospholipid (Dahlbäck *et al.*, 1990b) suggesting that this domain serves to anchor protein S to the phospholipid surface. Both free and bound protein S bind negatively charged phospholipid (Schwalbe *et al.*, 1990). Free protein S binds to platelet-derived microparticles in a calcium-dependent reaction probably involving the Gla domain but the bound form of the protein does not (Dahlbäck *et al.*, 1992). Moreover, protein S supports the binding of both protein C and APC to microparticles (Dahlbäck *et al.*, 1992) The requirement for negatively charged phospholipid in the inactivation of factor Va by APC and protein S suggests that, as with other coagulation reactions, the inactivator complex forms on the surface of platelets or endothelial cells.

Protein S is found in the α granules of platelets from where it is released by thrombin. Stimulated platelets each have about 400 protein S-binding sites per platelet. Protein S enhances the binding of APC to activated platelets; about 200 molecules of APC bind per platelet (Harris and Esmon, 1985). In one set of experiments, Suzuki *et al.* (1984) reported that protein S enhanced the rate of inactivation of platelet-associated factor Va by APC some 25-fold. However, Solymoss *et al.* (1988) found little enhancement of the rate of APC-mediated membrane-bound factor Va inactivation (two-fold). This result is now supported by data of Tans *et al.* (1991), Jane *et al.* (1992) and Bakker *et al.* (1992) who employed *in vitro* systems with purified human components. Since results generated are divergent from those derived from plasma coagulation systems, purified systems may provide a poor model of the physiological situation (Jane *et al.*, 1992).

In the study of Solymoss *et al.* (1988), the most striking effect of protein S was to abrogate the protective effect of factor Xa upon factor Va. It was thus concluded that the physiological function of protein S is to allow APC to attack factor Va in the presence of factor Xa. Factor IXa exerts a similar protective effect on factor VIIIa and protein S also abrogates this protective effect (Regan *et al.*, 1994). Thus one function of protein S is to make the cofactors Va and VIIIa vulnerable to APC-mediated inactivation. Since antibodies against the thrombin-sensitive region and the first EGF domain of protein S are the most potent inhibitors of APC cofactor function (Dahlbäck *et al.*, 1990b), it would appear as if these regions are involved directly in protein–protein interactions on the phospholipid surface.

Studies of the species specificity of the cofactor activities of protein S and amino acid sequence comparisons have suggested that various residues within the thrombin-sensitive region (Arg49, Gln52, Gln61) and the first EGF domain (Ser81, Ser92, Lys97, Thr103, Pro106) may be involved in binding to APC (Greengard *et al.*, 1995a).

Human protein S binds to factor Va in a calcium-dependent, saturable and reversible manner with a K_d of 33 nM (Heeb *et al.*, 1993). Further, since protein S at a concentration of 16 nM is capable of 50% inhibition of the activity of the prothrombinase complex at 1 nM factor Xa, 20 pM factor Va and 50 μM phospholipid mix (Heeb *et al.*, 1993), it would appear that protein S possesses an anticoagulant function which is independent of APC. The inhibition of prothrombinase activity is not, however, totally dependent on factor Va (Heeb *et al.*, 1993). Indeed, Heeb *et al.* (1994) have shown that protein S can also bind to factor Xa (K_d= 19 nM) and inhibits it, again in the absence of APC. These conclusions have been independently confirmed by Hackeng *et al.* (1994). Protein S bound to C4bBP was found to be incapable of interacting with factor V or Va. Whilst the protein S–C4bBP complex does inhibit factor Xa, thrombin-cleaved protein S does not, suggesting that the conformational change induced by cleavage of the thrombin-sensitive loop results in the loss of a factor Xa binding site (Hackeng *et al.*, 1994).

A further anticoagulant property of protein S is its ability, in the presence of phospholipid, to inhibit the intrinsic activation of factor X through a specific interaction with factor VIII (Koppelman *et al.*, 1995a). Interestingly, the protein S–C4bBP complex showed a five-fold higher affinity for factor VIII than both free protein S and thrombin-cleaved protein S, suggesting that C4bBP might also interact with factor VIII to bring about inhibition of factor X activation, a prediction subsequently confirmed by experiment (Koppelman *et al.*, 1995b).

Van Wijnen *et al.* (1996) have provided evidence of direct interactions between protein S and the phospholipid membrane surface that are essential for its role in inhibiting both tenase and prothrombinase complexes. The dissociation constant for the binding of protein S to phospholipid ranged from 7 to 74 nM depending upon protein S preparation.

Arnljots and Dahlbäck (1995) demonstrated the *in vivo* importance of protein S as a cofactor to APC by means of a rabbit model of microarterial thrombosis. Infusion of protein S and APC into the rabbits was significantly more potent in terms of its anticoagulant effect than protein C alone.

Endothelial cells synthesise and release protein S (Fair *et al.*, 1986) and the endothelial cell surface promotes protein S-dependent inactivation of factor Va by APC (Stern *et al.*, 1986). Indeed, the binding of protein S to endothelial cells, which requires calcium, is essential for the expression of its cofactor activity for APC (Hackeng *et al.*, 1993). Antibody binding experiments have shown that at least two regions in protein S, the γ-carboxyglutamic acid domain and the thrombin-sensitive region, are involved in the expression of cofactor activity (Hackeng *et al.*, 1993) Thus one of the ways in which endothelial cells oppose coagulation is via the local promotion of the APC anticoagulant pathway.

A 138 kDa plasma protein has been reported which binds to protein S (Walker, 1986). This protein comprises a 94 kDa heavy chain and a 46 kDa light chain and enhances both the anticoagulant activity of APC and the rate of factor Va inactivation by APC and protein S.

Protein S is a potent mitogen of smooth muscle cells with a time course similar to that of platelet-derived growth factor-AA (Gasic *et al.*, 1992; Benzakour *et al.*, 1995). Antibody-binding experiments suggested that the mitogenic activity resided in the third and fourth EGF-like domains. Protein S may thus be important in promoting smooth muscle cell proliferation, an important cause of arterial restenosis after balloon angioplasty or vascular injury.

6.2.6 Protein S receptor

Protein S binds and activates *Tyro* 3, a member of a family of receptor-like tyrosine kinases which is expressed both in the nervous system and in non-neuronal tissues such as kidney, ovary and testis (Godowski *et al.*, 1995; Stitt *et al.*, 1995). A protein S receptor detected on the surface of human smooth muscle cells (Benzakour *et al.*, 1995) appears not to be identical to *Tyro* 3 and may represent a second cellular receptor. This receptor may mediate the proliferative response of smooth muscle cells to protein S.

6.3 The protein S (*PROS*) gene

6.3.1 cDNA cloning

A full-length protein S cDNA clone was first isolated from a bovine cDNA library (Dahlbäck *et al.*, 1986). Human protein S cDNA clones have also been isolated and sequenced (Lundwall *et al.*, 1986; Hoskins *et al.*, 1987; Ploos van Amstel *et al.*, 1987a–c). The full-length cDNA encodes a 635-amino-acid mature protein (24-amino-acid signal peptide plus 17-amino-acid propeptide), a 41-amino-acid leader sequence, 108 bp 5′ non-coding region and, 1132 bp 3′ non-coding region. The derived human protein S amino acid sequence was found to be approximately 82% homologous to that of the bovine, but at the nucleotide sequence level, the two protein S coding sequences are approximately 87.5% homologous. He and Dahlbäck (1993) reported that the rabbit protein S cDNA is 89% identical to the human and bovine sequences at the nucleotide sequence level and 82% and 81% homologous, respectively, at the amino acid sequence level. The sequences of protein S cDNAs from the rhesus monkey and the pig have also been determined; the encoded proteins are, respectively, 93% and 83% homologous to the human sequence at the amino acid sequence level (Greengard *et al.*, 1995a).

At the 3′ end of the protein S mRNA-coding region, there appear to be two alternative polyadenylation sequences at nucleotides 3254–3259 and 3289–3294 (Hoskins *et al.*, 1987). Since electrophoretic resolution of the two possible mRNA species at approximately 3.5 kb has not been achieved, it remains unclear whether or not both polyadenylation sites are functional. At only 2.4 kb in length, bovine liver protein S mRNA is much shorter than its human counterpart (Lundwall *et al.*, 1986). Both mRNAs are reasonably homologous up until the rather unusual AGTAAA polyadenylation signal at nucleotide 2358 in the bovine gene (Dahlbäck *et al.*, 1986; Lundwall *et al.*, 1986). The human homologue of this motif is ACTAAA (at nucleotides 2459–2464; Hoskins *et al.*, 1987). Were this G↔C change sufficient for the motif to be no longer recognizable as a polyadenylation signal, then it could account for the dramatic mRNA size difference between human and bovine mRNAs.

Derivation of the amino acid sequence of protein S has turned up some interesting homologies. For example, amino acids 273–388 of human protein S exhibit 17%

homology with a portion of the β-subunit of DNA-directed RNA polymerase (Hoskins *et al.*, 1987). Considerable homology has also been reported with rat andro-gen-binding protein (Baker *et al.*, 1987; Ploos van Amstel *et al.*, 1987a), human SHBG (Gershagen *et al.*, 1987), a growth arrest-specific protein, *gas* (Manfioletti *et al.*, 1993), and two basement membrane proteins, laminin A and merosin, which have important roles in cell proliferation, migration and differentiation (Joseph and Baker, 1992).

6.3.2. Structure and evolution of the PROS gene

The structure of the *PROS* gene has been determined by several groups (Edenbrandt *et al.*, 1990; Ploos van Amstel *et al.*, 1990; Schmidel *et al.*, 1990). The gene comprises 15 exons and spans some 80 kb of chromosomal DNA (*Figure 6.2*). The structure of the *PROS* gene indicates that the domain structure of protein S correlates with the position of the introns in the coding sequence (*Table 6.1*). The introns are of the same phase and located in identical positions to those of the other vitamin K-dependent serine proteinases, factors VII, IX, X, prothrombin and protein C. Further, the exon/intron distribution at the 3′ end of the gene is identical to that exhibited by the human SHBG and rat androgen-binding protein genes. These similarities are consis-tent with the *PROS* gene having arisen by exon shuffling.

The existence of a second sequence, homologous with and physically linked to the *PROS* gene, was first shown by Ploos van Amstel *et al.* (1987b). This sequence is now known to represent a *PROS* pseudogene which possesses some 97% homology with the *PROS* gene. The latter clearly represents the active gene since all liver cDNAs isolated

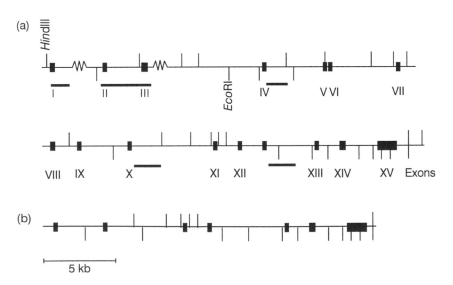

Figure 6.2. Structure of the human *PROS* gene and pseudogene. (a) Restriction map and exon/intron distribution of the human *PROS* gene after Edenbrandt *et al.* (1990) and Schmidel *et al.* (1990). Exons 1–15 are denoted by solid bars and Roman numerals. Restriction enzyme cleavage sites are denoted by vertical lines above (*Hind*III) and below (*Eco*RI) the gene. The approximate position of *Alu* repeat sequences is denoted by solid lines under the restriction map. Zig-zag lines denote unmapped regions of the gene. (b) Restriction map and exon/intron distribution of part of the *PROS* pseudogene. Derived from Edenbrandt *et al.* (1990).

to date are identical to the *PROS* sequence and no transcripts from the *PROS* pseudo-gene are detectable in liver, endothelial cells or platelets (Reitsma *et al.*, 1994). The pseudogene is homologous to exons 2–13 and the 3' untranslated region of the *PROS* gene (*Figure 6.2*; Edenbrandt *et al.*, 1990; Ploos van Amstel *et al.*, 1990; Schmidel *et al.*, 1990). The positions of the introns are virtually identical to those in the *PROS* gene but the nucleotide sequences have diverged by 3.4%. However, four Stop codons are present in the 'normal' reading frame as well as 33 missense mutations, 15 silent mutations and a dinucleotide deletion (which alters the reading frame). Southern blotting experiments have demonstrated that gorilla and chimpanzee both possess two copies of a *PROS* gene sequence per haploid genome whereas African green monkey, rhesus monkey and orangutan possess only one. This dates the duplication of the ancestral *PROS* gene to about 13–17 million years ago after the branching off of the orangutan from the other Old World apes. Sequence analysis confirmed the presence of one of the Stop codons identified in the human *PROS* pseudogene in the chimpanzee and gorilla.

Primer extension experiments identified two major products (Ploos van Amstel *et al.*, 1990), suggesting the existence of at least two distinct *PROS* mRNA species. These two forms are predicted to possess 5' untranslated regions of 175 and 285 bp respectively.

6.3.3 DNA polymorphisms

A TCC/CCC (Ser460/Pro) polymorphism has been detected in the Dutch population by Bertina *et al.* (1990). This polymorphism was detectable because the variant protein (Protein S Heerlen) was no longer recognized by the monoclonal antibody (S-12) used in an enzyme-linked immunosorbent assay (ELISA). The CCC allele occurs at a frequency of 0.0052 and is detectable by the presence/ absence of an *Rsa*I site in exon 13 of the *PROS* gene. The change occurs within the consensus sequence for the potential N-linked glycosylation of Asn458. Experimental data are consistent with the interpretation that glycosylation at this site is inhibited by the T→C base change. Duchemin *et al.* (1995) found that the Heerlen polymorphism occurred in, 19% of French patients with symptomatic protein S deficiency compared with only 0.8% of healthy controls. The patients bearing the Heerlen polymorphism were deficient in free protein S but exhibited normal levels of C4bBP. Duchemin *et al.* (1995) claimed that this could be caused by the ability of one molecule of C4bBP to bind two molecules of protein S Heerlen. Whatever the reason, the Heerlen polymorphism appears to be rare in Italian patients with protein S deficiency (Castaman *et al.*, 1996b).

A Lys/Glu polymorphism has been noted at residue 155 in 1.6% of the normal Japanese population (Yamazaki *et al.*, 1993). However, there is as yet no evidence that this polymorphism is associated in any way with lower biological activity or an increased risk of thrombosis. A T/A dimorphism has also been reported at position +54 of intron 11 with respective allele frequencies of 0.76 and 0.24 in a Caucasian population (Mustafa *et al.*, 1996).

A further neutral polymorphism has been detected at the third position of codon Pro626 (CCA/CCG); formatted for polymerase chain reaction (PCR), this much more common dimorphism occurs with frequencies of 0.52 and 0.48, respectively, in the Dutch population (Diepstraten *et al.*, 1991) and 0.57 and 0.43 in the French population (Gandrille *et al.*, 1995a). This polymorphism has great utility since it can be used to distinguish alleles at both genomic and cDNA level to confirm or exclude 'allelic exlusion' (see below). Another polymorphism useful in this regard is a C/A dimorphism (respective frequencies, 0.83, 0.17) in the 3' untranslated region of the *PROS* gene, 520 bp 3' to the Stop codon (Mustafa *et al.*, 1996).

Other possible DNA sequence polymorphisms have been detected at codons –31, –16, 35, 180, 303, 304 and 344 (Lundwall *et al.*, 1986; Hoskins *et al.*, 1987; Ploos van Amstel *et al.*, 1987a; Gandrille *et al.*, 1995a).

6.3.4 Chromosomal localization

The *PROS* gene was tentatively localized to chromosome 3 by somatic cell hybrid analysis (Ploos van Amstel *et al.*, 1987a, b). This was confirmed by somatic cell hybrid analysis with regional localization to 3p21→q21 (Long *et al.*, 1988; Watkins *et al.*, 1988). The location of the *PROS* gene was refined still further by *in situ* hybridization to the region 3p11.1→3q11.2 (Watkins *et al.*, 1988). The *PROS* pseudogene is also located on chromosome 3 (Ploos van Amstel *et al.*, 1987b). Segregation studies between the protein S deficiency state and a *PROS* pseudogene *Msp*I variant revealed no recombination events in 23 informative meioses, consistent with a distance of less than 4 cM between the two homologous sequences.

6.3.5 Expression

Radioimmunoassay and intrinsic labelling experiments have shown that protein S is synthesised and secreted by both human HepG2 hepatoma cells (Fair and Marlar, 1986) and human endothelial cells (Fair *et al.*, 1986). It is also synthesised by megakaryocytes (Ogura *et al.*, 1987) and is found in small amounts in platelets (Schwarz *et al.*, 1985). Northern blotting studies have now confirmed the expression of the *PROS* gene in human liver with the detection of a *PROS* mRNA of between 3500 and 4000 nucleotides (Lundwall *et al.*, 1986; Hoskins *et al.*, 1987; Ploos van Amstel *et al.*, 1987a, b). Data on the representation of *PROS* cDNA clones in a human liver cDNA library suggest that *PROS* mRNA comprises approximately 0.01% of total liver mRNA. He and Dahlbäck (1993) have shown by Northern blotting that a 2.4 kb rabbit *PROS* mRNA is present in liver, testis, ovary, uterus, lung and brain but absent or virtually absent in kidney, intestine, muscle, pancreas, heart and spleen. By contrast, in the rat, *PROS* mRNAs were expressed in all tissues examined (brain, heart, lung, diaphragm, liver, spleen, stomach, intestine, kidney, ovary, uterus, placenta, bladder; Jamison *et al.*, 1995); some evidence for tissue-specific alternative mRNA splicing was also reported.

The human *PROS* mRNA, at 3.5 kb, is considerably longer than the 2.4 kb bovine and rabbit mRNAs (Lundwall *et al.*, 1986; Hoskins *et al.*, 1987; He and Dahlbäck, 1993). This is due to the presence of a much longer 3' non-coding sequence in the human *PROS* mRNA. Minor mRNA species of 3100 and 2600 nucleotides have also been reported in human (Hoskins *et al.*, 1987) and could represent alternative processing products.

By means of Northern blotting, Western blotting and ELISA, protein S synthesis has been shown to be down-regulated by tumour necrosis factor-α (TNF-α) by about 70% in SV40-transfected human microvascular endothelial cells and by about 50% in primary human umbilical vein and dermal microvascular endothelial cells (Hooper *et al.*, 1994). However, no such effect was noted in HepG2 cells. Other cytokines (interleukins, 1 and 6, transforming growth factor-β) had no effect on protein S levels.

Human protein S has been expressed *in vitro* in murine C127 epitheloid cells

(Chang *et al.*, 1992). The recombinant protein was fully γ-carboxylated and β-hydroxylated and exhibited normal cofactor activity for APC in a clotting assay. However, the plasma and recombinant proteins differed with respect to N-linked carbohydrate side chains. Rabbit protein S has been expressed in the adenovirus-transfected Syrian hamster cell line AV12–664 (He and Dahlbäck, 1993). It was correctly γ-carboxylated and β-hydroxylated and contained the correct N-linked carbohydrate chains.

6.4 Protein S deficiency

Since its first description in 1984 (Comp *et al.*, 1984; Schwartz *et al.*, 1984), protein S deficiency has come to be recognized as a common cause of heritable thrombophilia. It has recently been reported that it is associated with premature arterial thrombosis as well as with venous thromboembolism (see below). Owing to difficulties in functional assay and reliance upon antigen assays, many cases have probably been missed. The classification of the deficiency state is somewhat confusing since it takes account of the proportions of protein S found in the free form and bound to C4bBP, the former fraction being the only portion functional as a cofactor to APC. This is discussed in more detail below. An added complexity is an acquired form of protein S deficiency caused by the formation of autoantibodies to protein S (Sorice *et al.*, 1996).

6.4.1 Laboratory and clinical features

Classificatory systems. Deficiency of protein S associated with recurrent thromboembolism was first reported by Schwarz *et al.* (1984) and Comp *et al.* (1984). The laboratory diagnosis of protein S deficiency is complicated by the fact that a proportion of circulating protein S is bound to C4bBP. In functional assays, only free protein S is measured whilst in immunological assays, either total free and bound protein S or free protein S may be measured (reviewed by Preissner, 1990). Comp *et al.* (1986) defined two types of protein S deficiency. Type I is characterized by very low levels of free protein S (i.e. functional activity) as a result of the bulk of protein S present being complexed with C4bBP (Dahlbäck and Stenflo, 1993). These individuals may exhibit either normal or reduced total protein S antigen levels, although usually it is reduced in some members of a kindred. Comp *et al.* (1986) also described a case of two brothers with very low levels of both free and total protein S which they termed type II deficiency. This phenotype is consistent with severe homozygous protein S deficiency since both parents had type I deficiency. The nomenclature described here is somewhat confusing since it conflicts with the nomenclature adopted for deficiencies of antithrombin III and protein C. In protein S deficiency, the concentration of free protein S is likely to be determined by other factors such as the plasma concentration of C4bBP (an acute phase reactant), in addition to the effect of any inherited lesion.

A considerable number of different cases of type I deficiency have now been reported (e.g. Comp and Esmon, 1984; Comp *et al.*, 1984, 1986; Schwarz *et al.*, 1984; Broekmans *et al.*, 1985; Sas *et al.*, 1985; Kamiya *et al.*, 1986; Engesser *et al.*, 1987b). Schwarz *et al.* (1989) reported a further case of protein S deficiency (type I by Comp's classification) with an antigen level of 17% and undetectable activity. The patient's mother exhibited normal activity/antigen levels but possessed a variant 65 kDa protein S molecule in her plasma. The patient's father (59% antigen,, 10% activity) possessed only molecules of the normal size (70 kDa). The patient possessed the 65 kDa variant protein and is therefore presumed to be a compound heterozygote for the two inherited defects.

Nishioka and Suzuki (1990) suggested that the protein S–C4bBP complex competitively inhibits the APC cofactor function of free protein S. This may contribute to low protein S activity in patients with low levels of free protein S.

One example of type II deficiency has been reported in a brother and sister (Chafa *et al.*, 1989). These cases, appeared to be homozygous (2–3% free protein S, 13–21% total protein S) since several children of the proposita exhibited a less severe form of the deficiency with total protein S levels characteristic of the heterozygous state.

A dysfunctional protein S molecule has been reported by Mannucci *et al.* (1989) in the plasma of a young female patient with deep vein thrombosis who was taking oral contraceptives. The patient's plasma had total and free protein S antigen levels at the lower limit of normal (63%, 65%) but markedly reduced functional protein S (15%). Five other family members (two siblings, mother and maternal uncle) exhibited a similar pattern but none had suffered venous thromboembolism at the time of the study. Several other examples of dysfunctional protein S molecules have now been reported and characterized at the molecular genetic level (Yamazuki *et al.*, 1993; Hayashi *et al.*, 1994; Gandrille *et al.*, 1995a; see Section 6.4.2, 'Single base-pair substitutions'.)

Studies such as these and data generated from DNA sequencing studies will increasingly highlight the complexities of individual cases and serve to render Comp's original classification (Comp *et al.*, 1986) obsolete. It is therefore likely to be superseded by that recommended by the Scientific and Standardization Committeee of the International Society of Thrombosis and Haemostasis (München, July, 1992): type I, protein S total and free antigen and activity levels reduced; type II, protein S activity reduced, total and free antigen levels normal; type III, normal total protein S, reduced free antigen and activity levels.

Free protein S levels are therefore reduced in both type I and type III deficiencies although total protein S is only reduced in type I. In an attempt to establish the nature of the difference between the two types of protein S deficiency, Zöller *et al.* (1995b) looked at the relationship between free and total protein S and isoforms of C4bBP. Coexistence of type I and type III deficiencies was found in 14/18 protein S-deficient families, suggesting that these apparently distinct types of the condition were actually phenotypic variants of the same disease. In clinically and phenotypically normal relatives from these families, protein S correlated well with the C4bBPβ level (only the β chain containing C4bBP binds protein S) and free protein S equalled the molar excess of protein S over C4bBPβ. In these individuals, the concentration of total protein S overlapped with the concentrations exhibited by those individuals with protein S deficiency. Thus some protein S-deficient family members possessed total protein S levels within the normal range, a characteristic of type III protein S deficiency. Because protein S and C4bBPβ correlate (possibly due to an age effect or because both are weak acute phase reactants), type III-deficient family members also possessed higher plasma concentrations of C4bBPβ than those with type I deficiency, resulting in a low level of free protein S. The take-home message is that the plasma level of free protein S is a more reliable marker of protein S deficiency than total protein S.

Clinical manifestations. As implied above, the majority of protein S-deficient patients are heterozygous for an inherited defect. However, cases of homozygous protein S deficiency have been reported and are associated with neonatal purpura fulminans (Mahasandana *et al.*, 1990a, b; Marlar and Neumann, 1990, 1996; Pegelow *et al.*, 1992).

The clinical manifestations of heterozygous protein S deficiency are almost exclusively superficial thrombophlebitis, deep vein thrombosis and pulmonary embolism (Broekmans *et al.*, 1985; Engesser *et al.*, 1987b; Pabinger and Schneider, 1996). However, a significant number of cases of premature arterial thrombosis associated with protein S deficiency have also been reported (Mannucci *et al.*, 1986b; Girolami *et al.*, 1989; Thommen *et al.*, 1989; Allaart *et al.*, 1990; reviewed in Allaart *et al.*, 1990, and Blanco *et al.*, 1994). Sanson *et al.* (1996) have reported a slightly higher incidence of natural abortion and stillbirth (1.5-fold) in protein S-deficient mothers compared with controls.

The mean age of onset of thrombotic disease is reported to be between 24 (Briët *et al.*, 1988), 28 (Engesser *et al.*, 1987b) and 32 (Zöller *et al.*, 1995a) years of age. Some 50% of protein S-deficient patients will have experienced a thromboembolic episode before the age of 25 although one-third of protein S-deficient individuals may still be free of symptoms by the age of 35. Blanco *et al.* (1994) have reviewed the literature concerning cases of hereditary protein S deficiency manifesting clinically before the age of 18. Zöller *et al.* (1995a) have reported that venous thrombotic events occurred in 47% of protein S-deficient patients and in 7% of their relatives who did not have a deficiency of protein S. Pabinger and Schneider (1996) reported that the probability of developing thrombosis by the age of 50–60 was between 80 and 90%, with a recurrence risk of 63%.

The prevalence of protein S deficiency in the general population (1 in 15 000–20 000) and its incidence in patients with juvenile and/or recurrent venous thrombosis are similar to those for protein C deficiency (Broekmans *et al.*, 1986; Gladson *et al.*, 1988). Briët *et al.* (1988) found that some 4% of their sample of 257 individuals with recurrent thromboembolism were suffering from protein S deficiency (this translates into a prevalence of about 3/100 000 in the Dutch population). Gladson *et al.* (1988) quoted the prevalence of protein S deficiency in a similar group of 141 patients to be 5%. Other prevalence data reported include 2.2% (Heijboer *et al.*, 1990), 6.9% (Awidi *et al.*, 1993) and 7.6% (Melissari *et al.*, 1992).

Bertina (1985) found that in the majority (67%) of patients with protein S deficiency, the first thrombotic episode occurred without an apparent cause. A similar proportion of spontaneous cases (71%) was reported by Henkens *et al.* (1993). Known risk factors, when present, were surgery (6%), immobilization (11%) and oral contraceptive use (16%) (Bertina, 1985). None of the protein S-deficient patients experienced a thrombotic episode before the age of 15. However, half had experienced one or more thrombotic episodes before they were 25. It was found that a decrease in the plasma protein S level below 0.67 U ml^{-1} was associated with a significant risk of venous thromboembolism. The same group screened 37 patients with arterial occlusion presenting before the age of 45 and identified three patients who were heterozygous for protein S deficiency (Allaart *et al.*, 1990). Family studies of the relationship between arterial thrombosis and protein S deficiency were, however, less convincing than similar studies on venous thrombosis. There is no evidence for a difference in the incidence of protein S deficiency-related thrombosis between the sexes (Briët *et al.*, 1988).

Pabinger *et al.* (1994a) performed a retrospective multicentre study of 34 patients with type I protein S deficiency and concluded that there was no excess thrombotic risk associated with oral contraceptive use. However, in this report, only total protein S was measured and there are a number of independent reports of thrombosis in individuals with low free protein S who have taken oral contraceptives (Mannucci *et al.*, 1989; Heistinger *et al.*, 1992; Koelman *et al.*, 1992; Villa *et al.*, 1996). It is thought that oral contraception may reduce the level of circulating plasma protein S (Malm *et al.*, 1988; Quehenberger *et al.*, 1996).

One further risk factor could be the co-inheritance of protein S deficiency with APC resistance (see below). Co-inheritance of the two deficiency states probably increases the likelihood of the individual concerned coming to clinical attention.

6.4.2 Molecular genetic analysis of protein S deficiency

The molecular genetic analysis of protein S deficiency has been recently reviewed (Aiach *et al.*, 1995) and a database of mutational lesions, many of which are as yet unpublished, is available (Gandrille *et al.*, 1997). We shall nevertheless review the molecular genetics of protein S deficiency in some detail here, although relying solely on the published literature.

Restriction fragment length polymorphism (RFLP) tracking. Little molecular genetic analysis of protein S deficiency has so far been carried out. In one large pedigree with type I (Comp's classification) protein S deficiency, Ploos van Amstel *et al.* (1989b) detected an extra 3.5 kb *PROS MspI* fragment which co-segregated with the protein S-deficient phenotype. The site of mutation was localized to the *PROS* pseudogene. Thus the mutation responsible for the protein S-deficient phenotype in this pedigree, which presumably lies within the *PROS* gene, was co-inherited with the pseudogene mutation, implying tight linkage between the two *PROS* sequences.

Marchetti *et al.* (1993) successfully used the Pro626 polymorphism within the *PROS* coding sequence as a marker to track protein S deficiency through a family affected by recurrent venous thrombosis. This analysis employed restriction enzyme digestion of PCR fragments derived from platelet *PROS* cDNA. Formstone *et al.* (1996) used the same approach to perform a prenatal exclusion of severe protein S deficiency in an at-risk family.

Deletions and insertions. Ploos van Amstel *et al.* (1989a) reported the case of an individual who exhibited an extra 4 kb *MspI PROS* fragment which co-segregated with the protein S-deficient phenotype in his two-generation family. Heterozygotes for this lesion suffered from recurrent thromboses. Hybridization of the proband's DNA with *PROS* cDNA probes indicated the reduced signal intensity of specific bands, consistent with a deletion in one of the patient's *PROS* genes within the SHBG-homologous domain.

Schmidel *et al.* (1991) reported a deletion of the *PROS* gene in two apparently unrelated families with protein S deficiency. The deletion was found to span 5.3 kb, including 90% of intron 12, the entire exon 13 and about 25% of intron 13. The mutations in the two families are almost certainly identical by descent since both originated from the same area of West Virginia some 150 years ago. Another study of nine protein S-deficient patients by Southern blotting had previously revealed no obvious deletions or rearrangements of the *PROS* gene (Tanimoto *et al.*, 1989).

A further *MspI* variant has been reported in a family with type I protein S deficiency but the variant did not segregate with the reduced plasma protein S phenotype (Ploos van Amstel *et al.*, 1989b). This suggests either that high-frequency recombination events have occurred between the *PROS* and *PROS* pseudogene loci or, much more likely, a second non-allelic cause of protein S deficiency (e.g. over-expression of C4bBP) must be considered.

Short deletions of the *PROS* gene have also been reported: Gómez *et al.* (1994) described a case of homozygous protein S deficiency causing neonatal purpura fulminans due to a deletion of an A within the AAA codon encoding Lys43. Both Borgel *et al.* (1994) and Gandrille *et al.* (1995a) independently reported a deletion of a T within the CCT codon encoding Pro82 in cases of type I protein S deficiency. Further frameshift deletions [T within the TTG codon encoding Leu261 (Gómez *et al.*, 1995); G within the GGG codon encoding Gly267 (Gómez *et al.*, 1995); AACAGGC spanning codons −14 to −16 (Duchemin *et al.*, 1996); A within the GGA codon encoding Gly448 (Andersen *et al.*, 1996); and CTTA spanning codons 44 and 45 (Duchemin *et al.*, 1996) have also been reported.

Short insertions have also been reported in codons −25, 565 and 578 (Reitsma *et al.*, 1994; Gómez *et al.*, 1995).

Single base-pair substitutions.

Mutation screening methodology. Using PCR primers specific for the *PROS* gene, Reitsma *et al.* (1994) screened a panel of eight unrelated type I-deficient patients from the Dutch population for *PROS* gene lesions. Exons, 1, 2 and 4–15 could be readily screened by this method since primers could be designed so as to avoid PCR amplification from the pseudogene. However, this was not possible in the case of exon 3 since its flanking sequences were identical to those in the pseudogene and both gene and pseudogene sequences were co-amplified. (Thus the PCR product containing exon 3 was indistinguishable from its pseudogene counterpart and any signal indicative of a heterozygous *PROS* mutation would have been diluted and possibly lost by the pseudogene sequence.) Two patients possessed a G→A transition at position +5 of the intron, 10 donor splice site, two patients possessed a Term→Tyr substitution at codon 636 and one patient possessed an insertion of a single base (T) at codon −25 (see below). In the remaining three patients, no pathological lesions were found.

In an attempt to resolve this question, analysis of platelet *PROS* mRNA from the patients was initiated (Reitsma *et al.*, 1994). The codon 626 G/A dimorphism was used as a marker for allele-specific transcription. When genomic DNA was tested, five of the patients were found to be informative for this marker including four with a defined lesion. However, when tested on *PROS* mRNA/cDNA, transcription from one allele was found to be abolished ('allelic exclusion') in two patients and markedly reduced in another. In normal wild-type individuals, both *PROS* alleles are codominantly expressed (as evidenced by analysis of *PROS* mRNA in an individual heterozygous for the protein S Heerlen dimorphism and subsequently in five control individuals; Formstone *et al.*, 1995). Using a restriction enzyme-based assay, Reitsma *et al.* (1994) demonstrated that no transcript from the *PROS* pseudogene was produced in liver, endothelial cells or platelets.

The absence of detectable lesions in the coding sequence of the *PROS* gene in three patients was intriguing. Either these patients were not actually protein S-deficient (unlikely since the phenotypic data were convincing) or the lesions could be located in a hitherto unknown 5′ exon or a non-coding portion of the gene. Another possibility to be considered is that there might be locus heterogeneity in protein S deficiency [i.e. another locus such as *C4BPA* or *C4BPB* (see Section 6.5.1) was responsible]. Alternatively, gene conversion events could have taken place between the closely linked *PROS* gene and pseudogene (see Mustafa *et al.*, 1995, for a possible example of this phenomenon) which could have abolished PCR amplification from the *PROS* allele which had experienced the conversion. Regardless of the correct explanation, the phenomenon of allelic exclusion limits the utility of a cDNA-dependent PCR [reverse transcriptase (RT)-PCR] analysis in a search for *PROS* gene lesions.

The phenomenon of allelic exclusion has been observed with various other genes (Cooper and Krawczak, 1993) and may reflect a poorly understood process which couples transcription with translation. The scale of the problem of allelic exclusion has yet to be fully evaluated. Nevertheless, a cDNA-based approach retains certain advantages which can commend its use. In principle, the complexity of the *PROS* gene (15 exons spread out over some 80 kb of genomic DNA) implies that cDNA analysis would avoid the PCR amplification of individual exon sequences. Certainly, cDNA analysis would detect the aberrant products of splicing mutations which might go undetected in genomic DNA analysis if they lay at some distance from the splice junctions. Finally, the presence of the homologous pseudogene prevents easy analysis of exon 3 via PCR amplification of genomic DNA and could also hinder mutation detection if it mediates gene conversion events.

The mRNA-based RT-PCR approach has also been tested by Formstone *et al.* (1995) in the characterization of seven novel *PROS* gene lesions. Although this approach was found to be inappropriate for use in the characterization of type I protein S deficiency (due to the phenomenon of allelic exclusion), it did provide a rapid and effective method for detecting mutations in type III deficiency (see below). The loss of mRNA from one *PROS* allele was observed for 6/8 protein S-deficient individuals who were heterozygous for the codon 626 G/A dimorphism, thus illustrating the limited utility of this methodology. Another potential problem for RT-PCR-based methodology is presented by the existence of alternatively spliced products in the *PROS* gene (Berg *et al.*, 1996b). This must be borne in mind when employing RT-PCR-based procedures since they could in principle interfere with both mutation detection and the characterization of splicing defects.

One alternative approach to the detection of *PROS* gene lesions is perhaps more promising: the screening of the 15 exons by denaturing gradient gel electrophoresis (DGGE) followed by selective DNA sequencing (Gandrille *et al.*, 1995a). Careful design of the PCR primers used for amplification allowed specific (and selective) amplification of most of the exons of the active *PROS* gene. Mutations detected (see below) were verified by family studies, evolutionary sequence comparisons and by screening 100 healthy controls for these lesions. SSCP has been successfully attempted by Beauchamp *et al.* (1996). On the other hand, the problem of the less than total reliability of single-strand conformation polymorphism (SSCP) and DGGE may be avoided by exon-by-exon sequencing (Formstone *et al.*, 1995; Gómez *et al.*, 1995; Mustafa et al., 1995; Simmonds *et al.*, 1996).

Quantitative protein S deficiency. Reitsma *et al.* (1994) performed the first study of 'type I deficiency' patients. Two patients possessed a G→A transition at position +5 of the intron, 10 donor splice site whilst two patients exhibited a Term→Tyr substitution at codon 636. The latter substitution abolishes the normal Stop codon and predicts a protein S molecule which is extended by 14 amino acids, teminating instead at codon 649. A further patient possessed an insertion of a single base (T) at codon –25, predicting a premature Stop codon at –4. Gómez *et al.* (1995) reported eight further single base-pair substitutions, deletions and insertions in their screen of

15 patients with type I protein S deficiency. However, while the mutation detection frequency (53%) was rather low, there is as yet no obvious explanation.

Single base-pair substitutions in a further 10 'type I' deficiency patients were characterized by Gandrille *et al.* (1995a). Only one was a nonsense mutation (Cys22→Term). The others (Glu26→Ala, Phe31→Cys, Thr37→Met, Arg49→His, Asp204→Gly, Glu208→Lys, Cys224→Trp, Cys224→Arg and Asp335→Asn) occur in residues that are evolutionarily conserved and presumably interfere with protein folding. This same group have recently reported two more point mutations causing type I protein S deficiency: Leu259→Pro and Cys625→Arg (Duchemin *et al.*, 1996). Further nonsense mutations (Gln522→Term, Yamazaki *et al.*, 1995; Arg410→Term, Andersen *et al.*, 1996) have been reported in cases of type I protein S deficiency characterized by markedly reduced platelet mRNA levels consistent with allelic exclusion. Four substitutions in Cys residues causing type I protein S deficiency have been proposed to disrupt disulphide linkages and hence protein stability (Simmonds *et al.*, 1996).

Of the mutations reported by Formstone *et al.* (1995), two were frameshift deletions (1908 del AC in Thr547/Pro548 and 927 del G in Gly220) found in type I deficiency patients; these lesions clearly influenced mRNA stability since they resulted in allelic exclusion. Two splice site mutations were also detected in other patients: the first, a G→A transition in the invariant GT dinucleotide at the donor splice site of exon 4, resulted in the generation of two abnormal yet in-frame mRNA species. The predominant mRNA resulted from utilization of a cryptic splice site 48 bp 3' to the donor splice site whereas the rarer mRNA species was generated by the skipping of exon 4. Abnormal protein was not detected in the plasma from this type I deficiency patient. The second splicing mutation, also in a type I deficiency patient, was in position –1 in the donor splice site of 10 and resulted in the production of an unstable mRNA as evidenced by allelic exclusion.

Mustafa *et al.* (1995) reported the identification of seven *PROS* gene lesions in a total of 9/10 patients studied. These mutations comprised four missense mutations, a short 2 bp deletion, a mutation in the donor splice site of intron 10 and a nonsense mutation which led to allelic exclusion.

One example of type I protein S deficiency (Arg474→Cys) has been shown by *in vitro* expression studies to be due to intracellular degradation and impaired secretion of the variant (Yamazaki *et al.*, 1996). Site-directed mutagenesis was used to show that substitution of Arg474 by Ala or Glu (but not Lys) also reduced the secretion of recombinant protein S, suggesting that a positively charged basic amino acid may be required at this residue within the SHBG-like domain.

A Cys145→Tyr substitution in the second EGF domain has been reported by Beauchamp et al. (1996). This lesion is predicted to abolish the disulphide link between Cys145 and Cys158 and may be predicted to result in a protein with reduced stability. Borgel *et al.* (1996) reported an example of the occurrence of a *de novo* mutation (GCC→CCC, Ala484→Pro) causing sporadic type I protein S deficiency. Non-paternity was ruled out by tracking *PROC* and *HBB* gene polymorphisms and by typing for four variable number tandem repeat (VNTR) polymorphisms.

Qualitative protein S deficiency. A dysfunctional protein S variant (Protein S Tokushima) has been characterized in several members of a family with thrombotic disease who exhibited normal total and free protein S antigen values but a low activity level (Hayashi *et al.*, 1994). The proband possessed an AAG→GAG transition converting Lys155 to Glu in the second EGF domain of protein S. Protein S Tokushima, both purified from the patient's plasma and expressed *in vitro*, exhibited an abnormal electrophoretic mobility and reduced affinity for two monoclonal antibodies, one recognizing the thrombin-sensitive domain of the protein and the other its Ca^{2+}-dependent conformation. Hayashi *et al.* (1995) further demonstrated that this variant protein was unable to interact with either APC or factor X. Indeed, the variant was unable to inhibit the activity of the prothrombinase complex. The Lys155→Glu variant has also been detected by Yamazaki *et al.* (1993) in a family exhibiting a dysfunctional protein S molecule. However, in this family, the variant did not co-segregate with either the protein S deficiency phenotype or the symptoms of thrombotic disease. The reason for this surprising result is at present unclear.

Gandrille *et al.* (1995a) reported several 'type II' mutations. A G→A transition at position +5 in the intron 5 donor splice site is thought to give rise to an abnormal mRNA either by exon skipping or cryptic splice site utilization. Two mutations (Arg2→Leu and Arg1→His) at the propeptide cleavage site are predicted to interfere with normal propeptide processing; the former, which was found in two apparently unrelated patients of African origin, also appears to be associated with abnormal γ-carboxylation. A further dysfunctional protein S molecule was found to result from a Thr103→Asn substitution in the first EGF domain: the longer side chain of asparagine might influence protein binding to the EGF domain without affecting domain folding.

Formstone *et al.* (1995) characterized the *PROS* gene lesions in two type III deficiency patients. The first, a Met570→Thr missense mutation, occurred in a residue which is conserved in human SHBG. The amino acid environment of Met570 appears hydrophobic which suggests that this region may be internal to the protein. The type III deficiency state indicates an effect on C4bBP interaction. Indeed, Lys571 is predicted to interfere with protein S binding to C4bBP (Greengard *et al.*, 1995a). The second lesion resulting in type III deficiency, Asn217→Ser, occurred in a residue located in the fourth EGF domain of protein S [*Figure 6.3* (see Frontispiece)]. By analogy with the calcium-binding pockets of the first EGF domain of factors IX and X, the mutated Asn217 residue is believed to play a significant role in the binding of calcium (Handford *et al.*, 1991; Selander-Sunnerhagen *et al.*, 1992; *Figure 6.3*). The substitution of Asn217 by Ser results in the loss of a γ carbon and the replacement of the apical amide and reactive carbonyl group by a less reactive hydroxyl group [*Figure 6.3* (see Frontispiece)]. The resultant reduction in both steric and polaric contact between the mutated residue and the calcium ion would be expected to lead to loss of calcium binding affinity. This proposal remains, however, to be tested by the biochemical characterization of an *in vitro*-expressed protein. In the meantime, other examples of this substitution continue to be reported (Beauchamp *et al.*, 1996).

Beauchamp *et al.* (1996) also reported a Ser624→Leu substitution in the SHBG domain causing type I/III deficiency. *In vitro* expression of mutant protein S molecules lacking the region 607–635 have implicated this region in C4bBP binding (Nelson and Long, 1992), an observation consistent with the type III deficiency state in the above patient.

Simmonds *et al.* (1996) reported two cases of type II protein S deficiency (Lys9→Glu and Gly340→Asp. The former was proposed to interfere with electrostatic interactions in the Gla domain.

Coinheritance of APC resistance and protein S deficiency. A number of cases of APC resistance co-occurring with either heterozygous or homozygous protein S deficiency were noted by Zöller *et al.* (1995b). Zöller *et al.* (1995a) found that the factor

V Leiden lesion was present in 38% of protein S-deficient patients ($n=16$), a much higher frequency than that found in normal controls. Some 72% of family members of the protein S-deficient individuals who possessed both gene defects had experienced thrombotic symptoms as compared with 19% of those with only protein S deficiency and 19% of those with only the factor V Leiden variant. Very similar data have been reported by Koeleman *et al.* (1995): 38% of protein S-deficient patients ($n=16$) possessed the factor V Leiden lesion. In those families in which both abnormalities were segregating, 80% of clinically symptomatic individuals possessed both types of defect. In three protein S deficiency pedigrees presented by Beauchamp *et al.* (1996), 12 individuals carried a *PROS* gene lesion, seven individuals carried the factor V Leiden lesion, and eight individuals carried both: in these three groups, thrombosis was reported in one, one and three individuals, respectively. Gurgey *et al.* (1996) have also reported combined factor V Leiden and protein S deficiency in several children presenting with thrombosis in the central nervous system. Taken together, these data suggest that co-inheritance of the two deficiency states appears to increase the likelihood of an individual coming to clinical attention.

Antenatal diagnosis. Formstone *et al.* (1996) performed molecular genetic and phenotypic analyses in a highly unusual case of combined protein S and protein C deficiency manifesting in a family in which a child had died perinatally from renal vein thrombosis. Antenatal diagnosis in a second pregnancy was initially performed by indirect RFLP tracking using a neutral dimorphism within the *PROS* gene and served to exclude severe protein S deficiency. An umbilical vein blood sample at 22 weeks gestation revealed isolated protein C deficiency. This pregnancy proceeded to a full-term delivery without thrombotic complications. Molecular genetic analysis of the *PROC* and *PROS* genes segregating in the family then yielded one *PROC* gene lesion in the father (Arg169→Gln) and two *PROS* gene lesions (Met570→Thr and 1908 delAC), one in each parent. These lesions were shown to segregate with the respective deficiency states through the family pedigree. Analysis of DNA from paraffin-embedded liver tissue taken from the deceased child revealed the presence of both *PROS* mutations as well as the *PROC* mutation. Genotypic analysis of the second child revealed a *PROC* mutation but neither *PROS* mutation, consistent with its possession of normal protein S levels and a low/borderline protein C level. Antenatal diagnosis was then performed in a third pregnancy by direct mutation detection. However, although the fetus carried only the paternal *PROS* and *PROC* gene lesions, the child developed renal thrombosis *in utero*. It may be that a further genetic lesion at a third locus still remains to be defined. Alternatively, the intrauterine development of thrombosis in this infant could have been caused at least in part by a transplacental thrombotic stimulus arising in the protein S-deficient maternal circulation. This analysis therefore serves as a warning against extrapolating too readily from genotype to phenotype in families with a complex thrombotic disorder.

6.5 C4bBP and its interactions with protein S

6.5.1 Physiology, structure and function

Protein S exists in two forms in plasma, 40% as free protein S, the remainder bound to C4bBP. This 570 kDa protein also serves to regulate the complement pathway and occurs in human plasma at a concentration of 150 μg ml^{-1}. Protein S is bound to C4bBP in a non-covalent 1:1 stoichiometric complex at a site distinct from the site of interaction with complement protein C4b. Calcium increases the rate of association of protein S and C4bBP: in the absence of calcium, the K_d equals 10^{-7} M whereas in its presence, the K_d is 5×10^{-10} M. C4bBP is an acute phase response protein which under certain conditions exhibits an increase in concentration to 400% of normal (Barnum and Dahlbäck, 1990).

The function of protein S as a cofactor to APC in the inactivation of factors Va and VIIIa is lost on binding to C4bBP (Dahlbäck, 1986). Thus, in protein S deficiency, it is the level of free protein S in the plasma which determines phenotypic severity. The complex of protein S and C4bBP may, however, also serve to inhibit the cofactor activity of protein S (Nishioka and Suzuki, 1990), possibly by interacting with APC, thereby acting as a competitive inhibitor of free protein S.

6.5.2 Molecular biology of C4bBP

C4bBP possesses a unique spider-like structure as visualized by electron microscopy; the seven legs (30 × 300 Å) corresponding to 70 kDa α-chain subunits, surround a smaller, eighth β-chain subunit which is thought to be the binding site for protein S. The primary structure of C4bBP α chain has been determined by amino acid sequencing and by cDNA cloning (Chung et al., 1985a,b). The 549-amino-acid residue subunit protein contains eight 60-amino-acid residue short consensus repeat (SCRe) units (also known as Sushi domains). SCRes have been found in a large number of complement and non-complement proteins (e.g. factor XIII β subunit), but their function remains unclear. Each of the α-chains in C4bBP contains a binding site for the activated form of complement protein C4b. The primary structure of the β chain has been determined from the cDNA (Hillarp and Dahlbäck, 1989).

The 235-amino-acid residue protein contains three SCRe repeats plus a 60-amino-acid long C-terminal non-repeat region. Sequence homology between the α- and β-chain SCRes varies from 17% to 35%. The amino acid sequence of the β chain predicts five potential glycosylation sites. The C-terminal regions of the α and β chains appear to be α-helical in structure and are linked to each other by disulphide bridges between cysteine residues found at homologous locations. The amino acid sequence similarity of the C-terminal regions of the α- and β-chains is 25–30% (Hillarp and Dahlbäck, 1989).

In addition to the sequence similarity, the genes for the α and β subunits of C4bBP (C4BPA and C4BPB) are closely linked on chromosome 1q32 in a head-to-tail orientation (Andersson et al., 1990; Padro-Manuel et al., 1990). The C4BPA gene comprises 12 exons spanning over 40 kb (Rodriguez de Cordoba et al., 1991) and contains a promoter which is responsive to hepatocyte nuclear factor-1 (Arenzana et al., 1995). The C4BPB gene contains eight exons spanning in excess of 10 kb; alternative processing from two different promoters yields two different transcripts which are encoded by eight and six exons respectively (Hillarp et al., 1993). The structures of the C4BPA and C4BPB genes are similar, consistent with these genes having evolved by a process of duplication and divergence.

The biosynthesis of C4bBP involves the assembly of six or seven α chains and one or no β chain (Dahlbäck, 1991). Although little is known as to the mechanisms underlying the regulated synthesis of these chains, Northern blotting experiments have demonstrated that the C4BPB mRNA is considerably less abundant than the C4BPA mRNA (Hillarp and Dahlbäck, 1989).

6.5.3 C4bBP binding to protein S

C4bBP is heterogeneous both in terms of subunit composition and protein S binding (Hillarp *et al.*, 1989); 30% of C4bBP in plasma occurs as a lower molecular weight species than the remainder. Approximately 50% of the lower molecular weight species binds protein S and contains the β chain but possibly only six α chains. The higher molecular weight species is thought to contain seven α chains but lacks the β chain and does not bind protein S (Hillarp *et al.*, 1989). These data are consistent with a location of the protein S binding site on the β chain (reviewed by Dahlbäck, 1991). This binding site was localized to one or more of the SCRe modules of the β chain (Hardig *et al.*, 1993). Experiments involving the competitive binding of synthetic peptides and monoclonal antibodies raised against them have pointed to the involvement of residues 31–45 of the β chain of C4bBP in the binding of protein S (Fernandez and Griffin, 1994; Fernandez *et al.*, 1994). The concentration of the β-chain-containing C4bBP is equimolar to the concentration of C4bBP–protein S complexes, suggesting that free protein S represents the molar excess of protein S (Griffin *et al.*, 1992).

6.5.4 Clinical aspects

It has been suggested that an increased concentration of C4bBP may serve to lower the concentration of free protein S in plasma (Comp *et al.*, 1986). Conversely, a lowered level of C4bBP (as found in a family with inherited C4bBP deficiency; Comp *et al.*, 1990) would result in an elevated level of free protein S. However, the concentration of free protein S during infectious disease appears to be both normal and fairly stable despite a dramatically raised C4bBP level (Hesselvik *et al.*, 1991). Garcia de Frutos *et al.* (1994) have demonstrated differential regulation of the α and β chains of C4bBP during the acute phase; the plasma concentration of C4bBP lacking the β chain increased by more than that containing the β chain. Since the total protein S level during the acute phase was increased to the same extent as the level of C4bBP containing the β chain, it appears that stable levels of free protein S during the acute phase are the result of differential regulation of C4bBP α- and β-chain expression. The free protein S concentration represents the molar excess of protein S over C4bBP containing the β chain.

The C4bBP–protein S complex is capable of binding to phospholipid vesicles and this may provide a suitable surface for interactions of this multi-protein complex. For example, the complex can interfere competitively with the cofactor function of free protein S in aiding the action of APC which occurs on phospholipid surfaces.

Although no familial elevation of C4bBP has so far been reported, this might be predicted to produce a similar phenotype to protein S deficiency. If so, it would be detected by finding raised levels of protein S–C4bBP complex accompanied by deficiency of free protein S in a patient with recurrent venous thromboembolism.

Another possible abnormality of C4bBP that would produce the same effect without increased total C4bBP could be a molecule with increased affinity for protein S. Indeed this may be the explanation for the type II protein S deficiency reported by Comp *et al.* (1984), in which patients have reduced free protein S but normal amounts of total C4bBP.

As an acute phase reactant, C4bBP is greatly elevated following surgery or in patients with inflammation or neoplasia (Taylor, 1992). In these situations, all associated with increased risk of thrombosis, the reduced free protein S consequent upon elevation of C4bBP may have a contributory role to play in the pathogenesis of thrombosis.

Factor V and activated protein C resistance

7.1 Properties and role in coagulation

Factor Va serves as a cofactor to factor Xa in the prothrombinase complex which, in the presence of phospholipid and calcium ions, activates prothrombin (reviewed in Mann et al., 1988). Together with protein S, factor V also acts as a cofactor to activated protein C (APC) in the degradation of factor VIIIa (Shen and Dahlbäck, 1994).

The purification of unactivated bovine procofactor V was first achieved in 1979 (Esmon, 1979; Nesheim et al., 1979a). Human procofactor V was isolated soon after and shown to be activated by thrombin (Dahlbäck, 1980). The protein was partially sequenced and this sequence was completed from cDNA clones. Sequence data confirmed the homology of the two coagulation cofactors V and VIII (reviewed by Kane and Davie, 1988; Mann et al., 1988; Jenny and Mann, 1989).

Factor V deficiency normally gives rise to a bleeding diathesis. The relevance of factor V to thrombotic disease lies in the very frequent mutation at Arg506 in the APC cleavage site. Factor Va is normally inactivated by this cleavage. Abolition of APC cleavage as a consequence of mutation within the cleavage site (APC resistance) leads to excess thrombin generation and predisposition to venous thrombosis. The importance of factor V is evident from a transgenic model: mice homozygous for a factor V defect die either embryonically, possibly as a result of an abnormality in the yolk sac vasculature, or neonatally from massive haemorrhage (Cui et al., 1996).

7.2 Structure and function of factor V

7.2.1 Physiology and physical properties

Factor V is synthesised in both hepatocytes and megakaryocytes. It circulates in the plasma as a single-chain glycoprotein, M_r 330 kDa, with a half-life of 11–15 h (Rand et al., 1995). The concentration of factor V in platelet-poor plasma is 7–10 μg ml^{-1} (20 nM) (Tracy et al., 1982), whilst platelets contain about 4 μg factor V per 10^9 platelets. Thus in whole blood, about 80% of the factor V is free in the plasma and 20% is stored in the platelets. Platelet factor V is stored in complex with multimerin, a multimeric platelet protein stored in α granules and expressed on the surface of activated platelets (Hayward et al., 1995). Platelet factor V is releasable upon platelet activation and may play an important role in haemostasis (Jenny and Mann, 1989).

Single-chain factor V has no intrinsic cofactor activity but circulates as a procofactor (Nesheim et al., 1979b). It is activated by thrombin which cleaves specific peptide bonds yielding light and heavy chains of factor Va. Factor Va may then perform its cofactor role of enhancing the rate of activation of prothrombin by factor X by more than, 1000-fold (Rosing *et al.*, 1980; reviewed by Mann *et al.*, 1988).

7.2.2 Primary structure, domains and sequence homologies

The 2196-amino-acid sequence of factor V was established on the basis of both amino acid and cDNA sequencing (Jenny et al., 1987; Kane et al., 1987). cDNA sequencing indicated the presence of a hydrophobic 28-amino-acid leader sequence. Factor V contains several types of internal repeats (*Figure 7.1*); the largest portion consists of a triplicated A domain of about 300 residues that is also found in ceruloplasmin, the major copper transport protein in plasma. Interestingly, factor V also contains a copper ion (Mann *et al.*, 1984).

The second and third A domains are separated by a long B domain of about 800 residues unrelated in sequence to any known protein (*Figure 7.1*); it consists of 31 tandem repeats of a nine residue motif with the consensus sequence, T/N/P, L, S, P, D, L, S, Q, T. The C-terminal region of factor V consists of a duplicated domain of about 150 amino acids that possesses homology to the mouse milk fat globule-binding protein (Stubbs *et al.*, 1990) and more distantly to the slime mould lectin, discoidin (Vehar *et al.*, 1984). This domain is important for phospholipid binding. The domain structure of factor V can thus be represented as A1–A2–B–A3–C1–C2. Factor V also possesses an acidic peptide between its A2 and B domains; residues 653–698 contain 17 Asp or Glu residues out of a total of 45.

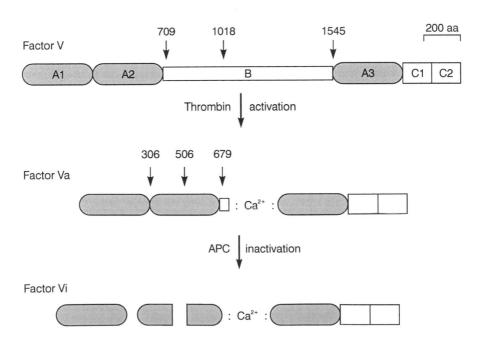

Figure 7.1. Structure and processing of human factor V. Factor V possesses the domain organization A1–A2–B–A3–C1–C2. It is activated by thrombin cleavage at positions 709, 1018 and 1545. The two chains of factor Va are non-covalently linked in the presence of calcium ions. Factor Va is inactivated by activated protein C cleavage at positions 306, 506 and 679. Reproduced from Hillarp *et al.* (1995) with permission from Churchill Livingstone.

7.2.3 Secondary structure and post-translational modifications

Homology with bovine factor Va suggests the following disulphide linkages in the human protein: C139–C165, C220–C301, C471–C498, C576–C656, C1697–C1723, C1879–C2033 and C2038–C2193 (Xue et al., 1993, 1994).

There are 37 potential N-linked glycosylation sites of which 25 are in the B domain. The B domain isolated from plasma factor V has a M_r of 70 kDa, suggesting that it is about 50% carbohydrate by weight when compared with the predicted mass from amino acid sequence. The B domain is entirely removed during proteolytic activation of the procofactor.

Factor Va is also phosphorylated by activated platelets (Kalafatis et al., 1993).

7.2.4 Proteolytic processing, activation and inhibition

The activation of single-chain procofactor V requires proteolytic cleavage by either thrombin or factor Xa (Marquette et al., 1995). Factor Va thus produced is stable. Thrombin cleaves human factor V at three scissile bonds, Arg709–Ser710, Arg1018–Thr1019 and Arg1545–Ser1546 (Figure 7.1), yielding chains of M_r 105 kDa, 70 kDa and 74 kDa; the 105 kDa and 74 kDa species form the heavy (A1–A2) and light (A3–C1–C2) chains of factor Va, respectively (Suzuki et al., 1982; Keller et al., 1995; factor 7.1). Factor Xa also cleaves the heavy chain at Arg348–Ser349 to activate human factor V, resulting in a 105 kDa/220 kDa (A1–A2/B1–A3–C1–C2) heterodimer (Monkovic and Tracy, 1990). During coagulation, factor V is first activated by factor Xa and then processed by thrombin (Monkovic and Tracy, 1990; Yang et al., 1990). Thus activation of factor V by factor Xa permits the initial assembly of the pro-thrombinase complex, thereby yielding thrombin which can then fully back-activate the procofactor. Factor V is also reportedly activated by meizothrombin (Tans et al., 1994). Factor V may also be activated by plasmin (Lee and Mann, 1989), platelet-associated proteinases such as calpain (Bradford et al., 1988), elastase and cathepsin G (Allen and Tracy, 1995).

Factor Va is inactivated by APC (Suzuki et al., 1983c; Figure 7.1). APC is thought to bind factor Va through residues 311–325 in the serine proteinase domain (Mesters et al., 1993b). Early studies demonstrated that APC cleaves human factor Va in both the heavy chain (Arg505–Arg 506) and the light chain (Arg1765–Leu1766) (Guinto and Esmon, 1984; Odegaard and Mann, 1987). Membrane-bound human factor V is now known to be cleaved by APC at four sites: Arg306, Arg506, Arg679 and Lys994 but, in the absence of membrane, no cleavage occurs (Kalafatis et al., 1994). Membrane-bound human factor Va is cleaved at three sites by APC (Arg306, Arg506 and Arg679; Kalafatis et al., 1994). Inactivation of both factors V and Va requires the presence of membrane and cleavage of Arg306 (Kalafatis et al., 1994). Cleavage at Arg506 precedes, and may promote subsequent cleavage at Arg306 and Arg679 (Kalafatis et al., 1994, 1995a). In the absence of membrane, APC cleaves the heavy chain of factor V only at Arg506 and Arg679 (Kalafatis et al., 1994). Thus whereas in membrane-bound factor Va, cleavage at Arg506 is required for exposure of the inactivating cleavage site at Arg306, the Arg306 cleavage site is immediately available in membrane-bound factor V and prior cleavage at Arg506 is not required. Bovine factor Va is cleaved by APC at three sites in the heavy chain (Arg residues 306, 505 and 662; Guinto et al., 1992; Kalafatis and Mann, 1993).

Impaired cleavage at Arg306 results in a cofactor that is cleaved at Arg506 and Arg679 and still possesses approximately 60% normal cofactor activity (Lu et al.,

1994b). Abolition of the cleavage site at Arg506 (in the Gln506 variant, see Section 7.5.2 below) slows the rate of inactivation of membrane-bound factor Va but does not abolish it (Kalafatis et al., 1995a).

The APC-mediated inactivation of platelet factor V is not completely analogous to that found with plasma factor V (Camire et al., 1995). APC is unable to inactivate platelet factor V completely even though the same cleavages apparently occur in platelet factor V as in plasma factor V. It is possible that this difference may be due to differential post-translational modification of the two forms of factor V.

The function of the B domain that is released from factor V upon activation has been studied by site-directed mutagenesis. A recombinant protein lacking residues 811–1441 was expressed in COS cells and found to possess only 38% of the activity of wild-type factor Va (Kane et al., 1990). Sequences within the B domain appear to be required for thrombin activation of factor V (Pittman et al., 1994).

7.2.5 Interactions with phospholipid, platelets and calcium ions

A single high-affinity calcium-binding site ($K_d=10^{-9}$ M) and two lower affinity sites ($K_d = 10^{-5}$ M) are present on factor V. Factor Va has a single low-affinity site which appears upon assembly of the heavy and light chains. A single functionally important calcium-binding site is present on the heavy chain of factor Va.

Phospholipid stimulates APC-catalysed factor Va inactivation (Bakker et al., 1992). Both factors V and Va bind to negatively charged phospholipids with a K_d in the nanomolar range. A phospholipid-binding site on bovine factor Va has been localized to amino acid residues, 1667–1765 in the A3 domain at the N-terminal end of the light chain (Kalafatis et al., 1990). Ortel et al. (1992) also identified a phospholipid-binding site within the C-terminal C2 domain of the light chain.

Platelets express about 1000 high-affinity sites for factor Va ($K_d= 10^{-10}$ M) and about 3500 lower affinity sites. Binding to platelets occurs through the light chain. Assembly of the prothrombinase complex on thrombin-activated platelets has revealed that about 3000 factor Va sites bind factor Xa at the high-affinity sites ($K_d =10^{-10}$ M) with a 1:1 stoichiometry. Similar prothrombinase complexes can form on the surface of neutrophils, monocytes, lymphocytes and damaged endothelial cells.

7.2.6 Interactions with factor Xa, APC and protein S

Both light and heavy chains of factor Va interact with factor Xa, although isolated light chains bound to platelet membranes will bind factor Xa (Tracy and Mann, 1983). It is thought that factors Va and Xa bind independently to the membrane surface before interacting with each other with high affinity (Ye and Esmon, 1995). Factor Va cleaved at Arg506 exhibits a reduced ability to interact with factor Xa (Nicolaes et al., 1995). Binding of prothrombin to factor Va is mediated via the heavy chain of factor Va (Guinto and Esmon, 1984). This reaction is calcium-independent, and shows 1:1 stoichiometry.

Interaction of factor Va with APC occurs exclusively via the light chain of factor Va (Krishnaswamy et al., 1986), involving residues, 1865–1875 in the A3 domain of the cofactor (Walker et al., 1990); a peptide corresponding to these residues (RAG-MQTPFLI) inhibits the anticoagulant activity of APC towards factor V. Mesters et al. (1993a) provided evidence for a binding site for factor Va on the light chain of protein C; specifically, a synthetic peptide corresponding to residues 142–155 of protein

C (in the connecting region) binds directly to factor Va and interferes with APC inactivation of factor Xa.

Factor Va performs a second role in relation to protein C. It serves to enhance the rate of protein C activation by thrombin/thrombomodulin some 50-fold (Salem *et al.*, 1983a, b). The K_m for factor Va in this reaction is 14 nM, 100 times higher than its K_m for accelerating prothrombin activation by factor Xa. Thus, factor Va does not merely exert a coagulant influence but it also helps to up-regulate the protein C anticoagulant pathway which in turn acts so as to inhibit factor Va. This feedback mechanism therefore serves to limit the dissemination of the clotting process.

Protein S increases the rate of APC inactivation of factor Va by some two-fold (Nesheim *et al.*, 1982; Solymoss *et al.*, 1988). Further, factor Xa protects factor Va from APC inactivation by lowering the effective concentration of factor Va available to interact with APC (Suzuki *et al.*, 1983c; Solymoss *et al.*, 1988). Since protein S abolishes the ability of factor Xa to protect factor Va from inactivation without affecting the interaction between factors Va and Xa, it would appear that one role for protein S is to permit the inactivation of factor Va by APC in the presence of factor Xa. Both free protein S and thrombin-cleaved protein S bind factor V and factor Va (Hackeng *et al.*, 1994). However, since protein S complexed with C4bBP does not bind these factors (Hackeng *et al.*, 1994), we may surmise that the factor Va binding site on protein S is masked by the interaction with C4bBP.

7.2.7 Role as cofactor to APC in the inactivation of factor VIIIa

The discovery that APC resistance, a major cause of hereditary thrombophilia (see below), could be corrected by factor V (Dahlbäck and Hildebrand, 1994) represented the first suggestion that factor V might possess anticoagulant properties and act as a cofactor to APC. Using a purified system, Shen and Dahlbäck (1994) demonstrated that in the presence of factor V and protein S, APC effectively inhibited factor VIIIa. However, APC alone or APC together with factor V were ineffective whilst APC together with protein S was less efficient than when factor V was present. Factor V activated by thrombin was however inefficient as a cofactor. The APC cofactor activity of factor V was maximal at a molar concentration of factor V that was similar to that of APC and protein S. Similar data have been reported by Varadi *et al.* (1996). Together with other kinetic data, these findings are consistent with the view that factor V and protein S act synergistically as cofactors to APC in its inactivation of factor VIIIa through the formation of a 1:1:1 stoichiometric complex of APC, factor V and protein S on phospholipid surfaces.

7.3 Molecular genetics of factor V

A number of cDNA clones for factor V have been isolated and sequenced (Kane and Davie, 1986; Jenny *et al.*, 1987; Kane *et al.*, 1987; Dahlbäck *et al.*, 1988). A full-length cDNA contains 6672 bp coding sequence, 163 bp 3' untranslated region, plus a 5' untranslated region of either 97 bp or 103 bp, depending upon which of two transcriptional initiation sites are used (Cripe *et al.*, 1992). The coding region comprises a 28-amino-acid hydrophobic leader peptide and a protein of 2196 amino acids (709-amino-acid heavy chain region, 836-amino-acid connecting region and a 651-amino-acid light chain region).

A cDNA for bovine factor V has been characterized (Guinto *et al.*, 1992) which is 88%, 59% and 88% homologous to its human counterpart in the heavy chain, connecting region and light chain respectively.

The human factor V (*F5*) gene contains 25 exons ranging in size from 72 bp to 2820 bp in length and spans in excess of 80 kb (Cripe *et al.*, 1992). The size of intron 2 is still unclear. The structure of the *F5* gene is very similar to that encoding human factor VIII (*F8*).

The sites of synthesis of factor V have been studied in various cells and tissues by means of bioassays for coagulant activity, radioimmunoassays or [^{35}S]-methionine incorporation. These methods have demonstrated the synthesis of factor V in human megakaryocytes and hepatocytes (reviewed by Tuddenham and Cooper, 1994). Northern blotting studies have demonstrated the presence of a 7000 bp mRNA in both human hepatocytes and HepG2 cells (Jenny *et al.*, 1987; Kane *et al.*, 1987; Mazzorana *et al.*, 1991).

An intragenic G/A polymorphism in exon 16 at nucleotide 5380 (codon 1771, Val/Met) has been described (Bayston *et al.*, 1994); formatted for polymerase chain reaction (PCR), the alternative alleles occur at frequencies of 0.69 (A) and 0.31 (G). An AGA/AAA polymorphism, predicting the alternative codons Arg and Lys at residue 485, has been reported with respective allele frequencies of 0.97 and 0.03 in the French population (Gandrille *et al.*, 1995b). Lunghi *et al.* (1996) have reported three further polymorphisms in the *F5* coding region (His/Arg1299, Leu/Ile1257, Ser/Ser1240); the Arg1299 (R2) allele was claimed to be associated with a reduced plasma factor V level.

An extragenic *Pst*I polymorphism has been reported in the human *F5* gene region (McAlpine *et al.*, 1990); alternative, 10.8 kb and 8.5 kb alleles occur at respective frequencies of 0.74 and 0.26 in a Caucasian population sample.

The *F5* gene has been mapped to chromosome 1q21–25 by somatic cell hybrid analysis and *in situ* hybridization (Dahlbäck *et al.*, 1988; Wang *et al.*, 1988).

7.4 Factor V deficiency

Factor V deficiency is an autosomal recessive bleeding diathesis characterized by low levels of plasma and/or platelet factor V. Symptoms include epistaxis, menorrhagia, excessive bleeding after dental extraction or surgery and occasionally haemarthroses. Factor V deficiency is normally characterized by a prolonged prothrombin time and activated partial thromboplastin time (APTT) which is corrected by the addition of barium-absorbed normal plasma but not by the addition of normal serum. Some 150 cases of factor V deficiency have been reported and its estimated prevalence is of the order of approximately $1/10^6$ (reviewed by Tuddenham and Cooper, 1994). Several cases of thrombotic disease have also been reported to be associated with the deficiency state (Miller, 1965; Reich *et al.*, 1976; Manotti *et al.*, 1989). Although perhaps prophetic, this association may be coincidental rather than causative.

7.5 Factor V variants, APC resistance and thrombotic disease

7.5.1 The phenomenon of APC resistance

Dahlbäck *et al.* (1993) first described a poor anticoagulant response to APC which appeared to suggest the possible existence of a novel cofactor to APC. These workers developed a novel APTT-based method to determine the anticoagulant response of patient plasma to added purified APC and used it to identify three unrelated families with a poor response to APC. In the first family, the anticoagulant response of the propositus' plasma to APC was consistently much lower than that of control plasma (i.e. the addition of APC to his plasma failed to elicit the normal prolongation of clotting time). That this was caused by an inherited abnormality was evidenced by the finding that plasma from 14 out of 19 members of his family (four of whom had experienced episodes of venous thrombosis) also responded poorly to APC. The other families studied contained individuals who manifested a very similar laboratory phenotype. Again, this trait appeared to be inherited as an autosomal dominant trait with incomplete penetrance. Griffin *et al.* (1993) then showed that plasma from APC cofactor-deficient individuals did not cross-correct each other indicating that the deficient factor was identical in the different patients studied.

Dahlbäck *et al.* (1993) proceeded to exclude the possibility of autoantibodies to protein C, a fast-acting proteinase inhibitor of APC, protein S deficiency and mutations in the genes encoding factor VIII (*F8*) and von Willebrand factor (*VWF*). When patient plasma was mixed with normal plasma, the normal plasma corrected the poor anticoagulant response of the patient plasma, indicating that the latter lacked a factor present in normal plasma. Since the poor anticoagulant response to APC could not at this stage be easily explained by the accepted model of the protein C anticoagulant pathway, Dahlbäck *et al.* (1993) initially postulated the existence of a novel cofactor to APC in normal plasma which was deficient in the patients studied. However, Dahlbäck and Hildebrand (1994) soon showed that their APC cofactor was factor V: purification of a protein possessing APC cofactor activity yielded factor V. Affinity-purified factor V was then shown to correct the poor anticoagulant response to APC of APC-resistant plasma in a dose-dependent manner. Since this plasma contained normal levels of factor V procoagulant activity, it was inferred that the phenomenon of APC resistance was due to a specific defect in the *anticoagulant* function of factor V.

Bertina *et al.* (1994) performed a series of experiments which identified the defect responsible for APC resistance in a majority of thrombotic patients exhibiting this phenotype. Firstly, they assayed a series of plasmas deficient in a single haemostatic factor and found that all contained a normal level of 'APC cofactor' except for those plasmas deficient in factor V. Further evidence implicating factor V came from the establishment of close genetic linkage between D1S61, located only 4 cM from *F5*, and APC resistance in a three-generation family. Close linkage between intragenic polymorphisms in the *F5* gene and APC resistance was also reported by Zöller and Dahlbäck (1994).

Bertina *et al.* (1994) then selectively sequenced the portion of the *F5* gene encoding the region around the N-terminal APC cleavage site. A CGA→CAA transition was noted, predicting the substitution of Arg506 by Gln (*Figure 7.2*). This substitution was then shown to prevent the APC-mediated inactivation of factor Va generated by factor Xa cleavage although it did not prevent inactivation of factor Va generated by thrombin cleavage (see Section 7.2.3). The other members of the family were screened for the CGA→CAA transition by oligonucleotide discriminant hybridization and co-segregation of the lesion with the APC-resistant phenotype was confirmed. Further, homozygosity for this lesion was confirmed in two other individuals previously classified as being homozygous for the APC resistance trait. Of 64 unrelated thrombotic patients who exhibited the APC-resistant phenotype, 56 (87%) were found to carry

Factor Va

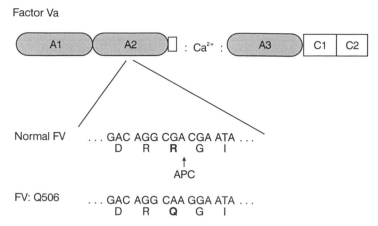

Figure 7.2. Mutation in factor V in a patient with APC resistance. APC inactivates factor Va by cleavages in the A2 domain (Arg506) as well as at Arg306 and Arg679. The nucleotide and translated-amino-acid sequences flanking the APC cleavage site at position 506 are presented for the wild-type sequence and for an individual with factor V Leiden. Reproduced from Hillarp *et al.* (1995) with permission from Churchill Livingstone.

the Arg506 mutation. So did three out of 64 (4.7%) healthy controls taken from the Dutch population, making this variant (now termed Factor V Leiden) the most common inherited abnormality of blood coagulation. These results were soon confirmed by others (Greengard *et al.*, 1994a; Sun *et al.*, 1994; Voorberg *et al.*, 1994; Zöller and Dahlbäck, 1994; reviewed by Dahlbäck, 1994, 1995b; Hillarp *et al.*, 1995).

In a study of 50 Swedish families with inherited APC resistance, Zöller *et al.* (1994) found the factor V Leiden mutation in 47 of them. Perfect co-segregation between a low APC ratio and the presence of the mutation was seen in 40 families. However, in seven families, co-segregation was not perfect: 12 out of 57 APC-resistant individuals lacked the factor V mutation. Further, in three families with APC resistance, the factor V Leiden mutation was absent, suggesting the possible existence of additional lesions underlying APC resistance.

The finding that factor V acts synergistically with protein S as a cofactor in the inactivation of factor VIIIa (Shen and Dahlbäck, 1994; Section 7.2.6) provides the explanation for the phenomenon of APC resistance. The factor V Leiden variant is not cleaved due to the mutation at Arg506 and therefore the subsequent inactivating cleavage at Arg306 does not occur (Kalafatis *et al.*, 1994). Factor V Leiden is therefore degraded at a lower rate and factor Va in APC-resistant plasma will give rise to a higher than normal rate of thrombin generation (Greengard *et al.*, 1994b). This thrombin then activates factors V and VIII, resulting in a loss of factor V with its APC cofactor activity. Prothrombin fragment 1+2 and thrombin/antithrombin (TAT) levels, both markers of activation of the coagulation system, are elevated in individuals bearing the factor V lesion (Simioni *et al.*, 1996a; Zöller *et al.*, 1996).

The simple conclusion that factor V acts as a cofactor for APC was initially challenged by Griffin *et al.* (1995). Varying the concentration of either normal or Leiden variant factor V did not appear to alter the APC resistance ratio in a typical clotting assay. However, the finding that factor V affects the apparent rate of inactivation of purified factor VIIIa by APC in the presence of protein S, phospholipid and calcium ions (Shen and Dahlbäck, 1994) still points to an important interaction between these proteins.

Unlike genetic analysis, however, the accuracy of the APC resistance test is critically dependent upon where the cut-off value for the APC sensitivity ratio is set (Kambayashi *et al.*, 1995). By increasing the ratio threshold to 2.76 which corresponds to the mean ratio minus one standard deviation (SD), Aillaud *et al.* (1995) increased the sensitivity to 100%, albeit with a parallel decrease in specificity (71%). A typical comparison of the APC sensitivity ratios found in patients and controls is presented in *Figure 7.3*.

Kalafatis *et al.* (1994) pointed out that APC-resistant patients might escape detection if factor V is not properly activated prior to the assay. This is because the inactivating cleavage at Arg306 on membrane-bound factor V or factor Xa-cleaved factor V does not require prior cleavage at Arg506. Misclassification of APC resistance as protein S deficiency due to interference with functional protein S assays can also occur (Faioni *et al.*, 1993). One way around this is to use a one-stage tissue factor-dependent factor V assay which can distinguish APC resistance from protein S deficiency and

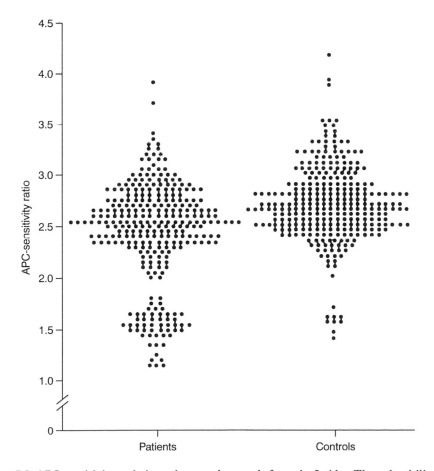

Figure 7.3. APC sensitivity ratio in patients and controls from the Leiden Thrombophilia Study. Reproduced from Koster *et al.* (1993) Venous thrombosis due to poor anticoagulant response to activated protein C: Leiden Thrombophilia Study. *Lancet*, vol. 342, pp. 1503–1506. © 1993 The Lancet Ltd.

possession of a lupus anticoagulant, and can be employed with patients who are receiving anticoagulant therapy (Le *et al.*, 1995). The various available assays for APC resistance vary markedly in their performance (Legnani *et al.*, 1996). Several novel assays remain to be rigorously tested. One of these is a chromogenic assay which is based upon the ability of APC to limit the generation of factor Xa by inactivating factor VIIIa in plasma (Varadi *et al.*, 1995). Engel *et al.* (1996) have also described a highly sensitive and specific assay which employs factor V-deficient plasma.

Henkens *et al.* (1995a) have shown that raised factor VIII:C levels can lead to a reduction in the APC sensitivity ratio and therefore yield false positives in tests for APC resistance. Misclassification of APC resistance as type II protein C deficiency may also occur (Ireland *et al.*, 1995). The use of both phenotypic and genotypic assays to screen for APC resistance has been reviewed by Dahlbäck (1995c, d) and Bertina *et al.* (1995).

7.5.2 Biochemistry of the Gln506 variant

Effects on inactivation of factor Va. Kalafatis *et al.* (1995a) purified factor V from the plasma of two APC-resistant patients and showed that it possessed normal procoagulant activity in both clotting time and thrombin formation assays. However, cleavage of the thrombin-activated membrane-bound factor Va differed between the normal molecule and the Gln506 variant. As we have seen (Section 7.2.3), inactivation of plasma factor Va normally occurs via a series of sequential cleavages. Cleavage at Arg506 is followed by cleavage at Arg306 and Arg679 which results in loss of cofactor activity. Membrane-bound plasma factor Va from patients exhibiting APC resistance (i.e. whose factor Va lacks the Arg506 cleavage site) is still inactivated following cleavage at Arg306 and Arg679: cleavage at Arg306 is responsible for the loss of approximately 70% of cofactor activity whilst cleavage at Arg679 is responsible for the loss of approximately 30% of cofactor activity (Lu *et al.*, 1994b; Kalafatis *et al.*, 1994, 1995a; Aparicio and Dahlbäck, 1996). However, the rate of inactivation of factor Va Gln506 is slower than in the normal molecule: after 5 min incubation with APC, membrane-bound normal factor Va is completely inactivated while factor Va Gln506 retains approximately 50% of its initial activity. Similar results have been reported by Heeb *et al.* (1995). Thus cleavage at Arg506 in factor Va has no effect on cofactor activity but cleavage at this site is nevertheless still necessary for optimal exposure of the inactivating cleavage sites at Arg306 and Arg679. Since APC can still inactivate the Gln506 variant of factor V, albeit at a decreased rate, this may go some way toward explaining why APC resistance is a comparatively mild risk factor for thrombosis.

However, the experiments of Kalafatis *et al.* (1995a) were performed using high concentrations of factor Va. Nicolaes *et al.* (1995) have shown that when conditions of low factor Va concentration are employed and when its cofactor activity is assayed in a prothrombin activation mixture containing a low concentration of factor Xa, the difference in APC-catalysed inactivation between wild-type factor V and its Leiden variant becomes more pronounced.

Antifibrinolytic effects. In addition to the effects on cleavage of factor Va, factor V Leiden is also associated with delayed fibrinolysis. APC is known to be profibrinolytic (Chapter 6) through its ability to attenuate thrombin generation. Bajzar *et al.* (1996c) have shown that approximately, 10-fold more APC is required to reduce clot lysis time from 140 min to 50 min in clots containing factor V Leiden compared with clots containing normal factor V. However, in the absence of thrombin-activatable fibrinolysis inhibitor (TAFI; see Section 5.2.8), APC did not affect clot lysis either with normal factor V or factor V Leiden. The impaired profibrinolytic response to APC in individuals with the factor V Leiden lesion is therefore TAFI-dependent. This phenomenon thus represents a second mechanism by which the factor V Leiden lesion contributes toward a prothrombotic tendency.

7.5.3 Prevalence of APC-resistant factor V variants

A number of *retrospective* studies of thrombotic patients have been performed. Griffin *et al.* (1993) determined the prevalence of APC resistance in a group of 25 patients with venous thrombophilia in whom no identifiable abnormality had been found. A total of 16 patients (64%) manifested an APC-induced APTT prolongation time that was below the values observed for 97% of controls. Of these 16 patients, eight had positive family histories of thrombosis whereas eight had none.

Koster *et al.* (1993) studied 301 consecutive patients with venous thrombosis and found APC resistance in 64 (21%) of them. Response to APC was measured by the 'APC sensitivity ratio', defined as the APTT + APC divided by the APTT – APC. Normal responders (APC sensitivity ratio, 2.25–3.17) could in this way be readily distinguished from poor responders (1.46–2.10). A clear inverse relationship was apparent between the degree of response to APC and the risk of thrombosis; a poor response to APC was associated with an almost seven-fold increase in the risk of thrombosis. Six patients exhibited a very low APC sensitivity ratio (less than 1.25); in one case, both the patient's parents were poor responders, consistent with the patient being homozygous for the Arg506→Gln mutation.

Svensson and Dahlbäck (1994) studied 104 consecutive patients with venous thrombosis (45% had a family history of thrombotic disease); 33% of patients exhibited an APC ratio below the 5th percentile of the control values. In addition, some 211 members of 34 families of individuals with APC resistance were studied; 41% of APC resistant individuals had experienced a thrombotic episode by the age of 45 compared with only 3% for relatives without APC resistance.

Cadroy *et al.* (1994) noted a frequency of APC resistance of only 19% in their study of French patients with a previous history of venous thrombosis, a finding supported by similar data: 13% (Leroy-Matheron *et al.*, 1996) and 17% (Manten *et al.*, 1996). Voorberg *et al.* (1994) reported a frequency of 37% among Dutch patients with idiopathic recurrent thromboembolism.

Rosendaal *et al.* (1995) found that 19.5% of 471 consecutive patients aged less than 70 with a first objectively confirmed episode of venous thrombosis possessed the factor V Leiden mutation (85 heterozygotes and seven homozygotes).

The prevalence of APC resistance as reported in the studies of Koster *et al.* (1993), Cadroy *et al.* (1994) and Rosendaal *et al.* (1995) is clearly lower than that noted in the study of Griffin *et al.* (1993). In the case of the study of Koster *et al.* (1993), this could have been due to the exclusion of Coumarin-treated patients. Such patients are more likely to suffer from recurrent thrombotic disease and if APC cofactor deficiency were preferentially associated with severe disease, then the prevalence of the condition in the population could have been underestimated. Similarly, it is likely that patients with an age of onset less than 45 years, recurrence and a family history of thrombotic disease will exhibit a higher frequency of APC resistance than consecutive patients with a single episode of deep vein thrombosis (DVT). (The median age at first thrombotic event for APC-resistant members of thrombophilia families has been determined to be 29 years; Lensen et al., 1996.) Some clearly anomalous results have been reported (e.g. Desmarais et al., 1996) and these have probably resulted from differences in patient selection or the way that the APC resistance assay was performed. It may be, however, that there are population differences in the prevalence of both APC resistance and the factor V Leiden variant.

Perhaps the most accurate measure of the prevalence of the factor V Leiden variant among thrombotic patients has come from a *prospective* study of male physicians (Ridker et al., 1995a). The frequency of the factor V Leiden mutation in individuals with DVT or pulmonary embolism (PE) was 11.6% ($n = 121$). This translates into a relative risk of 2.7 (95% confidence interval, 1.3–5.6). When men over the age of 60 were considered, the prevalence increased to 26% ($n = 31$) representing a relative risk of 7.0 (95% confidence interval, 2.6–19.1). Results from a second prospective study (relative risk = 2.4, 95% confidence interval, 1.3–4.5) are in agreement with this report (Simioni et al., 1997).

Factor V Leiden is a very important cause of thrombosis in childhood. Nowak-Göttl et al. (1996) reported that 10/19 (52%) children with venous thrombosis and 7/18 (38%) children with arterial thrombosis possessed the factor V Leiden variant. Interestingly, in 3/17 factor V Leiden-bearing children, an additional inherited coagulation disorder was found which presumably increased the likelihood of coming to clinical attention.

In the Caucasian population, between, 1% and 7% of normal healthy controls exhibit APC resistance and/or possess the factor V Leiden lesion [Dutch population: Koster et al. (1993), Bertina et al. (1994), Rosendaal et al. (1995); British population: Beauchamp et al. (1994), Emmerich et al. (1995), Hallam et al. (1995a), Rees et al. (1995); French population: Emmerich et al. (1995), Gandrille et al. (1995b); German population: März et al. (1995), Hallam et al. (1995a), Rees et al. (1995), Braun et al. (1996), Schröder et al. (1996); Italian population: Mannucci et al. (1996); Austrian population: Halbmeyer et al. (1994b); American population: Ridker et al. (1995a); Brazilian population: Arruda et al, (1995); Greek population: Rees et al. (1995), Chaida et al. (1996)]. In areas of southern Sweden, however, the frequency of the factor V Leiden lesion may be as high as 15% (Svensson and Dahlbäck, 1994; Dahlbäck, 1996)

By contrast, the frequency of the factor V Leiden variant is rather lower in the American black population (3/214; 1.4%; Pottinger et al., 1995) and in the Chinese population (1/618; 0.12%; Ko et al., 1996). Similarly in the Japanese population: Takamiya et al. (1995) reported no cases in a sample of 192 healthy volunteers. Takamiya et al. (1995) failed to find any individual with an APC ratio significantly below the lower limit of normal, whereas Fujimura et al. (1995) reported a prevalence of APC resistance of 18% in the Japanese population. However, Fujimura et al. (1995) did not find an example of factor V Leiden in their patients, implying the existence of a second cause of APC resistance in the Japanese population. Negative results have also been reported for the African population (0/612 chromosomes screened; Rees et al., 1995), the Greenland Inuit (0/266 chromosomes screened; de Maat et al., 1996), Hong Kong Chinese (0/586 chromosomes screened; Chan et al., 1996) and the indigenous Australian Aboriginal and Papua New Guinea peoples (0/336 chromosomes screened; Rees et al., 1995) while the mutation occurred with a frequency of only 0.6% in people from the Indian sub-continent (Rees et al., 1995).

Given the prevalence of the factor V Leiden mutation in the Caucasian population, it is estimated that the homozygous defect will have a prevalence of between 0.09% and 0.5% (Greengard *et al.*, 1994b). Indeed, after many surveys, numerous examples of symptomatic individuals homozygous for the factor V Leiden variant have now been reported (e.g. Koster *et al.*, 1993; Svensson and Dahlbäck, 1994; Villa *et al.*, 1995).

Why is this factor V variant so common? Its high frequency in the general population suggests that it confers, or has conferred, some selective advantage on bearers of the factor V Leiden variant. Dahlbäck (1994) speculated that a slight hypercoagulable state associated with possession of the factor V Leiden variant might be advantageous in certain situations such as traumatic injury and pregnancy. Nichols *et al.* (1996) have demonstrated that carriership of the factor V Leiden lesion can serve to ameliorate the clinical (bleeding) phenotype exhibited by haemophilia A patients. This explains at least some of the phenotypic variability associated with specific *F8* gene mutations causing haemophilia A and is consistent with the possibility that the factor V Leiden lesion may confer a selective advantage to some individuals under some circumstances.

Bertina *et al.* (1994) have shown that there is substantial linkage disequilibrium between the factor V Leiden variant and a *Hin*fI polymorphism within the *F5* gene. At least for the Dutch population, this argues against a recurrent mutation model (CG is a hypermutable site in the human genome; see Chapter 2) and in favour of one, or at least very few, identical-by-descent mutations spreading in the population. The process of mutational spread itself implies the influence of selection but it is as yet unclear how selection might have acted. Zivelin *et al.* (1997) have presented evidence based upon haplotype analysis for a single origin for the Caucasian factor V Leiden variant some 21 000–34 000 years ago. The factor V Leiden variant has, however, been found in diverse populations including Ashkenazi Jews and an African American (Greengard *et al.*, 1994a) suggesting that recurrent mutation is also a cause of the extreme geographical dispersal of this lesion.

Kirschbaum and Foster (1995) have described a PCR-based method which employs sequence-specific primers for the rapid detection of the factor V Leiden mutation. Gandrille *et al.* (1995c) have also reported a rapid non-isotopic detection method for screening large groups of subjects for the factor V Leiden lesion; this employs a modified oligonucleotide which introduces a *Hin*dIII restriction site into a PCR product. Many other PCR-based methods for the detection of the lesion have been reported: Rabès *et al.* (1995); van de Locht *et al.* (1995); Greengard *et al.* (1995b); Guillerm *et al.* (1996); Bellissimo *et al.* (1996); Blasczyk *et al.* (1996); Zotz *et al.* (1996); Reitsma *et al.* (1996); Corral *et al.* (1996); Margaglione *et al.* (1996); and Rees *et al.* (1996).

7.5.4 Clinical aspects

Thrombotic risk associated with APC resistance. A considerable body of work has now built up on the importance of APC resistance in clinical practice (Hillarp *et al.*, 1995). In their study of 308 individuals from 50 Swedish families with inherited APC resistance, Zöller *et al.* (1994) found that 30% of heterozygotes, 44% of homozygotes and, 10% of wild-type individuals had experienced thromboembolic symptoms. The mean age at first thrombotic event was 36 for heterozygotes, 25 for homozygotes and 36 for normals (Zöller *et al.*, 1994; *Figure 7.4*). By 33 years of age, 20% of heterozygotes, 40% of homozygotes and 8% wild-type individuals had experienced thromboembolic symptoms. In a retrospective study, Rosendaal *et al.* (1995) estimated the relative risks of coming to clinical attention for heterozygotes and homozygotes to be seven-fold

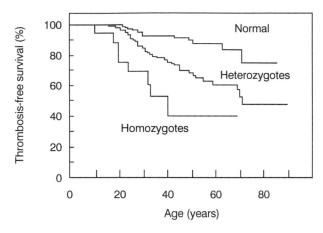

Figure 7.4. Thrombosis-free survival curves. The relationship between genotype and severity of thrombotic disease in 50 families with inherited APC resistance. The probability of being free of thrombotic events at a certain age is presented in the Kaplan–Meier analysis of 146 normal individuals, 144 heterozygotes and 18 homozygotes. Differences between normals and heterozygotes and between heterozygotes and homozygotes were highly significant. Reproduced from Dahlbäck (1995d) with kind permission from F.K. Schattauer Verlagsgesellschaft mbH.

and 80-fold respectively. In a prospective study of 77 (of the 22 071) participants in the Physician's Health Study who survived an episode of DVT, the factor V Leiden lesion was found to be associated with a four- to five-fold increased risk of recurrent venous thrombosis (Ridker *et al.*, 1995b).

Since APC resistance is fairly frequent in the general population, this phenomenon is unlikely to be the sole arbiter of whether or not a thrombotic tendency becomes clinically overt. Indeed, siblings homozygous for the factor V Leiden mutation have been found to vary quite markedly in their thrombotic manifestations (Greengard *et al.*, 1994b). Age, sex and other predisposing factors (e.g. pregnancy, surgery, trauma, contraceptive pill usage) appear likely to be important determinants of the APC sensitivity ratio in normal individuals (Zöller *et al.*, 1994b; Hellgren *et al.*, 1995; Bokarewa *et al.*, 1996; Rai *et al.*, 1996).

Vandenbroucke *et al.* (1994) reported that among 155 women who had developed DVT, 23% carried the factor V Leiden lesion (five homozygotes and 30 heterozygotes) compared with 3.6% of controls. This translates into a risk of venous thrombosis of 7.9 (95% confidence limits, 3.2–19.4). This value is rather higher than that reported for men (Ridker *et al.*, 1995a; see Section 7.5.3) When oral contraceptive use was considered, women who used these preparations and who possessed the factor V Leiden lesion were some 34-fold (confidence interval, 7.8–154) more likely to experience a thrombotic episode than women in whom both risk factors were absent. Thus the risk of venous thrombosis in women who used oral contraceptives is much higher when they bear the factor V Leiden mutation. Further, homozygosity for the factor V Leiden mutation has been shown to be a paricularly important risk factor for venous thrombosis in women using oral contraceptive preparations (Rintelen *et al.*, 1996b). However, the magnitude of this risk is in part dependent upon the type of oral contraceptive preparation used (Bloemenkamp *et al.*, 1995).

In reporting a case of portal vein thrombosis associated with the factor V Leiden mutation, Levoir *et al.* (1995) suggested that APC resistance should be sought in cases of thrombosis in unusual sites. This suggestion was echoed by Hillarp *et al.* (1995) and Dhote *et al.* (1995) who described cases of arterial and venous retinal thrombosis and Budd–Chiari syndrome in APC-resistant patients. Similarly, carriership for factor V Leiden occurs at a higher frequency in individuals who have experienced cerebral venous thrombosis than in controls (Brey and Coull, 1996; Deschiens *et al.*, 1996; Martinelli *et al.*, 1996a; Zuber *et al.*, 1996). By contrast, PE and superficial thrombophlebitis appear to occur less frequently among heterozygous carriers of the factor V Leiden mutation (Hillarp *et al.*, 1995). DVT occurs at comparable frequencies in the two groups (Hillarp *et al.*, 1995).

Co-inheritance of APC resistance and protein C deficiency. In principle, co-inheritance of the common factor V Leiden lesion with a protein C gene defect could explain the difference in thrombotic risk noted between families with clinically symptomatic (overt) protein C deficiency and those with the much more common clinically asymptomatic (covert) deficiency state (Milctich *et al.*, 1987; Tait *et al.*, 1995; see Section 4.2.6). In order to examine this postulate, the prevalence of the factor V Leiden variant among protein C-deficient patients has been examined.

Co-inheritance of the factor V Leiden mutation with protein C deficiency has been shown to increase the risk of coming to clinical attention, at least in the Dutch population. Koeleman *et al.* (1994) found that 19% of a sample of 48 Dutch symptomatic protein C-deficient probands (with known protein C gene lesions) possessed the factor V mutation as compared to 2% in the general population. When family studies were performed, a thrombotic episode had been experienced by 73% of family members with both a protein C (*PROC*) gene lesion and the factor V (*F5*) gene lesion. By contrast, 31% of family members bearing only the *PROC* gene lesion and 13% of family members bearing only the *F5* gene lesion had experienced a thrombotic episode. Finally, two-locus linkage analysis provided strong support (Lod = 19.6 at θ = 0.00) for the hypothesis that the *F5* and *PROC* gene loci were the two trait loci involved. Even so, at least in one family, not all cases of severe thrombotic disease could be accounted for by co-inheritance of the *PROC* and *F5* gene lesions, suggesting that still further factors may be involved.

The factor V Leiden variant has however been found to be less frequent among British (0.06) and Swedish/Danish (0.15) protein C deficiency patients (Hallam *et al.*, 1995a). Indeed, in the British population, the frequency of the factor V Leiden allele was not significantly elevated compared with healthy controls. Co-inheritance of the factor V Leiden variant is therefore unlikely to be the sole determinant of whether an individual with protein C deficiency will come to clinical attention.

Hallam *et al.* (1995a) also analysed patient data by type of protein C deficiency. It was noted that the frequency of the factor V Leiden variant was 2.8-fold higher in type II deficiency patients compared with type I patients. Hallam *et al.* (1995a) suggested the following possible explanation for this disparity: together with protein S and factor V, APC forms a 1:1:1 stoichiometric complex on the phospholipid surface that is capable of inactivating factor VIIIa (Shen and Dahlbäck, 1994). The factor V Leiden defect is likely to reduce the level of this inactivating complex. This is because thrombin generation is increased in APC-resistant plasma due to the excess of factor Va (Shen and Dahlbäck, 1994). Factor V activated by thrombin is, however, inefficient as a cofactor in the inactivation of factor VIIIa (Shen and Dahlbäck, 1994). Although the reduced synthesis of protein C, characteristic of a type I protein C deficiency state, would presumably reduce the level of factor VIIIa-inactivating complex still further, a normal level of factor V is still available to wild-type protein C molecules for complex formation. By contrast, a dysfunctional (type II) protein C molecule might still be able to interact with protein S and factor V to generate a non-functional complex. Once formed, this inactive complex could sequester factor V and protein S thereby reducing the amount of factor V accessible to wild-type protein C (it might also compete with its functional counterpart for access to factor VIIIa). This reduction might explain why carriership of the factor V Leiden variant increases the likelihood of clinical detection more dramatically for type II protein C deficiency than for type I deficiency patients. However, these authors did not detect all the *PROC* gene lesions responsible for the protein C deficiency and could conceivably have misclassified APC resistance as type II protein C deficiency (Ireland *et al.*, 1995; Faioni *et al.*, 1996).

Gandrille *et al.* (1995b) have also reported data on the frequency of association of protein C deficiency and the factor V Leiden variant: some 14% of the 113 protein C-deficient patients studied possessed the factor V variant compared with only 1% of controls.

Kalafatis *et al.* (1995b) re-examined their large Protein C Vermont pedigree (type II protein C deficiency caused by a Glu20→Ala substitution) and noted that 73% of family members who possessed both the protein C mutation and the factor V Leiden variant had experienced a thrombotic episode. However, whereas recombinant APC is able to cleave factor V R506Q and factor Va R506Q, albeit slowly, minimal cleavage occurs when recombinant APC E20A is employed (Kalafatis *et al.*, 1995b). In combination, therefore, these inherited variants serve to generate a stable procofactor (factor V) and a stable active cofactor (factor Va), thereby greatly increasing the probability of thrombosis.

A similar situation has been reported by Brenner *et al.* (1996): all 5/11 clinically symptomatic carriers of a Thr298→Met protein C mutation also carried the factor V Leiden variant whereas venous thromboembolism was not observed in those individuals who carried the *PROC* lesion but lacked the factor V variant. Co-inheritance of the two deficiency states appears therefore to increase the likelihood of an individual coming to clinical attention.

Co-inheritance of APC resistance and antithrombin III (ATIII) deficiency. The co-inheritance of factor V Leiden and inherited ATIII deficiency has been studied in a total of 128 families with ATIII deficiency (van Boven *et al.*, 1996). Altogether, 127 probands and 188 of their relatives were screened for the factor V Leiden variant: the variant was found in 18 families. Nine families were available for assessment of the mode of inheritance and clinical significance of the combined deficiency state. Co-segregation of both defects was observed in four families, a consequence of the presence of both the *AT3* (chromosome 1q23–q25.1) and the *F5* (1q21–q25) genes on the same region of the long arm of chromosome 1. This potentially provides an explanation for those cases of the inheritance of severe thrombosis through a family pedigree through several generations.

A total of 11/12 individuals possessing both a reactive site or pleiotropic effect ATIII mutation and the factor V Leiden variant had experienced thrombotic symptoms (van Boven *et al.*, 1996). By contrast, for individuals bearing both a heparin binding site defect of ATIII and the factor V Leiden variant, only 2/6 manifested thrombotic symptoms. The severity of the ATIII defect may therefore have a bearing on the likelihood of the combined deficiency state coming to clinical attention.

Martinelli *et al.* (1996b) reported finding the factor V Leiden variant in 2/16 (12.5%) symptomatic patients with ATIII deficiency compared with 0/18 asymptomatic deficient relatives. Lane *et al.* (1996b) described antenatal diagnosis in the case of a child at risk for homozygous ATIII deficiency and heterozygous factor V Leiden.

Co-inheritance of APC resistance and protein S deficiency. A number of cases of APC resistance co-occurring with either heterozygous or homozygous protein S deficiency were noted by Zöller *et al.* (1995b). Zöller *et al.* (1995a) found that the factor V Leiden lesion was present in 38% of protein S-deficient patients (*n*=16), a much higher frequency than that found in normal controls. Some 72% of family members of the protein S-deficient individuals who possessed both gene defects had experienced thrombotic symptoms compared with 19% of those with only protein S deficiency and 19% of those with only the factor V Leiden variant. Very similar data have been reported by Koeleman *et al.* (1995): 38% of protein S-deficient patients (*n*=16) possessed the factor V Leiden lesion. In those families in which both abnormalities were segregating, 80% of clinically symptomatic individuals possessed both types of defect. Martinelli *et al.* (1996a) reported finding the factor V Leiden variant in 2/13 (14%) symptomatic patients with protein S deficiency compared with 0/15 asymptomatic deficient relatives while Gurgey *et al.* (1996) reported combined factor V Leiden and protein S deficiency in children presenting with thrombosis in the central nervous system. Co-inheritance of the two deficiency states appears therefore to increase the likelihood of an individual coming to clinical attention.

Combined factor V Leiden and factor V deficiency. Three unrelated cases of combined factor V Leiden and factor V deficiency (pseudo-homozygous APC resistance) have been reported in individuals with thrombotic disease (Simioni *et al.*, 1996b; Zehnder and Jain, 1996). Such individuals appear to exhibit severe APC resistance compatible with homozygosity for the factor V Leiden lesion but are in fact only heterozygous for this lesion. The authors of these papers considered that the second factor V defect contributed toward a predisposition to thrombosis in these patients. Clearly in cases manifesting a heterozygous 'knockout' mutation of factor V (e.g. Zehnder and Jain, 1996), those individuals possessing a factor V Leiden mutation on the other allele are functionally hemizygous.

Combined factor V Leiden and factor XII deficiency? To investigate the possible co-occurrence of factor V Leiden and factor XII deficiency, Castaman *et al.* (1996a) reinvestigated 19 clinically symptomatic individuals from 13 families previously diagnosed as having factor XII deficiency. However, factor V Leiden was found in none of them. As yet, therefore, there is no evidence for a combination of factor V Leiden and factor XII deficiency being more likely to come to clinical attention.

Combined factor V Leiden and plasminogen deficiency. Züger *et al.* (1996) described a family with heterozygous plasminogen deficiency and the factor V Leiden variant. Six members of the family were deficient in plasminogen but only two of these individuals had suffered thromboembolic events. Both individuals in the family who were found to be heterozygous plasminogen-deficient and heterozygous for the factor V Leiden variant had experienced thromboembolic episodes whilst one of two individuals heterozygous for the factor V Leiden variant was affected.

Arterial thrombosis. Cases of arterial thrombosis have also been reported in individuals either heterozygous or homozygous for the factor V Leiden mutation (Holm *et al.*, 1994; Lindblad *et al.*, 1994; Zöller *et al.*, 1994b, 1995a). A possible association between APC resistance and arterial thrombosis (e.g. myocardial infarction, stroke) has proven to be rather contentious (Samani *et al.*, 1994; Biasiutti *et al.*, 1995; van Bockxmeer *et al.*, 1995; Catto *et al.*, 1995; Emmerich *et al.*, 1995; Forsyth and Dolan, 1995; Kontula *et al.*, 1995; Ma *et al.*, 1995; Simioni *et al.*, 1995; Ardissino *et al.*, 1996; Fisher *et al.*, 1996; Ganesan *et al.*, 1996; Montaruli *et al.*, 1996; Press *et al.*, 1996). Halbmeyer *et al.* (1994b) noted that APC resistance was significantly higher among stroke patients (20%) than among controls (2%) whilst März *et al.* (1995) noted a significantly higher frequency of the factor V Leiden mutation in patients with angiographically proven coronary artery disease (9%) than in healthy controls (4%). However, a prospective study of male physicians has unequivocally demonstrated that the factor V Leiden mutation is not associated with an increased risk of myocardial infarction ($n=374$) or stroke ($n=209$) (Ridker *et al.*, 1995a). This study is almost certainly the most reliable performed to date since it was prospective and therefore presumably free of the common types of epidemiological bias that can influence the results of retrospective studies.

Clinical management. In a retrospective study of 21 APC-resistant patients, the risk of recurrence of thrombosis after the first thrombotic event (4.8%) does not appear to be significantly different for factor V Leiden heterozygotes than for individuals with a history of venous thromboembolism but no APC resistance/factor V Leiden mutation (5%; Rintelen *et al.*, 1996a). Without properly controlled prospective studies, it is hard to make specific recommendations regarding clinical management of APC-resistant patients. Hillarp *et al.* (1995) recommend treating factor V Leiden homozygotes and symptomatic heterozygotes as if they were patients with a deficiency of protein C, protein S or ATIII. Prophylactic therapy for APC-resistant individuals should therefore be considered in risk situations (e.g. major surgery, trauma) and, in the absence of any further information, oral contraceptive use should be avoided. It is clear however that exposure of APC-resistant individuals to a risk situation does not always trigger a thrombotic event (Samama *et al.*, 1995; Girolami *et al.*, 1996).

Thrombomodulin and its deficiency state

8.1 Introduction

Esmon and Owen (1981) used a Langendorff heart preparation (perfusion through the coronary circulation and collection from the coronary sinus via a slit in the ventricle) to show that when protein C and thrombin were perfused together, protein C emerged fully activated in a 4 sec pass time. By contrast, no activation at all occurred during a 30 min incubation of the same concentration of thrombin and protein C in a test tube. The endothelial surface of the coronary microcirculation had increased the rate of activation by at least 20 000 fold.

The endothelial receptor, thrombomodulin, responsible for this dramatic effect has been purified and its sequence established from cDNA cloning. Intensive study of the structure and function of this protein has elucidated three distinct anticoagulant functions related to thrombin or its inhibition (reviewed by Esmon, 1989b, 1995; Dittman and Majerus, 1990). The physiology, biochemistry and genetics of thrombomodulin will be discussed and its clinical importance assessed.

8.2 Physiology, structure and function

8.2.1 Physiology, distribution and physical properties

Each endothelial cell possesses about 50 000–100 000 molecules of thrombomodulin on its surface. Immunohistochemistry has shown that most thrombomodulin is located in the microvasculature which accounts for 99% of the total endothelial surface presented to blood. The endothelial cell surface area : blood volume ratio is much higher in the capillaries than in the major vessels. As a result, the concentration of thrombomodulin is more than 1000-fold higher in the microvasculature. Thrombomodulin is also found in lymphatics, but hepatic sinusoids and post-capillary venules of lymph nodes lack thrombomodulin.

Plasma contains about 20 ng ml^{-1} of a soluble form of thrombomodulin (Ishii and Majerus, 1985) which may have physiological importance because, in blood depleted of thrombomodulin by specific antibodies, the inactivation of factor Va during clotting is delayed. Soluble thrombomodulin purified from human placenta has a molecular weight of 75 000 on unreduced sodium dodecyl sulphate polyacrylamide gel electrophoresis (SDS-PAGE). It is thought to be generated either by proteolytic cleavage of, or by the action of active oxygen species on, cellular thrombomodulin (Ishii *et al.*, 1991; Sawada *et al.*, 1992).

Thrombomodulin accelerates the activation of protein C by thrombin. Since this can be suppressed by active site-blocked thrombin, it may be concluded that thrombin interacts with thrombomodulin and acquires enhanced activity for its substrate protein C. The affinity of cultured endothelial cells for thrombin is high, with a K_d of 0.5 nM, far below the concentration at which effective clot formation can occur. However, the K_m of the complex (0.7 μM) is about 10 times higher than the concentrations of protein C in plasma (50–80 nM). Thus the physiologic rate of activation of protein C is proportional to its concentration in plasma.

If thrombin bound to thrombomodulin remained an active procoagulant, the effect of a high-affinity receptor in the microvasculature would be to concentrate thrombin therein and lead to rapid fibrin occlusion. This does not occur because all procoagulant

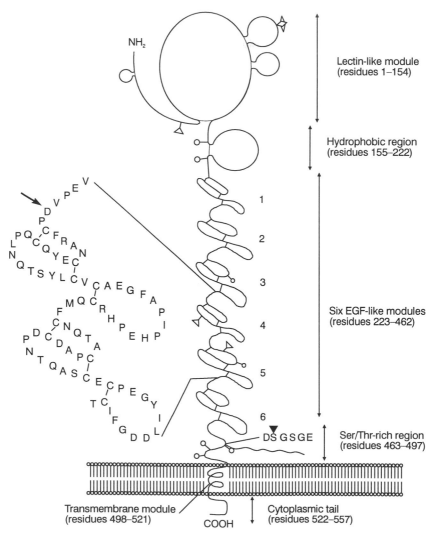

Figure 8.1. Secondary structure and membrane interactions of human thrombomodulin. The arrow denotes Asp349 which has been implicated in the interaction of thrombomodulin with the Gla domain of protein C. The arrowhead denotes the site of attachment of the glycosaminoglycan (wavy line). The O-linked and N-linked carbohydrate sidechains are marked by circles and triangles respectively. Reproduced from Dahlbäck and Stenflo (1994) by kind permission of Churchill Livingstone.

properties of thrombin including platelet activation, fibrinogen cleavage, the activation of factors V, VIII and XIII, the inactivation of protein S and the factor Xa-mediated activation of prothrombin (Thompson and Salem, 1986) are inhibited by thrombomodulin binding (Ohishi *et al.*, 1993).

Thrombomodulin also stimulates the rate of inactivation of thrombin by antithrombin III (ATIII) (Hofsteenge *et al.*, 1986; Preissner *et al.*, 1987; Aritomi *et al.*, 1993). Human thrombomodulin may lack this property because the relevant binding site is masked by vitronectin (Preissner *et al.*, 1990b).

8.2.2 Primary sequence, secondary structure and domains

The complete sequence of human thrombomodulin was established by cDNA sequencing. The structure of the 575-amino-acid 60.3 kDa unmodified protein is shown in *Figure 8.1*. The first 18 or 22 residues constitute a hydrophobic leader sequence that is probably cleaved by a signal peptidase. An N-terminal lectin-like domain (residues 1–154) and a hydrophobic region (residues 155–222) may interact with lipid membrane (*Figure 8.1*). This sequence is distantly related to the lectin-like region of asialoglycoprotein receptor (Patthy, 1988). By analogy, disulphide bridges may occur between Cys residues 12–17, 34–149, 78–115 and 119–140. Following this region are six repeats of an epidermal growth factor (EGF) domain (residues 223–462) and a short serine/threonine-rich O-linked glycosylation domain (residues 463–497). The next domain (residues 498–521) is highly hydrophobic and probably serves as a transmembrane linker to a C-terminal intra-cytoplasmic segment (residues 522–557) which contains a single cysteine. The EGF domains 4–6 and the Ser/Thr-rich domain region are required for cofactor activity (Tsiang *et al.*, 1992).

8.2.3 Post-translational modification

The predicted M_r from sequence is considerably less than that observed for purified protein; the difference is due to glycosylation which introduces a chondroitin sulphate group into at least a proportion of human thrombomodulin. The serine/threonine-rich region (*Figure 8.1*) is the site of O-linked glycosylation with attachment of the glycosaminoglycans to a primary site at Ser474 and a secondary site at Ser472 (Gerlitz *et al.*, 1993). Sites for N-linked glycosylation are present in the fifth EGF domain and in the lectin-like region (*Figure 8.1*). In recombinant human soluble thrombomodulin, the carbohydrate side chains and the chondroitin sulphate moiety contribute both to the direct anticoagulant activity and the ATIII-dependent anticoagulant activity of thrombomodulin (Nawa *et al.*, 1990; Parkinson *et al.*, 1990, 1992a; Koyama *et al.*, 1991). The glycosaminoglycan interacts directly with thrombin via the anion-binding exosite 2, an interaction which is distinct from that of thrombomodulin which interacts with thrombin via anion-binding exosite 1 (see Section 8.2.4). Thus two thrombin molecules may bind simultaneously to thrombomodulin (Ye *et al.*, 1993).

The sequence that acts as a signal for β-hydroxylation of aspartic acid or asparagine residues is present in EGF domains 3 and 6 in thrombomodulin, and a β-hydroxy-aspartate residue has been found in the bovine protein (Stenflo *et al.*, 1988). Such modified residues are involved in calcium binding, which is known to be important for some aspects of thrombomodulin function.

The cytoplasmic tail contains several residues that are potential sites for phosphorylation, including a tyrosine and a serine.

8.2.4 Anticoagulant properties and structure/function correlation

Interaction with thrombin. Thrombomodulin binds thrombin with high affinity and enhances the rate of activation of protein C by thrombin 20 000-fold (Esmon and Owen, 1980, 1981; reviewed by Esmon, 1989a, b; Dittman and Majerus, 1990; Sadler *et al.*, 1993). Thrombomodulin provides about 50–60% of thrombin-binding sites on the endothelial cell surface, approximately 10^5 thrombomodulin molecules per endothelial cell (Maruyama and Majerus, 1985). Thrombomodulin is located predominantly in the microvasculature, which accounts for 99% of the total endothelial surface presented to blood. The concentration of membrane-bound thrombomodulin is more than 1000-fold higher in the microvasculature than in the major vessels. Thus, whilst thrombin is free in the major vessels, it is bound by thrombomodulin as soon as it enters the microvasculature.

Upon binding thrombomodulin, the active site of thrombin is structurally altered (Ye *et al.*, 1991) and it loses its ability to activate factors V and VIII and to clot fibrinogen whilst its activity toward protein C is enhanced (reviewed by Walker and Fay, 1992). The half-life of the thrombin–thrombomodulin complex on the cell membrane is less than 15 sec (Esmon, 1987), the complex being internalized by endocytosis.

Thrombomodulin binds to thrombin near the active centre of the proteinase at the anion-binding exosite 1 (Sadler *et al.*, 1993; Liu *et al.*, 1994). Specifically, synthetic peptides corresponding to residues Arg89–Asn95 and Thr147–Ser158 inhibit the thrombin–thrombomodulin interaction (Nishioka *et al.*, 1993). The active site is not involved since diisopropylfluorophosphate-inhibited thrombin binds with the same affinity as native thrombin (Bar-Shavit *et al.*, 1989). Hirudin and fibrinogen compete with thrombomodulin for this anion-binding exosite, but this could be indicative of a common site or steric hindrance by large molecular ligands. Antibodies to a peptide corresponding to residues 62–73 of thrombin B chain inhibit interaction of thrombin with fibrinogen, hirudin and thrombomodulin.

A primary binding site for thrombin on thrombomodulin has been localized to EGF domains 5 and 6 (Kurosawa *et al.*, 1988). The region did not support activation of protein C, however, for which EGF domain 4 was also required (Kurosawa *et al.*, 1988; Stearns *et al.*, 1989; Zushi *et al.*, 1989; Hayashi *et al.*, 1990). Further localization of the binding site was obtained by Tsiang *et al.* (1990) who showed that a peptide corresponding to residues 426–444 in the EGF-5 domain inhibits binding of thrombin to thrombomodulin.

Multiple changes occur in the structure of the active site of thrombin upon binding to thrombomodulin (Musci *et al.*, 1988; Ye *et al.*, 1991). Presumably these changes are related to the marked alteration of substrate specificity that occurs after thrombin binds to thrombomodulin.

The thrombin–thrombomodulin complex is internalized (Horvat and Palade, 1993; Conway *et al.*, 1994). This process may represent another means to remove thrombin from the ciculation. After transport to the lysosomes and release of thrombin, thrombomodulin is returned to the cell surface. Protein C inhibits this endocytosis cycle.

Interactions with protein C and factor V. The thrombin–thrombomodulin complex binds protein C in a Ca^{2+}-dependent fashion, unlike the binding of thrombin which is Ca^{2+}-independent. The calcium dependence of protein C activation results from a conformational change in thrombin brought about by the occupancy of anion binding exosite 2 by chondroitin sulphate (Liu *et al.*, 1994). EGF4, and more specifically Asp349, is required for protein C activation (Zushi *et al.*, 1991). The thrombomodulin–thrombin complex appears to interact with both the Gla and EGF domains of protein C (Hogg *et al.*, 1992; Olsen *et al.*, 1992). Specific residues within EGF domains 4, 5 and 6 that are involved in the cofactor activity of thrombomodulin have been identified by site-directed mutagenesis (Zushi *et al.*, 1991; Parkinson *et al.*, 1992b; Clarke *et al.*, 1993; Lentz *et al.*, 1993; Nagashima *et al.*, 1993). Protein C binds calcium via its Gla domain and the β-hydroxylated aspartate residue 71 (Öhlin *et al.*, 1988b; see Chapter 5). Once activated, protein C is released from the complex (cf. other cofactor complexes; Mann *et al.*, 1990).

Factor V has been implicated in the modulation of the thrombomodulin–protein C interaction. At low concentrations (less than 50 nmol l⁻¹), factor Va or its isolated light chain stimulate activation of protein C on cell surfaces by about 50-fold (Salem *et al.*, 1993a, b; Maruyama *et al.*, 1984), but above this concentration of factor Va (or of its isolated light chain), inhibition of the reaction is observed. This may be a physiological negative feedback loop to autoregulate activated protein C (APC) production by substrate product inhibition. Thus APC cleaves factor Va releasing the light chain, which could control further APC production.

8.2.5 Regulation of thrombomodulin activity and expression

A variety of mechanisms have been implicated in the regulation of thrombomodulin expression by endothelial cells *in vitro* and *in vivo* (reviewed in Dittman and Majerus, 1990). Phorbol myristate acetate induces endocytosis of thrombomodulin with subsequent degradation of the receptor. Tumour necrosis factor (TNF) also promotes endocytosis and degradation of thrombomodulin whilst inducing tissue factor expression, a complete reversal of the endothelial cell properties towards a procoagulant state. Conversely, retinoic acid up-regulates thrombomodulin expression and down-regulates tissue factor expression (Koyama *et al.*, 1994). Oral contraceptive use appears to decrease circulating thrombomodulin (Quehenberger *et al.*, 1996).

Thrombomodulin expression in the endothelium is also influenced by shear stress (Malek *et al.*, 1994), suggesting that the receptor may play a protective role against thrombosis in regions of stasis or low flow rate. Thrombomodulin expression is also up-regulated by heat shock (Conway *et al.*, 1994).

Thrombomodulin functions as an antagonist of thrombin receptor-mediated endothelial cell activation (Parkinson *et al.*, 1993). Whilst it is not normally expressed in vessel wall smooth muscle cells *in vivo*, its expression is initiated by culture of smooth muscle cells *in vitro* (Soff *et al.*, 1991). This may be a reflection of an *in vivo* ability of luminal smooth muscle cells to express thrombomodulin after endothelial cell injury, thereby providing protection from thrombosis (Fink *et al.*, 1993).

8.3 The thrombomodulin (*THBD*) gene

8.3.1 Structure of the intronless gene

A number of cDNA clones encoding human thrombomodulin have been isolated from endothelial cell or lung cDNA libraries (Jackman *et al.*, 1987; Suzuki *et al.*, 1987b; Wen *et al.*, 1987; Shirai *et al.*, 1988). The full-length cDNA includes 1725 bp encoding an 18-amino-acid signal peptide and 557-amino-acid mature protein. The transcriptional initiation sites of the human *THBD* gene have been mapped to two positions 158 and 163 bp upstream of the ATG initiation codon (Fritze *et al.*, 1988). A TATAAA box occurs 32 bp and 27 bp upstream of the two transcriptional initiation sites whilst a possible CCAAT box (GCAAT) lies 83 bp further upstream. In addition, four GC boxes (GGCGGG or CCGCCC; possible Sp1-binding sites) are present in the 500 bp upstream of the initiation codon. Further upstream, retinoic acid- and *TNF*-responsive elements have also been characterized (Yu *et al.*, 1992; von der Ahe *et al.*, 1993; Dittman *et al.*, 1994) as well as several potential heat shock responsive elements (Conway *et al.*, 1994).

At the 3′ end of the gene, four potential polyadenylation sites (AATAAA) occur within the non-coding region. However, in three independent cDNAs (Jackman *et al.*, 1987; Wen *et al.*, 1987; Shirai *et al.*, 1988), it is the fourth AATAAA motif that is used. This gives rise to a 1779 bp 3′ non-coding region and thus a full-length cDNA of approximately 3660 bp consistent with the size of thrombomodulin mRNA as measured by Northern blotting studies. Various other motifs are apparent in the 3′ non-coding region: a total of six CACTG pentanucleotide sequences are present which may fulfil some role in 3′ end formation. In addition, the sequence TTATTTAT is present in two overlapping copies 55 bp upstream of the fourth AATAAA motif. This sequence has been previously observed to occur in the 3′ non-coding regions of mRNAs encoding proteins that are involved in the inflammatory response (Caput *et al.*, 1986). The 3′ untranslated region may also harbour a cAMP-responsive element (Tazawa *et al.*, 1994).

Van der Velden *et al.* (1991) have reported a GCC/GTC (Ala/Val) sequence polymorphism at residue 455 in the sixth EGF-like domain. This polymorphism occurs with frequencies of 0.82/0.18 in the Dutch population.

The human *THBD* gene is unusual in that it does not contain introns. It has been allocated to chromosome 20 by hybridization of a cDNA probe to flow-sorted human chromosomes (Wen *et al.*, 1987). *In situ* hybridization with a cDNA probe regionally localized the *THBD* gene to 20p12–cen (Espinosa *et al.*, 1989).

Murine and bovine thrombomodulin cDNAs have been isolated (Jackman *et al.*, 1986; Dittman and Majerus, 1989; Imada *et al.*, 1990). The mouse and bovine thrombomodulin sequences are 68% and 64% homologous to the human protein at the amino acid sequence level.

8.3.2 Gene expression

The expression of the *THBD* gene has been well reviewed (e.g. Dittman and Majerus, 1990; Tuddenham and Cooper, 1994). Northern blotting studies reveal the presence of a 3.7–3.8 kb *THBD* mRNA in umbilical cord epithelial cells, peripheral blood monocytes, lung and placenta but not in RNA derived from HepG2 cells, the monocyte cell line U937 or brain. Transcription of the *THBD* gene is increased in mouse haemangioma cells by treatment with cycloheximide and thrombin (Dittman *et al.*, 1989) and in human endothelial cells treated with phorbol esters or dibutyryl cyclic AMP (cAMP; Hirokawa and Aoki, 1991). cAMP and retinoic acid have a synergistic effect on *THBD* gene expression in F9 embryonal carcinoma cells (Weiler-Guettler *et al.*, 1992). Suppression of synthesis of *THBD* mRNA has been observed following hypoxia, and treatment with interleukin-1β, lipopolysaccharide and *TNF*.

Human *THBD* cDNAs have been successfully expressed *in vitro* (Suzuki *et al.*, 1987b; Lin *et al.*, 1994). These constructs directed the synthesis of immunoreactive and functionally active human thrombomodulin.

8.4 Clinical aspects

8.4.1 Thrombomodulin-deficient transgenic mice

Transgenic mice have been made deficient for thrombomodulin by targeted disruption of the *THBD* gene (Healy *et al.*, 1995; Rosenberg *et al.*, 1995). The heterozygous deficient (TM$^{+/-}$) mice exhibited 50% reductions in the levels of thrombomodulin mRNA and protein but developed normally, were fertile and did not manifest any thrombotic symptoms. By contrast, the homozygous deficient mice (TM$^{-/-}$) died before embryonic day 9.5 and exhibited marked retardation in growth and development. Thrombomodulin must therefore have a developmental function unrelated to its role in the regulation of coagulation.

8.4.2 Levels of plasma thrombomodulin in disease

Plasma levels of soluble thrombomodulin have been shown to be elevated in patients with venous and arterial thrombosis, cerebral infarction and disseminated intravascular coagulation (Takano *et al.*, 1990; Takahashi *et al.*, 1992; Seigneur *et al.*, 1993). However, recombinant human thrombomodulin exhibits an antithrombotic effect on thrombin-induced thromboembolism in rats and mice (Kumada *et al.*, 1988; Gomi *et al.*, 1990; Solis *et al.*, 1991, 1994; Gonda *et al.*, 1993; Mohri *et al.*, 1994; Ono *et al.*, 1994).

8.4.3 Lesions in the thrombomodulin gene

Only one lesion in the thrombomodulin gene has been reported to date. It was detected in a 45-year-old man who had experienced a pulmonary embolism but who exhibited normal plasma levels of ATIII, protein C, protein S, APC resistance, plasminogen, C4b binding protein, tissue plasminogen activator, plasminogen activator inhibitor and HCF2 (Öhlin and Marlar, 1995). The lesion, detected by a combination of single strand conformation polymorphism analysis and DNA sequencing, was a GAC→TAC transversion which is predicted to convert Asp468 to Tyr. This residue is located between the transmembrane domain and the sixth EGF domain, close to one of the putative O-linked glycosylation sites. Since there is currently no three-dimensional structure of thrombomodulin available, it is unclear how this lesion perturbs the structure/function of the thrombomodulin molecule. However, the patient exhibited a highly variable but sometimes lower than normal level of soluble thrombomodulin in his plasma (1.9–13.6 ng l^{-1}, cf. NR 2.2–4.8 ng l^{-1}). It is possible either that the patient exhibited reduced expression of the variant thrombomodulin on the endothelial cell membrane or that the mutation led to a conformational alteration in the molecule which makes it less susceptible to proteolytic degradation.

It is unclear how frequent *THBD* gene defects are likely to be. The above lesion was detected in a study of 28 patients with a family history of thrombosis but with normal levels of ATIII, protein C, protein S and APC resistance.

Plasminogen and its deficiency

9.1 Introduction

Plasminogen is the plasma zymogen precursor of plasmin, the serine proteinase which effects the proteolytic degradation of fibrin (fibrinolysis). Fibrinolysis is highly important in clinical therapeutics since plasminogen activators can be used to treat thrombosis. Clinical aspects of fibrinolysis other than the hereditary disorders of plasminogen are beyond the scope of this chapter but have been extensively reviewed elsewhere (de Bono, 1987; Sobel *et al.*, 1987). The structure and function of plasminogen have been reviewed by Ponting *et al.* (1992) whilst several reviews of the fibrinolytic pathway and its modulation may be consulted (Bachmann, 1987; Collen and Lijnen, 1987; Robbins, 1987; Vassalli *et al.*, 1991).

9.2 Physiology, structure and function

9.2.1 Physiology

Plasminogen is present in plasma at a concentration of 200 mg l^{-1} (2 μM). This represents the highest concentration of any component of the fibrinolytic system as would be expected for a terminal effector molecule. It is synthesised by hepatocytes and has a half-life in the circulation of 2.2 h.

Plasminogen is activated to plasmin by two major pathways, that dependent on tissue plasminogen activator (tPA; the 'extrinsic' pathway of plasminogen activation) and that involving contact factors of coagulation and urokinase (the 'intrinsic' pathway of plasminogen activation; *Figure 9.1*). Although plasmin is capable of digesting fibrinogen as well as fibrin, the activity of fibrinolysis is normally localized to the fibrin clot by the fact that plasminogen is activated efficiently by tPA only in the presence of fibrin. Moreover, the plasmin that diffuses away from the fibrin clot is rapidly neutralized by α2-antiplasmin. Fibrinolysis is further regulated by specific inhibitors of the activator molecules: plasminogen activator inhibitor-1 (PAI-1) which inhibits tPA and urokinase and C1 inhibitor which inhibits kallikrein. Other inhibitors that have been implicated in the modulation of fibrinolytic activity are histidine-rich glycoprotein (Chapter 13), the kringle 4-binding protein and thrombospondin.

9.2.2 Primary structure

Plasminogen was the first protein of fibrinolysis to be purified and its primary sequence was established by Edman degradation (Wiman, 1977; Sottrup-Jensen *et al.*, 1978). The mature plasma protein is a monomer of 791 residues with a computed molecular weight

Venous Thrombosis: from Genes to Clinical Medicine, D.N. Cooper and M. Krawczak.
© 1997 BIOS Scientific Publishers, Oxford.

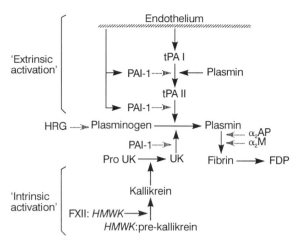

Figure 9.1. Activation of plasminogen via the 'extrinsic' and 'intrinsic' pathways.
tPA, tissue plasminogen activator; PAI-1, plasminogen activator inhibitor-1; HRG, histidine-rich glycoprotein; UK, urokinase; HMWK, high molecular weight kininogen.

(M_r) of 88 kDa. The actual M_r according to sodium dodecyl sulphate polyacrylamide gel electrophoresis (SDS-PAGE) is 92 kDa due to glycosylation (see Section 9.2.4). The N-terminal residue is glutamic acid. Internal autocatalytic cleavage occurs during activation yielding a species des-AA1–76 (Lys77–plasminogen). The latter is therefore often a component of plasma-derived plasminogen preparations.

The amino acid sequences of human, bovine, porcine, murine and rhesus macaque plasminogen have all been established (for alignment, see Ponting *et al.*, 1992).

9.2.3 Secondary structure and domains

There are 48 cysteine residues forming 24 disulphide bonds with no free sulphydryl groups (*Figure 9.2*). The exact disulphide bonds have not all been established, but are believed to be the following: 30–54; 34–42; 84–162; 105–145; 133–157; 168–296; 166–243; 187–226; 215–238; 256–333; 277–316; 305–326; 356–433; 377–418; 405–428; 460–539; 481–522; 510–534; 556–564; 546–664; 586–602; 678–745; 710–726; 735–765. Residue numbering here is that employed by Petersen *et al.* (1990). The N-terminal 77 residues that are readily released by plasmin, form a domain that contains two disulphide bonds and is strongly charged (E+D=15, K+R=9).

The next 463 residues (Lys77–Cys541) comprise a region containing five kringle domains. Each kringle has an approximate M_r of 10 kDa and three disulphide bonds.

Figure 9.2. Primary structure and disulphide linkage of human plasminogen. The positions of 18 introns are indicated by solid arrows. The signal peptide (–19 to –1) is cleaved at the Gly–1/Glu+1 bond. Conversion of plasminogen to plasmin occurs by cleavage of the Arg561–Val562 bond by plasminogen activators. PAP, preactivation peptide generated by the action of plasmin upon Glu–plasminogen; primary (but not sole) cleavage site is Lys77–Lys78 bond. K1–K5, kringles, located in the A chain. After cleavage of the Arg561–Val562 bond, the B chain remains attached to the A chain via the disulphide bonds Cys548–Cys666 and Cys558–Cys566. Carbohydrate attachment sites, Asn289 and Thr346, are depicted as diamonds. Reproduced from Petersen *et al.* (1990) with permission from the American Society for Biological Chemistry.

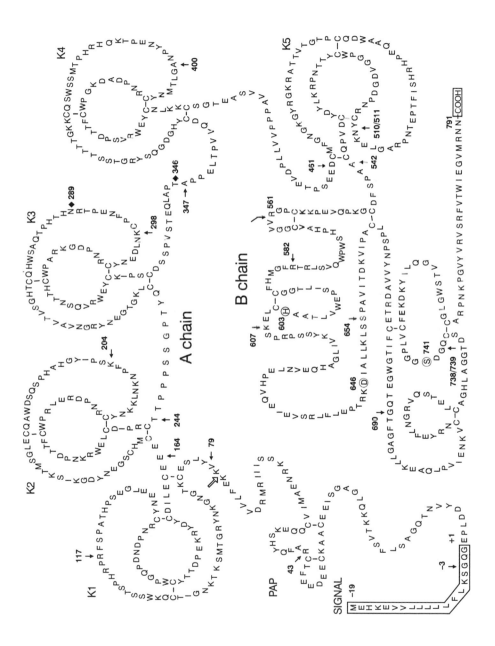

There is evidence that each kringle exists as an independent structure (Castellino *et al.*, 1981). Kringles are thought to mediate binding either to fibrin (Wu *et al.*, 1990), α2-antiplasmin (Wiman *et al.*, 1979), histidine-rich glycoprotein (Lijnen *et al.*, 1980) or to cellular receptors (Miles *et al.*, 1991). The remainder of the molecule, consisting of residues Ala542–Asn791, is the serine proteinase domain. The catalytic triad residues are located at His603, Asp646 and Ser741.

The structure of kringle 1 of human plasminogen has been determined by nuclear magnetic resonance (NMR; Rejante and Llinas, 1994). The crystal structure of kringle 4 has been determined to 1.9 Å resolution (Mulichak *et al.*, 1991).

9.2.4 Post-translational modifications

Carbohydrate accounts for 2% of the weight of plasminogen. There are two major types of plasminogen in plasma, differing in carbohydrate content, which can be separated by lysine affinity chromatography. The first type has only a trisaccharide sequence linked to Thr346 (approximately 67% of the total) but microheterogeneity is present in this fraction due to variable sialic acid attachment to *N*-acetyl galactosamine. The second major form of plasminogen has an additional biantennary oligosaccharide linked to Asn289 and comprises approximately 33% of the total. This modification influences the affinity of plasminogen for fibrin, α2-antiplasmin and cellular receptors as well as its rate of activation. The structure of this carbohydrate has also been established (see review by Castellino, 1984).

9.2.5 Binding sites

Plasminogen interacts with certain ω-aminocarboxylic acids which inhibit both its activation and the functional activity of plasmin. These compounds include lysine, 6-aminohexanoic acid (EACA) and trans-4-amino-methylcyclohexane-1-carboxylic acid (tranexamic acid). It is thought that these molecules mimic the binding characteristics of other molecules (e.g. fibrin and annexin II; see below).

The high-affinity lysine-binding site is located in kringle 1 ($K_d = 9$ μM). Both Lys and Glu plasminogen bind strongly and specifically to fibrin through this lysine-binding site, as is demonstrated by the fact that EACA abolishes such binding. The high-affinity binding site also mediates the interactions of plasmin with α2-antiplasmin and histidine-rich glycoprotein. Weaker binding sites are present in kringles 2 and 3. Kringle 4 has a weak affinity for fibrin whilst kringle 5 binds fibrin with moderate to high affinity (Wu *et al.*, 1990). Critical Arg, Asp, Tyr and Trp residues are present in kringles 1, 4 and 5 for the binding of ligand (Leech and Rickeli, 1980; Hochschwender and Laursen, 1981; Trexler *et al.*, 1982; McCance *et al.*, 1994; Menhart *et al.*, 1995).

The endothelial cell receptor for plasminogen is annexin II, a 40 kDa protein which also binds tPA (Hajjar *et al.*, 1994; reviewed by Redlitz and Plow, 1995). The kringle-associated lysine binding sites of plasminogen interact with the carboxyl terminal lysine residues of the receptor. Actin also serves as an endothelial cell-surface receptor for plasminogen (Dudani *et al.*, 1996).

9.2.6 Plasminogen activation

All mechanisms of plasminogen activation involve hydrolysis of the Arg561–Val562 peptide bond yielding two-chain plasmin. Thus the N-terminal A (heavy) chain is linked by two disulphide bonds to the smaller C-terminal B (light) chain which contains the catalytic site residues. 'Extrinsic' activation is effected by tPA. This activator is effective only in the presence of fibrin, due to the greatly increased affinity of fibrin-bound tPA for plasminogen. Kinetic data (Hoylaerts *et al.*, 1982) support a model in which tPA and plasminogen adsorb to fibrin, forming an ordered ternary complex. This allows efficient activation within a fibrin clot. 'Intrinsic' activation of plasminogen depends on the presence of contact factors, factor XII, high-molecular-weight kininogen and prekallikrein. Contact with negatively charged surfaces *in vitro* leads to conversion of prekallikrein to kallikrein which converts prourokinase to urokinase. Urokinase activates plasminogen but with much less fibrin specificity than tPA. The 'intrinsic' activation of plasminogen may not be physiologically important since most individuals with a contact factor deficiency have no problems with thrombosis.

9.2.7 Substrates of plasmin

Plasmin is a serine proteinase comprising two chains linked by Cys548–Cys666 and Cys558–Cys465. As with other members of the trypsin family of proteins, plasmin cleaves after arginine or lysine residues, with a preference for lysyl peptide bonds and basic amino acid esters and amides. However, unlike other proteinases of coagulation (e.g. factor Xa, thrombin, etc.), plasmin will attack a wide range of protein and peptide substrates. Its specificity for fibrin *in vivo* is determined purely by the fact that it binds to fibrin and that its primary activator, tPA, is fibrin-specific. Furthermore, free plasmin is rapidly neutralized by inhibitors. When plasmin is produced at a rate exceeding the inhibitory potential of blood (e.g. by infusion of exogenous non-fibrin-specific activators such as streptokinase), then fibrinogen and other proteins in blood are readily attacked by plasmin.

9.2.8 Inhibitors of plasmin

The most active inhibitor of plasmin is α2-antiplasmin, a single-chain protein of M_r 70 kDa, which very rapidly forms complexes with its target enzyme, giving rise to stable inactive plasmin–antiplasmin complexes. These complexes can be partially dissociated by SDS, yielding a non-disulphide-bonded peptide of M_r 8 kDa. Further dissociation with NH_4OH yields plasmin and modified inhibitor ($M_r = 60$ kDa) which is inferred to have been cleaved during the process of complex formation or subsequent dissociation (see Carrell *et al.*, 1987, for a more detailed discussion of the mode of action of serpins).

α2-Macroglobulin is a slow-binding plasmin inhibitor which forms active complexes in the circulation only after α2-antiplasmin has been depleted. α2-Macroglobulin–plasmin complexes are rapidly removed from the circulation and it is thought that α2-macroglobulin acts as a rather non-specific sink for free end peptidases. Other serpins [antithrombin III (ATIII), α1-proteinase inhibitor, C1-inhibitor] only neutralize plasmin in purified systems.

High-molecular-weight kininogen inhibits the binding of plasmin to platelets thereby inhibiting platelet activation by plasmin; this it does by forming a complex with plasmin in solution (Humphries *et al.*, 1994).

9.2.9 Protein polymorphisms

Isoelectric focusing has distinguished two alleles of plasminogen (plasminogen no. 1 = A and plasminogen no. 2 = B). In normal populations of Caucasians, the more common allele (1, A) had a frequency of 70% whereas in Orientals, the frequency was 96% (Hobart, 1979; Raum *et al.*, 1980b). The molecular basis of this polymorphism has not been established.

In the Japanese population, a non-functional variant (601Ala→Thr) reaches polymorphic frequency at 3.6% (Aoki *et al.*, 1984).

9.3 The plasminogen (*PLG*) gene

9.3.1 cDNA cloning

Plasminogen cDNA clones have been isolated from liver cDNA libraries (Malinowski *et al.*, 1984; Forsgren *et al.*, 1987). The full-length human plasminogen cDNA encodes all 790 amino acids of the mature protein plus the 20-amino-acid signal peptide sequence, and contains 64 bp and 252 bp of 5′ and 3′ non-coding sequence respectively. Inspection of the cDNA clones isolated reveals some heterogeneity in the site of poly(A) addition, either 17 bp or 46 bp downstream of the polyadenylation motif.

9.3.2 Structure and evolution

The exon/intron distribution of the human plasminogen (*PLG*) gene has been established with the sequencing of some 7853 bp of coding sequence and flanking regions (Petersen *et al.*, 1990). A total of 19 exons are present in the gene, with an average size of 146 bp. The gene comprises 19 exons, spanning 52.5 kb genomic DNA (*Figure 9.3; Table 9.1*).

Some 1 kb DNA sequence upstream of the Met translational initiation codon has been established (Petersen *et al.*, 1990). Malgaretti *et al.* (1990) determined the location of the transcriptional initiation site in human liver cells by primer extension: 161 bp upstream of Met. Two CTGGGA motifs, commonly found in acute phase response genes, are apparent in the putative promoter region. Potential binding sites for liver transcription factors HNF-1 and HNF-4 have also been noted (Malgaretti *et al.*, 1990).

The organization of the light chain-encoding region of the *PLG* gene is similar to that in the genes encoding tPA, urokinase and factor XII although the intron positions relative to the amino acid sequence differ slightly. Each of the five plasminogen kringles are encoded by two separate exons. The splice junctions of the introns within the kringles are type II whereas the introns between the kringles are type I as found in the other genes which encode kringle structures (i.e. those encoding tPA, urokinase, factor XII and the first kringle in prothrombin). This is consistent

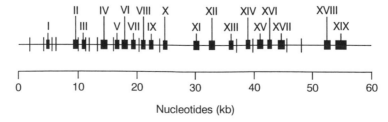

Figure 9.3. Exon/intron distribution in the human *PLG* gene (redrawn from data provided by Petersen *et al.*, 1990). The 19 exons are denoted by solid bars and are numbered I–XIX. Short vertical lines denote *Eco*RI restriction sites.

Table 9.1. Exons of the human *PLG* gene

Exon	Exon length (bp)	Protein domain
1	169	5′ non-coding
2	136	PAP1
3	107	PAP2
4	115	K1a
5	140	K1b
6	121	K2a
7	119	K2b
8	163	K3a
9	146	K3b
10	160	K4a
11	182	K4b
12	149	K5a
13	94	K5b
14	121	Act
15	75	His
16	141	Asp
17	107	Cnt
18	146	Loop
19	387	Ser + 3′non-coding region

Data from Petersen *et al.* (1990). Protein domains are PAP1 (N-terminal half of preactivation peptide), PAP2 (C-terminal half of preactivation peptide), K1a–K5a (N-terminal half of kringles), K1b–K5b (C-terminal half of kringles), Act (activation cleavage site), His, Asp and Ser (active site residues), Cnt (connecting region to the A chain) and Loop (disulphide loop).

with the hypothesis that these proteins have evolved at least in part by exon shuffling. The exception to this rule is the second kringle of prothrombin which is encoded by a single exon.

The *PLG* gene is also closely related to the lipoprotein (a) gene (*LPA*) and two apolipoprotein (a)-related genes or pseudogenes (Ichinose, 1992; Wade *et al.*, 1993; Byrne *et al.*, 1994). Apolipoprotein (a) is a plasma glycoprotein that is linked to apolipoprotein B100; it is highly homologous to plasminogen in both the kringle and proteinase domains (Eaton *et al.*, 1987; McLean *et al.*, 1987). Apolipoprotein (a) may have evolved from plasminogen by either of two mechanisms (Petersen *et al.*, 1990). Either exons II–IX were deleted followed by a recombination event that linked the signal peptide at intron A to kringle 4 at intron I or these missing exons are still present in an *LPA* gene intron and are removed during mRNA processing. Further evolution of the *LPA* genes would have involved multiple duplications of exons X and XI encoding plasminogen kringle 4 to generate the 37 kringles found in apolipoprotein (a).

Two plasminogen-related sequences (termed *PLGLA* and *PLGLB*) with approximately 95% nucleotide sequence homology to the *PLG* gene have also been characterized (Ichinose, 1992). These sequences differ from each other by only 1–2% in their nucleotide sequences. They probably represent pseudogenes since they possess both an in-frame Stop codon and a single base-pair substitution in the conserved GT dinucleotide of a donor splice site.

Isolation and characterization of a murine plasminogen cDNA clone (Degen *et al.*, 1990) should potentiate comparative evolutionary studies; the cDNA comprises a 24 bp 5′ non-coding region, a 2436 bp open reading frame and a 257 bp 3′ non-coding region. The sequence of the cDNA indicates that the mouse and human plasminogen sequences are 79% and 76% homologous at the protein and DNA levels respectively although the murine protein contains two extra amino acid residues (543 and 587).

9.3.3 DNA polymorphisms

Genetic polymorphism of the plasminogen protein has been known for some time (see Hobart, 1979, and Section 9.2.9). Comparison of human cDNA sequence (Malinowski et al., 1984; Forsgren et al., 1987) revealed a GTG/GTA silent substitution at Val564. Restriction fragment length polymorphisms (RFLPs) within the coding region have also been detected with restriction enzymes MaeII, XmnI and AvaII (Petersen et al., 1990); the point mutations responsible for these polymorphisms were located to amino acid residues Cys239, Phe295 and Gly743, respectively.

An RFLP search was conducted by Murray et al. (1987) using 22 restriction enzymes to screen DNA samples from 10 individuals. Three biallelic RFLPs have been found with restriction enzymes MspI, SacI and RsaI. These RFLPs are frequent in Caucasian populations (f = 0.33/0.67, 0.67/0.33 and 0.76/0.24, respectively) but are apparently not so common in Blacks and Asians. Significant linkage disequilibrium was noted between the three RFLPs and the previously reported protein polymorphism, PLG A/B (Raum et al., 1980b). A further RFLP detectable with TaqI (f = 0.94, 0.06) has also been reported (Candiani et al., 1989).

9.3.4 Chromosomal localization

The PLG gene was assigned to the long arm of chromosome 6 by somatic cell hybrid analysis (Murray et al., 1987). Using in situ hybridization, Murray et al. (1987) and Frank et al. (1989) confirmed the somatic cell hybrid-derived data and regionally localized the PLG gene to chromosome 6q26–q27. Frank et al. (1989) also noted a plasminogen-related sequence on chromosome 2q11–p11, possibly the same as that reported by Ichinose (1992) and partially sequenced by Petersen et al. (1990).

The LPA gene has also been mapped to chromosome 6q26–q27 by in situ hybridization (Frank et al., 1988; Ichinose, 1992). Indeed, two LPA-related genes have been shown to flank the LPA gene by pulsed field gel electrophoresis (PFGE) (Byrne et al., 1994). The apolipoprotein (a) and PLG gene sequences probably arose by duplication of a common ancestral gene (see Section 9.3.2) and have not subsequently been separated on the chromosome. The murine PLG gene maps to the proximal half of chromosome, 17, a region syntenic with the long arm of human chromosome 6 (Degen et al., 1990a).

9.3.5 Expression

Plasminogen has been shown by radioimmunoassay to be synthesised and secreted by primary cultures of rat hepatocytes (Bohmfalk and Fuller, 1980) and by perfused rat liver (Saito et al., 1980). In this latter report, the rate of plasminogen release was compatible with liver being the major site of synthesis. Liver is also thought to be the main site of synthesis of human plasminogen (Raum et al., 1980a). In monkey liver, two forms of plasminogen (1 and 2) are synthesised from two distinct mRNAs sized at 23S and, 18S respectively (Gonzalez-Gronow and Robbins,, 1984).

From the representation of PLG cDNA clones in human liver cDNA libraries (Malinowski et al., 1984; Forsgren et al., 1987), the abundance of PLG mRNA can be estimated at between 0.03% and 0.12% of total mRNA. Malgaretti *et al.* (1990) reported the highest levels of *PLG* mRNA in human liver and HepG2 cells with much lower levels detectable in Hep3B and HeLa cells. Several liver-specific DNase hypersensitive sites are present in the 40 kb region separating the *PLG* and *LPA* genes (Magnaghi *et al.*, 1994).

Human *PLG* cDNAs have been expressed *in vitro* in a baculovirus vector-infected insect cell system (Whitefleet-Smith *et al.*, 1989) and in mammalian cells (Busby *et al.*, 1991).

9.4 Plasminogen deficiency

Plasma plasminogen values vary in the normal population between 0.79 and 1.36 U ml^{-1} (Dolan *et al.*, 1994; Rodeghiero and Tosetto, 1996); although no sex difference was apparent, plasminogen levels decrease with age only in women. Dolan *et al.* (1994) noted that 20/752 (2.7%) healthy blood donors possessed plasminogen levels below 0.79 U ml^{-1} but these individuals were asymptomatic for thrombotic disease. Newborn children possess plasma plasminogen antigen levels about 50% of that of the adult with lowered plasmin generation rates (Estelles *et al.*, 1980; Corrigan *et al.*, 1989).

That plasminogen deficiency alone can be responsible for a clinical phenotype is evidenced by studies on transgenic mice (Bugge *et al.*, 1995). Although Plg$^{-/-}$ mice complete embryonic development and survive to adulthood, such animals are predisposed to severe thrombosis. Interestingly, transgenic mice deficient in both plasminogen and fibrinogen exhibit a much milder phenotype (Bugge *et al.*, 1995). Removal of fibrinogen from the extracellular environment thus alleviates the pathology associated with plasminogen deficiency. Further, transgenic mice deficient in both plasminogen and fibrinogen are phenotypically indistinguishable from fibrinogen-deficient mice (Bugge *et al.*, 1995). These authors concluded that the fundamental and possibly only essential physiological role of plasminogen is fibrinolysis.

9.4.1 Hypoplasminogenaemia and dysplasminogenaemia

Deficiencies of plasminogen so far described fall into two distinct categories:

(i) hypoplasminogenaemias are due to the decreased synthesis/secretion of an otherwise normal plasminogen. These cases are termed type I and are characterized by similarly reduced levels of activity and antigen.
(ii) Dysplasminogenaemias (type II) are due to the synthesis of a qualitatively abnormal plasminogen and are characterized by a decreased activity level but a normal level of antigen (reviewed by Robbins, 1988, 1990, 1992).

Cases of type I deficiency reported to date are summarized in *Table 9.2*. Activity levels are consistent with a heterozygous defect. The deficiency state is usually associated with deep vein thrombosis (DVT), pulmonary embolism (PE) and thrombophlebitis but the occasional episode of arterial thrombosis has been reported (Dolan *et al.*, 1988). In several cases, a possible trigger (e.g. injury, surgery, contraceptive pill) has been quoted but, with only one exception (Mannucci *et al.*, 1986a), no other cases of thrombosis were noted in the families even in phenotypically proven heterozygotes for the deficiency state. The age of onset varied from 21 to 33 (Dolan *et al.*, 1988) with a mean ($n=6$) of 27.

Table 9.2. Inherited hypoplasminogenaemias (type I deficiency)

Activity/antigen (% normal)	Clinical phenotype	Reference
55/49	DVT; father of propositus had PE	Mannucci et al. (1986a)
30	DVT, pulmonary hypertension	Lottenberg et al. (1985)
50	DVT	Hasegawa et al. (1982)
45	PE, thrombophlebitis	Ten Cate et al. (1983)
50	DVT	Girolami et al. (1986)
56/45	DVT	Dolan et al. (1988)
70/68	DVT/PE	Dolan et al. (1988)
58/60	Thrombophlebitis	Dolan et al. (1988)
57/60	Arterial thrombosis	Dolan et al. (1988)
48/55	DVT	Skoda et al. (1988)
54/54	Arterial thrombosis	Azuma et al. (1993)
63/66 'Frankfurt II'	Asymptomatic	Robbins et al. (1991a)

Reported examples of type II deficiency are summarized in *Table 9.3*. Activity levels vary from 20% to 80% whilst the antigen value is normal in most cases. The mode of inheritance of the deficient phenotype is usually autosomal dominant. As with type I deficiency, DVT, PE and thrombophlebitis are common in deficient patients; the age of onset varied from 15 (Aoki *et al.*, 1978) to 32 (Soria *et al.*, 1983a). The majority of patients so far described have a defect in the active site of the plasminogen molecule.

One patient (Tochigi I) suffered from recurrent thrombosis and exhibited a sharply reduced level of functional plasminogen (Aoki *et al.*, 1978). Active site titration studies and the absence of incorporation of diisopropylphosphorofluoridate into the light

Table 9.3. Inherited dysplasminogenaemias (type II deficiency)

Patient	Activity (%)	Antigen	Clinical phenotype	Nature of defect	Reference
Tochigi I	69	N	DVT, PE, TP	Ala601→Thr	Aoki et al. (1978)
Tochigi II	50	N	Asymptomatic	Ala601→Thr	Sakata and Aoki (1980)
Nagoya I	50	N	Asymptomatic	Ala601→Thr	Miyata et al. (1984)
Paris I	45	N	PE	Active site?	Miyata et al. (1984) Soria et al. (1983a)
Chicago I	41	N	DVT	Impaired tPA binding	Wohl et al. (1979)
Chicago II	88	N	DVT	Impaired tPA binding	Wohl et al. (1979)
Chicago III	80	N	DVT	Impaired tPA binding and impaired cleavage of Arg561–Val562 bond	Wohl et al. (1982)
Frankfurt I (1988a)	78	N	DVT/PE	Active site?	Scharrer et al. (1986)
Tokyo	10–20	N	DVT	Active site?	Kazama et al. (1981)
San Antonio	58	76	Axillary vein	Decreased tPA-mediated activation	Liu et al. (1988)
Jichi	48	N	DVT/PE	?	Manabe and Matsuda (1985)
Maywood I	26	N	?	Impaired cleavage of Arg561–Val562 bond	Robbins et al. (1991a)

chain of the abnormal molecule (Sakata and Aoki, 1980) suggested that the defect lay at or near the active site. The absence of proteolytic activity was then shown to be due to a Ala601→Thr substitution, two residues N-terminal to the active site His603 residue (Miyata *et al.*, 1982). The niece of this patient was identified as being a homozygote by virtue of her possessing very low (approximately 5% normal) plasminogen activity but normal antigen; she was, however, asymptomatic. Miyata *et al.* (1984) have since identified two further asymptomatic heterozygotes for this mutation in the same family. This variant is now known to account for some 94% of cases of dysplasminogenaemia in the Japanese popoulation (Tsutsumi *et al.*, 1996). It also appears to be fairly frequent in the general Japanese population (3.6%; Aoki *et al.*, 1984), an observation probably best explained in terms of a founder effect since the variant was not found in a Caucasian population sample. The Ala601→Thr substitution has, however, also been found in the Chinese population at a frequency of 2.9% (Li *et al.*, 1994) and so it may represent a very ancient mutation. Apart from the index cases, no other members of the type II families studied exhibited symptoms of thrombotic disease, not even those with markedly reduced plasminogen levels.

The actual prevalence of plasminogen deficiency in the general population has been variously estimated as between 0.29% (Tait *et al.*, 1996) and approximately 0.4% (Tait *et al.*, 1991a). By contrast, some 2–3% of unexplained DVTs in young patients may be due to plasminogen deficiency (Mannucci and Tripodi, 1987; Gladson *et al.*, 1988). However, 20 individuals identified by Tait *et al.* (1991a) in their survey of blood donors, possessed plasminogen levels persistently more than two standard deviations below the reference range, yet exhibited no history of thrombosis. Similarly, Shigekiyo *et al.* (1992) studied 40 members of two unrelated families with type I plasminogen deficiency; only 3/21 heterozygotes were symptomatic, a frequency not significantly different from that found in healthy controls. These authors suggested that an inherited defect of plasminogen might not on its own be sufficient to account for a predisposition to thrombosis. Finally, Tait *et al.* (1996) showed that the observed prevalence rate (0.29%, 95% CI 0.19–0.42%) for familial plasminogen deficiency in a cohort of 9611 blood donors was not significantly different from that calculated from reports of plasminogen deficiency in thrombotic cohorts (0.54%).

Various other studies, however, have arrived at entirely different conclusions. A study of 20 kindreds with type I plasminogen deficiency has shown that the prevalence of thrombotic events in heterozygous patients was 23.6% (22/93) compared with 0/95 unaffected siblings (Sartori *et al.*, 1994). Analysis of thrombosis-free survival curves also showed that plasminogen-deficient patients exhibited a significantly higher probability of developing thrombosis than unaffected siblings (Sartori *et al.*, 1994). Consistent with this finding, Girolami *et al.* (1994) showed that 26% of a group of 112 patients (from 28 kindreds) with plasminogen deficiency had experienced thrombotic manifestations as compared to 1% of unaffected family members. It may thus be concluded that inherited plasminogen deficiency is indeed an independent risk factor for venous thrombosis.

9.4.2 Molecular genetics

Studies of plasminogen deficiency at the DNA level are still in their infancy. As described above, the pathological change underlying plasminogen Tochigi was determined by amino acid sequencing to be an Ala601→Thr substitution (Miyata *et al.*,

1982); the causative lesion may be inferred to be a CCT→ACT transversion. This substitution occurs two residues N-terminal to the His603 active site residue. This variant is now known to account for some 94% of cases of dysplasminogenaemia in the Japanese population (Tsutsumi *et al.*, 1996).

The Ala601→Thr lesion has also been found in homozygous form in a Japanese patient (Plasminogen Kagoshima) with 8% activity/98% antigen, confusingly termed '*type I*' (Ichinose *et al.*, 1991). A daughter of the propositus was shown to be heterozygous for the mutation as were several members of a third family (Nagoya II); the presence of the lesion could be readily demonstrated since it removes a *Fnu*4HI restriction site. The second mutation ('*type II*') was found in PLG Nagoya I (51% activity/65% antigen); a GTC→TTC change converting Val355 to Phe. This lesion leads to the loss of an *Ava*II site. It occurs in the connecting region between the third and fourth kringles and may be important for either the structure or stability of the plasminogen protein. Two further members of this family were found to be heterozygous for the *type II* defect, two heterozygous for the *type I* mutation and two compound heterozygous for *type I* and *type II* lesions. Although no functional analysis was attempted, both mutations segregated with the predisposition to thrombosis in the families examined; 2/6 *type I* heterozygotes had a history of thrombosis compared with, 1/2 *type II* heterozygotes and 2/2 *type I/type II* compound heterozygotes. The frequency of homozygous severe plasminogen deficiency is much rarer in the Japanese population than would be expected on the basis of the frequency of the *type I* allele. It may thus be inferred that homozygosity for a plasminogen gene defect militates against the chances of survival.

A further Japanese mutation in the *PLG* gene, a TCC→CCC transition converting Ser572 to Pro, was found in a family with type I plasminogen deficiency (54% normal activity and antigen) and arterial thrombosis (Azuma *et al.*, 1993). This lesion occurs within the activation cleavage region between the kringle and serine proteinase domains in a residue which is conserved between different serine proteinases. The mutation may alter the conformation of the plasminogen protein, thereby either impairing its secretion or reducing its stability.

Other *PLG* gene lesions include a GCT→ACT transition in an essentially asymptomatic Japanese patient with type I plasminogen deficiency (60% normal activity and antigen) which is responsible for converting Ala675 to Thr (Mima *et al.*, 1996) and a GAC→AAC transition converting Asp676 to Asn in two clinically asymptomatic individuals with type I plasminogen deficiency (Tsutsumi *et al.*, 1996).

Züger *et al.* (1996) have described a family with heterozygous plasminogen deficiency and the factor V Leiden variant. Six members of the family exhibited a plasminogen deficiency but only two of these individuals had suffered thromboembolic events. Both individuals in the family who were heterozygous plasminogen-deficient and heterozygous for the factor V Leiden variant had experienced thromboembolic episodes, whilst one of two individuals heterozygous for the factor V Leiden variant was affected.

Heparin cofactor II

10.1 Introduction

There are two thrombin inhibitors in plasma whose activity is stimulated by heparin, antithrombin III (ATIII) and heparin cofactor II (HCF2). HCF2 was first described by Briginshaw and Shonberge (1974a,b) who distinguished it from ATIII on the basis that it inhibited thrombin but not factor Xa. HCF2 was then isolated by Tollefsen and Blank (1981) and Wunderwald *et al.* (1982).

HCF2, a member of the serpin superfamily (Carrell *et al.*, 1987), is synthesised in the liver. It is specifically responsive to dermatan sulphate (Tollefsen *et al.* 1983) whereas this proteoglycan has no effect on the ATIII–thrombin interaction. Dermatan sulphate comprises 60–70% of the glycosaminoglycan of the intima and media of large blood vessels. This property of dermatan sulphate responsiveness permitted assays of HCF2 in plasma to be devised which could then be used to screen for the predicted deficiency state, expected to be associated with thrombophilia. Although several families with HCF2 deficiency and thromboembolism have been described (see Section 10.4), the relatively high frequency of covert HCF2 deficiency in the general population casts doubt on whether this association is causal or merely coincidental.

10.2 Physiology, structure and function

10.2.1 Physiology

HCF2, which is synthesised in the liver, circulates as a single-chain glycoprotein M_r 65.6 kDa, at a concentration of approximately 80 μg ml^{-1} (1.2 μM), about half the molar concentration of ATIII. About 10-fold higher concentrations of heparin are required to stimulate HCF2 inhibition of thrombin compared with those required to stimulate ATIII (Tollefsen *et al.*, 1983). Dermatan sulphate stimulates HCF2 1000-fold but has almost no effect on ATIII. However, this effect of dermatan sulphate is inhibited by fibrinogen (Zammit and Dawes, 1995). Dermatan sulphate prolongs the activated partial thromboplastin time of plasma (Ofosu *et al.*, 1984) and upon infusion into experimental animals is antithrombotic (Fernandez *et al.*, 1986a). HCF2 exerts a regulatory role *in vivo* through its interaction with proteoglycans of the vessel walls (Liu *et al.*, 1995; Shirk *et al.*, 1996).

HCF2 thus inhibits thrombin in both a 'progressive' reaction and in an accelerated reaction catalysed by dermatan sulphate. Deletion mutagenesis studies have helped to identify residues in HCF2 which are essential to the progressive reaction (Sheffield and Blajchman, 1995): $\Delta106$ and $\Delta169$ mutants failed to react with thrombin even in the presence of dermatan sulphate whereas $\Delta58$ and $\Delta82$ mutants retained an ability

Venous Thrombosis: from Genes to Clinical Medicine, D.N. Cooper and M. Krawczak.
© 1997 BIOS Scientific Publishers, Oxford.

to complex with thrombin. The biochemistry of HCF2 and its mode of action have been reviewed by Tollefsen (1995).

10.2.2 Primary structure and secondary modifications

The amino acid sequence of HCF2 was deduced by characterization of cDNA clones and by amino acid sequencing. The protein (calculated $M_r=54\,996$ Da) contains 480 amino acids plus a signal peptide of 19 residues (Inhorn and Tollefsen, 1986; Ragg, 1986; Blinder et al., 1988). Three potential Asn-linked glycosylation sites are located at Asn30, Asn169 and Asn368. Since the plasma-derived protein possesses about 10% carbohydrate, one or more of these are likely to be occupied. Two tyrosine residues are sulphated in HCF2 derived from a human hepatoma line (Hortin et al., 1986).

10.2.3 Structure–function relationships

HCF2 is homologous to other serpins with about 30% sequence identity to ATIII. HCF2 is also likely to be structurally homologous to, and to have a similar mode of action to, other serpins. HCF2 contains an N-terminal acidic domain rich in negatively charged amino acids that is similar in composition to the C-terminal of hirudin, a potent thrombin inhibitor. This acidic domain is thought to bind to the anion-binding exosite I of thrombin (reviewed by Church and Hoffman, 1994).

HCF2 also contains a region between Lys165 and Phe195 which is highly homologous to ATIII. This region contains many positively charged amino acid residues that can interact with glycosaminoglycans. The binding sites for heparin and dermatan sulphate are located in this region. Both heparin and dermatan sulphate bind to thrombin via the anion-binding exosite II of the proteinase. Unlike the case of ATIII, however, in which heparin promotes the interaction of proteinase and serpin in a *ternary complex* via a template mechanism, glycosaminoglycans promote the HCF2-mediated inhibition of thrombin via an allosteric mechanism (i.e. the binding of glycosaminoglycans to HCF2 serves to accelerate the formation of HCF2–thrombin complexes; Sheehan et al., 1994).

The reactive site of HCF2 has been identified by isolating and sequencing the peptide after complex formation with thrombin (Griffiths et al., 1985). After release from complex with its proteinase target, the Leu444–Ser445 bond of HCF2 is cleaved, thereby identifying these residues as P1–P1'. Proline, present in the P2 position (Pro443), may be important in specifying thrombin as a target proteinase since Pro is conserved at this location in other thrombin substrates (Inhorn and Tollefsen, 1986).

A particular feature of HCF2 and ATIII which distinguishes them from other serpins is their ability to interact with glycosaminoglycans. This property has been localized to a region containing negatively charged residues by site-directed mutagenesis and through the study of a naturally occurring variant. Thus Blinder and Tollefsen (1990) mutated Lys185 to Met, Asn or Thr and showed that all these variants possessed reduced affinity for heparin, but normal activity against thrombin in the absence of glycosaminoglycan. Furthermore, these variants were not stimulated by dermatan sulphate. Conversely, mutation of Arg103 to Leu had no effect on the functional properties of HCF2 with or without addition of glycosaminoglycan. These results were consistent with the conclusion drawn from the study on the naturally occurring variant, HCF2 Oslo (Arg189→His), which exhibited a reduced affinity for dermatan sulphate (Blinder et al., 1988).

A synthetic peptide corresponding to residues 54–75 of HCF2 was found to inhibit cleavage of fibrinogen by thrombin (50% inhibition at 28 μM). This effect was achieved without blockage of the active site of thrombin (Hortin *et al.*, 1989). The polyanionic peptide formed by residues 54–75 may interact with a non-catalytic extended binding site of thrombin, similar or identical to that occupied by hirudin. Such a binding site may contribute to the efficiency of the inhibition of thrombin by HCF2.

10.3 The heparin cofactor (*HCF2*) gene

10.3.1 cDNA cloning

Several cDNA clones encoding HCF2 have now been isolated from libraries made from mRNA derived from human liver (Ragg, 1986; Inhorn and Tollefsen, 1986; Ragg and Preibisch, 1988; Blinder *et al.*, 1988). The full-length *HCF2* cDNA includes at least 60 bp 5′ non-coding region, 1497 bp encoding a 19-amino-acid hydrophobic signal peptide and a 480-amino-acid mature protein, and a 654 bp 3′ non-coding sequence.

A murine *Hcf2* cDNA has been sequenced (Zhang *et al.*, 1994); it contains a 1434 bp open reading frame (encoding a 23 residue signal peptide and 455 residue mature protein), a minimum of 90 bp 5′ untranslated sequence and a 580 bp 3′ untranslated region. The rabbit HCF2 cDNA contains a 1440 bp open reading frame (encoding a 19 residue signal peptide and 461 residue mature protein), a 77 bp 5′ untranslated region and a 661 bp 3′ untranslated region (Sheffield *et al.*, 1994). The predicted amino acid sequences of the mouse and rabbit proteins are 87% identical to that of human HCF2.

10.3.2 Structure and evolution

The mapping of the *HCF2* gene comprises five exons spread out over 14.5 kb of DNA (Ragg and Preibisch, 1988). Nearly 16 kb of DNA sequence from the *HCF2* gene region, including 1.7 kb 5′ flanking region has been established (Herzog *et al.*, 1991). A restriction map of the human *HCF2* gene is shown in *Figure 10.1*. A total of 11 *Alu* sequences have been found within the *HCF2* gene (Herzog *et al.*, 1991); their distribution is highly non-random (two 5′ to the gene, six in intron 1, two in intron 2 and one in intron 3).

Several regions of clustered transcriptional start sites occur over a 90 bp stretch of 5′ flanking sequence (Ragg and Preibisch, 1988). This may reflect heterogeneity of transcriptional initiation *in vivo*. An inverse CCAAT box element (ATTGG) is present in the proximal 5′ region of the *HCF2* gene. Three sequences matching the acute

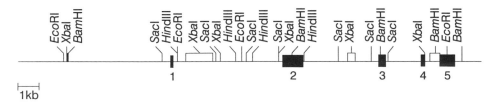

Figure 10.1. Exon/intron distribution and restriction map of the human *HCF2* gene. Exons are denoted by solid bars and are numbered 1–5. Reproduced from Ragg and Preibisch (1988) with permission from the American Society for Biological Chemistry.

phase responsive element consensus sequence in inverted orientation (TCCCAG) occur in an approximately 1 kb region upstream of the gene (Herzog *et al.*, 1991).

The number and distribution of exons in the *HCF2* gene closely parallel that seen in the human α1-proteinase inhibitor (*PI*) and rat angiotensinogen genes. For all three genes, the first exon consists entirely of 5' non-coding sequence, whereas the second exon contains the remaining few nucleotides of 5' non-coding sequence and encodes the signal peptide and N-terminal amino acid sequence. At both the nucleotide and amino acid sequence levels, homologies between HCF2 and α1-proteinase inhibitor (46% and 29% respectively) and between HCF2 and angiotensinogen (36% and 19% respectively) are apparent (Ragg and Preibisch, 1988). Although the HCF2, α1-proteinase inhibitor and angiotensinogen genes are clearly derived from a common ancestral sequence, they do differ markedly from each other at their N-terminal ends.

The murine *Hcf2* gene is approximately 7 kb in length and contains at least four exons (Zhang *et al.*, 1994). The exon–intron organization of the murine gene is thus essentially homologous to its human counterpart although it may lack a large intron in the 5' untranslated region.

10.3.3 DNA polymorphisms

Comparison of the available *HCF2* cDNA and genomic sequences yields three possible sequence polymorphisms; two are silent GGC/GGT and CAC/CAT nucleotide substitutions at codons Gly149 and His463, respectively, and the third is an AGA/AAA change at codon 218 which gives rise to an Arg/Lys substitution (Ragg *et al.*, 1986; Blinder *et al.*, 1988; Ragg and Preibisch, 1988; Herzog *et al.*, 1991).

An RFLP detectable with the restriction enzyme *Bam*HI has been found in the 5' half of the human *HCF2* gene (Blinder *et al.*, 1988); alternative 3.1 and 2.6 kb alleles occur with frequencies of 0.57 and 0.43 in the US population. An *Msp*I RFLP with alternative alleles of 2.9 kb and 1.2 kb occurring at frequencies of 0.58 and 0.42 has also been reported (Turner *et al.*, 1990). A third RFLP, detectable with *Hind*III and exhibiting alternative alleles of 20 kb and 17.5/2.5 kb, has been shown to occur with frequencies of 0.55 and 0.45, respectively (Herzog *et al.*, 1991).

10.3.4 Chromosomal localization

The *HCF2* gene was first localized to chromosome 22 by hybridization of flow-sorted human chromosomes spot-blotted on to nitrocellulose filters (Blinder *et al.*, 1988). Somatic cell hybrid analysis then further localized the gene to 22p–q11.2 (Herzog *et al.*, 1991). The murine gene maps to chromosome 16 (Zhang *et al.*, 1994).

10.3.5 Expression

From the representation of *HCF2* clones in human liver cDNA libaries, it may be estimated that *HCF2* mRNA comprises approximately 0.005% of total mRNA (Inhorn and Tollefsen, 1986; Ragg, 1986; Blinder *et al.*, 1988; Ragg and Preibisch, 1988). Ragg (1986) demonstrated the presence of a 2.3 kb *HCF2* mRNA in human liver. This 2.3 kb mRNA species is also present in HepG2 hepatoma cells together with a much fainter band at around 3.5 kb (Ragg and Preibisch, 1988).

A 2.3 kb *Hcf2* mRNA has been noted in murine liver by Northern blotting but no signal was detectable in heart, brain, spleen, lung, skeletal muscle, kidney, testis, placenta, pancreas or intestine.

Human recombinant HCF2 has been successfully expressed in *Escherichia coli* (Blinder *et al.*, 1988, 1989). Human HCF2 variants generated by site-directed mutagenesis have been expressed transiently in COS cells (Ragg *et al.*, 1990; Derechin *et al.*, 1990).

10.4 HCF2 deficiency

10.4.1 Clinical aspects

In the normal population, HCF2 activity levels cluster around a mean of 93 U dl^{-1} (90% confidence limits, 65–128; Rodeghiero and Tosetto, 1996). Relatively few examples of a HCF2 deficiency state have so far been reported. Tran *et al.* (1985) studied a patient who presented with intracranial arterial thrombosis. Two other members of her family had experienced episodes of either arterial or venous thrombosis. Measured activity (39–47% normal) and antigen (42–51% normal) levels in these individuals were consistent with heterozygous HCF2 deficiency. Sié *et al.* (1985) investigated a second individual with deep vein thrombosis after an ankle injury. The propositus and three relatives possessed 50% of normal levels of HCF2 activity/antigen although only the propositus had experienced symptoms of thromboembolic disease. Weisdorf and Edson (1990) described a patient with recurrent venous thrombosis, thrombophlebitis, pulmonary embolism and HCF2 activity (47–53% normal) and antigen (38% normal) values consistent with a heterozygous deficiency state. Simioni *et al.* (1990) reported two families with members that exhibit low levels of HCF2 activity and antigen. In both families, thrombotic events occurred in 2/6 affected members. Interestingly, a high incidence of spontaneous abortion in affected females was noted.

Three further cases of HCF2 deficiency were reported by Bertina *et al.* (1987) who screened 277 patients with a history of unexplained venous thrombosis. In the first case, the propositus (45% normal HCF2 antigen) had suffered from recurrent deep vein thrombosis. Although six heterozygous relatives were identified phenotypically, none were clinically affected. The second case, who had suffered a deep vein thrombosis, exhibited a 47% normal HCF2 antigen level. Although this individual had a family history of thrombosis, there was also evidence of concurrent protein S deficiency. The remaining case concerned an individual with a 50% normal HCF2 antigen level who had experienced pulmonary embolism and thrombophlebitis. Although his daughter also appeared to be deficient phenotypically (60% normal HCF2 antigen), she was asymptomatic. Other screening studies of consecutive thrombotic patients have also picked up cases of HCF2 deficiency (e.g. Awidi *et al.*, 1993).

A number of other studies have attempted to determine the frequency of HCF2 deficiency among individuals presenting with a history of venous thrombosis. No examples of HCF2-deficient individuals were found in studies of 74 (Sié *et al.*, 1985), 31 (Ezenagu and Brandt, 1986), 74 (Chuansumrit *et al.*, 1989) and 70 (Andersson *et al.*, 1987) such patients. However, in the study of Bertina *et al.* (1987) quoted above, three patients were described with less than 60% of normal HCF2 antigen (1.1%). Combining these data, HCF2 deficiency may thus be expected to occur at a frequency of 0.6% among patients presenting with thrombotic disease. At such a low frequency, it is obviously important to consider the frequency of the deficiency state in clinically normal individuals. Bertina *et al.* (1987) reported two cases (one inherited) in a screen of 107 healthy volunteers (0.9%). Similarly, Andersson *et al.* (1987) found two HCF2-deficient individuals in a screen of 379 blood donors (0.5%); both were subsequently shown to exhibit abnormal immunoelectrophoretic patterns. Thus, the similar prevalence of HCF2 deficiency in patients with recurrent thrombosis and apparently healthy controls suggests that it would be premature to conclude that HCF2 deficiency is *ipse facto* an important risk factor for thrombosis. It is possible that, under certain conditions (e.g. co-inheritance of a second deficiency state), HCF2 deficiency may increase the likelihood that a thrombotic event will occur and that the patient will come to clinical attention.

One of the cases reported by Andersson *et al.* (1987), a heterozygote with approximately 50% normal plasma HCF2 activity in the presence of dermatan sulphate, has been studied in further detail (Blinder *et al.*, 1989); the variant protein (HCF2 Oslo) was purified and shown to exhibit a dramatically reduced (at least 60-fold) affinity for dermatan sulphate. The molecular basis for this variation has been determined to be an Arg→His substitution at residue 189 (discussed below). The substitution of the homologous residue Arg129 in ATIII Geneva is associated with loss of heparin binding (Gandrille *et al.*, 1990). The fact that all the individuals known to possess the Arg189 mutation are asymptomatic indicates either that the binding of dermatan sulphate to HCF2 is physiologically unimportant or, perhaps more likely, that halving the normal HCF2 activity is insufficient to elicit pathological symptoms. Further studies of HCF2-deficient individuals and their families (perhaps including homozygous affected patients) should serve to shed some light on this question.

10.4.2 Molecular genetics

A single base-pair substitution has been detected in one of the heterozygous patients described by Andersson *et al.* (1987) whose HCF2 (HCF2 Oslo) exhibited normal binding to heparin but not to dermatan sulphate; the CGC→CAC transition results in the replacement of Arg189 by His (Blinder *et al.*, 1989). *In vitro* expression studies of a *HCF2* gene altered by site-directed mutagenesis at Arg189 confirmed that this mutation was the underlying cause of the observed phenotype (Blinder *et al.*, 1989). Site-directed mutagenesis has also been used to demonstrate that Lys185 is involved in heparin–glycosaminoglycan binding (Blinder and Tollefsen, 1990), consistent with the Arg189 mutant phenotype. Identical G→A transitions at Arg189 were also found (Blinder *et al.*, 1989) in the other apparently unrelated individual with HCF2 deficiency reported by Andersson *et al.* (1987) and in two other patients described by Borg *et al.* (1991) and Toulon *et al.* (1991). This may be yet another example of recurrent mutation at a CpG dinucleotide mediated by deamination of 5-methylcytosine. However, without restriction fragment length polymorphism haplotyping data, the possibility of identity by descent cannot be excluded.

By contrast to the above asymptomatic deficiency state, Matsuo *et al.* (1992) reported the case of a Japanese patient with angina pectoris and coronary artery disease who exhibited a reduced level of plasma HCF2 antigen. The basis of the HCF2 deficiency in this patient, termed HCF2 Awaji, was found to be the insertion of a T after the GAT codon encoding Asp88 (Kondo *et al.*, 1996). The propositus' sister was also found to harbour the lesion which co-segregated with the deficiency state. Although this might have predicted the presence of a protein prematurely truncated at amino acid 107, no such protein was found by crossed immunoelectrophoresis and Western blotting (Kondo *et al.*, 1996). *In vitro* expression of the mutant protein product in human kidney 293 cells demonstrated that the mutant protein is secreted normally (Kondo *et al.*, 1996) suggesting that it is subsequently degraded in the circulation.

In a screen of 305 patients with juvenile thromboembolic episodes, two unrelated individuals of Italian origin were found to possess a dinucleotide deletion (TT) in exon 5 (codons 457/458) of the *HCF2* gene (HCF2 Rimini; Bernardi *et al.*, 1996). One of these individuals also possessed the factor V Leiden variant while the other manifested type I protein c deficiency; perhaps the second defects increased the chances of these patients coming to clinical attention.

The phenotypic effects of other mutations may be assessed by *in vitro* mutagenesis/expression experiments (e.g. substitution of Arg for Leu in the Leu444–Ser445 thrombin cleavage site enhances the rate of thrombin inhibition; Derechin *et al.*, 1990).

Factor XII deficiency

11.1 Introduction

Of the four proteins involved in contact activation of plasma coagulation, two (factors XI and XII) were first identified through the study of patients exhibiting deficiency states. Factor XII is an activator not only of coagulation but also of fibrinolysis and of the kinin system. Thus in order to understand the physiological role of factor XII and to understand the likely effect of its deficiency, it is necessary to consider all three enzymatic cascades. Some of the known interactions of factor XII are depicted in *Figure 11.1*. Emphasis is placed upon those most likely to be of physiological importance. The possible interactions of factor XII with these pathways have been recently reviewed (Fujikawa and Saito, 1989).

11.2 Structure and function

11.2.1 Molecular structure, domains and biochemical properties

Factor XII purified from human plasma has an M_r of approximately 80 kDa, containing 16.8% carbohydrate (Fujikawa *et al.*, 1980). It is a single-chain species with no detectable proteinase activity (*Figure 11.2*). Conversion of the single-chain zymogen to an active serine proteinase is accomplished by cleavage of a single Arg–Val peptide bond producing α-factor XIIa. During contact activation of plasma, this initial cleavage is performed by kallikrein to yield an active serine proteinase which retains its surface binding property. Further cleavage by kallikrein at an Arg–Asp bond releases

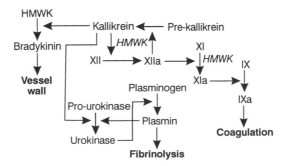

Figure 11.1. Interactions of factor XII. Factor XII reciprocally activates pre-kallikrein on contact with negatively charged surfaces. Kallikrein then activates prourokinase, initiating a cascade leading to fibrinolysis. Factor XIIa activates factor XI leading to thrombin generation and coagulation, and also releases bradykinin from high-molecular-weight kininogen (HMWK), causing vasodilation.

Venous Thrombosis: from Genes to Clinical Medicine, D.N. Cooper and M. Krawczak.
© 1997 BIOS Scientific Publishers, Oxford.

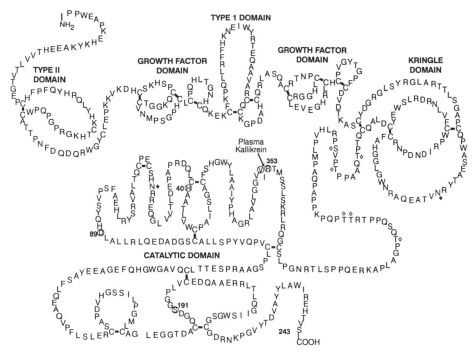

Figure 11.2. Primary sequence, putative secondary structure and functional domains of human factor XII. The site of proteolytic cleavage by kallikrein is shown by an arrow. The catalytic triad residues are circled. Disulphide linkages are tentative being based on homology to other protein domains. N-linked and O-linked glycosylation are denoted by closed and open diamonds, respectively. Reproduced from Seligsohn and Griffin (1995) with kind permission from McGraw-Hill Inc., New York.

the catalytic domain of factor XII from the N-terminal heavy chain in which the surface binding property resides. The catalytic domain thus freed is referred to as β-factor XIIa. Subsidiary cleavages also occur such that the final product has a light chain of only nine amino acids (Asn33–Arg343).

The amino acid sequence of factor XII was determined by automated Edman degradation (Fujikawa and MacMullen, 1983; McMullen and Fujikawa, 1985) and confirmed by the sequencing of cDNA clones (Cool et al., 1985). The linear sequence of the mature protein consists of 596 amino acid residues beginning with two proline residues. The first cleavage by kallikrein at Arg353–Val356 yields a disulphide-linked dimer of which the heavy chain comprises the first 353 residues and the light chain the C-terminal 243 residues. The former has been predicted on the basis of sequence homology to contain five domains. Type II and type I fibronectin domains are separated by a growth factor-like domain, followed by another growth factor-like domain and then a kringle domain. The kringle is followed by a connecting sequence that has six O-linked glycosylation sites. The light chain consists of a trypsin-like serine proteinase domain with its catalytic triad residues at His40, Asp89 and Ser191. The domain organization of factor XII more closely resembles those of tissue plasminogen activator (tPA) and prourokinase than it does the other coagulation factors, suggesting that factor XII may be, or may have been, more important in fibrinolysis than in coagulation.

Since βXIIa does not bind to negatively charged surfaces, the binding domain(s) must lie within the N-terminal portion of the molecule. Possible candidates for the binding site(s) are the type II domain (which in fibronectin is responsible for collagen binding) and the type I and kringle domains which in other proteins bind collagen. However, the actual surface binding site(s) may be much more localized since treatment of the N-terminal heavy chain with lysyl endopeptidase cleaves a single bond at Lys73 or Lys74 and destroys the ability of the whole fragments to bind to sulphatide (Fujikawa and Saito, 1989). The region surrounding this site is fairly basic (four Lys and two His in 10 residues), somewhat resembling the histidine-rich surface binding region of high-molecular-weight kininogen (HMWK). Indeed, factor XII and HMWK may compete for the same binding site on human umbilical vein endothelial cells (Reddigari *et al.*, 1993). Another surface binding site has been localized to the extreme N-terminal of factor XII: residues 1–14, by defining the epitope of two monoclonal antibodies that specifically inhibit this property (Clarke *et al.*, 1989).

Factor XII may also play a role in processes other than haemostasis; Schmeidler-Sapiro *et al.* (1991) have shown that both factor XII and factor XIIa exert a mitogenic effect on HepG2 hepatoma cells.

11.2.2 Post-translational modification of factor XII

The carbohydrate of factor XII (16.8% by weight) consists of 4.2% hexose, 4.7% hexosamine and 7.9% neuraminic acid (Fujikawa *et al.*, 1980). Attachment sites have been identified by sequencing as follows: Asn-linked chains at residue 230 in the heavy chain and residue 414 in the light chain. Six O-linked carbohydrate chains are present, all in the connecting region of the molecule between the kringle and catalytic domains (open diamonds in *Figure 11.3*) at Thr280, Thr286, Ser289, Thr309, Thr310 and Thr318). The function of these modified residues is unknown at present.

11.2.3 Interactions with other proteins

Kallikrein-dependent activation. The activation of factor XII by kallikrein is greatly accelerated by a negatively charged surface and by HMWK. The latter acts as a co-factor binding kallikrein to the surface. The overall rate enhancement provided by the accessory factors is about 50 000-fold. Part of the rate enhancement provided by negatively charged surfaces may be due to a conformational change in factor XII making it more susceptible to proteolysis.

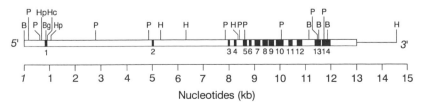

Figure 11.3. Structure of the human *F12* gene (after Cool and MacGillivray, 1987). Exons 1–14 are denoted by solid bars. Restriction sites mapped are: B, *Bam*HI; Bg, *Bgl*I; Hc, *Hinc*II; H, *Hind*III; Hp, *Hpa*II; P, *Pst*I.

Factor XIIa-dependent pre-kallikrein activation. Pre-kallikrein is activated by factor XIIa in a reaction that is accelerated modestly by HMWK in the presence of negatively charged surfaces.

Autoactivation of factor XII. When purified factor XII is incubated with negatively charged surfaces, after a short delay, factor XIIa begins to appear and after over an hour apparently fully activates itself (Tans *et al.*, 1983). Since the kinetics of this 'autoactivation' indicate that the activator is factor XIIa, this process cannot be the initial trigger of contact activation.

Factor XIIa activation of factor XI. This process is greatly enhanced by the presence of negatively charged surfaces and HMWK. It is thought that the latter acts by binding to surfaces together with factor XI, making the substrate available to α-factor XIIa. Supporting this is the observation that β-factor XIIa activation of factor XI is not enhanced by negatively charged surface nor by HMWK. Detailed kinetic studies on these interactions have not been performed.

Factor XIIa and fibrinolysis. The euglobulin fraction of plasma contains a readily demonstrated negatively charged surface-dependent fibrinolytic activator system, which is absent from factor XII deficient and from HMWK-deficient plasmas and delayed in pre-kallikrein-deficient plasma (see review in Fujikawa and Saito, 1989). It is now clear that this 'intrinsic activator' of fibrinolysis depends on the activation of prourokinase to urokinase by kallikrein, or more slowly by factors XIIa or XIa. Urokinase then activates plasminogen (*Figure 11.1*). Fibrinolysis is impaired in factor XII-deficient patients (Levi *et al.*, 1991), an observation which could account for the association of venous thromboembolism with factor XII deficiency.

Inhibitors of factor XIIa. Of the five plasma serine proteinase inhibitors (serpins) that have inhibitory activity against contact activation (ATIII, C1 inhibitor, α2-antiplasmin, α1-proteinase inhibitor and α2-macroglobulin), only C1 inhibitor and ATIII (the latter in the presence of heparin) are efficient against factor XIIa in contact-activated plasma (Pixley *et al.*, 1985); C1 inhibitor accounts for 74–92% of this inhibition in plasma (De Agostini *et al.*, 1984). These inhibitors are not active against surface-bound αXIIa, hence only βXIIa in free solution is neutralized by complex formation with serpins.

11.3 The factor XII (*F12*) gene

11.3.1 cDNA cloning

Human factor XII cDNA clones have been isolated from liver cDNA (Cool *et al.*, 1985; Que and Davie, 1986; Tripodi *et al.*, 1986a,b; Clarke *et al.*,1989). The factor XII protein comprises a signal peptide of 19 amino acids and a mature protein of 596 amino acids. Thus, together with a short (up to 50 bp) 5′ untranslated sequence and a 150 bp 3′ untranslated sequence, a full-length cDNA would measure 2045 bp.

11.3.2 Structure

The *F12* gene comprises 14 exons spanning a total of 12 kb (Cool and MacGillivray, 1987; *Figure 11.3*). The exons range in size from 57 to 320 bp (*Table 11.1*). The introns

Table 11.1. Exons and introns within the human *F12* gene

Exon	Exon length (bp)	Amino acid residues	Domain
1	105	−19 to −1	5′ UTR, signal peptide
2	57	1–19	Unknown homology
3	99	20–52	Fibronectin type II
4	70	53–76	Fibronectin type II
5	110	77–113	EGF-like
6	131	114–157	Fibronectin type I
7	104	158–192	EGF-like
8	165	193–247	Kringle
9	217	248–320	Kringle
10	231	321–397	Serine proteinase
11	136	398–443	Serine proteinase
12	143	444–491	Serlne proteinase
13	148	492–541	Serine protcinase
14	320	542–596	Serine proteinase/3′ UTR

are a little unusual in their distribution; exons 3–14 all occur in approximately 4.1 kb DNA whereas exons 1 and 2 are split by two large introns of 4.5 and 3.0 kb, respectively. Cool and MacGillivray (1987) sequenced in excess of 350 bp of 5′ flanking region and determined the position of the major transcriptional initiation site (i): a C nucleotide 49 bp upstream of the ATG initiation codon. Since the size of the S1 nuclease and primer extension products were in broad agreement, it would appear that the 49 bp 5′ untranslated region is not split by an intron. Other weaker transcriptional initiation sites were noted at +17, +23, +28 and +36. An ATGGTAGT sequence, similar to the SV40 enhancer core consensus sequence, is located in the human *F12* gene 330bp upstream of i.

Three *Alu*I repeat elements have been identified in the *F12* gene region by sequence analysis (Cool and MacGillivray, 1987); one is present at nucleotide −370 5′ to the gene and two others occur in intron B.

The 13 introns which divide the human *F12* gene separate structural domains of the molecule. Thus the signal peptide (exon 1), two regions of fibronectin homology (exons 3, 4 and 6), two EGF-like domains (exons 5 and 7) and a kringle (exons 8 and 9) are encoded by distinct exons or pairs of exons. The remaining exons (10–14) encode the serine proteinase domain and the 3′ untranslated region. Both the domain structure and the exon/intron organization of the *F12* gene bear a close resemblance to that exhibited by the tissue plasminogen activator (*PLAT*) and urokinase (*PLAU*) genes. Thus, consistent with the model of Patthy (1985), factor XII is more closely related evolutionarily to the plasminogen activators than to the true coagulation factors.

11.3.3 DNA polymorphisms

Comparison of different cDNA sequences has led to the identification of various possible DNA sequence polymorphisms within the human *F12* gene coding region. Most are neutral changes (at codons 160, 180, 209, 279 and 482) but one (CCT/TCT, Pro/Ser314) changes the amino acid specified. However, most of these variant bases were detected by Que and Davie (1986) and so far only appear in this one clone.

A *Taq*I polymorphism in intron B of the human *F12* gene has been reported (Bernardi *et al.*, 1986; Hofferbert *et al.*, 1996). A total of 45 unrelated Italians were originally screened for this RFLP but so far it has been found only in families with factor XII deficiency (Bernardi *et al.*, 1987, 1988), possibly because it occurs in association with a second lesion just upstream of the transcriptional initiation site (Hofferbert *et al.*, 1996).

11.3.4 Expression

Factor XII circulates in plasma with a concentration of 30 mg ml^{-1} and a half-life of 50–70 h (Saito, 1987). Based on organ perfusion studies, its site of synthesis is the liver where experimentally the rate of production was 0.2% that of albumin.

Northern blotting studies using a *F12* cDNA probe have detected a single mRNA species in human liver RNA variously measured at 2.1 (Que and Davie, 1986) and 2.4 kb (Cool *et al.*, 1985). No signal was seen, however, with RNA from human kidney or pancreatic carcinoma cells (Que and Davie, 1986). Estimates of the relative abundance of *F12* mRNA in total human liver poly(A)$^+$ mRNA can be obtained from the representation of *F12* cDNA in liver cDNA libraries. However, these estimates [0.018% (Cool *et al.*, 1985), 0.00017% (Tripodi *et al.*, 1986a) and 0.000057% (Que and Davie, 1986)] vary quite widely.

The enhanced expression of factor XII at both transcriptional and translational levels in response to oestrogen has been reported in the isolated rat liver (Gordon *et al.*, 1991). Factor XII has been shown to induce the expression of monocyte interleukin-1, a key mediator of the inflammatory response which is known to play a role in induction of the acute phase response and stimulation of endothelial cell procoagulant activity (Toossi *et al.*, 1992).

11.3.5 Chromosomal localization

The *F12* gene was allocated to human chromosome 5 by somatic cell hybrid analysis (Citarella *et al.*, 1988; Royle *et al.*, 1988). This was confirmed by *in situ* hybridization, the localization being refined to the region 5q33–qter (Royle *et al.*, 1988).

11.4 Factor XII deficiency

The first patient with this plasma defect to be studied was John Hageman, a man with no history of bleeding and whose blood was studied routinely as a preoperative screening procedure (Ratnoff and Colopy, 1955). He underwent surgery without bleeding complications as have numerous other patients with the same deficiency (Ratnoff and Saito, 1979). The *minimum* prevalence of the severe form of the deficiency has been estimated at 1/800 000 (Ratnoff and Saito, 1979). Since these patients are homozygous, the gene frequency for a non-functioning allele would be predicted to be about 1:900. Since neither the homozygous nor the heterozygous state is symptomatic for a bleeding disorder, those figures are probably underestimates.

Some individuals with factor XII deficiency have been reported to suffer from thrombosis. For example, Mr. Hageman died of pulmonary embolism and at least seven examples of myocardial infarction have been reported in factor XII-deficient individuals (Ratnoff *et al.*, 1968; Lämmle *et al.*, 1991). A review of 121 cases showed an 8.2% incidence of venous thromboembolism (Goodnough *et al.*, 1983) and in addition several cases of myocardial infarction at a relatively young age. This association might be explained on the basis that factor XII is implicated in the activation of fibrinolysis via prourokinase. Since factor XII-deficient individuals are generally asymptomatic, there is no indication for replacement therapy.

The majority of cases of factor XII deficiency are associated with equal amounts of factor XII activity and antigen (CRM⁻) but a few dysfunctional variants (CRM⁺) have been found (Saito et al., 1979, 1981; Berrettini et al., 1985; Takahashi and Saito, 1988; Miyata et al., 1989b; Lämmle et al., 1991; Wuillemin et al., 1991, 1992).

Mannhalter et al. (1987) studied a cohort of 107 patients with recurrent venous thrombosis and found 11 (10.3%) to possess factor XII activity (FXII:C) levels consistent with a heterozygous deficiency state. Halbmeyer et al. (1992) reported an incidence of factor XII deficiency of 8% in a study of 103 patients with recurrent venous thrombosis. Lämmle et al. (1991) studied 18 'homozygous', 20 'heterozygous' and 25 'possibly heterozygous' patients in the Swiss population: none of the homozygous patients (FXII:C less than 0.01 U ml⁻¹) exhibited an abnormal bleeding tendency whilst only one of the possible heterozygotes exhibited a history of mild bleeding and her homozygous sister was asymptomatic. These data confirmed the view that factor XII deficiency is not associated with a haemorrhagic diathesis. Two of the 18 homozygotes had suffered from venous thromboembolism at a young age, indicating that homozygous factor XII deficiency may be a risk factor in thrombotic disease. However, no clear evidence for an association between the heterozygous deficiency state and thrombosis emerged. This finding contrasted with those of Mannhalter et al. (1987). It is possible that additional defects were present at other loci in the latter group of heterozygous patients, with factor XII deficiency being a contributory rather than a causative factor in thrombotic disease.

Rodeghiero et al. (1992) studied 13 factor XII-deficient families and reported that the incidence of thrombosis was significantly higher than in age-matched controls. A likelihood analysis on first-degree relatives of the propositi yielded an odds ratio of 4.7 ($p=0.12$) for a thrombotic event between those classified as heterozygotes and those classified as normals. Although not statistically significant, this result did suggest the possible existence of an association. For individuals ($n=41$) who were obligatory heterozygotes, 17% experienced a thrombotic episode (odds ratio 8.2, $p=0.03$). The frequency of thrombotic episodes in obligate homozygotes (1 in 16 individuals) was too low for comparison with the data of Lämmle et al. (1991)

Halbmeyer et al. (1994a) screened 300 healthy blood donors and found seven individuals (2.3%) deficient in factor XII. Six exhibited a moderate deficiency state (FXII:C, 20–45%) whilst one manifested a severe deficiency state (FXII:C = 0.3%); all seven were associated with a CRM⁻ phenotype. Although 26 of the donors (8.7%) had a positive family history of thromboembolism, none of the factor XII-deficient individuals had such a family history.

Koster et al. (1994a) reported that 6% of patients with an objectively confirmed episode of venous thrombosis possessed factor XII levels consistent with a heterozygous deficiency. Since the corresponding frequency for a similarly sized sample of healthy controls was 5%, it is clear that there is no evidence for an increased prevalence of heterozygous factor XII deficiency in individuals presenting with venous thrombosis. Clearly, other factors must play a role in determining whether or not an individual with factor XII deficiency develops overt clinical symptoms of thrombotic disease.

To investigate the possible co-occurrence of factor V Leiden and factor XII deficiency, Castaman et al. (1996a) re-investigated 19 clinically symptomatic individuals from 13 families previously diagnosed as having factor XII deficiency. However, factor V Leiden was found in none of them. As yet, therefore, there is no evidence for a combination of factor V Leiden and factor XII deficiency being more likely to come to clinical attention.

To date, no molecular lesions in the *F12* gene have been described in patients with factor XII deficiency and a thrombotic tendency.

Fibrinogen and the dysfibrinogenaemias causing venous thrombosis

12.1 Introduction

Blood coagulates when soluble fibrinogen is converted to fibrin monomers which polymerize spontaneously to form a physical meshwork, the coagulum or clot. The enzyme responsible for this conversion is thrombin, the final effector enzyme of the intrinsic and extrinsic coagulation pathways. Fibrinogen has been studied intensively by a variety of different methods in an effort to construct a model of its structure and to elucidate its function. Such studies have been facilitated by the fact that fibrinogen is by far the most abundant coagulation protein in plasma with a concentration of 2–4 g l^{-1}. Since it is impossible in a short space to cover adequately the vast body of literature relating to this protein, this chapter attempts to summarize those aspects of its stucture and function relevant to the inherited defects of fibrinogen that cause thrombotic disease. The interested reader is referred to several comprehensive reviews (Mossesson and Doolittle, 1983; Doolittle, 1994; Hantgan *et al.*, 1994) for further detail.

12.2 Structure and function of fibrinogen

12.2.1 Primary sequence and covalent linkage of the fibrinogen chains

Fibrinogen is a disulphide-linked hexamer, containing three pairs of chains having the general formula (Aα, Bβ, γ). These chains are arranged so that there is an axis of two-fold symmetry through the centre of the molecule (*Figure 12.1*). The length of the bars corresponds to the length of the polypeptide chains Aα (610 residues), Bβ (461 residues), γ (411 residues). About 10% of γ chains in plasma have an extra 20 residues at their C termini due to differential mRNA processing (see below).

There are 8, 11 and 10 cysteine residues in the Aα, Bβ and γ chains, respectively, and these residues are all involved in a total of 29 disulphide linkages in the fibrinogen molecule. The disulphide linkages have been established by conventional methods using cyanogen bromide fragmentation (Henschen *et al.*, 1983). There are 17 inter-chain disulphide bonds (straight lines in *Figure 12.1*) connecting Aα, Bβ and γ chains in each symmetrical half of the molecule, and three disulphide bonds linking

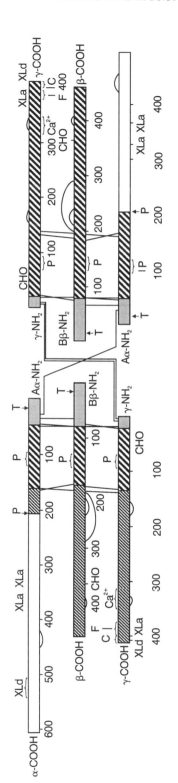

Figure 12.1. Diagram of the six chains of fibrinogen showing disulphide linkage; straight lines represent the 17 interchain bonds, curved lines the 12 intrachain bonds. T↓, sites of thrombin cleavage; P↓, the plasmin cleavage that releases the polar portion of the α chain; P↿, the plasmin-sensitive region of the coiled-coils; XLd, XLa, factor XIII cross-linking sites; Ca⁺⁺, calcium-binding sites; F, sites involved in fibrin monomer assembly; C, platelet-binding site; CHO, carbohydrate attachment. Stippled region, central domain; heavy hatched region, coiled-coil region; light hatched region, outer domain; open box, polar region. Reproduced from Hantgan *et al.* (1987) with permission from Lippincott-Raven Publishers.

the two halves in an antiparallel fashion (Doolittle, 1984). The 11 disulphide bonds nearest the N termini of the six chains are referred to as the N-disulphide knot. The alignment of the three chains is maintained by the N-disulphide knot and by the outer disulphide triplets, approximately 100 residues towards the C termini of each chain. There are 12 intra-chain disulphide bonds (curved lines in *Figure 12.1*), which presumably serve to maintain the tertiary structure of the individual chains. The disulphide rings flanking the coiled-coil region play an important part in dimer assembly (Zhang and Redman, 1994; *Figure 12.2*) but the symmetrical disulphide bonds between the Aα (28/28) and γ (8,9/8,9) chains appear not to be critical for fibrinogen assembly and secretion (Zhang *et al.*, 1993).

The three chains are strongly similar in sequence, suggesting evolution from a common ancestral protofibrinogen. However, the different properties of the three chains are conferred by sequences specific to each chain. Thus, fibrinopeptides A and B are located at the N termini of chains Aα and Bβ, and are released by thrombin cleavage of specific Arg–Gly bonds. Specific residues involved in the cross-linking of fibrin by factor XIIIa are found in the α and γ chains only (*Figure 12.2*). In particular,

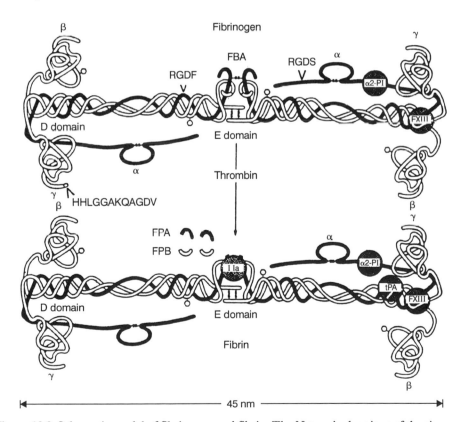

Figure 12.2. Schematic model of fibrinogen and fibrin. The N-terminal regions of the six fibrinogen chains are joined in the central domain (E domain) by disulphide bonds and form a dimeric structure. The E domain contains the FpA and FpB sequences that are cleaved by thrombin. Carbohydrate groups are located at one site on each of the γ and Bβ chains. The binding sites for thrombin (IIa), tissue plasminogen activator (tPA), factor XIII (FXIII) and α2-antiplasmin (α2-PI) are indicated. The platelet-binding motifs, HHLGGAKQAGDV and RGDS, are shown. Reproduced from Mosesson (1993) with permission from Thieme New York.

the γ chains have Glu (acceptor) and Lys (donor) pairs near their C termini, eight residues apart. The Aα chain has a long polar region of 403 residues at its C terminus that is highly susceptible to proteolytic attack, especially by plasmin. Release of this fragment is the first effect of plasmin on fibrinogen, and generates fragment X. Plasmin also readily attacks a central region of all three chains approximately half way between the N- and C-terminal disulphide knots.

12.2.2 Secondary structure

Fibrinogen has extensive helical structure. A sequence that occurs twice in each of the three chains is Cys–Pro–X–X–Cys and this is found at the two places in each half molecule where all three chains are disulphide-linked (Doolittle et al., 1978). The intervening 111 residue sequence of each chain has features strongly suggesting a helical conformation (absence of proline, polar and non-polar residues regularly repeated). Doolittle et al. (1978) proposed the now widely confirmed model of triple inter-chain disulphide linkages at either end of a coiled-coil of three strands much like a rope. The length of the coiled-coil region, calculated from extensive model building, is 160 Å, in excellent agreement with the connector region length seen in electron microscopy (see below). Doolittle et al. (1978) also proposed that the mid-region of the coiled-coil has a short non-helical segment that is susceptible to plasmin.

12.2.3 Tertiary structure

Evidence from physical, chemical and electron microscopic studies can be interpreted best in terms of a trinodular structure. All six N-terminal domains meet centrally in a globular region, cross-linked and stabilized by disulphide bonds (*Figure 12.2*). The three chains of each symmetrical half adopt an extended coiled-coil form connecting to C-terminal globular domains. At finer resolution, it is apparent that the Bβ and γ chains form independent distinct nodules. The C terminus of the Aα chain contains many hydrophilic residues and its position and fold are conjectural at present. With an M_r of 340 kDa, the molecule is 475 Å in length, connected by threads (seen in electron microscopy) corresponding to the coiled-coils which are 8–15 Å in diameter (Krakow et al., 1972; Fowler and Erikson, 1979). The three-dimensional structure of fibrinogen is still unclear, but ongoing crystallographic studies should shed some light on this question (reviewed by Doolittle et al., 1996).

Assembly of fibrinogen proceeds through the formation of αγ and βγ heterodimers (Huang et al., 1996) followed by the generation of αβγ half molecules which then dimerize to yield the mature six-chain molecule. Initial formation of the αγ and βγ complexes is potentiated by hydrophobic interactions between amino acids in the coiled-coil region of the molecule (Xu et al., 1996).

Fibrin monomers assemble in a staggered overlapping manner by non-covalent interactions between the E and D domains to form two-stranded fibrils that undergo trimolecular and tetramolecular branching (*Figure 12.3*). The fibrils also undergo non-covalent lateral associations to form a branched network of thick fibres. Fibrinogen has some flexibility in the long axis, and the coiled-coils with their disulphide rings may allow some twisting of the 'rope'.

Plasmin digests fibrinogen characteristically through the susceptible regions mentioned, producing fragments corresponding to the larger nodules (fragment D) and

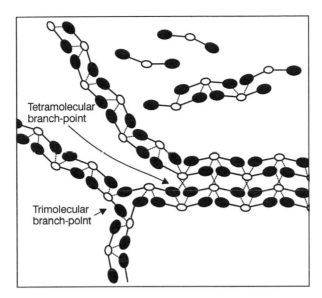

Figure 12.3. Fibrin assembly and matrix structure. Fibrin molecules are represented by tri-domain structures having a central E domain and two outer D domains. Fibrin monomers assemble in a staggered overlapping manner by non-covalent interactions between the E and D domains (–) to form two-stranded fibrils that undergo trimolecular and tetramolecular branching. The fibrils also undergo non-covalent lateral associations to form a branched network of thick fibres. Reproduced from Mosesson (1993) with permission from Thieme New York.

the central smaller nodule (fragment E). The assymetric combination of one lateral nodule connected to the central nodule is designated fragment Y.

Using the novel method of carrier protein-driven crystallization, Donahue *et al.* (1994) determined the structure of the C-terminal end of the γ chain. This portion of fibrinogen includes the GPIIb/IIIa receptor (see Section 12.2.10) and the donor and acceptor sites for factor XIIIa-catalysed cross linking (see Section 12.2.7).

12.2.4 Post-translational modifications

The human Aα chain is phosphorylated at Ser3 and Ser345 (Seydewitz and Witt, 1985). The longer γ-chain splicing variant contains sulphated tyrosines at residues 418 and 422 (Farrell *et al.*, 1991). The human Bβ-chain is hydroxylated at Pro31 (Henschen *et al.*, 1991) and glycosylated at residue Asn364 (Töpfer-Petersen *et al.*, 1976). The γ chain is also glycosylated at position Asn52 (Blombäck *et al.*, 1973).

12.2.5 Conversion of fibrinogen to fibrin

This occurs in three steps. Firstly, thrombin releases the two fibrinopeptides, A and B (FpA and FpB). Non-covalent assembly of fibrin monomers then proceeds rapidly to form protofibrils of two staggered strands which also assemble laterally. Finally, the coagulum is cross-linked by factor XIIIa (see Section 12.2.7). These steps will now be considered in more detail.

Fibrinopeptides A and B and their interactions with thrombin. Thrombin binds to fibrinogen in the region of the N-terminal peptides of the Aα and Bβ chains (reviewed by Binnie and Lord, 1993). Thrombin initially binds to the Aα chain between residues 30–44 (Hofsteenge and Stone, 1987; Binnie and Lord, 1991). High-resolution nuclear magnetic resonance (NMR) of peptides has shown that the thrombin-binding domain also spans residues Asp7–Arg16 of the Aα chain (Ni *et al.*, 1988a) with a non-polar cluster consisting of the side chains of residues Phe8, Leu9 and Val15 being directly involved in binding to thrombin (Ni *et al.*, 1988b).

The scissile bond is Arg16–Gly17 which is very rapidly cleaved when fibrinogen is exposed to thrombin, leading to release of FpA (Ala1–Arg16) at a rapid rate (Scheraga, 1986). The interaction between FpA and thrombin is highly specific, hydrophobic contacts providing the majority of interactions (Stubbs and Bode, 1993a, b; *Figure 12.4*). The small number of polar interactions noted probably serve to anchor and/or orient the peptide correctly.

Evolutionary conservation studies are useful to explore the structure and function of individual residues in FpA (Stubbs and Bode, 1993b). Asp7, by initiating the short α-helical turn, is responsible for the internal structural integrity of FpA but this residue can be replaced by Ser, Thr or Glu. Phe8 and Leu9 fill the aryl-binding site of thrombin; Phe is found predominantly in position 8 whereas a bulky hydrophobic residue is usually found in position 9. Ala10 is evolutionarily very variable; it makes no contact with thrombin. Glu11 makes a salt bridge with Arg173 of thrombin but since this residue is not invariant, this interaction may not be conserved between species. Gly12 is invariant because its left-handed helical conformation would cause strain with any other residue. Gly13 is virtually invariant because a larger residue would result in steric hindrance with Phe8. Val15 is highly conserved evolutionarily; this position always contains a branched-chain hydrophobic residue. The P1 residue (Arg16) is invariant,

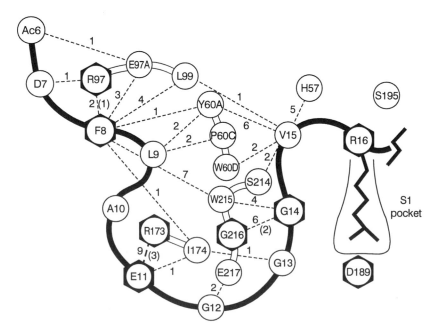

Figure 12.4. Structure and environment of fibrinopeptide A (thick line) in the active site cleft of thrombin (thin line). Broken lines indicate intrapeptide hydrogen bonds. Reproduced from Stubbs and Bode 1993b with permission from Thieme New York.

indicating the specificity of thrombin cleavage. Recent crystallographic studies have demonstrated a direct interaction via a salt bridge between Aα Arg16 and residue Asp189 in the catalytic site of thrombin (Martin *et al.*, 1992; Stubbs *et al.*, 1992).

Starting simultaneously, but proceeding more slowly, thrombin cleaves the scissile bond Arg14–Gly15 in the Bβ chain, releasing fibrinopeptide B (Gly1–Arg14). As polymerization proceeds to formation of protofibrils, the rate of fibrinopeptide B release increases, suggesting that its release is hastened by conformational changes (Hanna *et al.*, 1984). The release of fibrinopeptide B is not sufficient for normal clot formation since a venom, from the snake *Ancistrodon contortrix*, which only releases fibrinopeptide B, leads to fibrin polymerization at 37°C but only at low temperature. Stubbs and Bode (1993b) have made an evolutionary comparison of the FpB sequence which appears much more variable than that of FpA. There is some evidence for thrombin binding to the Bβ chain: Des-Bβ 15–42 fibrin exhibits severely diminished thrombin binding (Siebenlist *et al.*, 1990).

In addition to binding fibrinogen at its substrate site, thrombin binds to fibrin at various non-substrate sites. Stubbs and Bode (1993b) have proposed candidate hydrophobic/acidic regions in all three fibrinogen chains which might be capable of binding thrombin: these sequences, comprising residues 35–49, 150–165 (Aα), 66–80, 182–197 (Bβ) and 8–23, 124–139 (γ), all contain a Cys–X–X–X–Cys consensus at their C-terminal ends. Meh *et al.* (1996) have presented evidence for a high-affinity thrombin binding site in the D domain whilst two low-affinity thrombin binding sites reside in the central E domain.

The fibrinogen-binding exosite. Thrombin cleavage of fibrinogen requires not only fibrinopeptide recognition but also the possession of a fibrinogen-binding exosite (Kaminski and McDonagh, 1987; Fenton *et al.*, 1988; Vali and Scheraga, 1988). This exosite is centred around the Lys70–Glu80 loop of the proteinase (Stubbs and Bode, 1993a, b) whose structure is stabilized by a number of buried salt bridges. The exosite is capable of binding to residues 27–50 of the Aα chain, residues 15–42 of the Bβ chain and residues 8–23 of the γ chain (reviewed by Binnie and Lord, 1993; Stubbs and Bode, 1993a). The interactions of fibrinogen with thrombin are summarized schematically in *Figure 12.5*. Thrombin bound to fibrin via the fibrinogen-binding exosite is able to promote the factor XIII-mediated cross-linking of fibrin molecules in order to strengthen the clot (Greenberg *et al.*, 1987; Kaminski *et al.*, 1991; see Section 12.2.7).

Fibrin-bound thrombin is protected from inactivation by heparin–antithrombin III (Hogg and Jackson, 1989) which limits the effectiveness of heparin-mediated thrombin inhibition as a means of reducing fibrin-bound thrombin (Mirshahi *et al.*, 1989).

Fibrin assembly. Fibrin monomers formed by release of the fibrinopeptides have a structure virtually identical to that of fibrinogen but the very limited proteolysis exposes binding domains in the central E domain which interact with sites on the γ chain in the D domain (outer nodule). Hence DE contacts are formed which impose a half-staggered overlap mode of building the fibrin polymer chain. In this pattern, contacts are also established lengthwise between adjacent D domains (DD-long). Time-lapse studies show that double-stranded half-staggered protofibrils lengthen to about 800 nm before thickening by side-to-side contact begins (Hantgan *et al.*, 1980).

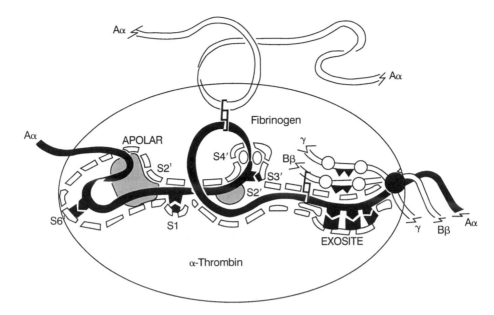

Figure 12.5. Major interactions of fibrinogen with thrombin. The boundary of the active site cleft is shown as a broken line. Hydrophobic interactions are shown by broken lines, acidic groups are marked by ● and basic groups by ■. The cleavage site on Aα fibrinogen is arrowed. Disulphide bridges are indicated by ⌇. The disulphide knot (●) lies in the vicinity of the fibrinogen-binding exosite. Reproduced from Stubbs and Bode (1993b) with permission from Thieme New York.

Thrombin cleavage of the fibrinopeptides releases FpA and FpB and exposes poly-merization sites, 'A' and 'B', located in the N-terminal portion of the molecule (Kudryk *et al.*, 1974). Complementary polymerization sites (a and b) are located on the γ chain of the molecule. Non-covalent assembly of fibrin monomers then proceeds rapidly to form protofibrils of two staggered strands which also assemble laterally (Mosesson, 1990). The two-stranded fibrils are formed by interaction between the N-terminal 'A' poly-merization sites on the central E domain and their complementary 'a' sites in the C-ter-minal region of the γ chain in the outer D domain (Olexa and Budzynski, 1980; Budzynski *et al.*, 1983). Similarly, the 'B' polymerization sites on the central E domain form an association with 'b' sites on the γ chain of the D domain.

The 'A' site is thought to comprise residues 17–20 of the Aα chain (Gly–Pro–Arg; Laudano and Doolittle, 1978). The 'a' site is variously reported as being located between residues 356 and 411 (Varadi and Scheraga, 1986), 303 and 405 (Southan *et al.*, 1985b) or 337–379 (Shimizu *et al.*, 1992) on the γ chain. The 'B' and 'b' polymer-ization sites may be at residues Gly15–Pro18 in the Bβ chain and in the C-terminal region of the Aα chain, respectively (Laudano and Doolittle, 1980; Hasegawa and Sasaki, 1990; Pandya *et al.*, 1991; Medved *et al.*, 1993).

Two further constitutive self-association sites have been recognized in the D domain, 'D:D' and 'γXL' (Mosesson *et al.*, 1995a,b). The D:D sites promote end-to-end alignment in assembling fibrils whilst the γXL sites promote intermolecular carboxyl terminal γ chain association that facilitates factor XIIIa-mediated cross-linking.

12.2.6 Calcium and fibrinogen

There are three Ca^{2+}-binding sites on human fibrinogen, one of which has been local-ized to γ chain residues 311–336 accounting for two sites (Dang et al., 1985). The third binding site is thought to be in the central domain of fibrinogen. Since the K_d for cal-cium binding to these three sites is 100-fold lower than the concentration of calcium ions in plasma, the sites must be fully saturated in fibrinogen in the circulation. The γ chain of fibrinogen bound with calcium ions has been shown to be less susceptible to plasmin degradation and to be more stable to heating and high pH (Haverkate and Timan, 1977). Although calcium is not essential for either fibrinopeptide release or for polymerization of fibrin monomers, the physical assembly of the monomers is accelerated in the presence of calcium (Brass et al., 1978). Evidence for the involve-ment of the C-terminal portion of the β chain in calcium-mediated interactions comes from the fact that the terminal tetrapeptide (QHRP) binds 10-fold more tightly to fibrinogen in the presence of 2 mmol calcium.

12.2.7 Interaction of fibrin/fibrinogen with factor XIII

Factor XIIIa catalyses cross-linking of the polymerized fibrin by creating isopeptide bonds between lysine and glutamine side chains present in the Aα and γ chains. This makes the clot stronger and more resistant to fibrinolysis by plasmin. In vitro, the action of cross-linking can be detected by resistance of the clot to dissolution by 6 M urea. The first pairs of glutamyl-lysyl bonds to form are just eight residues apart (398–406) at the C-terminal of the γ chain producing DD-long cross-links. Subsequently multiple cross-links form between adjacent Aα chains (the most impor-tant residues being glutamines 221, 237, 328; lysines 418, 448, 508, 539, 556, 580, 601; Matsuka et al., 1996; Sobel and Gawinowicz, 1996). This latter process may protect the plasmin-sensitive region of the coiled-coils.

Thrombin cleaves factor XIII after which active α2 subunits are released. Fibrinogen facilitates this process by lowering the effective calcium concentration required to the range of the plasma concentration of calcium. Fibrin polymers are able to accelerate thrombin-mediated factor XIII activation (Janus et al., 1983) after which factor XIIIa becomes bound to fibrin (Greenberg et al., 1985). Thrombin also serves as a cofactor in fibrin polymerization in a manner independent of its catalytic activity (Kaminski et al., 1991). These observations can be accommodated in a model in which factor XIII activation depends on the formation of a ternary complex between throm-bin, factor XIII and fibrin (Greenberg et al., 1987).

12.2.8 Interaction with fibronectin

Factor XIIIa cross-links fibronectin to the Aα chains of fibrin(ogen), both in plasma and in experiments with purified proteins (Mosher, 1976). The physiological role of this may be to increase clot strength according to studies using purified proteins. However, similar studies in plasma clotted with and without fibronectin showed only minimal effects (Chow et al., 1983).

12.2.9 Interaction of fibrin with the fibrinolytic pathway

Activation of plasminogen by tissue plasminogen activator (tPA) is dependent on the presence of fibrin (Rånby, 1982; Hoylaerts et al., 1982). Peptides corresponding to the

Aα chain residues 149–161 accelerate tPA-mediated plasminogen activation whilst Lys157 is essential for this reaction (Voskuilen *et al.*, 1987). Residues 311–379 at the C-terminal end of the γ chain of fibrinogen are also involved in the acceleration of tPA-catalysed activation of plasminogen (Yonegawa *et al.*, 1992). Plasminogen activation on the surface of fibrin serves to localize fibrinolysis.

Plasminogen binds to unmasked Lys residues at the C-terminal end of fibrinogen chains. Specifically, residues Leu121 and Lys122 of the Bβ chain of fibrinogen appear to be important in the binding of plasminogen (Varadi and Patthy, 1984). Plasminogen thus bound is rapidly converted into plasmin by fibrin-bound tPA; this mechanism thus represents the most important mechanism for the acceleration of fibrinolysis. Plasmin has a far higher affinity for fibrin as compared to fibrinogen (Lucas *et al.*, 1983).

α2-Antiplasmin inhibits plasmin release but has no effect on fibrin-bound plasmin. Indeed, fibrin blocks formation of the inhibited complex of plasmin with the serpin (Wiman *et al.*, 1979). The C-terminal region of α2-antiplasmin is rich in Lys residues and can therefore bind to the lysine-binding sites present on the kringles of plasminogen, thereby preventing the association of plasminogen with fibrin. When cross-linked to fibrin, α2-antiplasmin retains its ability to inactivate plasmin thereby preventing early clot lysis. A glutamine residue at the N terminus of α2-antiplasmin is known to be covalently linked to Lys303 of the Aα chain of fibrinogen (Kimura and Aoki, 1986; Tamaki and Aoki, 1982).

12.2.10 Interaction of fibrin/fibrinogen with platelets

Activated platelets bind fibrinogen in a specific saturable manner (Marguerie *et al.*, 1979) and the interaction between fibrinogen and the integrin regulates platelet aggregation. The receptor responsible for this binding is the platelet membrane glycoprotein complex IIb/IIIa (GpIIb/IIIa). The major recognition site for this binding is within residues 400–411 (HHLGGAKQAGDV; *Figure 12.2*) of the γ chain (Plow *et al.*, 1984). Other regions of the Aα chain of fibrinogen are also involved in binding to GpIIb/IIIa (e.g. RGDF 95–98, RGDS 572–575). The process of clot retraction and the stability of the haemostatic plug depend on these interactions which physically link the platelet in and to the fibrin meshwork of the clot.

12.2.11 Interaction with thrombospondin

Fibrinogen forms a reversible non-covalent complex with thrombospondin, a glycoprotein released from the α granules of platelets. Thrombospondin binding appears to involve residues 113–126 of the Aα chain and residues 243–252 of the Bβ chain (Bacon-Raguley *et al.*, 1987, 1990; Walz *et al.*, 1987).

12.3 The fibrinogen (*FGA, FGB, FGG*) genes

12.3.1 Structure of the α-fibrinogen (FGA) gene

Several cDNA clones encoding human Aα fibrinogen have been isolated and characterized (Imam *et al.*, 1983; Kant *et al.*, 1983; Rixon *et al.*, 1983; Uzan *et al.*, 1984). The full-length cDNA includes 1875 bp encoding the 625-amino-acid residue mature protein plus between 48 and 57 bp DNA encoding a signal peptide of either 16 or 19 amino acids in length (methionines occur at both −16 and −19). There is thus a minimum of

54 bp 5′ non-coding sequence (Kant *et al.*, 1983). The cDNA clones isolated possess 3′ non-coding regions of either 217 bp or 234 bp. Two alternative polyadenylation sites occur preceded, between 17 and 22 bp upstream, by AATAAA motifs.

The α-fibrinogen cDNA codes for 15 amino acids at the C terminus which are not found in the circulating plasma protein and which are proteolytically removed during the maturation of the protein. This amino acid extension is not required for fibrinogen assembly or secretion (Farrell *et al.*, 1993).

At the amino acid sequence level, human α-fibrinogen coding region is approximately 65% homologous to its bovine counterpart as inferred from cDNA sequence data. These sequences are approximately 57% homologous at the nucleotide sequence level, a much lower figure than that found when the genes encoding Bβ or γ fibrinogens are compared between the two species.

The *FGA* gene is situated in the centre of the human fibrinogen gene cluster, 13.4 kb downstream of the γ-chain (*FGG*) gene and approximately 13 kb upstream of the β-chain (*FGB*) gene (Hu *et al.*, 1995; Kant *et al.*, 1985; *Figure 12.6*). A detailed genomic map of the human *FGA* gene is lacking although the human gene is known to consist of five exons, analogous to that exhibited by the rat gene (see below). An alternatively processed *FGA* mRNA contains a sixth exon whose inclusion results in the production of an extended α chain (αE) (Fu *et al.*, 1992). Since the extended α chain is assembled into fibrinogen and its synthesis is enhanced by interleukin-6 (IL-6), this form of α-fibrinogen may well participate both in the acute phase response and in fibrinogen metabolism.

The rat *FGA* gene has also been cloned (Crabtree and Kant, 1981) and sequenced (Crabtree *et al.*, 1985). It consists of five exons spanning 8 kb. The first intron occurs within the signal peptide coding region. The second occurs three amino acids upstream of the first of the two disulphide knots, from which the inter-chain disulphide bonds originate. The third and fourth introns occur in the relaxed midpoint of the coiled-coil and nine amino acids before the second disulphide knot at the end of the coiled-coil, respectively. The positions of the second and third introns are conserved for all three rat fibrinogen genes. The α chain of rat fibrinogen is 531 amino acids in length, shorter than the 625-amino-acid residue human protein. This is due

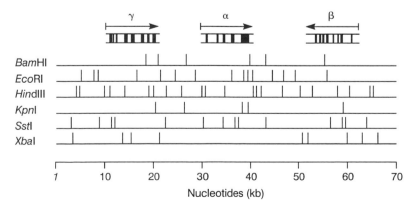

Figure 12.6. Restriction map of the human fibrinogen locus at chromosome 4q31. The arrows denote the direction of transcription. Derived from Kant *et al.* (1985) with permission from Dr G. Crabtree.

to a deletion of the repeated region seen in the human α-fibrinogen chain which comprises 10 repeats of 13 amino acids. Comparison of the remaining human and rat α-chain amino acid sequences show them to be 64% homologous. The cloning of the α-chain gene of lamprey fibrinogen has also been reported (Wang *et al.*, 1989; Pan and Doolittle, 1992): the C-terminal half of the protein is homologous to the C termini of the β and γ chains, suggesting that this gene could be a direct descendent of the primordial fibrinogen gene. Murakawa *et al.* (1993) have reported the evolutionary comparison of α-fibringens from five mammals.

The transcriptional initiation site (i) in the rat gene has been located 58 bp upstream of the Met initiation codon (Crabtree *et al.*, 1985). A TATA box (TTTAAG) is apparent 26 bp further upstream of i. In the human gene, i is located 55 bp upstream of Met (Hu *et al.*, 1995). The human *FGA* gene promoter contains six potential IL-6 responsive sequences but only one (CTGGGA; –122 to –127) is functional (Hu *et al.*, 1995). Other elements characterized include an hepatocyte nuclear factor 1 (HNF1) binding site between –47 and –59, a CCAAT/enhancer-binding protein site (C/EBP; –134 to –142) and positive (–1393 to –1133) and negative (–1133 to –749) regulatory elements (Hu *et al.*, 1995).

12.3.2 Structure of the FGB gene

cDNA clones for human β-fibrinogen have been isolated from human liver cDNA libraries (Chung *et al.*, 1983a; Kant *et al.*, 1983; Plaisancié *et al.*, 1984; Uzan *et al.*, 1984). The full-length cDNA thus comprises 1383 bp encoding the 461-amino-acid coding region followed by a 431 bp 3′ non-coding region. The mature protein is preceded by a signal peptide whose length would appear to be no more than 30 amino acid residues (methionine codons occur at –16 and –27).

One polyadenylation site lies 430 bp 3′ to the Stop codon (Chung *et al.*, 1983a). However, two-thirds of cDNA clones analysed possess rather shorter 3′ non-coding regions of 100 and approximately 166 bp. This 3′ end heterogeneity is consistent with the observation of two distinct mRNA species seen in Northern blotting experiments (see below). Huber *et al.* (1987) reported a cDNA clone with a 229 bp 3′ non-coding region.

The human *FGB* gene (Chung *et al.*, 1983a) comprises eight exons spaced out over a distance of about 8 kb (Chung *et al.*, 1983a; *Table 12.1*; *Figure 12.6*). Its transcriptional orientation is opposite to that exhibited by the *FGA* and *FGG* genes (Kant *et al.*, 1985).

Huber *et al.* (1987) sequenced 1500 bp DNA upstream of the putative initiation codon at Met30. Three different transcriptional initiation sites were located very close

Table 12.1. Exons in the human *FGB* gene

Exon	Amino acid residues	Protein domain
1	1–8	Signal peptide/portion of fibrinopeptide B
2	9–72	Fibrin polymerization/thrombin cleavage site
3	73–133	} Connecting region
4	134–209	
5	210–248	
6	249–289	
7	290–384	
8	385–461	

together, 2, 5 and 8 bp upstream of Met30. Putative TATA and CCAAT boxes are located 33 bp and 71 bp, respectively, upstream of the translational initiation codon. A high degree of homology was observed between the human and rat *FGB* genes in a 140 bp stretch upstream of the TATA box. CAT reporter gene and mobility shift assays have provided evidence for the presence of an IL-6-responsive element upstream of the human *FGB* gene (Huber *et al.*, 1990; Anderson *et al.*, 1993). Anderson *et al.* (1993) have also described a C/EBP binding site which is required for maximal induction of *FGB* gene transcription by IL-6. Glucocorticoid treatment is known to increase *FGB* gene expression in cultured human hepatocytes some five-fold (Mazzorana *et al.*, 1991) whilst IL-6 increases the rate of *FGB* gene transcription in cultured rat hepatocytes (Nesbitt and Fuller, 1991). Dalmon *et al.* (1993) reported that *in vitro* mutagenesis of the *FGB* IL-6-responsive element served to abolish IL-6 stimulation of transcription. Interestingly, the proximal HNF1 binding site was found to be necessary for IL-6 stimulation. This suggests not only a complex inter-play between the various proteins which bind to the IL-6-responsive region but may also indicate the existence of tight coupling between its tissue-specific and inducible elements.

β-Fibrinogen cDNA clones have been isolated from four other vertebrate species. A nearly full-length bovine *FGB* cDNA clone has been sequenced (Chung *et al.*, 1981). Comparison of the human and bovine β-fibrinogen sequences revealed more than 80% homology at the amino acid sequence level (Chung *et al.*, 1981) and 85% homology at the nucleic acid sequence level (Chung *et al.*, 1983a). β-Fibrinogen cDNA clones have also been isolated from rat (Crabtree and Kant, 1981; Eastman and Gilula, 1989), chicken (Weissbach *et al.*, 1991), *Xenopus* (Bhattacharya *et al.*, 1991; Roberts *et al.*, 1995) and lamprey (Bohonus *et al.*, 1986). The deduced amino acid sequences for lamprey and *Xenopus* β-fibrinogens are 49% and 66% homologous to that of human β-fibrinogen.

12.3.3 Structure of the FGG gene

cDNAs encoding γ-fibrinogen have been isolated from human liver cDNA libraries (Chung *et al.*, 1983b; Imam *et al.*, 1983; Kant *et al.*, 1983; Uzan *et al.*, 1984). The full-length cDNA encodes a 411-amino-acid mature protein, a 26-amino-acid signal peptide and includes 5′ and 3′ non-coding regions of 71 and 207 bp, respectively.

The human *FGG* gene is situated 13.4 kb upstream of the *FGA* gene and about 35 kb upstream of the *FGB* gene (Kant *et al.*, 1985). The human *FGG* gene comprises 10 exons spanning 10 kb genomic DNA (Fornace *et al.*, 1984a; Rixon *et al.*, 1985; *Figure 12.6*; *Table 12.2*). The transcriptional initiation site (i) has been mapped to 50 bp upstream of the Met initiation codon (Rixon *et al.*, 1985). This is only 8 bp 5′ to the corresponding site for i in the rat (Crabtree and Kant, 1982a; Fowlkes *et al.*, 1984). A TATA box (CATAAA) was located between 25 and 19 bp 5′ to i and a CCAAT box lies a further 30 bp upstream. Mizuguchi *et al.* (1995) have identified an upstream stimulatory factor (USF) between −66 and −77, a functional IL-6-responsive element between −301 and −306 and a negative regulatory element between −348 and −390.

The molecular basis for the two different forms of γ-fibrinogen (γA and γB) was demonstrated by analysis of the 3′ non-coding region (Fornace *et al.*, 1984a; Rixon *et al.*, 1985). The major form of human γ-fibrinogen, γA, is produced from an *FGG* mRNA spliced at nucleotide 7764 (codon 407) in exon IX and joined to exon X which

Table 12.2. Exons in the human *FGG* gene

Exon	Exon length (bp)	Amino acid residues	Protein domain
I	129	−26 to −1	Signal peptide
II	45	1–15	Disulphide bridge cysteines
III	184	16–76	Portion, coiled-coil, 1st disulphide ring
IV	94	77–108	Coiled-coil
V	131	109–151	Coiled-coil, 2nd disulphide ring
VI	134	152–196	D domain
VII	185	197–258	D domain
VIII	278	259–350	D domain
IX	170	351–407	} Cross-linking, fibrin polymerization,
X	219	408–411	} platelet receptor recognition

encodes amino acids 408–411. A 202 bp 3′ non-coding region includes an AATAAA motif 25 bp upstream of the polyadenylation site. The minor form of human γ-fibrinogen, γB, is generated by alternative processing and polyadenylation of the *FGG* gene transcript within intron I, resulting in the readthrough of the exon IX/intron I splice junction. A further 20 codons are thus included in exon IX in this transcript before the Stop codon is reached. Exon X is not used and polyadenylation occurs after 250 bp 3′ non-coding region and 17 bp downstream of an AATAAA motif. Alternative splicing and use of polyadenylation sites is also found in the rat *FGG* gene (Haidaris and Courtney, 1992).

A novel type of duplication has occurred within the human *FGG* gene; the majority of exon IX (158 bp) has been duplicated and inserted within intron H (Fornace *et al.*, 1984b; Rixon *et al.*, 1985). This sequence, termed a 'pseudoexon', lacks the splice recognition sites but is bounded on either side by almost perfect inverted repeat elements 102 bp long. It is thought that the duplication event occurred between 10 and 20 million years ago.

FGG cDNA clones have also been reported from the rat (Crabtree and Kant, 1981, 1982a) and from the lamprey (Strong *et al.*, 1985). These sequences show 83% and 50% homology with human γ-fibrinogen at the amino acid sequence level. Some homology is evident between human and rat γ-fibrinogen and mouse κ-casein at the amino acid sequence and at the nucleotide sequence levels (32%) (Thompson *et al.*, 1985). A bovine *FGG* cDNA has also been reported (Brown *et al.*, 1989).

The rat *FGG* gene exhibits an an identical exon/intron distribution to that exhibited by the human *FGG* gene but also a very similar alternative splicing mechanism at its 3′ end (Crabtree and Kant, 1982a; Crabtree *et al.*, 1985; Morgan *et al.*, 1987). Extensive inter-species homology within the highly conserved intron C is also evident (Rixon *et al.*, 1985).

12.3.4 DNA polymorphisms in the human fibrinogen genes

The sequencing of numerous cDNA and genomic clones for the human fibrinogen genes has yielded a considerable number of putatively neutral DNA sequence variants (Henschen, 1993; Tuddenham and Cooper, 1994); 80% of these variants are silent mutations in that they do not alter the amino acid being encoded. Some may be artefacts of cloning or sequencing but the authenticity of many is supported by amino acid sequencing data.

A number of restriction fragment length polymorphisms (RFLPs) have also been reported (reviewed by Tuddenham and Cooper, 1994). These are potentially useful in family studies for disorders involving either the absence of fibrinogen or the production of defective fibrinogen.

12.3.5 Chromosomal localization/organization of the fibrinogen genes

Using *in situ* hybridization and somatic cell hybrid analysis, the fibrinogen genes have been localized to 4q26–qter (Humphries *et al.*, 1984), to 4q23–q32 (Kant *et al.*, 1985), and even more precisely to 4q31 (Marino *et al.*, 1986). Close physical linkage of the fibrinogen genes is implied by the finding of linkage disequilibrium between RFLPs associated with the human *FGA*, *FGB* and *FGG* genes (Murray *et al.*, 1986; Humphries *et al.*, 1987). The relative extent of linkage disequilibrium favoured the order *FGG–FGA–FGB* on the chromosome. Both the close physical linkage and the gene order were confirmed by Kant *et al.* (1985); all three genes lie within 55 kb of each other. Whilst the *FGG* and *FGA* chain genes are transcribed in the same direction, the *FGB* gene is transcribed in the opposite direction.

In the rat, the three fibrinogen genes are also closely linked on chromosome 2q31–q34 (Kant and Crabtree, 1983; Marino *et al.*, 1986).

12.3.6 Regulation of fibrinogen biosynthesis

The liver is the main site of fibrinogen synthesis, with the hepatocyte identified by immunofluorescence as the cell type responsible. However, platelets contain a releasable pool of fibrinogen in their α granules and guinea-pig megakaryocytes synthesise fibrinogen (Leven et al., 1985). Regulation of fibrinogen levels at between 2 and 4 g l^{-1} is achieved by means of an indirect feedback loop involving the reticuloendothelial system. In humans, fibrinogen turnover takes between 3 and 4.5 days (MacFarlane et al., 1964). Up to a quarter of fibrinogen is extravascular but catabolic breakdown occurs intravascularly. When fibrinolysis is induced, the rate of synthesis can increase some 25-fold (Harker and Schlichter, 1972).

Factors affecting fibrinogen turnover. Fibrinogen is an acute phase protein whose rate of synthesis is markedly stimulated by inflammation or injury. Ritchie *et al.* (1982) demonstrated that fibrinogen degradation products (FDPs) were involved in a feedback loop that up-regulates not only fibrinogen but also other acute phase proteins. Fragments D and E from plasmin degradation of fibrin/fibrinogen act on Kupffer cells and macrophages, causing them to release IL-1 and -6 (hepatocyte-stimulating factor). The hepatocyte is thereby stimulated to increase synthesis of fibrinogen as well as other acute phase proteins.

Snake venoms, which produce disseminated fibrinogen breakdown, also elicit very rapid increase in fibrinogen mRNA production, along with a strong acute phase response (Crabtree and Kant, 1982b).

Regulation of fibrinogen gene expression. Some degree of coordinate control of expression of the three fibrinogen genes is apparent. This is in part an acute phase response (Green and Humphries, 1989). During this response, transcription of the three fibrinogen genes and the stability of their mRNAs are coordinately regulated (Nesbitt and Fuller, 1991); both IL-6 and phorbol esters are known to increase fibrinogen gene expression coordinately (Liu and Fuller, 1995). However, Roy *et al.* (1990) showed that overexpression of Bβ fibrinogen chains transfected into HepG2 cells enhanced the synthesis of all three fibrinogen chains. These authors have further demonstrated that transfection with any of the individual fibrinogen chain cDNAs specifically increased the synthesis of all three chains two- to four-fold (Roy *et al.*, 1994). This could be attributed directly to increases in the transcriptional activity of the three genes. Although the mechanism responsible is not yet understood, these experiments provide firm evidence for the coordinated expression of these genes.

A variety of experiments have now been carried out in order to characterize the DNA sequence elements upstream of the fibrinogen genes thought to control their expression. A combination of deletion mutagenesis chloramphenicol acetyltransferase (CAT) reporter gene and mobility shift assays and deoxyribonuclease I (DNase I) footprinting have demonstrated the presence of functional binding sites in the promoters of the *FGA* and *FGB* genes for the liver-specific transcription factor, HNF-1 (Courtois *et al.*, 1987, 1988; Kugler *et al.*, 1988). HNF-1 may play a role in determining the hepatic phenotype but it is probably only one of several elements necessary for hepatocyte-specific transcription of the fibrinogen genes.

CAT reporter gene and mobility shift assays have provided evidence for the presence of IL-6-responsive elements upstream of the human *FGA*, *FGB* and *FGG* genes (Huber *et al.*, 1990; Anderson *et al.*, 1993; Hu *et al.*, 1995; Mizuguchi *et al.*, 1995; Zhang *et al.*, 1995).

Transforming growth factor β induces a decrease in the basal level of *FGA*, *FGB* and *FGG* mRNAs and antagonizes the IL-6 induction of these mRNAs at late but not early times probably by a post-transcriptional mechanism (Hassan *et al.*, 1992). A body of evidence has now built up implicating the major late transcription factor (MLTF), originally described in adenovirus, in the expression of the rat *FGG* gene (Chodosh *et al.*, 1987; Morgan *et al.*, 1988). MLTF binding sites are present in the promoters of both the *FGG* and *FGB* genes. However, MLTF has been found in a wide variety of different cell types in which the *FGG* and *FGB* genes are not expressed. MLTF is thus unlikely to be the sole regulator of the coordinated transcription of the fibrinogen genes and probably acts as a constitutive transcription factor in the establishment of the basal level of expression of the *FGG* gene. Further promoter elements, capable of binding Sp1 and the CCAAT binding protein, CB2, have been characterized in the *FGG* gene promoter (Chodosh *et al.*, 1988; Morgan *et al.*, 1988)

12.3.7 In vitro *expression*

The human *FGA* gene has been expressed in *Escherichia coli* (Lord, 1985) as have the human *FGB* and *FGG* genes (Bolyard and Lord, 1988; 1989). The assembly and secretion of biologically active recombinant human fibrinogen has been achieved by co-transfection of COS cells (Roy *et al.*, 1991) and yeast (Roy *et al.*, 1995) with three expression vectors, one for each chain.

12.4 Evolution of the fibrinogen proteins

Sequence data on fibrinopeptides and on fibrinogen are available from a large number of vertebrate species. Using these data, Doolittle (1980) estimated the age of the fibrinogen gene and the time at which a primordial gene duplicated, at about the time of the emergence of plants and animals 1.5 billion years ago. Data from phylogeny and from the comparison of present day sequences, indicate that the primordial gene diverged into the precursor of the present *FGA* gene and into a precursor of *FGB* and *FGG* genes. The further divergence of this pre-*FGB/FGG* gene to *FGB* and *FGG* genes coincides with the emergence of the vertebrates about 500 million years ago. Fibrinogen is made up of three chains in all extant vertebrates, including the lamprey whose last fossil ancestor diverged from ours about 450 million years ago.

A feature that often remains constant in different members of a gene family is intron position. Only two intervening sequences are shared at homologous positions between the three fibrinogen genes, but an additional intron position is shared between *FGB* and *FGG* genes, confirming their closer evolutionary relationship. With the discovery of the extended α chain (Fu *et al.* 1992; Section 12.3.1), it has become clear that the C-terminal extension of αE shares the same degree of similarity with the β and γ chains as these chains do with each other, and that all three chains share a common ancestry.

12.5 The molecular genetics of fibrinogen disorders predisposing to venous thrombosis

12.5.1 Dysfibrinogenaemias causing thrombosis

Some 1% of patients presenting with venous thrombosis are reported to have an iden-tifiable abnormality of fibrinogen (Gladson *et al.*, 1988; Haverkate and Samama, 1995). Fibrinogen variants are detected on the basis of either a prolonged thrombin time, a prolonged Reptilase™ time or a lower chronometrically determined fibrino-gen level than the immunologically determined value. Ebert (1991a, b) has indexed a total of 242 families exhibiting variant fibrinogens that have been reported in the lit-erature. The majority are clinically asymptomatic, 28% are associated with a bleeding tendency, 18% with a thrombotic tendency and 2% with both. Some 55% exhibited no clinical symptoms, having been picked up by routine screening. Most individuals concerned were heterozygous for a gene defect but 5% were homozygous. Seventy-one per cent of variants resulted in abnormal polymerization, 37% in abnormal fib-rinopeptide release, 6% in abnormal stabilization and 30% a combination of these. Most of the characterized fibrinogen variants are associated with a bleeding tendency.

Individuals heterozygous for a dysfunctional fibrinogen molecule probably possess both homodimeric and heterodimeric fibrinogen dimers since, at least in the case of two Aα16 variants, both Aα chain alleles were expressed and the resulting fibrinogen dimers assembled randomly (Meh *et al.*, 1996).

The various dysfibrinogen variants associated with thrombosis (reviewed in Bertina, 1988; Galanakis, 1993; Henschen, 1993; Koopman and Haverkate, 1994; McDonagh *et al.*, 1994; Tuddenham and Cooper, 1994; Haverkate and Samama, 1995) may be classi-fied by reference to the type of functional defect exhibited. Fibrinogen Paris V (Dusard) exhibits decreased binding of plasminogen but, although tPA binding to fibrin is nor-mal, fibrin-mediated enhancement of plasminogen activation is greatly increased. Various other fibrinogen variants associated with familial/recurrent thrombosis have been reported. They include fibrinogens Baltimore I (defective fibrinopeptide release), Marburg (disordered monomer polymerization), Copenhagen (disordered peptide release and monomer polymerization), Charlottesville (disordered peptide release), Haifa (defective aggregation) and Naples (defective thrombin binding). Gene lesions that give rise to defective fibrinogen polymerization leading to denial of access to either plasminogen/plasmin or tPA and resulting in the inhibition of fibrinolysis, are likely to lead to an increased likelihood of recurrent thrombosis.

Fibrinogen New York I does not bind either thrombin or plasminogen properly and exhibits defective plasminogen activation by tPA (Liu *et al.*, 1985). This phenotype results from the absence of exon 2 from the *FGG* mRNA transcript, leading to the loss of amino acids 9–72 from the protein. However, it is unclear whether this is due to a gene deletion or to a splicing defect.

Haverkate and Samama (1995) collated 51 cases of dysfibrinogenaemia with thrombophilia and studied 27 cases which fulfilled certain strict criteria for inclusion. Propositi exhibited thrombosis of the lower limbs, sometimes with thrombophlebitis or pulmonary embolism. The mean age of onset of thrombotic symptoms for these individuals was 27 years (range 12–50). Venous thrombosis (24 cases) predominated over arterial thrombosis (three cases arterial thrombosis only, two cases venous and arterial thrombosis). Severe bleeding was also noted in two propositi post-partum whilst mild bleeding occurred in five cases. Relatives of the propositi were also investigated: 20/99 individuals with dysfibrinogenaemia had experienced thrombosis whereas no cases of thrombotic disease were noted among 88 relatives without dysfibrinogenaemia. Seven out of 15 female propositi had suffered from thrombosis *post partum* and the number of spontaneous abortions was reportedly rather high.

12.5.2 Molecular biological studies

An increasing number of molecular genetic studies have been performed on the dysfibrinogenaemias causing thrombotic disease. One example is fibrinogen γ Vlissingen, a variant which causes abnormal monomer polymerization and which manifests a defective interaction with calcium (Koopman *et al.*, 1991). A 6 bp deletion encompassing codons Asn319 and Asp320 of the *FGG* gene is responsible, thus implicating this site in the fibrin polymerization process (Koopman *et al.*, 1991). The same mutation has been found in a relative of this patient (fibrinogen Frankfurt IV; Haverkate and Samama, 1995).

Of the characterized dysfibrinogens causing thrombophilia, the Aα chain variants Chapel Hill II (Carrell and McDonagh, 1982) and New Albany (Henschen *et al.*, 1981b) both result from heterozygous Arg16→His (CGT→CAT) substitutions which prevent thrombin cleavage and subsequent release of FpA (see Section 12.5.3). Another variant causing delayed FpA release is fibrinogen γ Baltimore I caused by a Gly292→Val (GGC→GTC) substitution (Bantia *et al.*, 1990). Fibrinogen Aα Aarhus, associated with delayed polymerization, is caused by an AGG→GGG transition converting Arg19 to Gly (Blombäck *et al.*, 1988). The abnormal polymerization associated with fibrinogens γ Bergamo 2 and Barcelona III results from a CGC→CAC transition converting Arg275 to His (Reber *et al.*, 1986; Borrell *et al.*, 1995).

The mutation responsible for fibrinogen Naples is a [C]GCT→[C]ACT transition which converts Ala68 to Thr in the fibrinogen Bβ chain (Koopman *et al.*, 1992a). This type of dysfibrinogenaemia is clinically recessive since thrombotic manifestations only occur in homozygous family members. The mutation prevents thrombin binding to fibrin and illustrates the importance of this binding *in vivo* as a means to limit the presence of free thrombin in the circulation (see Section 12.5.3, 'Variants causing defective fibrinopeptide release').

Fibrinogen Marburg, another homozygous dysfibrinogenaemia, is caused by an AAA→TAA (Lys→Term) substitution at amino acid residue 461 leading to loss of 149

C-terminal amino acids from the Aα chain (Koopman *et al.*, 1992b). Although FpA release and fibrin stabilization were not affected, fibrin polymerization was dramatically decreased and binding to endothelial cells virtually abolished (Koopman *et al.*, 1992b; Sobel *et al.*, 1995). The C terminus of the Aα chain of fibrinogen is known to be important in stimulating tPA-induced plasminogen activation and its loss may be responsible for precipitating the thrombotic events.

Fibrinogens IJmuiden and Nijmegen are both heterozygous Bβ chain variants caused by Arg14→Cys and Arg44→Cys substitutions respectively (Koopman *et al.*, 1992c). Both variants exhibit defective fibrin polymerization and some of the abnormal molecules were bound to various proteins including albumin. It is thought that this may be due to the additional cysteine residues created forming disulphide-linked complexes with other proteins (see Section 12.5.3, 'Variants exhibiting abnormal disulphide linkages'). Another example of this phenomenon is fibrinogen Aa Paris V (Dusard) in which a CGT→TGT transition converts Arg554 to Cys (Koopman *et al.*, 1993). A substantial portion of fibrinogen Paris V (Dusard) molecules were found to be disulphide-linked to albumin.

Fibrinogen Milano III is associated with impaired fibrin polymerization and was caused by the insertion of a T after the ATT codon encoding Ile451 of the Aα chain which leads to a premature termination at residue 453 (Furlan *et al.*, 1994). The proposita, who suffered recurrent venous thrombosis, was homozygous for this lesion. Immunoblotting revealed the presence of albumin covalently linked to fibrinogen; this is presumably made possible by the loss of Cys472 which leaves Cys442 free to form a bridge with the free cysteine of albumin.

12.5.3 Structure–function relationships in variant fibrinogens causing thrombotic disease

Most patients with dysfibrinogenaemia exhibit a prolonged thrombin clotting time and usually delayed clotting induced by snake venoms such as Reptilase™. (One exception to this rule is fibrinogen Naples which exhibits a normal Reptilase™ clotting time.) All variants so far detected are characterized by delayed fibrinopeptide release and/or impaired fibrin polymerization which account for the prolonged thrombin clotting time. Many variants also manifest defective binding of calcium, thrombin, tPA or plasminogen and may not interact properly with platelets or endothelial cells. The majority of known variants have been detected by protein sequence analysis which accounts for the great preponderance of dysfunctional variants known. DNA sequence analysis is now increasingly used; it is much quicker and easier and allows a heterozygous lesion to be readily distinguished from its homozygous counterpart.

One difficulty in interpreting the nature of fibrinogen variants is to know whether the functional abnormality is a primary or a secondary consequence of the original defect. For example, the primary consequence of an Aα Arg16→His substitution is delayed release of FpA but secondary consequences include impaired fibrin formation and defective binding to thrombin and plasmin. Known variants of fibrinogen which are associated with thrombotic disease (summarized in *Table 12.3*) will now be described according to their functional consequences for fibrinogen.

Variants causing defective fibrinopeptide release. Thrombin cleavage of the Aα chain at Arg16–Gly17 yields FpA, thrombin interacting with the region between

Table 12.3. Inherited variants of human fibrinogen associated with thrombotic disease

Name	Structural defect	Functional defect	Reference
Hershey II[a]	Aα Arg16→Cys	FpA release	Galanakis et al. (1990)
Frankfurt V	Aα Arg16→Cys	?	Haverkate and Samama (1995)
Chapel Hill II[b]	Aα Arg16→His	FpA release	Carrell and McDonagh (1982)
New Albany	Aα Arg16→His	FpA release	Henschen et al. (1981b)
Aarhus[c]	Aα Arg19→Gly	FpA release, fibrin polymerization	Stenbjerg et al. (1983), Hessel et al. (1986), Blombäck et al. (1988)
Milano III[c]	Aα ins T in Ile451	Fibrin polymerization	Furlan et al. (1994)
Marburg[b,c]	Aα Lys461→Term	Fibrin polymerization, binding to endothelial cells	Fuchs et al. (1977), Koopman et al. (1992b)
Caracas V[d]	Aα Ser532→Cys	?	Haverkate and Samama (1995)
Paris V (Dusard, Dusart)	Aα Arg554→Cys	Fibrin polymerization, plasminogen binding	Soria et al. (1983b), Lijnen et al. (1984b), Koopman et al. (1993), Siebenlist et al. (1993)
Chapel Hill III	Aα Arg554→Cys	Fibrin polymerization, fibrinolysis	Carrell et al. (1983), Wada and Lord (1994)
New York I	Bβ (9–72) deletion	FpA and FpB release, fibrin polymerization, thrombin binding, tPA binding, plasminogen activation	Al-Mondhiry et al. (1975), Liu et al. (1985; 1986)
IJmuiden[d]	Bβ Arg14→Cys	FpB release, fibrin polymerization	Koopman et al. (1992c)
Nijmegen	Bβ Arg44→Cys	Fibrin polymerization, tPA binding, plasminogen activation	Engesser et al. (1988b), Koopman et al. (1992c)
Naples (Milano II)[c]	Bβ Ala68→Thr	FpA and FpB release, thrombin binding	Quattrone et al. (1979), Haverkate et al. (1986), Koopman et al. (1992a)
Bologna	γ Arg275→Cys	Abnormal D:D self-association site	Haverkate and Samama (1995), Mosesson et al. (1995b)
Bergamo II	γ Arg275→His	Fibrin polymerization	Reber et al. (1986)
Barcelona III	γ Arg275→His	Fibrin polymerization	Borrell et al. (1995)
Haifa	γ Arg275→His	Fibrin polymerization, Ca²⁺ effect on plasmin degradation	Soria et al. (1987), Kehl et al. (1984), Siebenlist et al. (1989)
Baltimore I[b]	γ Gly292→Val	Fibrin polymerization	Mosesson and Beck (1969), Beck et al. (1971), Brown and Crowe (1975), Bantia et al. (1990)
Bicêtre II	γ Asn308→Lys	?	Grailhe et al. (1993)
Giessen IV	γ Asp318→Gly	?	Haverkate and Samama (1995)
Melun	γ Asp364→Val	?	Haverkate and Samama (1995)
Vlissingen, Frankfurt IV[a]	γ (319–320) deletion	Fibrin polymerization, Ca²⁺ binding	Koopman et al. (1991), Haverkate and Samama (1995)

[a]Arterial thrombosis only; [b]bleeding symptoms as well as thrombosis; [c]homozygous; [d]venous and arterial thrombosis.

Asp7 and Arg16 (see Section 12.2.5). Three of the mutations causing thrombosis reside in the Aα Arg16 residue. Two (fibrinogens New Albany and Chapel Hill II) involve the substitution of Arg16 by His. Thrombin cleavage of the Aα His16–Gly17 bond proceeds much more slowly than the normal bond (Higgins *et al.*, 1983; Southan *et al.*, 1983, 1985a; De Cristofaro *et al.*, 1994). The His16 variant exhibits impaired fibrin formation and defective interactions with thrombin, factor XIIIa and plasmin. However, these functional defects are secondary to the delayed release of FpA.

Fibrinogen Hershey II (Aα Arg16→Cys) also results in defective FpA release; indeed, the Aα Cys16–Gly17 bond is not cleaved by thrombin at all (Matsuda *et al.*, 1986). The observation that some examples of this variant are associated with reduced FpB release has been held to imply that, at least at low thrombin concentrations, FpB release depends upon FpA release (Koopman and Haverkate, 1994). Fibrinogen Aarhus (Aα Arg19→Gly) is also characterized by defective FpA release (Blombäck *et al.*, 1988) consistent with Arg19 being a critical residue in the substrate binding site for thrombin.

Thrombin cleavage of the Bβ chain at Arg14–Gly15 yields FpB. Fibrinogen IJmuiden (Bβ Arg14→Cys) exhibits normal FpA release but reduced FpB release by thrombin.

Fibrinogen Naples (Bβ Ala68→Thr) exhibits a reduced release of FpA and FpB by thrombin but normal FpA release by Reptilase™. This indicates that the FpA cleavage site must be intact and that decreased FpA release could be due to a defective interaction between the variant fibrinogen and the fibrinogen binding exosite of thrombin (see following section 'Variants with defective interaction with thrombin or tPa'.).

Variants which impair fibrin polymerization. The polymerization of fibrin monomers occurs via the interaction of polymerization sites on different molecules (Section 12.2.5). The polymerization sites in the N-terminal portion of the molecule ('A' and 'B') are exposed by thrombin cleavage of FpA and FpB. Complementary polymerization sites 'a' and 'b' are located in the C-terminal region of the fibrinogen molecule.

Both fibrinogen IJmuiden (Bβ Arg14→Cys) and fibrinogen Nijmegen (Bβ Arg44→Cys) manifested a delayed clotting time with Reptilase™ and impaired fibrin polymerization after FpA removal by Reptilase™. This could be regarded as evidence for Arg14 and Arg44 contributing to the A polymerization site. However, the creation of novel cysteine residues suggests that abnormal disulphide linkages may be involved (see following section 'Variants exhibiting abnormal disulphide linkages').

Fibrinogen Aarhus (Aα Arg19→Gly) exhibited defective fibrin polymerization as well as defective FpA release (Blombäck *et al.*, 1988). This is consistent with Arg19 of the Aα chain being part of the N-terminal polymerization site, exposed after FpA removal. Fibrinogen Milano III (frameshift mutation leading to loss of 156 amino acids at the C-terminal end of the Aα chain) also results in defective fibrin polymerization (Furlan *et al.*, 1994).

Fibrinogens Bergamo II and Haifa (γ Arg275→His) exhibited defective fibrin polymerization after removal of FpA by Reptilase™ or FpA and FpB by thrombin. Fibrinogen Baltimore I (γ Gly292→Val) exhibited normal release of FpA and FpB but impaired fibrin polymerization (Bantia *et al.*, 1990).

Fibrinogen Vlissingen (γ 319–320 deletion) exhibits impaired fibrin polymerization but this impairment was lower or absent at high calcium concentrations which are known to stimulate fibrin polymerization. The loss of the high-affinity calcium binding site in the D domain may account for this impairment but defective fibrin polymerization in the absence of calcium suggests that the variant also affects calcium-independent polymerization (see the following section 'Variants manifesting a defective interaction with calcium).

Further, hitherto uncharacterized fibrinogen variants associated with defective fibrin polymerization and thrombotic disease include those reported by Hansen *et al.* (1980), Carrell *et al.* (1983), Ieko *et al.* (1991) and Hessel *et al.* (1995).

Variants with defective interaction with thrombin or tPA. The extended fibrinogen recognition site of thrombin appears to interact with residues 27–50 of the Aα-chain and residues 15–42 of the Bβ chain (reviewed by Binnie and Lord, 1993). Fibrinogen New York I (Bβ 9–72 deletion) is characterized by thrombin binding that is half that of normal (Liu *et al.*, 1985). It also manifests reduced tPA binding and a decreased stimulatory effect on tPA-induced plasminogen activation. This mutation clearly has multiple effects on fibrin(ogen) structure and function which are difficult or impossible to disentangle. One effect of the deletion is to remove residue Bβ Cys65 which normally forms a disulphide linkage to Aα Cys36.

Thrombin binding is perturbed in fibrinogen Milano IV (DeCristofaro *et al.*, 1994), a variant identical to fibrinogens Chapel Hill II and New Albany. De Cristofaro *et al.* 1994) postulated that this might be due to the salt bridge formed between Aα Arg16 and the Asp189 residue of thrombin (Section 12.2.5).

Thrombin binding is also impaired in the more subtle variant, fibrinogen Naples (Bβ Ala68→Thr); binding of active site-inhibited thrombin to fibrin generated by Reptilase™ from fibrinogen Naples was reduced to less than 10% of normal. Koopman and Haverkate (1994) have suggested that either the Bβ Ala68 residue participates in a non-polar interaction with thrombin and that this is disrupted by Thr or that the larger Thr side chain could lead to incorrect folding. Either way, it would appear that the N-terminal domain of fibrin(ogen) is required for a normal interaction with thrombin.

Fibrin normally binds tPA and plasminogen; indeed, the activation of plasminogen by tPA is greatly enhanced in the presence of fibrin (Section 12.2.9). Fibrinogen Nijmegen (Bβ Arg44→Cys) also exhibits reduced kringle 2-mediated tPA binding and a decreased stimulatory effect on tPA-induced plasminogen activation (Engesser *et al.*, 1988b). Reduced binding of plasminogen resulting in decreased stimulation of tPA-induced plasminogen activation was also found in fibrinogen Paris V (Dusard; Aα Arg554→Cys; Soria *et al.*, 1983b; Lijnen *et al.*, 1984b). The fibrin fibres formed with fibrinogen Paris V (Dusard) are much thinner than those found in normal fibrin (Koopman *et al.*, 1993), consistent with a defect which impairs the lateral association of fibres. This region of the Aα chain may contain the 'b' polymerization site and therefore could play an important role in lateral fibre association (Hasegawa and Sasaki, 1990)

Fibrinogen IJmuiden (Bβ Arg14→Cys) also exhibits decreased tPA-induced plasminogen activation. These functional abnormalities may be directly caused by the amino acid substitution or arise indirectly as a result of the impairment of fibrin polymerization. The latter explanation is consistent with fibrin polymerization reducing the stimulatory effect of fibrin on plasminogen activation (Suenson and Petersen, 1986).

The mutation responsible for fibrinogen Bicêtre II (γ Asn308→Lys) apparently does not affect the binding of tPA or plasminogen to fibrin (Grailhe *et al.*, 1993) despite the proximity of the mutation to the tPA binding site localized to residues 311–379 on the γ chain (Yonegawa *et al.*, 1992). The patient who had presented with thrombophlebitis and pulmonary embolism, was originally diagnosed on the basis of prolonged thrombin and Reptilase™ times (Grailhe *et al.*, 1993).

Variants manifesting a defective interaction with calcium. Fibrinogen Vlissingen (γ 319–320 deletion) exhibits a reduced number of high-affinity calcium-binding sites (Koopman *et al.*, 1991). The protective effect of calcium on the plasmin degradation of the C-terminal region of the γ chain was only partially present in this fibrinogen variant. These data indicate that the two-amino-acid deletion alters the high-affinity binding site in such a way as to decrease the protection of the γ 302–303 peptide bond against plasmin degradation. Fibrinogen Haifa (γ Arg275→Cys) also exhibits a reduced protective effect of calcium against plasmin degradation but no decrease in the number of high-affinity calcium binding sites was evident (Soria *et al.*, 1987). Consistent with this finding, Furlan *et al.* (1996) found normal calcium binding to five abnormal fibrinogens with mutations in the C-terminal portion of the γ chain.

Variants manifesting a defective interaction with endothelial cells. The binding of immobilized fibrinogen Marburg (Aα Lys461→Term) was almost completely abolished compared with wild-type fibrinogen. This variant clearly lacks amino acids 461–610 of the Aα chain. This region contains the RGD motif at residues 572–574 which is known to be essential for binding to endothelial cells (Cheresh *et al.*, 1989)

Variants exhibiting abnormal disulphide linkages. Several variant fibrinogens introduce a novel cysteine residue into the chain [e.g. IJmuiden, Nijmegen (Bβ), Hershey II, Paris V (Dusard), Chapel Hill III (Aα)]. These cysteines can form novel disulphide linkages analogous to those first reported for antithrombin III Northwick Park (Erdjument *et al.*, 1987). Indeed, 20% and 13% respectively of total plasma fibrinogen in heterozygous patients IJmuiden and Nijmegen was found to be disulphide-linked to albumin via the latter's free sulphydryl group. Disulphide-linked fibrinogen–albumin complexes were also noted in the case of fibrinogen Marburg (Aα Lys461→Term) presumably because Aα Cys472, which is normally linked to Aα Cys442, is absent. Fibrinogen Milano III (insertion of a T after the ATT codon encoding Ile451 of the Aα chain leading to premature termination at residue 453; Furlan *et al.*, 1994) also lacks Cys472, which leaves Cys442 free to form a bridge with the free cysteine of albumin. Finally, fibrinogen–albumin complexes have been found in both fibrinogens Paris V (Dusard) and Chapel Hill III (both Aα 554 Arg→Cys; Haverkate and Samama, 1995; Wada and Lord, 1994).

Other high-molecular-weight complexes, probably disulphide-linked fibrinogen dimers, were also noted in the patients with fibrinogens IJmuiden and Nijmegen at between 5 and 7% total plasma fibrinogen. Fibrinogen dimers may also be possible with the Aα Arg16→Cys variant (Hershey II) since these have been reported in at least one other case of such a variant (Miyata *et al.*, 1987). These dimers arise as a result of the formation of an extra inter-chain disulphide bond between two mutant Aα chains.

Characteristics of fibrinogen variants associated with thrombotic disease. As noted above (Section 12.5.1), some 20% of variant fibrinogens are associated with thrombophilia. According to Koopman and Haverkate (1994), thrombophilia occurs most often in fibrinogens with a defect in the C-terminal portion of the Aα and γ chains as well as the N-terminal portion of the Bβ chain. There are, however, exceptions to this emerging rule (fibrinogens Hershey II, Chapel Hill II, New Albany, Aarhus) and it remains to be seen whether the rule stands the test of time.

Fibrinogens with defects in the C-terminal portions of the Aα and γ chains may exhibit thrombotic manifestations as a result of defective fibrinolysis due to the decreased binding of plasminogen and the reduced stimulatory effect of these fibrin variants on tPA-induced plasminogen activation (Koopman *et al.*, 1986; Suenson and Petersen, 1986). Indeed, the C-terminal region of the γ chain is known to contain a site which stimulates plasminogen activation by tPA (Yonegawa *et al.*, 1992). This explanation may also apply to the N-terminal Bβ variants New York I and Nijmegen.

Two of the known Aα 554 Arg→Cys mutations, Paris V (Dusard) and Chapel Hill III, are associated with thrombotic disease (Wada and Lord, 1994). Studies of clot structure associated with fibrinogen Paris V (Dusard) have shown that the clot is more rigid than normal and less accessible to fibrinolytic enzymes (Collet *et al.*, 1993). This not only impairs fibrinolysis but also increases the likelihood that a clot will break, resulting in a high incidence of thromboembolism. Mosesson *et al.* (1996) have shown that Paris V (Dusard) fibrinogen, in the presence of factor XIIIa, exhibits a γ-chain cross-linking rate twice that of normal; the Aα 554 Arg→Cys substitution leads to increased γXL self-association and γ-chain cross-linking, one cause of the observed thrombophilic phenotype.

Defective thrombin binding to fibrin may also lead to thrombosis; the Bβ N-terminal variants New York I and Naples exhibit this characteristic and may result in an excess of thrombin in the circulation. Fibrinogens Malmö (Soria *et al.*, 1985) and Pamplona II (Fernandez *et al.*, 1986b), whose underlying defects remain to be determined, also exhibit abnormal thrombin binding and result in thrombotic disease.

12.5.4 Plasma fibrinogen levels and their relationship with thrombosis.

Elevated plasma fibrinogen levels are known to correlate strongly with various indicators of thrombin activation (Ceriello *et al.*, 1994) and represent an important risk factor for arterial thrombosis. Many RFLP association studies have been used to investigate the postulated association between plasma fibrinogen levels and the incidence of ischaemic heart disease (IHD; reviewed by Ernst *et al.*, 1994; Humphries *et al.*, 1994). Some genotypes exhibit higher mean plasma fibrinogen levels than others and the frequencies differ between racial groups (Iso *et al.*, 1995). This finding has been interpreted in terms of linkage disequilibrium (Baumann and Henschen, 1994) and may imply the presence in the vicinity of the fibrinogen genes of DNA sequences, perhaps regulatory elements, which play a role in determining plasma fibrinogen concentrations.

Elevated plasma fibrinogen levels have also been found to represent a risk factor for venous thrombosis: Koster *et al.* (1994b) demonstrated that a plasma fibrinogen greater than 5 g l^{-1} was associated with a four-fold increase in thrombotic risk. A deficiency of an H1H2 genotype for a *Hae*III polymorphism was noted in patients with deep vein thrombosis (Koster *et al.*, 1994). Possession of an H1H2 genotype was reportedly associated with a 40% reduction in thrombotic risk.

Defects of the fibrinolytic pathway associated with thrombotic disease

13.1 Introduction

The fibrinolytic system, like the other enzyme cascades of haemostasis, is virtually inactive in the resting state but is rapidly triggered to full activity by the event it exists to control, namely fibrin clot formation. Thrombi contain significant amounts of fibrin and are protected from fibrinolysis by incorporation of inhibitors [e.g. α2-antiplasmin, plasminogen activator inhibitor-1 (PAI-1)]. Defective fibrinolysis may play a role in both thrombus formation and persistence. The molecular biology of fibrinolysis has been well reviewed (Gerard *et al.*, 1986; Bachman, 1987; Collen and Lijnen, 1987). Here we briefly review the molecular biology and molecular genetics of known and suspected defects of the fibrinolytic pathway which are associated with thrombotic disease.

The terminal enzyme of the fibrinolytic system, plasmin, circulates as an inactive zymogen (plasminogen; see Chapter 8) that is activated by two important physiological activators: tissue-type plasminogen activator (tPA) and urokinase-type plasminogen activator (uPA). tPA is a serine proteinase with several unusual features that uniquely equip it for its role as primary fibrinolytic activator. Thus its main site of synthesis (vascular endothelium), its possession of significant activity in the single-chain form (usually an inactive zymogen) and the marked stimulation of its activity by fibrin are properties which ensure that tPA is perfectly designed for initiating clot lysis. The structural basis for its selective properties has now begun to be unravelled largely by molecular genetic methods. The prime stimulus to these studies has come from the development and use of recombinant tPA as a clot-selective fibrinolytic therapy, potentially applicable to all forms of thromboembolism. Striking success has been achieved in the management of coronary thrombosis by tPA infusion, if given at an early stage after the event (de Bono, 1987; Sobel *et al.*, 1987).

Defective activation of plasminogen has long been held to be an aetiological factor in thromboembolism, based on the finding of delayed *in vitro* clot lysis in a high proportion of patients. In most cases, the defect in clot dissolution appears to be an acquired characteristic mediated by or associated with other adverse prothrombotic factors such as age, obesity, smoking, hypertriglyceridaemia, surgery, malignancy, trauma or physical inactivity. The balance of fibrinolytic potential (an inferred property of blood based upon clot lysis observed *in vitro*) depends critically on the relative

Venous Thrombosis: from Genes to Clinical Medicine, D.N. Cooper and M. Krawczak.
© 1997 BIOS Scientific Publishers, Oxford.

amounts of tPA and of its main inhibitor, PAI-1 in blood (reviewed by Wiman and Hamsten, 1990). Some individuals with delayed clot lysis *in vitro* have a low resting level of tPA and/or poor release of tPA upon stimulation by various agonists such as venous cuffing, desamino-D-arginylvasopressin (DDAVP), exercise, etc. A number of reports of patients with defective fibrinolysis and thromboembolism have appeared (e.g. Harbourne *et al.*, 1991; Grimaudo *et al.*, 1992). A few kindreds have been reported in which defective fibrinolysis associated with venous thromboembolism is transmitted in an autosomal dominant manner.

Although no inherited disease caused by a mutation in the gene encoding tPA has yet been reported, it seems highly likely that mutations adversely affecting either tPA synthesis or the regulation of its release could cause a thrombotic phenotype. The physiology, structure/function and molecular genetics of tPA are therefore reviewed (Section 13.2) in anticipation of the discovery of its associated genetic disorders.

PAI-1 specifically and rapidly inactivates tPA (Kruithof *et al.*, 1984; Sprengers and Kluft, 1987; Andreasen *et al.*, 1990; Loskutoff *et al.*, 1988). Inactivation coincides with the formation of a tPA–PAI complex of M_r 110 kDa. PAI-1 has been purified to homogeneity (van Mourik *et al.*, 1984; Wagner and Binder, 1986) and shown to be distinct from a second plasminogen activator inhibitor, PAI-2, derived from placenta. PAI-1 belongs to the serpin superfamily of proteinase inhibitors and is 34% homologous to antithrombin III (ATIII). Its possible clinical importance lies in the fact that a high proportion of individuals with thrombotic disease exhibit elevated levels of PAI-1. Whether this has a genetic basis or not remains to be elucidated. The molecular biology and genetics of PAI-1 are reviewed in Section 13.3.

13.2 Tissue-type plasminogen activator (tPA)

13.2.1 Physiology, structure and function

tPA is synthesised and secreted by vascular endothelial cells as a single-chain glycoprotein of molecular weight 70 kDa. The plasma concentration of tPA varies in response to stress and a range of physiological and pharmacological stimulants (Prowse and Cash, 1984). Stored tPA is rapidly released into the circulation upon infusion of adrenalin, nicotinic acid or DDAVP. Exercise and venous occlusion also cause short-term release of tPA but the precise mechanism of action of these various agonists is presently unclear.

tPA is rapidly cleared from plasma ($T_{1/2}$ = 3.5 min) mainly by the liver by both PAI-1-dependent and -independent mechanisms (Camani *et al.*, 1994) both of which involve internalization mediated by the low-density lipoprotein receptor-related protein/α2-macroglobulin receptor (Orth *et al.*, 1994).

tPA binds specifically and with high affinity to fibrin (K_m = 0.16 μM) (Rånby, 1982; Rijken *et al.*, 1982) at a molar ratio of 0.88 moles tPA per mole fibrin. At the same time, tPA demonstrates greatly increased activity towards plasminogen without much change in the catalytic rate constants for the enzyme (K_{cat} increased from 0.06 to 0.1 sec⁻¹). The increased activity is explained by increased affinity of fibrin-bound tPA for its substrate plasminogen (K_m decreased from 65 to 0.16 μM; Hoylaerts *et al.*, 1982).

tPA is synthesised with a 35-amino-acid pre-pro-leader sequence followed by the 527 residues of the mature protein; 35 cysteine residues form 17 disulphide bridges with one free cysteine (residue 83) in the C-terminal loop of the growth factor-like domain. Complex oligosaccharide groups are added to Asn residues. Potential N-linked glycosylation sites occur at residues Asn 117, 184 and 448 whilst a single fucose group is attached to Thr61. The pre-pro-sequences are removed by endopeptidase cleavage.

Sequence homology and NMR studies show that tPA has a domain structure consisting of a 'finger' region (residues 4–50 of the mature protein), as found in fibronectin, followed by an epidermal growth factor (EGF)-like domain (residues 51–87) with three disulphide bonds typical of such domains. This is followed by two kringle domains (residues 88–175 and 176–263) one of which (kringle 2) has been analysed structurally by nuclear magnetic resonance spectroscopy (Byean et al., 1991). The kringles are connected to the serine proteinase domain (residues 276–527) which contains the catalytic site residues, His322, Asp371 and Ser478. The serine proteinase domain is disulphide-linked to kringle 2 (Cys264–Cys395) so that the heavy chain remains linked to the light chain after plasmin (or trypsin) cleavage at Arg275–Ile276 generates two-chain tPA. This cleavage occurs rapidly when tPA is adsorbed to a fibrin surface. tPA activates its substrate plasminogen by cleaving the peptide bond between residues Arg560–Val561 (see Chapter 8).

Both single-chain and two-chain tPA bind to and are stimulated by fibrin (Rånby et al., 1982; Rijken et al., 1982); in the presence of fibrin, the efficiency of tPA is enhanced some 500-fold through the formation of a ternary complex between tPA, plasminogen and fibrin. This serves to restrict plasmin generation by tPA to sites of fibrin deposition, thereby preventing systemic activation of plasminogen. Two-chain tPA is more active than single-chain tPA in the absence of fibrin (Bennett et al., 1991).

The finger, EGF, kringle 1 and kringle 2 domains may all be involved in fibrin binding (Van Zonneveld et al., 1986; Markland et al., 1989; Stern et al., 1989). Kringle 1 is thought to be involved in the initial binding of tPA to fibrin (Kaczmarek et al., 1993). Studies of deletion mutants suggest that the finger and kringle 2 domains serve to orientate tPA correctly on the fibrin polymer (Horrevoets et al., 1994).

The finger and EGF domains are also implicated in rapid clearance of tPA by the liver (Browne et al., 1988). Glycosylation mutants exhibit altered clearance by the liver (Lau et al., 1987).

The endothelial cell receptor for tPA is annexin II, a 40 kDa protein which also binds plasminogen (Hajjar, 1991; Hajjar et al., 1994; reviewed by Redlitz and Plow, 1995); tPA binding to the receptor occurs via a RDEKTQMIYQQ sequence within the finger domain.

Both single-chain and two-chain tPA are rapidly inhibited by PAI-1 (Hekman and Loskutoff, 1988). Since the concentration of PAI-1 in human plasma is higher than that of tPA, tPA circulates almost exclusively as a complex with PAI-1 (Lucore and Sobel, 1988). The region of the serine proteinase domain of tPA around residues 296–300 appears to interact with PAI-1 (Madison et al., 1989; Bennett et al., 1991).

An extensive analysis of tPA structure/function has been undertaken by Bennett et al. (1991); groups of charged residues were systematically altered to alanine. Mutations in the non-proteinase domains had little effect on fibrin stimulation whereas several regions of the proteinase domain were strongly implicated in fibrin binding or stimulation (specifically residues 408, 432, 434, 460, 462). Mutations in the proteinase domain decreased the activity of single-chain tPA (residues 426, 427, 429, 430). Binding of tPA to lysine-Sepharose was only affected by mutations in kringle 2. Clot lysis was affected by mutations in all domains except kringle 2. Many similar studies have been performed using site-directed mutagenesis to examine the relationship between structure and function in the tPA protein (Madison, 1994). Such data are very important for the design of second-generation tPA therapeutics (see Keyt *et al.*, 1994; Madison, 1994) with improved characteristics in terms of clot selectivity and resistance to PAI-1.

13.2.2 *Molecular genetics of tPA*

The molecular biology of tPA has been extensively reviewed (Cederholm-Williams, 1984; Blasi *et al.*, 1986; Gerard *et al.*, 1986). cDNA clones for human tPA have been isolated by a number of groups (e.g. Edlund *et al.*, 1983; Pennica *et al.*, 1983; Fisher *et al.*, 1985; Wei *et al.*, 1985; Harris *et al.*, 1986). Sequencing of cDNA clones established the amino-acid sequence of the 527-amino-acid residue mature protein and the 35-amino-acid residue leader sequence. The full-length cDNA contains 208 bp and 759 bp 5' and 3' non-coding regions, respectively.

The gene encoding human tPA (*PLAT*) comprises 14 exons spread out over nearly 40 kb of chromosomal DNA (Ny *et al.*, 1984; Browne *et al.*, 1985; Degen *et al.*, 1986). The entire region, including in excess of 3.5 kb 5' to exon 1, has been sequenced (Degen *et al.*, 1986). A considerable proportion of the human *PLAT* gene region (approx. 22%) is made up of repetitive DNA sequence; a total of 28 different *Alu* repetitive elements are present in the introns and 5' flanking sequence. The *PLAT* gene has been localized to chromosome 8p12–q11.2 (Yang-Feng *et al.*, 1986).

The various domains of the tPA protein are encoded by different exons. Thus, exon 3 encodes the fibronectin-like domain and exon 4 the EGF domain. The two kringles are bounded by introns E, G and I and split at identical locations by introns F and H. The amino acids that make up the active site within the light chain of the molecule are each encoded by different exons; the positions of three of the four introns (J, L and M) which divide the light chain-coding region are homologous with those found in the genes encoding the other serine proteinases, chymotrypsin, trypsin and elastase.

Sadler *et al.* (1991) have reported a CA repeat polymorphism within intron 1 of the human *PLAT* gene; formatted for polymerase chain reaction (PCR), 18 different alleles were distinguished corresponding to a size range of between 105 and 149 bp (allele frequencies varied between 0.003 and 0.418, yielding an overall heterozygosity of 77%). A similar $(TG)_{14} (TA)_{12}$ repeat polymorphism in intron 1 has been reported by Thomas and Drayna (1992); heterozygosity was 69% in a Caucasian population. A common *Alu* repeat insertion/deletion polymorphism in intron 8 does not appear to affect tPA synthesis (van den Eijnden-Schrauwen *et al.*, 1995; Tishkoff *et al.*, 1996).

The *PLAT* gene has been shown to be expressed in a wide variety of different tissues including the Bowes melanoma cell line, epithelial cells and endothelial cells (reviewed by Danø *et al.*, 1985; Tuddenham and Cooper, 1994). A wide variety of different compounds and substances have been found to influence the level of tPA secreted by cultured cells (reviewed in depth by Danø *et al.*, 1985; Tuddenham and Cooper, 1994). In summary, good evidence exists for a stimulatory effect on tPA secretion resulting from exposure to cyclic adenosine monophosphate (AMP), retinoic acid and phorbol esters. Exposure to thrombin and follicle-stimulating hormone also leads to increased secretion of tPA whereas interleukin-1 appears to inhibit secretion. Up-regulation of *PLAT* gene transcription has been reported after treatment with dexamethasone, endotoxin, α-thrombin, vascular endothelial growth factor, EGF and retinoic acid. Various *cis*-acting elements have been identified in the human *PLAT* gene promoter which are responsible both for the constitutive and inducible expression of the gene (Medcalf *et al.*, 1990; Fujiwara *et al.*, 1994; Sandler *et al.*, 1994; reviewed by Tuddenham and Cooper, 1994). Retinoic acid induction of *PLAT* gene expression has been shown to be mediated by a direct repeat element 7 kb upstream of the gene (Bulens *et al.*, 1995).

Herbert *et al.* (1994) have reported that tPA is a potent mitogen for human aortic smooth muscle cells *in vitro*. No difference was observed between double-chain and single-chain tPA in terms of their mitogenic effect. tPA may therefore contribute to smooth muscle proliferation after vascular injury as a result of percutaneous transluminal coronary angioplasty (PTCA) or as a consequence of atherosclerosis.

Carmeliet *et al.* (1994) reported the creation of mice with deficiencies of tPA and uPA by gene 'knockout' techniques. tPA-deficient mice exhibited decreased thrombolytic potential whilst uPA-deficient mice occasionally developed spontaneous fibrin deposition. Both tPA and uPA-deficient mice exhibited an increased incidence of thrombosis in response to endotoxin injection. Mice deficient in both factors exhibited extensive spontaneous fibrin deposition and an almost total loss of fibrinolytic potential.

13.2.3 Defects of tPA synthesis or regulation associated with thromboembolism

Normally, only a small proportion of the circulating tPA antigen is active, the remainder being complexed with PAI-1. Thus, whilst no correlation exists between the basal plasma levels of tPA antigen and activity, the concentration of tPA antigen may parallel the concentration of PAI-1 quite closely. An elevated level of tPA antigen has been found to be associated with an increased risk of cardiovascular disease in a study of patients with severe angina pectoris and angiographically proven coronary artery disease (Jansson *et al.*, 1991) and the level of tPA antigen was predictive of mortality in the longer term (Jansson *et al.*, 1993). Finally, in the prospective Physicians Health Study, higher tPA antigen concentrations were noted among individuals who later suffered myocardial infarction (Ridker *et al.*, 1993).

It seems very likely that failure to produce and/or release tPA normally would cause thromboembolic disease. Numerous studies have demonstrated decreased fibrinolytic activity in vein biopsies of patients with deep vein thrombosis (DVT). Most, however, have been found to be due to increased plasma levels of PAI-1 rather than decreased release of tPA (reviewed by Bachman and Kruithof, 1984; Francis, 1989; Wiman and Hamsten, 1990; Booth and Bennett, 1994). In most of these studies, there is little evidence of familial occurrence of impaired fibrinolysis.

Familial hypofibrinolysis has been reported in several families and defects in tPA release were initially suspected: for example, in a family reported by Johansson *et al.* (1978), 12/13 members who had experienced thromboembolism also had low euglobulin fibrinolytic activity after venous stasis and/or DDAVP infusion. Fibrinolytic activity of vessel wall biopsy was normal in these individuals, which suggested normal tPA synthesis but defective release. Other families with similar defects have been described by Jørgensen *et al.* (1982) and Stead *et al.* (1983). However, after the discovery of PAI-1, re-investigation of these families has shown that an increased level of PAI-1 was responsible for the observed defect (Nilsson *et al.*, 1985; Pizzo *et al.*, 1986; Jørgensen and Bonnevie-Nielsen, 1987).

One of the few cases of a potential defect in tPA synthesis and/or release is that of Petäjä *et al.* (1989, 1991) who have reported a family with low fibrinolytic capacity and low tPA activity with apparently normal PAI-1 levels.

Families or sporadic cases with thrombophilia and abnormal fibrinolysis in which a low level of tPA has been found should therefore be further studied by molecular genetic analysis of the *PLAT* locus: restriction fragment length polymorphism tracking and ultimately DNA sequencing if the *PLAT* locus is shown to co-segregate with the defect through the family pedigree. Mouse models of homologous gene 'knockout' or transgenic over-production of tPA are beginning to shed light on the likely clinical effects of disorders of tPA synthesis or regulation in man (reviewed by Carmeliet and Collen, 1994).

The over-expression of tPA in endothelial cells may in principle be considered as a means of boosting their thromboresistance. Thus Dichek *et al.* (1996), using a baboon model, transduced endothelial cells with a human tPA cDNA borne by a retroviral vector and demonstrated enhanced local antithrombotic activity. Such a strategy may prove therapeutically useful by prevented restenosis after angioplasty or the introduction of implanted devices.

13.3 Plasminogen activator inhibitor-1 (PAI-1)

13.3.1 Physiology, structure and function of PAI-1

PAI-1, a member of the serpin family, has a molecular weight of 52 kDa. It is synthesised in endothelial cells and hepatocytes and is also present in the α granules of platelets (Kruithof *et al.*, 1986) and in plasma (reviewed in Sprengers and Kluft, 1987). PAI-1 reacts with, and thereby inhibits, tPA with a second-order rate constant of more than 10^7 M^{-1} sec^{-1}. PAI-1 is released from endothelial cells in an active form but *in vitro* undergoes rapid inactivation to a latent form that does not inhibit tPA. This 'latent' PAI-1 can be reactivated *in vitro* by treatment with protein-denaturing agents followed by their removal and renaturation of PAI-1 in the presence of non-ionic detergents (Hekman and Loskutoff, 1985). In human plasma, the bulk of PAI-1 occurs in the active form.

The concentration of PAI-1 in healthy human blood ranges from undetectable to 1.3 nmol l^{-1}. The half-life of PAI-1 after infusion in rabbits is 7 min and it is probably cleared by the liver. The half-life of tPA in plasma due to inactivation by PAI-1 is about 100 sec, which suggests that the inhibitor is physiologically important. Perhaps the most compelling direct evidence for the importance of PAI-1 is the recent demonstration that overexpression in a transgenic mouse model leads to venous thrombosis (Erickson *et al.*, 1990; see below).

Owing to differential processing at the N terminus, plasma PAI-1 has either 379 residues (N terminus Val–His–His) or 381 residues (N terminus Ser–Ala–Val–His–His ...). The two forms are present in approximately equimolar amounts (Andreasen *et al.*, 1986). The reactive centre bond (P1–P1') is Arg346–Met347 (Andreasen *et al.*, 1986a, 1990), numbered from the distal N-terminal amino acid (Ny *et al.*, 1986).

In common with other serpins, PAI-1 forms a complex with its target proteinases (tPA and two-chain uPA) that is stable to moderate denaturation. The 'bait loop' presented to the proteinase is the reactive centre loop P15–P2' (Glu332–Ala 348). This region is strongly homologous to the bait loop regions of other serpins. The reactive loop is thought to complex with its proteinase target through formation of a covalent bond between Met347 and the catalytic site Ser478 residue of tPA, thereby forming a stabilized tetrahedral intermediate (Carrell and Travis, 1985).

The structure of intact latent human PAI-1 has been determined by X-ray crystallography to a resolution of 2.6 Å (Mottonen et al., 1992). The entire N-terminal side of the bait loop (residues Ser333–Val343) is inserted as the central β strand (4A) into β-sheet A. This β strand forms H bonds and hydrophobic interactions with strands 3A and 5A which probably contribute to the stability of the latent structure. The remainder of the reactive centre loop (Ser344–Arg356; P2–P10') emerges from sheet A and forms a loop which extends from the C terminus of strand 4A and the N terminus of strand 4B. Latent PAI-1 is inactive either because part of its bait loop is inaccessible to, or lacks the conformation necessary to bind to, its cognate proteinases. The latent form of PAI-1 differs from the active form in that it possesses a salt bridge between Arg30 and the P4' residue (Glu350) of the bait loop. This salt bridge is thought to be responsible for the significantly higher stability exhibited by the latent form of PAI-1 (Lawrence et al., 1994).

The transition from the active to the latent form of PAI-1 is prevented in vivo by binding to vitronectin (Wiman et al., 1988) which binds with high affinity to PAI-1 (Mimuro and Loskutoff, 1989). Vitronectin is a 75 kDa protein found in plasma and a member of the family of adhesive proteins including fibrinogen, fibronectin and von Willebrand factor which support platelet adhesion/aggregation. All possess the tripeptide sequence Arg–Gly–Asp which interacts with the platelet membrane glycoprotein complex GPIIb/IIIa. The structural requirements for the interaction of vitronectin with PAI-1 have been studied by Preissner et al. (1990a) and the binding site appears to involve residues 110–145 of the serpin (van Meijer et al., 1994). PAI-1 bound to vitronectin is functionally active. Although the functional significance of the PAI-1/vitronectin interaction has not been established, it may be that stabilization of PAI-1 is important.

In common with its fellow serpin, ATIII, PAI-1 binds heparin (Ehrlich et al., 1991), an interaction which may involve residues Lys65, Lys69, Arg80 and Lys88 (Ehrlich et al., 1992). Heparin appears to promote a thrombin inhibitory function of PAI-1 (Gebbink et al., 1993).

Mutational analysis of PAI-1 has indicated that the selectivity of this serpin for tPA is not conferred simply by the structure of the bait loop. Thus, Lawrence et al. (1990) created chimaeric reactive loop mutants in which residues P17–P2' of PAI-1 were replaced with the corresponding regions of PAI-2 and ATIII. These chimaeric proteins retained strong and selective reactivity with tPA, indicating that the sequence of the reactive loop is not critical. The PAI-2 loop chimaera had 200-fold greater reactivity with tPA than native PAI-2. Even the ATIII loop chimaera was a potent inhibitor of two-chain tPA ($K_i = 3.3 \times 10^6$), only eight-fold less potent than wild-type PAI-1 ($K_i = 2.3 \times 10^7$). However, replacement of the active centre with Met346–Ser347 resulted in an inactive inhibitor (Shubeita et al., 1990). The extended domains of PAI-1 responsible for its selective interaction with tPA remain to be established.

One newly understood role for PAI-1 is as an inhibitor of smooth muscle cell migration; this function is mediated by binding to the $\alpha_v\beta_3$ attachment site on vitronectin, thereby competitively inhibiting binding of the integrin $\alpha_v\beta_3$ to the same site (Stefansson and Lawrence, 1996). PAI-1 is inactivated by protein C (see Chapter 5) and neutrophil elastase (Wu et al., 1995).

13.3.2 Molecular genetics of PAI-1

Human PAI-1 cDNA clones have been isolated from cDNA libraries made from mRNA derived from umbilical vein endothelial cells (Ginsburg *et al.*, 1986; Pannekoek *et al.*, 1986) and placenta (Ny *et al.*, 1986; Wun and Kretzmer, 1987). These clones fell into two size classes of approximately 2.0 kb and 3.0 kb which differed from each other with respect to the lengths of their 3′ non-coding regions; 800 bp and 1800 bp, respectively. The PAI-1 cDNA encodes a 23-amino-acid residue hydrophobic leader peptide and a 379-amino-acid residue mature protein. cDNA clones encoding rat and bovine PAI-1 have also been characterized (Mimuro *et al.*, 1989). The amino acid sequences of the encoded proteins are predicted to be 81% and 87% homologous, respectively, to that of human.

Genomic clones spanning the human *PAI1* gene have been isolated by several groups (Loskutoff *et al.*, 1987; Bosma *et al.*, 1988; Strandberg *et al.*, 1988; van Zonneveld *et al.*, 1988; Follo and Ginsburg, 1989). The gene is split into nine exons spanning more than 12 kb. Bosma *et al.* (1988) sequenced the entire gene region including over 1500 bp 5′ flanking sequence and 2200 bp 3′ flanking sequence. The *PAI1* gene has been allocated to human chromosome 7q21.3–q22 by *in situ* hybridization (Klinger *et al.*, 1987). The major site of transcriptional initiation in the human *PAI1* gene is 145 bp upstream of the Met initiation codon (van Zonneveld *et al.*, 1988).

Several polymorphisms have been reported in the *PAI1* gene region: a HindIII polymorphism (Klinger *et al.*, 1987), a $(CA)_n$ repeat polymorphism (Dawson *et al.*, 1991) and a 4G/5G polymorphism in the promoter region (Dawson *et al.*, 1993). Since specific alleles of these polymorphisms correlate with the plasma concentration of PAI-1, at least some of the variance of PAI-1 concentration is likely to be due to genetic causes.

The 5′ flanking regions of the human *PAI1* and *PLAT* genes exhibit a remarkable degree of sequence homology. The sequence of the non-coding strand of the *PAI1* gene between −1520 and −1008 is 81% homologous to the *PLAT* coding strand between −3491 and −2977 with only six gaps required (Bosma *et al.*, 1988). This homology suggested the presence of sequence elements which could mediate the coordinate expression of the *PLAT* and *PAI1* genes. Bosma *et al.* (1988) pointed out that the expression of both genes is increased *in vivo* after myocardial infarction, surgery, or severe trauma and *in vitro* after treatment with thrombin, histamine or dexamethasone (Kluft *et al.*, 1986). Similarly, the expression of both genes is decreased after stanozolol treatment (Verheijen *et al.*, 1984). However, in apparent contradiction, Lucore *et al.* (1988) have shown that the expression of *PAI1* mRNA in various human cell lines bears an inverse relationship to the distribution of *PLAT* mRNA. Moreover, no homology has been noted between the rat *PAI1* gene promoter and that of the human *PLAT* gene (Bosma *et al.*, 1991).

Van Zonneveld *et al.* (1988) have demonstrated the presence of a glucocorticoid-dependent enhancer element within the first 305 bp of the promoter region; this finding complements the *in vivo* studies which have already documented glucocorticoid inducibility of the *PAI1* gene.

In various human cell lines, PAI-1 activity is induced by endotoxin, thrombin, interleukin-1, and glucocorticoids whilst being inhibited by cAMP (reviewed by Tuddenham and Cooper, 1994).

The expression of the *PAI1* gene has been well reviewed (Andreasen *et al.*, 1990; Loskutoff, 1991; Tuddenham and Cooper, 1994). Vascular endothelial cells appear to be the major site of synthesis for plasma PAI-1. Northern blotting experiments have demonstrated the presence of two distinct *PAI1* mRNA species (approx. 2.4 and 3.2 kb) in endothelial cells (Ginsburg *et al.*, 1986; Pannekoek *et al.*, 1986). Increases in the steady-state *PAI1* mRNA level have been reported after treatment with gluco-corticoids, transforming growth factor β (TGFβ), EGF, vascular endothelial growth factor, tumour necrosis factor α (TNFα), thrombin, interleukins-1 and -6, lipopolysaccharide, endotoxin, insulin and insulin-like growth factor. *PAI1* gene expression is also up-regulated by tPA itself (Fujii *et al.*, 1990). In cell culture, *PAI1* gene expression is influenced by serum concentration and substrate adhesion (Ryan *et al.*, 1996). PAI-1 gene expression is therefore tissue-specific but modifiable by a range of different cellular factors.

13.3.3 PAI-1 and cardiovascular disease

The plasma concentration of PAI-1 is a critical determinant of free tPA available for plasminogen activation and fibrinolysis. A high PAI-1 level might therefore be expected to inhibit fibrinolysis and could, in principle, exert a prothrombotic effect. Increased levels of PAI-1 activity have been reported in retrospective studies of patients with a variety of pathological conditions [e.g. coronary heart disease (Paramo *et al.*, 1985a; Mehta *et al.*, 1987) and acute myocardial infarction (Hamsten *et al.*, 1985, 1986; Almér and Öhlin, 1987; Verheugt *et al.*, 1987; reviewed by Wiman and Hamsten, 1990; Juhan-Vague *et al.*, 1995)]. Moreover, in a prospective study, high levels of plasma PAI-1 have been shown to constitute an important risk factor for recurrent myocardial infarction (Hamsten *et al.*, 1987). These observations are consistent with delayed thrombolysis of the coronary occlusion and suggest that a high plasma PAI-1 level could be a significant risk factor in arterial thrombosis (reviewed by Juhan-Vague and Collen, 1992). Studies of the association between raised PAI-1 levels and the incidence of post-operative or recurrent venous thrombosis have, however, generally been more equivocal. A case–control study of 953 patients with atherosclerotic disease has demonstrated that higher PAI-1 levels correlated with the future likelihood of myocardial infarction, ischaemic attacks and peripheral arterial thrombosis (Cortellaro *et al.*, 1993).

A mechanistic model is therefore beginning to emerge: the endothelial cells of atherosclerotic arteries synthesise increased amounts of PAI-1 (Schneiderman *et al.*, 1992; Raghunath *et al.*, 1995) and the release of this excess PAI-1 is enhanced by macrophages (present in atherosclerotic plaques) through the stimulatory influence of TNFα or TGFβ (Hakkert *et al.*, 1990; Tipping *et al.*, 1993). PAI-1, once released, is incorporated into fibrin clot to a much higher concentration (up to 200 nmol l^{-1}) than in plasma (approx. 0.4 nmol l^{-1}) (Robbie *et al.*, 1993; Stringer *et al.*, 1994) thereby stabilizing the clot and increasing fibrinolytic resistance (Levi *et al.*, 1992; Biemond *et al.*, 1995; Handt *et al.*, 1996).

Fibrinolytic capacity appears to be a useful parameter for determining the risk of recurrence in patients with spontaneous or recurrent DVT (Korninger *et al.*, 1984; Pizzo *et al.*, 1985). However, a low fibrinolytic response may not simply be due to an elevated PAI-1 level but could also be due to defective release of tPA. Juhan-Vague *et al.* (1987) found that fibrinolytic capacity was reduced in 35% of patients with DVT; 25% of these were deficient in tPA release whereas 75% exhibited raised PAI-1 levels. This conclusion was consistent with previous findings from other groups (Wiman *et al.*, 1985; Stalder *et al.*, 1985).

PAI-1 is increased in patients with a recent episode of thrombosis (Juhan-Vague *et al.*, 1984; Nilsson *et al.*, 1985; Wiman and Chmielewska, 1985; Sloan and Firkin, 1989; reviewed by Juhan-Vague *et al.*, 1995). Patients with high pre-operative PAI-1 levels have been reported to exhibit an increased risk of developing post-operative venous thrombosis (Paramo *et al.*, 1985b; Aranda *et al.*, 1988). However, no differences in PAI-1 levels have been noted between patients with or without DVT in prospective studies (Mellbring *et al.*, 1985; Kluft *et al.*, 1986; Sue-Ling *et al.*, 1987; Sørensen *et al.*, 1990).

Arnman *et al.* (1994) have reported that PAI-1 mRNA levels are significantly higher in thrombotic arteries and veins compared with their healthy counterparts. However, in all these studies, it is still unclear whether increased PAI-1 levels represent a risk factor for thrombosis or whether PAI-1, an acute phase reactant, is increased as a consequence of the thrombotic event. Other problems include the considerable variability of PAI-1 activity levels exhibited by normal subjects ($0.5–47$ U ml^{-1}) and possible variation in sampling procedures.

Several families in which decreased fibrinolytic capacity and elevated levels of PAI-1 activity/antigen co-segregated with thrombophilia have been reported (Alexandre *et al.*, 1980; Nilsson *et al.*, 1985; Jørgensen and Bonnevie-Nielsen, 1987; Patrassi *et al.*, 1992). However, the data were purely correlative and a causal relationship could not be confirmed. More recently, Engesser *et al.* (1989) studied 203 patients presenting with venous thromboembolism. PAI-1 activity and antigen, measured 3 months after their last thrombotic episode, were found to be elevated in 19 (9%) patients. However, in only eight patients did this elevated level persist when re-tested after a further 12 months. Although seven of these patients had a family history of thrombosis, none of the family members tested who had a history of thrombosis exhibited an abnormally high PAI-1 level. Despite one report claiming that the children of men with 'premature' myocardial infarction exhibited elevated levels of PAI-1 (Rallidis *et al.*, 1996), currently available data suggest that if high PAI-1 levels are associated with thrombosis, an inherited and/or persistent elevation of plasma PAI-1 is the exception rather than the rule.

Elevated levels of plasma PAI-1 have been found to be associated with diabetes mellitus. Schneider and Sobel (1991) have demonstrated that both insulin and insulin-like growth factor 1 influence the synthesis of PAI-1 and have claimed a synergistic effect on PAI-1 gene expression. These results are consistent with the hypothesis that hyperinsulinaemia may attenuate fibrinolytic activity, thereby contributing to accelerated atherosclerosis.

One approach to examining the cause and effect relationship between raised PAI-1 levels and thrombosis is through the development of mice transgenic for the *PAI1* gene. Erickson *et al.* (1990) studied such mice and reported that the mice exhibited subcutaneous haemorrhage at the tips of their tails at 3 days after birth which necrosed by 12 days. This necrosis and swollen hind feet resulted solely from venous rather than arterial occlusions. *PAI1* mRNA and PAI-1 protein levels, however, declined to control levels by 28 days and this decline paralleled the resolution of the venous occlusions. This study therefore provides strong evidence for a causal relationship between elevated PAI-1 levels and venous occlusion. The opposite experiment, the 'knockout' of the *PAI1* gene by homologous recombination, resulted in mice that were able to lyse clots more efficiently and which were relatively resistant to endotoxin-induced venous thrombosis (Carmeliet *et al.*, 1993a, b; reviewed by Carmeliet and Collen, 1994).

As yet, no molecular genetic studies have been carried out on individuals with raised PAI-1 levels. On the basis of PAI-1 activity measurements and the analysis of pedigrees in which thrombophilia is segregating (Engesser *et al.*, 1989), the determinants of raised PAI-1 levels appear not to reside at the PAI-1 locus itself. Dawson *et al.* (1991, 1993) have, however, demonstrated that genetic variation at the *PAI1* locus can influence the level of plasma PAI-1. Specifically, the 4G allele of the 4G/5G polymorphism ($f = 0.47$), 675 bp upstream of the transcriptional initiation site in the *PAI1* gene, was associated with a 21% higher plasma PAI-1 level than the alternative insertion allele ($f = 0.53$). Moreover, whereas the 5G allele exhibits enhanced NF-κB-like protein-binding potential, the 4G allele produces six-fold more mRNA than the 5G allele in response to interleukin-1. This polymorphism is not, however, associated with an elevated risk of myocardial infarction (Ye *et al.*, 1995).

If increased levels of PAI-1 are indeed directly responsible for a thromboembolic tendency then, whether or not the defect responsible is inherited, various potential therapies may be considered. One such approach involves treatment with antisense oligonucleotides. Sawa *et al.* (1994) treated human endothelial and smooth muscle cells with a 20-base oligonucleotide targeted to the 3' untranslated region of the *PAI1* gene. This resulted in a decrease in *PAI1* mRNA transcription, PAI-1 synthesis and a concomitant increase in cell-associated plasmin activity. Down-regulation of PAI-1 synthesis is therefore at least theoretically feasible even if problems with the delivery and stability of the oligonucleotide may preclude its use *in vivo*.

Other potential causes of familial thrombotic disease

14.1 Introduction

A number of other causes of venous thrombotic disease are either known or suspected. Some of these (e.g. prekallikrein or factor VII deficiency) may either represent chance association or a rare cause of thrombotic disease as a consequence of the deficiency state representing only a very minor risk factor. Other causes (e.g. factor VIII overexpression or hyperhomocysteinaemia) are undoubtedly very important but have yet to be studied in any detail. For this reason, this chapter attempts to present a hotch-potch of mini-reviews which are intended to serve the reader with gateways to wider reading in the area.

14.2 Prothrombin

14.2.1 Introduction

Fibrinogen is converted to fibrin clot by thrombin, which is itself derived from its zymogen prothrombin by factor Xa cleavage. Thrombin also has many other roles in haemostasis ranging from the positive-feedback activation of factors V and VIII, through activation of factors XI and XIII and platelets to negative-feedback regulation via the protein C pathway. The structure and function of thrombin have been comprehensively reviewed by Stubbs and Bode (1993a).

Prothrombin is synthesised by hepatocytes and is released into the circulation as a single-chain glycoprotein of M_r 71 600. The concentration of prothrombin in human plasma is 100 mg ml^{-1} and it exhibits a half-life in plasma of 2–4 days. Prothrombin is activated by factor Xa on the surface of activated platelets and is thus localized to sites of vascular injury. Thrombin then converts fibrinogen to fibrin monomer by specific cleavages, releasing fibrinopeptides A and B.

14.2.2 Physiology, structure and function

Prothrombin is a single-chain glycoprotein of 579 residues. Pre-pro-prothrombin has a leader sequence targeting the protein for secretion (residues –43 to –18) and a pro-sequence (residues –17 to –1) highly homologous to the pro-sequences found in other vitamin K-dependent γ-carboxylated proteins. The pre-segment of the protein is cleaved by a signal peptidase and the pro-segment by a processing proteinase revealing the mature N terminus. The presence of the pro-segment (residues −17 to −1) is essential for modification of the first 10 N-terminal glutamate residues to γ-carboxyglutamic acid. Prothrombin is glycosylated; N-linked carbohydrate chains are attached to Asn residues 78, 100 and 373.

The Gla domain occupies the N-terminal 32 residues, with six of the 10 Gla residues being located in pairs. Immediately following the Gla domain is a short segment (residues 33–46) containing three residues with aromatic side chains (Phe40, Trp41 and Tyr44), the 'aromatic stack'. A tetradecapeptide disulphide loop (residues 47–60) connects the helix to the first kringle domain. A second kringle follows and is connected to the segment which becomes the light (A) chain of thrombin (residues 285–320). The remainder of the molecule constitutes the heavy (B) chain of thrombin. The B chain has the typical catalytic triad of a serine proteinase located at His363, Asp419 and Ser525. Human α-thrombin, complexed stoichiometrically (1:1) with D–Phe–Pro–Arg chloromethylketone, formed an orthorhombic crystal suitable for X-ray crystallography (Bode *et al.*, 1989). Thrombin is an almost spherical molecule; the B domain is largely composed of two barrel-like domains, connected by twin structures and four helical regions. Notably different from other serine proteinase structures is the series of elongated and exposed loops, particularly around the active site cleft (Stubbs and Bode, 1993a).

Conversion of prothrombin to thrombin with release of a large activation peptide occurs as a result of the cleavage of two peptide bonds (Arg284–Thr285 and Arg320–Ile321) by factor Xa. The first cleavage releases a fragment consisting of the Gla domain, helix, loop and both kringles, designated fragment 1+2. The remainder of the molecule is not yet enzymatically active and is termed prothrombin 1. The second cleavage occurs between the light and heavy chains of thrombin but does not release them as separate fragments due to the single disulphide bond Cys22–Cys439. The resulting two-chain molecule is α-thrombin, the active enzyme. Alternative hydrolysis of the Arg320–Ile321 bond yields meizothrombin. Subsequent cleavage of Arg284–Thr285 yields fragment 1+2 and α-thrombin. Efficient conversion of prothrombin to thrombin by factor Xa only occurs in the presence of activated factor V, phospholipid and calcium ions. The activator complex (sometimes called prothrombinase) thus consists of an enzyme (Xa) a protein cofactor (Va) and phospholipid in a configuration whose assembly requires millimolar concentrations of calcium.

The most important inhibitor of thrombin in blood is antithrombin III (ATIII; Chapter 4). This serpin forms inactive stoichiometric complexes with thrombin, a process enhanced by heparan sulphate present on the endothelium. Other thrombin inhibitors present in blood are heparin cofactor II (Chapter 10) and protease nexin I.

14.2.3 Molecular genetics

Degen *et al.* (1983) isolated a human cDNA clone for prothrombin. Bancroft *et al.* (1990) mapped the major transcriptional initiation site which predicts an exon 1 of 110 bp in length and a full-length cDNA of 2.1 kb, consistent with the size of human prothrombin mRNA as measured by Northern blotting of hepatocyte mRNA. The human gene for prothrombin (*F2*) has been isolated and its structure determined (Degen *et al.*, 1983; Degen and Davie, 1987). The *F2* gene, which is located at chromosome 11p11–q12 (Royle *et al.*, 1987), covers about 19 kb and is split into 14 exons. Some 40% of the gene is composed of repetitive DNA sequences; altogether 30 *Alu* and two long interspersed nuclear elements (LINEs) are present within a 20 kb region. The promoter region of the human *F2* gene has been functionally character-ized (Chow *et al.*, 1991; Bancroft *et al.*, 1992). Several restriction fragment length poly-morphisms (RFLPs) have been reported (Tuddenham and Cooper, 1994).

14.2.4 A prothrombin variant and thrombotic risk

Prothrombin deficiency is a rare autosomal incompletely recessive disorder associ-ated with easy bruising, menorrhagia and haemorrhage following surgery, trauma or childbirth. Patients with more than 50% normal activity normally do not suffer bleed-ing problems, probably on account of the relatively long half-life of plasma pro-thrombin.

In the context of venous thrombosis, our interest in prothrombin results from a study of a DNA sequence polymorphism in the prothrombin gene which appears to confer thrombotic risk upon heterozygous carriers (Poort *et al.*, 1996). These workers detected a G→A transition at nucleotide 20210 of the prothrombin gene within the 3′ untrans-lated region. The polymorphism is located in the last base of the mRNA some 97 bp 3′ to the TAG-Term codon. The 20210A variant was found in heterozygous form in 2.3% of controls (95% CI, 1.4–5.6%). By contrast, it was present in 18% of selected patients with a personal and family history of venous thrombosis and in 6.2% of unselected patients. This translates into an increased risk of venous thrombosis of 2.8 (95% CI, 1.4–5.6). The risk of thrombosis increased for all ages and for both sexes. An association was also found between the presence of the 20210A allele and elevated prothrombin levels: indi-viduals homozygous for 20210G had a mean plasma prothrombin level of 1.05 U ml^{-1} as compared with 1.32 U ml^{-1} for heterozygotes. Poort *et al.* (1996) suggested that this lesion could exert a pathological effect by influencing mRNA cleavage and polyadeny-lation and hence conceivably mRNA stability.

14.3 Histidine-rich glycoprotein

14.3.1 Introduction

Histidine-rich glycoprotein (HRG) was first isolated and characterized from human plasma (Lijnen *et al.*, 1980) and subsequently from platelets (Leung *et al.*, 1983). It has several *in vitro* properties that make it appear likely to be important in the regulation of coagulation and fibrinolysis. It competes for heparin with ATIII and heparin cofac-tor II and also inhibits the binding of plasminogen to fibrin. Therefore increased lev-els of the glycoprotein would be predicted to promote thrombosis by up-regulating thrombin formation and down-regulating fibrinolysis

14.3.2 Physiology, structure and function

HRG is a 75 kDa monomeric glycoprotein which is synthesised in the liver. It is a negative acute phase protein (Smith *et al.*, 1989), which occurs in plasma at a concen-tration of 100 ± 45 mg l^{-1}. It interacts with heparin (Lijnen *et al.*, 1983, 1984a) and competes with ATIII and heparin cofactor II for heparin binding. Thus Lijnen *et al.* (1984a) reported a negative correlation between the plasma level of HRG and the anti-coagulant activity of heparin.

HRG also modulates fibrinolysis by interacting with the high-affinity binding site of plasminogen (Lijnen *et al.*, 1980), thereby inhibiting plasminogen binding to fib-rin. Binding studies demonstrated that HRG at a concentration of 1.8 μM reduced the binding of plasminogen to a fibrin clot by 50% (Lijnen *et al.*, 1980). From the dissociation constant of the reaction, about 50% of HRG is predicted to circulate as a reversible complex, thus reducing the effective concentration of plasminogen and exerting an anti-fibrinolytic effect. HRG also binds thrombospondin (Leung *et al.*, 1984) and fibrinogen (Leung, 1986) and may enhance plasminogen activation by tissue-type plasminogen inactivator.

The amino acid sequence of HRG was inferred from cDNA sequencing studies by Koide *et al.* (1986). The 18-amino-acid hydrophobic leader sequence is removed by cleavage of the Ala–Val dipeptide bond by a signal peptidase. The 507-amino-acid mature protein is calculated to have a molecular weight of 57.6 kDa, free of carbohydrate chains. Four potential carbohydrate attachment sites of sequence Asn–X–Thr/Ser were found at residues Asn45, Asn107, Asn326 and Asn327. In plasma, two alternative glycosylated forms of HRG are found with apparent molecular weights of 77 kDa and 75 kDa, respectively; these forms represent a polymorphism in the general population with respective frequencies of 0.35 and 0.65 (Hennis *et al.*, 1995a). The molecular basis of the polymorphism is a CCC/TCC dimorphism generating either Pro or Ser at amino acid residue 186 (Hennis *et al.*, 1995a). The presence of Ser at residue 186 creates another glycosylation site leading to the introduction of an extra carbohydrate group at Asn184.

The HRG protein is highly hydrophilic and the structure is predicted to consist of 8% helix, 14% β sheet, 46% β turn and 32% random coil (Koide *et al.*, 1986). HRG is characterized by regions rich in histidine and proline. The histidine-rich region is located between residues 330 and 389 and is arranged as 12 tandem repeats of five amino acids. This region shows about 50% sequence homology with the histidine-rich region of human high-molecular-weight kininogen (Koide *et al.*, 1986) and may be responsible for the binding of haem and metal ions to HRG (Morgan, 1985). In all, five different types of repeat are found in the HRG protein which make up more than half the protein (Koide *et al.*, 1986). This type of organization is consistent with the evolution of the HRG gene by duplication and divergence.

14.3.3 Molecular genetics

cDNA clones for HRG have been isolated from a human liver cDNA library (Koide *et al.*, 1986). A full-length cDNA sequence comprises a 121 bp 5′ non-coding region, 2067 bp encoding an 18-amino-acid hydrophobic leader peptide and a 507-amino-acid mature protein and a 352 bp 3′ non-coding region. The gene for human HRG comprises nine exons spread out over 11 kb of genomic DNA (Koide *et al.*, 1986).

A *Kpn*I RFLP has been reported at the *HRG* locus; alleles A1 (19.3 kb) and A2 (16.8 kb) occur at frequencies of 0.61 and 0.39, respectively, in a Caucasian population sample (Hennis and Kluft, 1991). A dinucleotide (CT/CA) repeat polymorphism has been noted in intron G: 15 alleles of length 233–267 bp contribute to an overall observed heterozygosity of 82% (Hennis *et al.*, 1992).

The human *HRG* gene has been localized to chromosome 3q28–29 (Oldenburg *et al.*, 1989; van den Berg *et al.*, 1990; Hennis *et al.*, 1994).

14.3.4 Clinical aspects

From what we know of its function in the control of haemostasis, an excess of HRG might be expected to give rise to thrombotic complications. Engesser *et al.* (1987a) reported a case of elevated plasma HRG in patient with recurrent venous thrombosis and myocardial infarction. Four other members of the family, which had a history of thrombosis, also possessed elevated HRG levels consistent with this trait having an hereditary basis. Similar reports of patients with thrombophilia and an elevated plasma HRG level have appeared (Engesser et al., 1988a; Shigekiyo et al., 1991; Castaman et al., 1993). Three families with venous thrombosis and raised levels of both HRG and plasminogen activator inhibitor 1 (PAI-1) have also been reported (Schved et al., 1991; Falkon et al., 1992; Angeles-Cano et al., 1993).

One intriguing case is HRG Eindhoven, an HRG variant associated with an abnormally high plasma level in a patient with recurrent arterial thromboembolism (Hoffman et al., 1993). The proposita consitently exhibited a plasma level 2.7 times that of normal. This variant also displayed abnormal heparin binding as measured by crossed affinity immunoelectrophoresis. Eight relatives of the proposita were found to possess high plasma HRG levels, consistent with an autosomal dominant mode of inheritance; all, however, were clinically asymptomatic. Hoffman et al. (1993) suggested that this 'syndrome' of increased HRG associated with reduced heparin binding was caused by an abnormality which is somehow involved in abnormal regulation of HRG concentration either through its synthesis or through its clearance.

Engesser et al. (1988) reported a prevalence of elevated plasma HRG of 5.9% in their study of 203 unrelated patients with idiopathic thrombophilia. In their more extensive study of 695 patients in whom other causes of thrombophilia had been excluded, Ehrenforth et al. (1993) noted abnormally high HRG levels (>148%) in 10.8% of patients. Familial elevation of plasma HRG was found in 3/10 families investigated, one of which possessed several members with episodes of venous and arterial thrombosis. By means of a twin study, Boomsma et al. (1993) demonstrated that genetic factors could explain 69% of the variance in plasma HRG levels. The basis for this genetic influence now appears to be the Pro/Ser polymorphism at residue 186: the Ser form contains an extra carbohydrate group (Section 14.2.2) and is associated with a markedly higher plasma level of HRG (Hennis et al., 1995b). Some 59% of the variance in plasma HRG levels can be attributed to the effect of this polymorphism (Hennis et al., 1995b).

The problem with the above studies is that it has been unclear whether the association between elevated levels of HRG and venous thrombosis are causal or merely coincidental. In an attempt to overcome this problem, Hennis et al. (1995c) performed a genetic analysis on a Dutch family with both elevated HRG levels and venous thrombosis. Using the dinucleotide repeat polymorphism as a genetic marker, linkage was assessed between the HRG gene and the high-expressing phenotype; a Lod_{max} of 4.17 at a recombination fraction of zero was strongly supportive of linkage. Hennis et al. (1995c) also demonstrated that possession of allele 6 of the dinucleotide repeat polymorphism was associated with the possession of a higher HRG level (mean = 149%, n =13) than that found in individuals lacking allele 6 (109%, n =25). However, only 3/6 of the family members with a history of thrombosis had an elevated HRG level. Further, elevated HRG levels were also noted in seven family members with no history of thrombosis.

HRG deficiency (21% normal plasma level) was reported in a patient who experienced a right transverse sinus thrombosis while taking the oral contraceptive pill (Shigekiyo et al., 1993). Four other family members possessed HRG levels between 20% and 35% normal (NR 109 ± 51%) but were asymptomatic. No other abnormality was found in any of the affected individuals (Shigekiyo et al., 1995). A further family with HRG deficiency segregating with venous thrombosis has recently been reported (Souto et al., 1996). It is not known whether HRG deficiency represents a predisposing factor to thrombotic disease. Although Ehrenforth et al. (1993) reported a prevalence of HRG deficiency (defined as below 52% normal) of 3.4% in their 695 patient study, a very similar prevalence was noted in their sample of healthy controls.

14.4 Factor VII

14.4.1 Introduction

Factor VII has a key role in coagulation. Indeed, activation of factor VII is the primary event in blood coagulation. The circulating zymogen factor VII has no activity, but upon contact with tissue factor exposed by vascular injury, it becomes converted to factor VIIa which is capable of activating both factor IX and factor X. Factor VII circulates at a very low concentration (10 nM) and has a short half-life in the circulation. Elevated factor VII levels have been associated with increased risk of cardiovascular disease (Meade, 1983; Meade et al., 1986; reviewed by Morrissey, 1995) whereas reduced factor VII levels appear to inhibit thrombus formation (Barstad et al., 1994). Factor VII deficiency is associated both with bleeding and thrombotic manifestations. The physiology, structure and function of factor VII are therefore briefly reviewed.

14.4.2 Physiology, structure and function

Factor VII is synthesised in the liver (Wion et al., 1985) and is present in plasma in low amounts (0.5 μg ml^{-1}; Fair, 1983) and has the shortest half-life of the classical coagulation factors at 3–4 h. Synthesis of active factor VII requires the presence of vitamin K. For these reasons, levels of factor VII are a sensitive index of liver function and vitamin K status. Levels of the zymogen fall rapidly in liver disease or on commencement of oral anticoagulation with vitamin K antagonists. The predominant form present in normal plasma is a single-chain glycoprotein zymogen of M_r 50 kDa. This single-chain form is now considered to have no proteinase activity towards its substrates factor IX or factor X. Activation of factor VII occurs by cleavage of a single bond. However, activated factor VIIa requires tissue factor and calcium ions to exhibit significant proteolytic activity against its physiological substrates factor IX and factor X. Although classically shown as the direct activator of factor X, it has become clear that the more important physiological substrate of the tissue factor–factor VIIa complex is factor IX. Further amplification of factor X activation then proceeds via the IXa/VIII complex (see below). The factor VIIa–tissue factor complex is inhibited with high efficiency by tissue factor pathway inhibitor (TFPI) complexed to factor Xa, forming a quaternary complex .

Factor VII is synthesised with a 38-amino-acid pre-pro-leader sequence containing a hydrophobic transmembrane domain targeting the protein for secretion (residues −36 to −24). Residues −1 to −17 contain highly conserved residues found in other vitamin K-dependent proteins. The mature protein is produced by cleavage of an arginyl–alanine bond and consists of a single chain of 406 residues with an M_r of 50 kDa. The region containing the γ-carboxyglutamic acid residues, of which there are

10, extends from residue 1 to residue 35 and includes the pro-sequence thought to be responsible for targeting the protein for γ-carboxylation. Two epidermal growth factor (EGF) domains are separated from the Gla domain by a short amphipathic helix (residues 37 – 46). Following the EGF domains is a connecting region in the middle of which is the preferred cleavage site for factor Xa, an Arg-Ile bond at residues 152/153. The catalytic domain is a typical serine proteinase containing the catalytic triad at His193, Asp242 and Ser344.

The arginyl–isoleucine bond referred to above is cleaved by factor Xa in the presence of tissue factor. The data of Bauer et al. (1989) measuring activation peptides of factor IX and factor X suggest that there is a continuous low-level generation of active factors, a process that may occur in the extravascular space.

Activated factor VII consists of a light chain, containing the Gla domain, amphipathic helix and growth factor domains, which is disulphide-linked via a bond from cysteine 135 to cysteine 262 to the heavy-chain catalytic domain. In vitro activation of factor VII by factor XIIa is markedly augmented in the cold in plasma from women taking oral contraceptives (Gordon et al., 1982) and from survivors of myocardial infarction (Gordon et al., 1987).

Factor VII has no or virtually no activity in the absence of tissue factor, which is an essential cofactor (Nemerson, 1988). Upon binding to tissue factor, which is the cell-surface receptor for factor VII, factor VII becomes activated in the presence of factor X (Rao and Rapaport, 1988). A reciprocal activation occurs in which the primary enzyme is likely to be factor Xa. The trimolecular complex undergoes a conformational change such that factor VIIa becomes more tightly bound to tissue factor, as factor Xa is released from the activator complex. The enzyme kinetics of this system have been studied in great detail by Nemerson and colleagues (for review see Nemerson, 1988). It was long held that the primary substrate of the factor VIIa–tissue factor complex physiologically was factor X. However, detailed re-evaluation of the kinetics of activation of factor IX versus factor X by VIIa–tissue factor suggests that, in fact, factor IX is the more important substrate (Komiyama et al., 1990; Repke et al., 1990). The recent determination of the crystal structure of the factor VIIa–tissue factor complex to 2.0 Å (Banner et al., 1996) has yielded insights into various structural and functional aspects of factor VII activation. Tissue factor binds to factor VII in three regions of the molecule, thereby removing much of the serine proteinase's internal and rotational flexibility.

Studies using synthetic peptides, predicted from the primary sequence of factor VII, by two groups have yielded partly conflicting results. Wildgoose et al. (1990) found that a peptide matching residues 195–206 of factor VII inhibited the interaction of factor VII with cell surface tissue factor at 1–2 mM. A modified peptide including an aspartamide residue was five- to 10-fold more potent. Flanking peptides were without effect in this study. However, Kumar et al. (1991) reported that the peptide matching residues 206–218 inhibited factor VII–tissue factor interaction. The variant factor VII 304 Gln isolated by O'Brien et al. (1991) has reduced affinity for human tissue factor and none at all for rabbit tissue factor, further implicating a heavy-chain region of the protease VIIa in binding to its cofactor. A third region of factor VII important for binding to tissue factor is the Gla domain, since Sakai et al. (1990) found that a factor VII derivative lacking the Gla domain failed to bind cell-surface tissue factor (Sakai et al., 1990). Binding sites for substrates of factor VIIa are presumed to involve the active site cleft.

Regulation of factor VII activity is provided by a recently discovered inhibitor associated with the lipoprotein fraction of plasma, TFPI (see below). TFPI is a Kunitz-type inhibitor which binds first to factor Xa and then forms a tetramolecular complex with factor VII–tissue factor.

14.4.3 Molecular genetics

cDNA clones encoding human factor VII have been isolated and characterized (Hagen et al., 1986); the 1398 bp coding region encoded a 60-amino-acid leader peptide followed by a 406-amino-acid factor VII mature protein with 5′ and 3′ non-coding regions of more than 35 and 1026 bp respectively. The human factor VII (F7) gene comprises nine exons spanning 12 kb (O'Hara et al., 1987) and is located at chromosome 13q34, some 2.8 kb upstream of the factor X (F10) gene (Miao et al., 1992). Exon 1b is absent in one of the cDNA clones characterized by Hagen et al. (1986) and its frequent removal from factor VII mRNA transcripts is suggested by Northern blotting studies of human liver mRNA which show that more than 90% of factor VII mRNA does not contain exon 1b (Berkner et al., 1986). Five imperfect tandemly repeated minisatellite sequences (MLEs) are located in the gene region (O'Hara et al., 1987). The MLE in intron G exhibits a copy number polymorphism (O'Hara and Grant, 1988). A further variable number tandem repeat (VNTR) polymorphism occurs in the same intron (de Knijff et al., 1994).

14.4.4 Factor VII deficiency and thrombotic disease

Inherited factor VII deficiency is a rare autosomal recessive disorder. The clinical features are quite variable, with a rather poor correlation between reported coagulant activity and clinical bleeding tendency. Several families have been reported in whom reduced factor VII activity was apparently associated with a thrombotic tendency.

Fair et al. (personal communication) studied a series of 68 patients from 10 kindreds; most of those patients with less than 10% normal plasma factor VII activity and antigen levels had severe bleeding problems. However, two individuals with homozygous deficiency in one family (factor VII levels of less than 5% using non-human thromboplastin but around 30% using human thromboplastin with normal levels of factor VII antigen) had a clinical history of thrombosis. At least 10 other individuals with factor VII deficiency and thrombosis, arterial and/or venous, have been described (Godall et al., 1962; Hall et al., 1964; Cooke, 1968; Gershwin and Gude, 1973; Girolami et al., 1977, 1978; Solanki and Corn, 1980; Shifter et al., 1984; Triplett et al., 1985; Escoffre et al., 1995). For a general review of thrombosis in patients with coagulation defects, see Goodnough et al. (1983).

Green et al. (1991) described an interesting variant of factor VII which is associated with a lower than normal level of plasma factor VII. The variant M2 allele differs from the wild-type (M1) by a G→A transition causing an Arg→Gln substitution at residue 353. The alleles are characterized by the presence or absence of an MspI restriction site. The M1/M2 and M2/M2 genotypes are associated with 23% and 67% reductions in the level of plasma factor VII activity. The M2 variant occurs with a frequency of 10% in the general population. This high frequency, suggestive of a balanced polymorphism, could indicate that this variant confers some protection against thrombosis and/or myocardial infarction.

Heywood *et al.* (1996) have shown that FVII: C levels are significantly associated with the *Msp*I polymorphism. However, Koster *et al.* (1994b) conducted a case–control study for 199 patients and found that neither the factor VII plasma level nor the *Msp*I genotype were related to a higher risk of venous thrombosis.

14.5 Hyperhomocysteinaemia

The association of homozygous homocystinuria with vascular and venous thrombotic disease was first reported over 30 years ago (Gibson *et al.*, 1964; reviewed by Rees and Rodgers, 1993). However, studies of patients with venous thromboembolism initially failed to find any association with plasma homocysteine levels (Brattström *et al.*, 1991; Beaumont *et al.*, 1992; Amundsen *et al.*, 1995).

Hyperhomocysteinaemia is an uncommon metabolic disorder associated with both vascular disease and thrombosis (reviewed by Rees and Rodgers, 1993). Homocysteine is derived by the metabolic conversion of methionine and is metabolized either through remethylation to methionine (catalysed by methionine synthase and 5,10-methylenetetrahydrofolate reductase) or transulphuration to cysteine [catalysed by cystathionine β-synthase (CBS)].

Hyperhomocysteinaemia has recently been shown to be a risk factor for venous thrombosis in people younger than 40 years of age (Falcon *et al.*, 1994) with hyperhomocysteinaemia being found in 19% of patients and 2% of controls. Subsequently, den Heijer *et al.* (1995) demonstrated that hyperhomocysteinaemia was a risk factor for recurrent venous thrombosis in patients between 20 and 70 years of age. In a cross-sectional study of consecutive unrelated patients with a history of venous or arterial occlusive disease, Fermo *et al.* (1995) found that moderate hyperhomocysteinaemia was present in 13.1% and 19.2% of patients, with venous and arterial patients, respectively. Den Heijer *et al.* (1996) performed a case–control study and showed that plasma homocysteine levels above the 95th percentile confer an increased risk of venous thrombosis (matched odds ratio 2.5, 95% CI 1.2–5.2). This association appears to be rather stronger for women than for men and increases with age. A very similar result has been reported by Simioni *et al.* (1996d): the odds ratio in this case was 2.6 (95% CI 1.1–5.9).

One cause of hyperhomocysteinaemia is the deficiency of CBS. CBS catalyses the condensation of homocysteine, an intermediate sulphydryl amino acid, and serine to form cystathionine using vitamin B6 as a cofactor. Homozygous CBS deficiency, which has an incidence of 1/335 000 live births, results in severe homocystinuria, a condition characterized by premature vascular disease and thrombosis (Mudd *et al.*, 1989). Heterozygous CBS deficiency, manifesting as mild hyperhomocysteinaemia, is however much more common with a prevalence of between 0.3% and 1.0% of the general population (Mudd *et al.*, 1989). Presumably, one cause of CBS deficiency would be defects in the *CBS* gene, located on chromosome 21q22.3. However, although hyperhomocysteinaemia occurs in approximately 30% of patients with premature occlusive arterial disease (POAD), no mutations were identifiable in the *CBS* genes of four POAD patients with hyperhomocysteinaemia and reduced CBS activity (Kozich *et al.*, 1995).

Other causes of hyperhomocysteinaemia are defects in vitamin B6 or B12 absorption/metabolism or the deficiency of 5,10-methylenetetrahydrofolate reductase (MTHFR; Malinov, 1994). The *MTHFR* gene is located on chromosome 1p36.3.

Frosst *et al.* (1995) have described a C→T polymorphism at nucleotide 677 in the *MTHFR* gene responsible for an Ala→Val substitution resulting in the synthesis of a thermolabile form of MTHFR. The T allele was found to occur with a frequency of 0.38 in the French-Canadian population and was associated with reduced enzyme activity and increased plasma homocysteine levels. Homozygosity for this polymorphism is associated with a three-fold increase in risk of premature cardiovascular disease (Kluijtmans *et al.*, 1996).

There may also be a combinatorial effect of hyperhomocysteinaemia and the possession of the factor V Leiden variant although such an effect is hard to measure (den Heijer *et al.*, 1996). Mandel *et al.* (1996) studied a total of 45 members of seven unrelated families in which hyperhomocysteinaemia was segregating. Six individuals with hyperhomocysteinaemia and factor V Leiden had experienced either venous or arterial thrombotic episodes whereas those patients with hyperhomocysteinaemia who lacked the factor V Leiden variant had not experienced any thrombotic manifestations.

A clear mechanistic explanation for the association of homocysteinaemia with venous thrombosis is not yet available (reviewed by De Stefano *et al.*, 1996). Ratnoff (1968) reported that homocystine, the oxidized dimer of homocysteine, could activate factor XII and it was suggested that deposition of homocysteine in the vessel wall could initiate contact activation and so promote thrombosis in individuals with hyperhomocysteinaemia. Factor VII deficiency has been found in four patients with homozygous homocysteinaemia (Munnich *et al.*, 1983). This might be an indirect consequence of homocysteine inducing tissue factor activity (Fryer *et al.*, 1993) since any enhancement of the extrinsic pathway of coagulation could lead to a reduction in factor VII. (Some patients with hyperhomocysteinaemia could thus have been misclassified as factor VII-deficient.)

Other causes of a heritable hyperhomocysteinaemia might include methionine synthase deficiency. The methionine synthase (*MTR*) gene has now been cloned, chromosomally localized (1q43) and the first pathological lesions reported (Leclerc *et al.*, 1996). It is, however, as yet unclear whether lesions in this gene could be responsible for a predisposition to venous thrombosis.

Homocysteine has also been shown to suppress protein C activation which it may bring about by acting as a competitive inhibitor of thrombin (Rodgers and Conn, 1990). This could down-regulate the protein C anticoagulant pathway, resulting in an increased likelihood of thrombosis. Finally, homocysteine may promote the proliferation of smooth muscle cells at the same time as inhibiting endothelial cell growth (Tsai *et al.*, 1994), a combination which could be sufficient to explain homocysteine-induced atherosclerosis.

Transgenic mice deficient in CBS have been made which are either moderately or severely homocysteinemic (Watanabe *et al.*, 1995). This may provide an animal model for the study of the *in vivo* role of elevated levels of homocysteine in the aetiology of cardiovascular disease.

14.6 Prekallikrein

Several cases of prekallikrein deficiency have been reported to be associated with either myocardial infarction (Hathaway, 1965) or venous/arterial thrombosis (Goodnough *et al.*, 1983; Harris, 1985) but since these are rare isolated cases, few general conclusions can be drawn.

14.7 Tissue factor and TFPI

Tissue factor levels are significantly increased in patients with disseminated intravascular coagulation (Wada *et al.*, 1994). This effect may be mediated by thrombin and fibrin degradation products which stimulate tissue factor production by endothelial cells (Brox *et al.*, 1984; Hamaguchi *et al.*, 1991). Tissue factor expression is also known to be stimulated by lipopolysaccharide and various cytokines (reviewed by Tuddenham and Cooper, 1994). Interestingly, homocysteine, a risk factor for premature vascular disease and thrombosis, also induces tissue factor synthesis in endothelial cells (Fryer *et al.*, 1993). Thus increased tissue factor production by endothelial cells could be an important risk factor in thrombotic disease (reviewed by Wada *et al.*, 1995; Diquélou *et al.*, 1995).

TFPI is the major inhibitor of the tissue factor–factor VII complex (reviewed by Broze 1995a, b). Recombinant TFPI has been shown to reduce fibrin deposition and inhibit thrombus formation on procoagulant extracellular matrix in a factor Xa-dependent manner (Lindahl *et al.*, 1994). Not surprisingly, the possible role of the defective synthesis or release of TFPI in thromboembolic disease is beginning to be explored (reviewed by Abildgaard, 1995). Novotny *et al.* (1991) found no difference in TFPI antigen levels between patients with thrombotic disease and controls. However, Goodwin *et al.* (1993) demonstrated significantly higher levels of plasma TFPI in a thromboembolic group than in a control group using a chromogenic assay. The authors postulated that this difference might be due to increased proteolytic cleavage and subsequent release from the vessel wall. Conversely, TFPI activity has been found to be lower in stroke patients than in controls (Abumiya *et al.*, 1995).

14.8 Factor VIII

Factor VIII is a cofactor for the activation of factor X by factor IXa. It circulates in plasma as a pro-cofactor bound to its carrier protein, von Willebrand factor (vWF). Raised levels of both vWF and factor VIII have been implicated in the aetiology of arterial thrombosis (Cucuianu *et al.*, 1983; Meade *et al.*, 1986; Blann and McCollum, 1994) whilst the low level of plasma factor VIII manifested by haemophilia A patients is associated with an 80% reduction in mortality from coronary heart disease (Rosendaal *et al.*, 1989).

Koster *et al.* (1995a) studied the association between the plasma levels of vWF and factor VIII in a group of 301 consecutive patients with an objectively confirmed episode of deep vein thrombosis. In univariate analysis, the risk of venous thrombosis increased with increasing vWF or factor VIII concentration and was higher in individuals with non-O blood groups than in those with group O. However, in multivariate analysis, only factor VIII appeared to be a risk factor. Thus blood group and vWF may be involved in causing thrombosis but they may only exert their effects through factor VIII. For individuals with plasma factor VIII concentrations more than 35% above the mean (>1500 IU l^{-1}), the adjusted odds ratio was 4.8 (95% CI 2.3–10.0). The prevalence of high factor VIII level was 11% in the healthy controls (cf. 25% in the thrombotic patients tested). Elevated plasma factor VIII may therefore represent yet another common risk factor for venous thrombosis which, when combined with an inherited deficiency state (e.g. protein C, ATIII, protein S, activated protein C resistance, etc.), may increase the likelihood that a patient will come to clinical attention.

Elevated factor VIII:C values could in principle be caused by a failure to inactivate factor VIII:C. This could arise as a result of mutation in one of the two activated protein C cleavage sites in factor VIII, a situation analagous to that of factor V Leiden (Chapter 7). However, Roelse *et al.* (1996) screened residues Arg336 and Arg562 of factor VIII for mutation in 125 consecutive patients with venous thrombosis and failed to detect any lesions. Thus this does not appear to be a common cause of venous thrombosis.

The epidemiology of venous thrombosis

15.1 Introduction

In many respects, venous thrombosis serves as a paradigm for any multifactorial disorder. Its complexity was already apparent to the early Rudolf Virchow who, in 1856, was among the first to realize that the pathophysiology of thrombosis involves three interrelated factors ('Virchow's triad'): damage to the vessel wall, slowing down of the blood flow (stasis), and increase in blood coagulability. On their own, none of these factors are capable of causing thrombotic disease but it is their temporal and spatial coincidence which threatens millions of people each year with what is a major cause of disablement and death, at least in the Western world.

Blood must stay fluid to provide organ nutrition but, at the same time, must be able to form blood clots to plug significant breaches in the vessel wall. These requirements are balanced by the highly complex coagulation system which, upon vessel damage, coordinates the action of both prothrombotic clotting factors and their inhibitors. Since both types of protein must remain in perfect equilibrium during the coagulation process in order to avoid distal ischaemia, it is not surprising that direct vessel damage is the strongest risk factor for venous thrombosis (Carter, 1994a). However, unlike in arterial thrombosis where gross damage is often evident at the site of thrombus formation, most changes observed in venous thrombosis are anatomically more subtle.

Blood flow, the second component of Virchow's triad, is an important factor in the thrombotic process because flow conditions affect both the function of clot formation substrates (i.e. platelets) and the assembly of coagulation protein complexes. Furthermore, shear forces not only remove excess activated clotting factors but may also remove solid matter, such as the thrombus itself. Since flow rates and shear forces are lowest in the leg veins, these vessels are not surprisingly the most frequent site of deep venous thrombosis (DVT; Carter, 1994a).

Whilst the first two components of Virchow's triad in most instances represent acquired conditions, blood hypercoagulability has both intrinsic and extrinsic causes. Acquired conditions such as immobilization, trauma, oral contraceptive use and malignancy, induce higher blood viscosity indirectly through stasis or activation of the coagulation mechanism. Therefore, they are called 'secondary hypercoagulable states' (Schafer, 1994). The precise mechanism of thrombosis is unknown for most secondary hypercoagulable states. These states not only predispose apparently normal individuals to thrombosis but are also likely to trigger disease episodes in patients with an inherited prothrombotic tendency.

Venous Thrombosis: from Genes to Clinical Medicine, D.N. Cooper and M. Krawczak.
© 1997 BIOS Scientific Publishers, Oxford.

In this book, we have concentrated mainly upon genetic factors responsible for blood hypercoagulability and thrombosis. Because of their predetermined nature in at-risk individuals, the pathological conditions resulting from defects in the corresponding genes (mostly through a deficiency of their encoded products) are termed 'primary hypercoagulable states' (Schafer, 1994). Many of these states have been intensively studied and the underlying prothrombotic mutations characterized in a number of genes. Apart from their importance for understanding the pathophysiology of venous thrombosis, the corresponding genotypes allow a more refined assessment of thrombotic risk, at least in those individuals threatened by hereditary thrombophilia.

Epidemiological studies have consistently demonstrated, however, that the hitherto defined genetic defects explain only a certain proportion of the average life-time risk for venous thrombosis. This disparity implies that many predisposing genetic factors are still unknown and/or that the way in which risks, issuing from the three components of Virchow's triad, integrate is not sufficiently well understood. In this chapter, we shall review the current state of genetic epidemiology of venous thrombosis and make an attempt to exploit the relative and combined contributions of its known risk factors. Although still at the level of a 'back of the envelope calculation', this computational exercise is intended to highlight how far our current knowledge about thrombotic risk is from being totally comprehensive.

The potential of a given risk factor to predispose to venous thrombosis is usually measured by means of the *relative risk*. This figure reflects how much more likely an individual exposed to a risk factor is to develop the disease than a non-exposed individual. The relative risk can only be estimated properly in cohort studies where a given number of exposed and non-exposed participants is followed-up for a given period of time, and the proportions of those clinically symptomatic determined. Such studies are often difficult to perform and they are usually expensive and rather inefficient. Alternatively, the relative risk can be approximated in case–control studies by means of the so-called *odds ratio*, given by:

$$\text{odds ratio} = \frac{F_T \cdot (1-F_P)}{(1-F_T) \cdot F_P}$$

(Hennekens and Buring, 1987). Here, F_T and F_P denote the frequencies of exposure among thrombotic and non-thrombotic individuals, respectively. It can be shown mathematically that the odds ratio is a stable and unbiased estimate of the relative risk (Hennekens and Buring, 1987). In the following sections, most of the studies reviewed were of case–control design. However, because of the above statistical equivalence, relative risk estimates will be referred to as 'relative risks' even if they were obtained via odds ratios.

15.2 General risk factors

The validity of any epidemiological study of venous thrombosis is critically dependent upon initial population selection and the diagnostic methods employed. A critical review of the literature indicates that there is substantial variation in the performance of such tests in terms of their sensitivity and specificity. Although the diagnosis of pulmonary embolism (PE), the major final event in the course of the disease, is not always made accurately (Alpert and Dalen, 1994), a rough estimate of its incidence may nevertheless be possible by reference to death certificates and registers. The lack of a uniformly accepted, cheap and reliable diagnostic procedure for DVT, however, should caution against the uncritical acceptance of results from epidemiological studies in this field.

Early population studies (Gjores, 1956; Coon *et al.*, 1973) arrived at prevalence figures for DVT of 2–5%, but since these studies had to rely upon clinical diagnoses rather than objective tests, the results are unlikely to be accurate. In 1991, a retrospective American study was published (Anderson *et al.*, 1991) that was based upon individuals examined during short-term hospital stays. The average annual incidence of first episodes of DVT was extrapolated to be around 50 per 100 000. Although more than 80% of DVT cases had been confirmed by objective testing in this study, it was nevertheless hampered by the fact that asymptomatic DVT could not be addressed. Furthermore, hospital patients represent a preselected group with multiple pathologies and an age structure different from that of the general population. Since bedridden home-care patients and individuals living in non-hospital facilities such as residential homes were also not included, figures arrived at among hospitalized patients are likely to underestimate the true incidence of DVT. Another study of DVT reliant upon hospital-based diagnoses was performed in Sweden (Nordstrom *et al.*, 1992). Since ascending venography was used for diagnosis, the estimated incidence of 160 per 100 000 per year appears to be reliable within the limitations of the study. This number includes recurrent cases of DVT as well as patients with symptoms of PE and so the two studies are mutually consistent.

Regional variability in prevalence is apparent for thrombotic disorders. In Europe, a three-fold difference exists between low-risk Mediterranean populations (Italy, France and Spain) and northern European countries (Tuddenham and Cooper, 1994). However, experience with immigrant populations as well as comparative studies have clearly demonstrated that these discrepancies are likely to be due mainly to different environmental and behavioural patterns, including diet and physical morphology and activity (Tuddenham and Cooper, 1994).

15.2.1 Age

In the study of Anderson *et al.* (1991), annual incidence rates for DVT varied between 17 per 100 000 for age 40–49 to 232 per 100 000 for age 70–79. The risk for DVT showed a dramatic increase with the fifth decade and was found to double with each successive decade (Carter, 1994b). However, it should be remembered that most risk factors predisposing for DVT (e.g. immobilization, malignancy and thrombotic history) also correlate with age. It is thus not immediately clear whether age is an independent risk factor for DVT. Certainly, studies controlling for confounding variables within multivariate designs are rare (Carter, 1994b).

15.2.2 Oral contraceptive use

The temporal coincidence of the first mass prescription of oral contraceptives with the occurrence of unusually early cases of thrombosis prompted a possible disease-association for the pill and other hormonal drugs. Indeed, initial case–control studies performed in the late 1960s and early 1970s revealed relative risks of between 4 and 11 for idiopathic DVT, and of 3 for recurrence of the disorder (Carter, 1994b). Large prospective studies in the USA (Porter, 1982) and the UK (Royal College of General Practitioner's Oral Contraception Study, 1978) later confirmed these preliminary results. Although both prospective studies arrived at similar figures to the case–control studies (7 and 4.2), a bias leading to an overestimate of the true relative risk associated with oral contraceptive use could not be excluded. Diagnosis was not performed blindly in both studies so that the knowledge of a proband's medication status could have influenced diagnostic scrutiny.

Thorogood et al. (1992) observed a relative risk of only 1.6 for oral contraceptive use, but this finding was obtained in a case–control study based upon fatal venous thromboembolism, a rare cause of death among young women. An extensive case–control study recently performed by the World Health Organization (1995) was largely confirmative of the results from the 1960s and 1970s. The odds ratio of contraceptive use for venous thromboembolism was 4.15 in European countries and 3.25 in non-European ('developing') countries. Risk estimates were generally higher for DVT than for PE. Interestingly, the thrombotic risk was found to increase within 4 months of start of use, remained unaffected by duration, and decreased to its previous level within 3 months after cessation of use.

Some studies provided indirect evidence for a dosage effect upon the thrombotic risk of oestrogen intake via oral contraceptives (Jeffcoate et al., 1968; Coronary Drug Project Research Group, 1970). However, since oestrogen was administered in these cohorts to suppress lactation and to reduce myocardial reinfarction risks, respectively, the hormone dosage was probably too high to allow meaningful extrapolation to the level prevailing in contraceptive pills. Stolley et al. (1975) claimed that reducing the oestrogen content to below 100 μg cuts the risk of venous thrombosis by about 50%. Although this and other findings have been challenged because of possible confounding effects (Kierkegaard, 1985), the same result was consistently obtained. This has resulted in the withdrawal of preparations containing more than 100 μg oestrogen from the market (Samsioe, 1994). Introduction of contraceptives with 30 μg oestrogen served to reduce the risk for venous thrombembolism even further, probably by another 50% (Samsioe, 1994). A reduction of oestrogen content from 50 μg to 30 μg does not, however, influence haemostatic variables [antithrombin, fibrinogen and plasminogen activator inhibitor-1 activity (PAI-1)] to any considerable extent (Scarabin et al., 1995).

There is no epidemiological evidence for an increased thrombotic risk for progestin-only contraceptives (Samsioe, 1994). In combined preparations, however, higher risks are associated with third-generation progestagens (e.g. desogestrel, gestodene) as compared to first- and second-generation progestagens (World Health Organization, 1995). This implies that the beneficial effect of third-generation progestagens, which lies in enabling a reduction of oestrogen content without loss of efficiency, is somewhat offset. The thrombotic risk conferred by the two older types of steroids appears to be lower only in combination with low (less than 50 μg) but not with high oestrogen doses.

15.2.3 A previous history of thrombosis

Previous thrombotic events are clearly a risk factor for recurrent DVT. In the study of Nordstrom et al. (1992), 14% of cases of DVT had a history of the disease. If related to a population prevalence of 2% to 5%, this figure is in accord with the relative risk of 7.9 for previous thrombotic episodes emerging from a French community-based study (Samama et al., 1993). Similarly, Thorogood et al. (1992) found a significantly higher prevalence of a history of thrombotic episodes in their retrospective case–control study of women who died from fatal venous thromboembolism. The relative risk was 4.0. In a prospective cohort study among patients of a university outpatient thrombosis clinic, Prandoni *et al.* (1996) noted an 18% recurrence risk for venous thromboembolism after 2 years, 25% after 5 years, and 30% after 8 years. Whilst cancer and impaired coagulation were associated with an increased risk, surgery and recent trauma or fracture were not.

15.2.4 Malignancy

An association between carcinoma and DVT has been recognized for some time (Carter, 1994b). In their 30 month follow-up of suspected cases of DVT from a thrombosis clinic, Goldberg *et al.* (1987) found that DVT-positive probands under the age of 50 developed cancer some 20 times more often than their age-matched, DVT-negative counterparts. This association was not, however, observed among individuals over the age of 70. A comparison of idiopathic and recurrent cases of DVT performed in a 2-year follow-up by Prandoni *et al.* (1992) revealed a five-fold higher cancer rate in the idiopathic as compared to the group of secondary thromboses. Both studies, although of different designs, clearly emphasize the association between malignancy and DVT.

15.2.5 Surgery

Orthopaedic procedures or trauma of the lower limbs are major risk factors for DVT. Thus, severe trauma or surgery of the knee region have been shown to be associated with a venographically confirmed DVT rate of 50% (Carter, 1994b). A significantly increased incidence of DVT has also been recognized after hip replacement or hip fractures (Carter, 1994b). The most likely explanation for these associations probably lies in the direct vessel injuries involved which increase blood coagulability.

Non-orthopaedic surgery accounts for the majority of in-patient surgical procedures, but these patients are generally younger than those undergoing orthopaedic surgery. Furthermore, risk factors such as trauma of the lower limbs or prolonged immobilization can be expected to be a rarer ocurrence. Although these differences might suggest a lower DVT risk for non-orthopaedic surgery, liability to thrombosis is still apparent. Thus, post-operative DVT was found in 10–30% of cases of abdominal surgery (Bergqvist, 1983), 10–50% of prostatectomies, a higher risk bound up with open rather than transurethral surgery (Mayo *et al.*, 1971; Nicolaides *et al.*, 1972), and 12% for abdominal as compared to 7% in vaginal hysterectomies (Walsh *et al.*, 1974). Thorogood *et al.* (1992) noted a relative risk of 11 for recent operation or accident in a case–control study of women who died from venous thromboembolism. It is, however, unclear whether these findings could be extrapolated to non-fatal thromboembolism.

15.2.6 Pregnancy

DVT is a recognized complication of pregnancy and the puerperium. However, owing to concerns about using radiation-based diagnostic tests during pregnancy, accurate data on DVT incidence are rare. The available data nevertheless suggest that the monthly DVT incidence is 0.3–2% among post-partum women, considerably higher than the pre-partum monthly risk of 0.01% (Drill and Calhoun, 1968). A later retrospective study arrived at risk figures at least one order of magnitude lower (Kierkegaard, 1983), indicating that the earlier studies had yielded considerable overestimates. The relative ratio of pre-partum and post-partum risk was however found to be comparable. No evidence has been found for a non-uniform distribution of DVT risk over the pre-partum period (Carter, 1994b).

15.3 Genetic risk factors

It is at present unclear whether venous thrombosis can occur in the absence of an obvious genetic predisposition. However, for a substantial proportion of cases, the primary hypercoagulable state underlying pathogenesis can be traced back to one or more specific prothrombotic mutations. The vast majority of these lesions result in a deficiency or processing irregularity of one of the proteins involved in either blood clotting [e.g. antithrombin III (ATIII), protein S, protein C] or fibrinolysis (e.g. fibrinogen, plasminogen, tissue plasminogen activator). Thus, Heijboer *et al.* (1990) observed an inherited deficiency state either of the coagulation-inhibiting or fibrinolytic proteins in 8.3% of outpatients with DVT, half of whom had a family history of thrombosis. This figure, although in line with previous estimates, was nevertheless likely to represent a serious underestimate since resistance to activated protein C (APC resistance), the most frequent known single gene defect in a venous thrombotic disorder (Dahlbäck *et al.*, 1993), was not included.

In striking contrast to the study of Heijboer *et al.* (1990), Melissari *et al.* (1992) found a congenital deficiency state known to predispose to thrombosis in 27% of 393 patients with a history of acute venous thromboembolism (DVT/PE). No satisfying explanation for this discrepancy was provided. Interestingly, however, the estimate of Melissari *et al.* (1992) is in keeping with the result of a study among Dutch patients with inherited thrombophilia (Bertina, 1988). Hereditary thrombophilia is defined in terms of the early ocurrence (less than 41 years) of thrombotic episodes in the patient and at least one of his or her closest relatives. Screening for essentially the same deficiency states as Melissari *et al.* (1992), hereditary thrombophilia was found to be explicable in terms of one of these deficiency states in 35 of 113 independent families (31%). Similarly, Hirsh *et al.* (1989) concluded that 20% of thrombotic patients under the age of 45 years may be expected to have a deficiency of ATIII, protein C or protein S.

This coincidence of prevalence figures suggests an ascertainment bias operating in the study of Melissari *et al.* (1992) and others arriving at similarly high estimates. Thus, if familial clustering or younger age led to a higher referral probability in the study of Melissari *et al.* (1992), a higher than average proportion of cases would have been due to hereditary thrombophilia. Indeed, since a positive family history was noted among patients with a primary hypercoagulable state (40–50%) as frequent as by Heijboer *et al.* (1990), the relative proportion of familial cases among all patients analysed must have been some three-fold higher.

15.3.1 Protein C deficiency

Protein C is a vitamin K-dependent protein but, unlike vitamin K-dependent factors II (thrombin), VII, IX and X, which are procoagulants, protein C is an anticoagulant (Chapter 5). Together with protein S (Chapter 6), protein C forms a complex that inactivates factors Va and VIIIa. Before it can exert this anticoagulant function, protein C has to be activated by thrombin cleavage on the endothelial cell surface.

Although an important genetic risk factor for venous thromboembolism, heterozygous protein C deficiency accounts for only 3% of cases of DVT (Heijboer *et al.*, 1990).

Prevalence of the deficiency is estimated to be 0.1–0.5% in the general population (Miletich *et al.*, 1987; Tait *et al.*, 1995) but only a minority of heteroyzgotes actually develop any thrombotic symptoms. Indirect evidence for the validity of such a high 'covert' prevalence comes from the close coincidence of the predicted (some 1 in 640 000) and observed incidence of homozygote or compound heterozygote carriers with severe protein C deficiency (1 in 500 000–750 000 births; Marlar and Neumann, 1990).

Since only a small proportion (three of 70) of heterozygous relatives of homozygous patients suffer any thrombotic events (Tuddenham *et al.*, 1989), thrombosis due to protein C deficiency could be thought of as following a recessive mode of inheritance in their families. On the other hand, the disease risk of protein C-deficient heterozygotes is at least 50% in families with a history of thrombotic disease (Allaart *et al.*, 1993). A large collaborative study among the kindreds of 75 patients of Austrian, German and Swiss origin (Pabinger and Schneider, 1996) revealed a life-time thrombosis risk of 80% or higher for heterozygous relatives, a finding which was in keeping with the annual incidence rate of 2.5% for protein C heterozygous family members noted before (Pabinger *et al.*, 1994b). Since index patients were heterozygous rather than homozygous for protein C deficiency in all these studies, the deficiency-associated disease state was shown to segregate as an autosomal dominant trait with reduced penetrance. This type of protein C deficiency makes up 6–9% of cases of hereditary thrombophilia defined in the above fashion (Lane *et al.*, 1996c).

The dual character of thrombosis consequent to inherited protein C deficiency is best explained by the influence of the genetic background. Heterozygous protein C deficiency is obviously not sufficient for thrombosis to manifest, although the dramatically increased risk of heterozygous relatives of heterozygous patients points towards a major role for protein C deficiency in these families. The heritability of the thrombotic disease state, on the other hand, argues in favour of the existence of other genetic factors necessary for a predisposition to thrombosis. These factors may be comparatively rare in the population, but have accumulated in affected families so as to bring them to clinical attention. The increased risk of 10% for non-deficient family members (Allaart *et al.*, 1993) further supports this view. By contrast, in families where the index patient is homozygous or compound heterozygous for protein C deficiency, the simultaneous occurrence of genetic factors predisposing heterozygotes to thrombosis is relatively unlikely. Therefore, heterozygous relatives of homozygotes will have a disease risk not much different from that of heterozygotes in the general population.

Based upon the analysis of unselected outpatients, a relative thrombotic risk of 7 has been reported for general heterozygous protein C deficiency by Koster *et al.* (1995b). This figure is surprisingly close to the relative risk of 5, observed in thrombotic families segregating heterozygous protein C deficiency (Allaart *et al.*, 1993). The similarity of the two estimates implies that, in the thrombotic families, deleterious genetic background effects must have combined with the deficiency state in an additive or sub-additive rather than a multiplicative fashion. For the general population, the data of Koster *et al.* (1995b) suggest an average life-time risk of approximately 25% for heterozygous protein C deficiency. Tait *et al.* (1995), by contrast, noted 14 heterozygotes among 9648 blood donors from western Scotland, none of whom had experienced thrombotic symptoms or showed a strong family history of the disease. Their study population, however, was likely to be highly selected for high average health and young age and therefore most probably against symptomatic protein C deficiency.

15.3.2 ATIII deficiency

ATIII inhibits blood coagulation through the inactivation of thrombin and other clotting factors such as factors XIIa, XIa, IXa and Xa (Chapter 4). ATIII deficiency is an autosomal dominant disorder with reduced and highly variable penetrance. Thus, between 15% and 100% of the ATIII-deficient members of affected families suffer from clinical thrombotic episodes (Demers et al., 1992). Based upon the disease history of 69 patients from 25 families segregating hereditary ATIII deficiency, Pabinger and Schneider (1996) estimated the life-time thrombotic risk for heterozygous members to be 80–90%.

Symptomatic ATIII deficiency has a prevalence of between 0.02% and 0.05% in the general population (De Stefano et al., 1996), but asymptomatic deficiency is probably up to 10 times more frequent (0.17%; Tait et al., 1994). In their survey of 277 unselected outpatient patients with DVT, Heijboer et al. (1990) observed three (1.1%) individuals with ATIII deficiency. This prevalence figure is identical to that arrived at by Koster et al. (1995b) in another Dutch study of similar design. Among patients with hereditary thrombophilia, however, the frequency of ATIII deficiency appears to be somewhat higher. With the exception of Tabernero et al. (1991), who found only 0.5% of thrombophiliacs to be ATIII-deficient, most studies of patients selected for young age and family background revealed prevalence rates of between 4% and 7% (Lane et al., 1996c). In the study of venous thromboembolism by Melissari et al. (1992), presumably biased towards the inclusion of cases of hereditary thrombophilia, a frequency of 5% was observed.

15.3.3 Factor V Leiden (APC resistance)

Factor V is a plasma glycoprotein synthesised in the liver and in megakaryocytes (Chapter 7). Upon activation by thrombin or factor Xa, it serves as a cofactor in the prothrombinase complex, increasing its catalytic activity approximately 2000-fold (Lane et al., 1996c). Factor Va is inactivated by APC through the degradation of the factor Va heavy chain, the first step being cleavage at Arg506.

Resistance to the anticoagulant activity of APC has been reported to be a major cause of hereditary thrombophilia (Chapter 7). It is inherited as an autosomal trait and is associated with the disease in some 40% of thrombophilic families. The frequency of APC resistance in unselected white patients with venous thromboembolism is generally found to be 20% (De Stefano et al., 1996; Lane et al., 1996c). This figure appears to apply not only to DVT but also to cerebral venous thrombosis (Brey and Coull, 1996). A population-based case–control study of patients with objectively confirmed DVT yielded a relative risk of 6.6 for APC resistance (Koster et al., 1993). APC resistance is not a strong risk factor for PE. Among 146 patients with a confirmed history of this complication, Desmarais et al. (1996) found eight APC-resistant individuals (5.5%). Comparison with a prevalence of 14 among 348 patients (4.0%) where PE could be ruled out yielded a relative risk estimate of 1.4.

There have been some reports of exceptionally high frequencies of APC resistance among unselected patients with venous thrombosis (Griffin et al., 1993; Svensson and Dahlbäck, 1994). To a considerable extent, however, the apparent discrepancies are likely to reflect differences in diagnostic methodology rather than population-based effects. APC resistance is assessed by the in vitro measurement of the relative delay in blood coagulation time upon APC addition (clotting time ratio, CTR). A strong correlation holds for the published studies between the average CTR in healthy controls and the estimated frequency of APC resistance among thrombotic patients. This relationship suggests that assay design has been critical for the presumed discriminative power of a study. Finally, coagulation time can be assumed to be influenced in different thrombotic patients by different factors, both causal and therapeutic (Kraus et al., 1995).

A single G to A transition, changing Arg506 to Gln, thereby creating a inactivation-resistant procoagulant factor Va, has been shown to be the molecular basis for APC resistance in the vast majority of affected individuals (Bertina et al., 1994; Greengard et al., 1994a; Voorberg et al., 1994). An average 17% of patients with venous thromboembolism were found to be carriers of this lesion, called factor V Leiden (De Stefano et al., 1996). A similar prevalence figure (18.2%) was arrived at by Ridker et al. (1995b) in their study of idiopathic venous thromboembolism among otherwise healthy US males.

The same study suggested that carriership for factor V Leiden would also confer a four- to five-fold increase in recurrence risk. This finding, however, contrasts with that of Rintelen et al. (1996a) who reported a lack of association between heterozygous factor V Leiden carriership and increased recurrence rates in unselected patients with venous thromboembolism. During an average follow-up of 5 years, some 25% of both carriers and non-carriers experienced a recurrent thrombotic episode; this rate was identical to that generally reported, disregarding factor V Leiden genotype (Prandoni et al., 1996). Since the recurrence rate among idiopathic cases studied by Ridker et al. (1995b) was 29% for carriers, but only 11% for non-carriers, the latter group of patients was probably selected for the avoidance of risk factors predisposing to recurrent venous thrombosis.

In unaffected European controls, the frequency of factor V Leiden heterozygotes is approximately 4–5% (Rees et al., 1995) which compares well with estimates of the general prevalence of APC resistance of 5% (Rosendaal et al., 1994; Svenson and Dahlbäck, 1994). The mutation is found at a frequency of 0.6% in Asia Minor and is virtually absent from southeast Asian, Australasian, South American and African populations (Rees et al., 1995).

Legnani et al. (1996) demonstrated that even with optimal data interpretation using two different coagulation assays, a certain proportion of cases with APC resistance among their patients (mainly with juvenile thromboembolism) was not explicable in terms of the factor V Leiden mutation. Leroy-Matheron et al. (1996) also reported on an incomplete co-segregation of factor V Leiden with APC-resistance in the families of 297 patients with unexplained thrombosis. In a group of 63 patients with ischaemic stroke studied by Fisher et al. (1996), six were diagnosed APC-resistant, but none of them had factor V Leiden. These findings imply that factor V Leiden is the predominant, but not the only cause of APC resistance. It is estimated that, when all experimental variables are carefully controlled, the APC test is some 85–90% sensitive and specific for the factor V mutation (Dahlbäck, 1995c). A study of Olivieri et al. (1995) provided an indication as to which factors might confound the assessment of APC resistance. Of eight apparently healthy women diagnosed APC-resistant among 50 users and 50 non-users of oral contraceptives, seven were on the pill but only two were heterozygous for factor V Leiden.

Because of its high population frequency, homozygosity for factor V Leiden can be expected to be quite common. Indeed, Rosendaal et al. (1995) identified seven homozygotes among 471 consecutive patients with DVT; however, no homozygote was found in 474 healthy controls. The clinical course of the homozygous patients showed no remarkable features, indicating that the homozygous abnormality is much less severe than, for example, homozygous protein C or protein S deficiency. This view is also supported by the findings of Greengard *et al.* (1994b) who reported that two out of four homozygous offspring of a heterozygous couple had remained symptom-free until the age of 30 whilst a heterozygous brother had severe thrombotic episodes that first occurred around the age of 20. Based upon the prevalence expected in their control group under Hardy–Weinberg conditions, Rosendaal *et al.* (1995) calculated the relative thrombosis risk to be 79.4 for homozygous carriership of factor V Leiden. When controlled for age, the relative risk was highest (140) in individuals less than 30 years of age, decreasing (98) in those between 30 and 50 years, and falling to 30 for individuals more than 50 years old. No such age effect was observed for factor V Leiden heterozygosity.

15.3.4 Protein S deficiency

Protein S is the principal cofactor of activated protein C in the inactivation of factors Va and VIIIa (Chapter 6). It is synthesised in the liver and, in smaller quantities, in other cells. Protein S exists in free form (30–40%) and, alternatively, in a stoichiometric complex with C4b-binding protein. Only the free form, however, has APC cofactor activity (Lane *et al.*, 1996c).

Protein S deficiency was noted in six out of 277 consecutive patients (2.2%) with venographically proven DVT by Heijboer *et al.* (1990). This figure is substantially smaller than the 7.6% prevalence figure reported by Melissari *et al.* (1992) among 393 adult patients with a history of acute venous thromboembolism. However, due to the potential ascertainment bias affecting the study of Melissari *et al.* (1992), their estimate is probably more representative of the frequency of protein S deficiency among inherited thrombophilias. In families segregating heterozygous protein S deficiency, the thrombotic risk profile of deficient individuals was not notably different from that noted for protein C and ATIII deficiency (Pabinger and Schneider, 1996).

Data on the prevalence of protein S deficiency in the general population are lacking (De Stefano *et al.*, 1996; Lane *et al.*, 1996c) so that relative risks are difficult to determine. Extrapolation from cohorts of thrombotic patients suggest a prevalence of symptomatic protein S deficiency of 1 in 33 000 (De Stefano *et al.*, 1996) but, like with protein C deficiency, the existence of a much more frequent, clinically 'covert' deficiency state cannot be excluded. Indeed, according to unpublished data of Drs Walker and Tait, six of 5000 Scottish blood donors had both total and free protein S levels 2 SD below the mean, suggesting a population frequency of approximately 0.1% (I. Walker, personal communication). Furthermore, at least one close relative of these individuals had a similar phenotype. Additional evidence for the common occurrence of protein S deficiency was provided by the study of Halbmeyer *et al.* (1994b). Two of 50 controls had a protein S activity of 60%, indicative of a mild deficiency state. One of these individuals, however, was also APC resistant.

15.3.5 Hyperhomocysteinaemia

Homocysteine is a sulphydryl amino acid which is derived from methionine through metabolitic conversion; it occurs in free and protein-bound form. Total homocysteine has a concentration in normal plasma of between 5 and 16 μmol l^{-1} (De Stefano *et al.*,

1996). It is normally metabolized in the cell to either methionine or cysteine, and although defects in this pathway represent an important risk factor for venous thrombosis (Chapter 14), the precise pathogenic mechanisms underlying the disease association are not yet fully understood. Animal studies have suggested that homocysteine is toxic to endothelial cells and that its persistence could cause vessel wall injury. Other mechanisms may involve factor V activation and interference with protein C activation (De Stefano et al., 1996).

The most frequent cause of severe hyperhomocysteinaemia (plasma concentration >100 µmol l^{-1}) is homozygous deficiency for cystathionine β-synthase, an enzyme which catalyses one step in the transulphuration reaction leading to cysteine. This abnormality occurs at a frequency of between 1 in 200 000 and 1 in 300 000 in the general population (Mudd et al., 1989; Malinov, 1994). The heterozygous deficiency state, causing mild (16–25 µmol l^{-1}) or moderate (25–100 µmol l^{-1}) hyperhomocysteinaemia, has a prevalence of 0.3–1.4% (Mudd et al., 1989), suggesting either strong preclinical selection or low penetrance for the homozygous deficiency state. Another cause of mild or moderate hyperhomocysteinaemia is homozygous thermolability of methylenetetrahydrofolate reductase (MTHFR), a component of the remethylation pathway leading to methionine. This condition has a population frequency of 5%. Finally, malnutrition and age are confounding factors in the causation of hyperhomocysteinaemia, leading to an acquired form of the disease.

Mild and moderate hyperhomocysteinaemia occurred in 28 (10%) of 269 patients with a first episode of venous thrombosis (Den Heijer et al., 1996). Together with the observed prevalence of 13 cases among 269 age- and sex-matched controls, this figure yielded a relative risk of 2.3. The disease association was stronger in females than in males and increased with age. In a study of 107 patients who had experienced their first venous thromboembolic event before the age of 45, 13.1% were diagnosed to have moderate hyperhomocysteinaemia (Fermo et al., 1995). These individuals were also found to have experienced a two-fold increase in recurrence risk as compared to normohomocysteinaemic patients. Inheritance of hyperhomocysteinaemia was confirmed by Fermo et al. (1995) in the vast majority of families analysed. This result corroborated the findings of Falcon et al. (1994) that most index cases of early venous thrombosis (less than 40 years) with hyperhomocysteinaemia (19% of total) had at least one first-degree relative with the same condition.

15.3.6 Elevated factor VIII level

Factor VIII is synthesised in the liver and secreted to the plasma where it serves as a cofactor in the activation of factor X by activated factor IX. Factor VIII is noncovalently bound to its carrier protein, von Willebrand factor (vWF).

Owing to its procoagulant potential, the heritable overexpression of factor VIII could represent a potent prothrombotic genetic factor. Indeed, Koster et al. (1995a) have demonstrated that an elevated factor VIII level is a substantial risk factor for DVT. Of 301 patients with an objectively diagnosed episode of venous thrombosis, 76 (25%) had a plasma factor VIII coagulant activity higher than 1500 IU l^{-1}. Only 34 age and sex-matched controls (11%) fell into this high-risk stratum, yielding a relative risk of 2.7. Heritability of elevated factor VIII levels has, however, not yet been demonstrated. Elevated vWF also appeared to be a risk factor for DVT but, when corrected for factor VIII levels in a multivariate analysis, relative risks were not notably different from unity.

15.3.7 Plasminogen deficiency

Plasminogen is the plasma zymogen of plasmin, a serine protease which, upon activation, effects the degradation of fibrin. It is synthesised in the liver and normally occurs in plasma at high concentrations (approximately 200 mg l^{-1}).

The role of plasminogen deficiency in venous thrombosis (Chapter 9) is still unclear. Families segregating the deficiency state have been reported, but often the index case was the only family member suffering from thrombosis (Dolan and Preston, 1988). In a retrospective study of 20 families, Sartori et al. (1994) reported that 9.5% of plasminogen-deficient relatives of a deficient, thrombotic index patient had also experienced a thrombotic event. All 95 non-deficient siblings were asymptomatic.

Heijboer et al. (1990), on the other hand, observed only four plasminogen-deficient individuals among 277 consecutive thrombotic patients (1.4%) and one deficient individual among 138 controls (0.7%). A similar proportion of symptom-free cases was found by Tait et al. (1992). Of 9611 blood donors, 61 (0.6%) had a plasminogen level lower than 65%, demarcating the lower end of the normal range in that study. In patients with early onset venous thrombosis, plasminogen deficiency was reported to occur at a frequency of 0.9% in Italy (Fermo et al., 1995) and Jordan (Awidi et al., 1993). These findings suggest that any association of plasminogen deficiency with thrombotic disease would be much dependent upon the presence of further genetic risk factors, and that the deficiency state represents a strong risk factor only in families with at least one deficient individual suffering thrombosis.

15.3.8 Increased prothrombin level

Prothrombin (or factor II) is the zymogen of thrombin, a key enzyme in the blood coagulation cascade which converts fibrinogen to fibrin. It is activated by factor Xa on the surface of activated platelets in the presence of factor V (Chapter 14).

In a recent study of 28 probands with documented hereditary venous thrombophilia, Poort et al. (1996) failed to find any variations other than known polymorphisms in the exons and 5' untranslated region of the prothrombin gene. At nucleotide position 20210 in the 3' untranslated region, however, a G → A transition was found in five patients (18%). None of the five healthy controls showed this variant. The following case–control study (Poort et al., 1996) identified 29 heterozygous carriers among 471 unselected thrombotic patients (6.2%) and 11 carriers among age- and sex-matched controls (2.3%). This polymorphism, which was also demonstrated to be associated with an increased plasma prothrombin level, thus confers a relative risk of thrombosis of 2.8.

15.3.9 Other presumed genetic risk factors

Kawasaki et al. (1995) investigated the possible association between hyperlipidaemia and DVT. When judged on the basis of serum total cholesterol and triglyceride, 59% of consecutive patients with DVT were diagnosed hyperlipidaemic as opposed to 29% of controls. If these results are corroborated by others, hyperlipidaemia would represent a major risk factor for venous thrombosis in the Japanese population.

Koster et al. (1994b) confirmed an important role for hyperfibrinogenaemia but not for factor VII level as a risk factor for venous thrombosis. Among 199 patients with objectively confirmed DVT and the same number of age- and sex-matched controls, individuals with a plasma fibrinogen level higher than 5 g l⁻¹ had an almost four-fold increased disease risk as compared to those with levels less than 3 g l⁻¹. Adopting a threshold of 4 g l⁻¹ for the definition of hyperfibrinogenaemia, this phenotype was found among 8% of controls and 15% of patients (relative risk 2.0). The distribution of factor VII levels in the same cohorts, by contrast, did not show any significant differences.

PAI-1 (Chapter 13) is unlikely to represent a strong genetic risk factor for venous thrombosis. An increased PAI-1 activity, which might be assumed to predispose to thrombotic episodes through excessive masking of plasminogen activator, was found by Engesser *et al.* (1989) in 14/113 (12%) unrelated thrombotic patients with a familiy history of the disease, but only in 5/90 non-familial cases (6%). Since the latter figure was not notably different from the prevalence in controls (3/100), this was regarded as evidence against a strong disease association in the general population. Furthermore, in half of the patients, the increased PAI-1 activity was only transient and had disappeared after 1 year (Engesser *et al.*, 1989). In the seven families of patients with persistently elevated PAI-1 levels and a familial history of disease, the phenotype was either not inherited or failed to co-segregate with thrombosis.

Factor XII deficiency (Chapter 11) has been implicated in thrombosis on the basis of a study by Halbmayer *et al.* (1992). Of 103 patients with thrombosis, 15 had factor XII deficiency. However, the study cohort included a variety of disease states and when confined to venous thromboembolism, the prevalence of factor XII deficiency was only 8%. This figure is in keeping with the results of Koster *et al.* (1994a) who observed a frequency of 6% among 350 unselected patients with DVT and 5% among 350 healthy age- and sex-matched controls. Both studies therefore imply that factor XII deficiency is not a strong risk factor for venous thromboembolism.

15.4 Integration of risk factors

As with other multifactorial diseases, each case of venous thrombosis marks the end-point of a highly individual history of exposure to a plethora of risk factors. Some risk factors, mainly those of genetic type, may be present at birth or even before, but in the vast majority of cases, they will not inevitably lead to thrombosis in later life. With the exception of some rare genetic defects, causing, for example, neonatal purpura fulminans, known risk factors are not associated with venous thromboembolism in an all-or-nothing fashion. For every condition observed in patients, at least a small number of healthy individuals can be found with the same condition, with the same age and sex, but without any history of thrombosis.

This lack of unequivocal association is explicable in terms of a large number of risk factors integrating during an individual's lifetime, rendering identical exposure profiles in different individuals very unlikely. Each exposure contributes to the liability of an individual to disease, most often in a poorly understood way, and only when this liability exceeds a certain threshold will thrombosis occur. Critical for the relative impact of a combination of risk factors upon an individual's disease risk is the way in which the risk factors integrate.

Complete lack of interaction is unlikely to hold for risk factors for venous thrombosis. For example, in a study among patients after major trauma, an age-related relative risk of 1.05 per year was arrived at by means of logistic regression (Geerts *et al.*,

1994). This result would imply an increase of DVT risk by 63% within a decade, almost half of that found in unselected thrombotic patients (Section 15.2.1). One meaningful explanation for this substantial discrepancy in age-related relative risk would be that, in terms of thrombotic liability, injury implies a much more dramatic physiological change in the young as compared to old people. Therefore, age would be a less important thrombotic risk factor in those experiencing a major trauma.

A model commonly adopted in epidemiology for the integration of risk factors is logistic regression. If n risk factors are to be considered, let $x = (x_1,...,x_n)$ denote their indicator function (i.e. $x_i = 1$ if the individual in question has been exposed to the ith risk factor, $x_i = 0$ otherwise). The disease risk is then assumed to be:

$$r(x) = \frac{e^{h(x)}}{1+e^{h(x)}}$$

where $h(x) = a + b_1 \cdot x_1 +...+ b_n \cdot x_n$ is a linear combination of the x_i. This model implies that risk factors contribute independently on the so-called 'logit'-scale, and that they do not interact. For two risk factors, parameters a, b_1, and b_2 can be determined from the disease risks bound up with the three risk profiles other than mutual exposure [$x = (1,1)$]. Deviation from the assumption of no interaction can then be measured by the ratio of observed over expected risk for mutual exposure. If the two risk factors do not interact, this ratio should be unity.

15.4.1 APC resistance and protein C deficiency

The integration of risk factors has been explored most rigorously for primary hyper-coagulable states and, because of its comparatively high population frequency, hereditary APC resistance has been the most important one. In light of the direct functional relationship, protein C deficiency is a good candidate for a risk factor strongly interacting with APC resistance. Furthermore, the equivocal inheritance pattern of thrombosis as a consequence of protein C deficiency requires a second predisposing trait segregating in families of heterozygous protein C-deficient thrombosis patients.

Koeleman et al. (1994) studied the co-segregation of the two traits in six clinically dominant protein C-deficient families from The Netherlands. When propositi were excluded, a thrombotic episode had been experienced by 12/19 family members (63%) with both mutations, 9/31 family members (29%) with the protein C mutation only, 2/20 members (10%) with the factor V Leiden mutation only, and by two of the remaining 28 individuals (7%). Therefore, the relative risk of protein C deficiency was 1.6 times higher among carriers of the factor V Leiden variant than among non-carriers. This difference points to an interaction of the two risk factors in the respective families. In fact, assuming a logistic model, the observed risk for carriers of both mutations (0.63) would be 1.7 times higher than expected (0.38) under the assumption of no interaction. The combined risk, relative to mutual non-carriership, was 8.8 and approximates to the product of the two individual relative risks (5.0 for protein C and 1.9 for factor V Leiden).

Brenner et al. (1996) reported upon an even more dramatic example of interaction. In a large Arab family segregating factor V Leiden and the T298M mutation of protein C, venous thrombosis was exclusively confined to compound carriers of both lesions (five of 11 were affected). By contrast, all respective heterozygotes for T298M (13) and factor V Leiden (10) alone as well as the 12 family members bearing neither mutation were symptom-free by the time of the study.

The results of Gandrille *et al.* (1995b) appear to contradict the above family data. Among 113 French unrelated symptomatic patients with protein C deficiency, 15 (14%) were found to be carriers of factor V Leiden. Hallam *et al.* (1995a) observed the mutation in only 6% of British protein C-deficient thrombotic patients. Since these prevalence figures are lower than the frequency among unselected thrombotic patients (Section 15.3.3), presence of the mutation did not appear to increase the relative thrombotic risk of protein C-deficient individuals.

Various explanations are possible for this apparent discrepancy. Firstly, ascertainment bias could have affected the two family studies. The probability of a family being recognized as segregating both risk factors increases with the number of affected family members bearing both risk factors. Secondly, additional genetic variants may be necessary for an interaction between APC resistance and protein C deficiency to become effective. Carriership for such variants would, however, not only increase the likelihood for a family to come to clinical attention but also the chance of simultaneous occurrence of the two risk factors in affected family members. Finally, population differences in the frequency of factor V Leiden among protein C-deficient thrombotic patients have been documented (Hallam *et al.*, 1995a) and must be taken into account.

15.4.2 APC resistance and protein S deficiency

Protein S acts as a cofactor of activated protein C and its deficiency might therefore also interact with APC resistance as a thrombotic risk factor. Indeed, Koeleman *et al.* (1995) found six carriers of the mutation among 16 protein S-deficient symptomatic patients, admitting, however, that this high prevalence (36%) was likely to reflect selection of probands for familial thrombosis. Nevertheless, an identical proportion was noted by Zöller *et al.* (1995a) who also studied the co-segregation of protein S deficiency and factor V Leiden in seven families. Of 18 family members with both defects, 12 (72%) had thrombosis. A history of the disease was noted in 4/21 individuals carrying either mutation alone (19%), and in 1/44 non-carriers (2%). As with protein C deficiency, the combined risk (0.72) was 1.4 times higher than expected (0.50) in a logistic model without interaction. The combined risk, relative to mutual non-carriership, was 29.3 and was thus again very close to the product of the two individual relative risks (5.3 for both protein S deficiency and factor V Leiden).

15.4.3 APC resistance and ATIII deficiency

Data related to the interaction of factor V Leiden and ATIII deficiency as risk factors for venous thrombosis are more scarce. The only available study to date was performed by van Boven *et al.* (1996) in nine families segregating factor V Leiden together with an ATIII mutation. Although the sample size was too small to allow any quantitative assessment of their interaction, the fact that 11/12 carriers of both defects suffered thrombosis, compared with 15/30 carriers of the ATIII mutation only, and 1/5 carriers of factor V Leiden, indicates that the co-occurrence of both lesions enhances their individual potential to predispose to venous thrombosis.

15.4.4 Genetic risk factors, oral contraceptives and pregnancy

Oral contraceptive use is a strong thrombotic risk factor in the female half of the population (Section 15.2.2) and, because of its high prevalence, may represent an epidemiologically relevant enhancer of other risk factors. Vandenbroucke *et al.* (1994) studied the predisposing interaction of oral contraceptive use and factor V Leiden among 155 consecutive premenopausal women (aged 15–49) with DVT and 169 controls. The relative risk for oral contraceptive use was 3.8, a value which was in perfect agreement with that observed in European populations in a large WHO study (1995). Factor V Leiden yielded a relative risk of 7.9. After the exclusion of five homozygotes, all of whom had experienced DVT, this risk figure reduced to 6.8 for factor V Leiden heterozygosity and thus matched the result of Koster *et al.* (1993) for unselected DVT patients.

Interaction of the two conditions was apparent. Factor V Leiden turned out to be a stronger risk factor among users of oral contraceptives than among non-users (relative risks 9.4 and 6.9). Contraceptive pill usage presented a greater risk to heterozygotes for the factor V mutation than to wild-type homozygotes (5.0 and 3.7). The combined relative risk was 34.7, compared with mutual non-exposure. This value was again close to the product of the two individual relative risks. A later study by the same authors (Bloemenkamp *et al.*, 1995) provided some evidence that, among carriers of the factor V Leiden mutation, the risk of third-generation progestagens in combined preparations (Section 15.2.2) is not higher than that of older makes.

If a multiplicative relationship holds for factor V Leiden homozygosity as well, its relative risk combined with oral contraceptive use could be as high as 300. Indeed, Rintelen *et al.* (1996b) claimed an important precipitating role for oral contraceptives in factor V Leiden-homozygous females on the basis of their observation that, of 17 female thrombosis patients with the homozygous genotype, 12 (70%) had taken the pill for 6–150 months prior to thrombosis. Since a similar proportion of pill-takers has been observed by Vandenbroucke *et al.* (1994) among both heterozygous and wild-type homozygous DVT patients, however, the data of Rintelen *et al.* (1996b) were of limited affirmative power.

Because of their relative rareness, other genetic risk factors have not been analysed in much detail for their potential synergistic effects with oral contraceptives. One study (Pabinger *et al.*, 1994a) has focused upon the thrombosis-free period following a first pill prescription to females with inherited ATIII, protein C and protein S deficiency. The length of this time interval was compared with the symptom-free period of age-matched controls suffering from the same pro-coagulant deficiency state. It turned out that the annual hazard risk (i.e. risk of first thrombotic episode occurring in a given time interval) was increased by oral contraceptives only in ATIII-deficient females (27.5% per annum in pill takers as compared to 3.4% per annum in non-users). For protein C and protein S deficiency, hazard rates were not statistically different (protein C, 12.0% vs. 6.9%; protein S, 6.5% vs. 8.6%).

Hellgren *et al.* (1995) determined the APC response in 34 women who experienced venous thrombotic complications in connection with pregnancy. Since 20 of these were APC resistant (59%), a much higher proportion than observed among unselected thrombotic patients, this was held to indicate a synergistic effect. The number of positive diagnoses was the same among non-pregnant female controls (6/57, 10%) and normal pregnancies (2/18, 11%) but, unfortunately, no data were included on the prevalence of APC resistance among non-pregnant female thrombosis patients. Previous studies by the same group, however, suggest a comparatively high frequency of 40% (Zöller *et al.*, 1994). Adopting this value, APC resistance was bound up with a relative risk of 11.4 in pregnant as compared to 5.7 in non-pregnant women. The relative risk of pregnancy among carriers and non-carriers, as well as the combined risk, could not be calculated in this way.

15.5 Conclusions

In this chapter, we have attempted to review the current state of the epidemiology of venous thrombotic disorders. Major risk factors encountered are summarized in *Table 15.1* together with their frequencies in the thrombotic and non-thrombotic parts of an archetypal Caucasian population. Included in *Table 15.1* are prevalences and relative risks considered in previous sections as well as data from other, sometimes unpublished sources. Although very much based upon average figures, this compilation can nevertheless guide our thinking as to the relative contribution of each individual factor to the overall risk for venous thrombosis.

A rough estimate of the percentage of cases of venous thrombosis attributable to a given risk factor (percentage attributable risk, PAR) in the general population is provided by:

$$\text{PAR} = 100 \cdot \frac{F_{\text{P}} \cdot (\text{RR}-1)}{F_{\text{P}} \cdot (\text{RR}-1) + 1}$$

where F_{P} is the frequency in the general population and RR is the relative risk (Hennekens and Buring, 1987). Inspection of *Table 15.1* reveals that the epidemiologically most important genetic risk factor is factor V Leiden. Carriership for this

Table 15.1. Relative venous thrombosis risks in an archetypal Caucasian population

Hypercoagulable states	F_{T}	F_{P}	RR	PAR
Primary				
Protein C deficiency	2.1%	0.3%	7.1	1.8%
ATIII deficiency	1.1%	0.2%	5.6	0.9%
Factor V Leiden	20%	4%	6.0	16.7%
Protein S deficiency	2.2%	0.2%	11.2	2.0%
Hyperhomocysteinaemia	10%	4.8%	2.2	5.4%
Increased prothrombin level	6.2%	2.3%	2.8	4.0%
Hyperfibrinogenaemia	15%	8%	2.0	7.4%
(elevated factor VIII)	25%	11%	2.7	15.6%
Secondary				
Oral contraceptives	21%	6%[a]	4.2	16.1%
History of thrombosis	14%	2%	8.0	12.3%
Pregnancy	6.2%	1.3%[b]	5.0[c]	4.9%
Recent surgery or trauma	23%	4%[d]	7.2[e]	19.9%

F_{T}, frequency among patients with venous thrombotic disorder; F_{P}, frequency in the general population; RR, relative risk (or odds ratio at times); PAR, percentage attributable risk.
[a]Estimate by Welsh Family Planning Association for 1996 (43% of Welsh women aged 16–49, which comprise 25% of all Welsh females).
[b]Estimate communicated by Dr Angus Clarke (2.6% of women in the Cardiff area become pregnant in a given year).
[c]Assuming a thrombosis risk of 0.8% for the total pregnancy period and a general risk of 1.6×10^{-3} per annum.
[d]Estimate based on annual referral rate to hospitals in the Cardiff area (provided by Dr Angus Clarke).
[e]Adopting two-thirds of the estimate of Thorogood *et al.* (1992) for fatal venous thromboembolism in females.

mutation explains 16.7% of all cases of venous thrombosis in the population. Since segregation of elevated factor VIII levels has not yet been demonstrated in affected families, it is unclear whether this condition also represents a genuine genetic risk factor or is instead a secondary phenomenon associated with other prothrombotic states. Simple summation in *Table 15.1* reveals, however, that, increased factor VIII level excluded, the known primary hypercoagulable states together account for only 40% of thrombotic risk.

Unfortunately, the combined frequencies of primary and/or secondary hypercoagulable states as well as the nature of their higher-order interaction are not sufficiently well known to allow assessment of the overall PAR of the risk factors listed. Nevertheless, a first approximation can be arrived at by assuming that both the frequency and the relative risk of a combination of genetic risk factors equals the product of the individual prevalences and risks, respectively. Under this simplistic assumption, the primary hypercoagulable states in *Table 15.1* would yield a combined PAR value of 46.1% (factor VIII level excluded). Thus, if a small number of major gene defects underlie venous thrombosis in the vast majority of cases (a scenario bearing the most optimistic perspectives for future epidemiological work), then half of these defects would still remain to be characterized. Major gene effects are, however, comparatively easy to detect and to map by means of family studies.

Alternatively, those 50% of cases without a known genetic basis could be entirely due to exogenous causes. This scenario is unlikely even though the PAR values for the four most important secondary hypercoagulable states listed in *Table 15.1* sum up to 53%. First, there is substantial exposural overlap between different environmental risk factors, rendering simple summation of PARs inadmissible. For example, a substantial proportion of individuals subsumed under 'Recent surgery or trauma' would also be included in the 'History of thrombosis' group. Second, all relative risks are moderately high, implying that a considerable proportion of individuals exposed to a given environmental risk factor will not suffer from venous thrombosis. Inter-individual differences in liability must therefore exist and their most parsimonious explanation would be in terms of genetic variation. Finally, venous thrombosis is the result of an endogenous physiological process; its occurrence reflects the specificity and efficiency of the blood coagulation system to fulfil its natural biological role. Since the regulation of this system is under strong genetic control, a purely exogenous thrombotic pathway is scarcely conceivable.

The final scenario to be considered here is that of a large number of minor genetic risk factors exerting their predisposing effects in a concerted, not necessarily additive, fashion. Discrimination between this and a mainly environmental pattern of causation should be feasible through twin studies or heritability studies among close relatives. Mapping of the underlying genes, however, will be difficult since the number of families or pairs of affected relatives required in a genome scan would be negatively correlated with the relative contribution of a gene in question to the family-specific risk profile. A more promising approach should be the mutation analysis of candidate genes in thrombotic patients selected for the absence of confounding, especially monogenic and environmental risk factors.

In the light of their major impact on public health in industrialized countries, a comprehensive characterization of the factors predisposing to venous thrombosis is of prime importance. At the same time, a deeper understanding is required of the interaction of individual risk factors. Such knowledge would allow us not only to pin-point risk profiles with high likelihoods of disease but also to defuse them by avoiding or eliminating their most crucial components. For the unforseeable future venous thrombosis will thus remain a major field of work in genetic epidemiology.

References

Abildgaard U (1968) Highly purified antithrombin III with heparin co-factor activity prepared by disc electrophoresis. *Scand. J. Clin. Lab. Invest.* **21**: 89–91.

Abildgaard U (1969) Binding of thrombin to antithrombin III. *Scand. J. Clin. Lab. Invest.* **24**: 23–27.

Abildgaard U (1981) Antithrombin and related inhibitors of coagulation. In: Pollar L (ed.) *Recent Advances in Blood Coagulation*, Vol. 3, pp.151–173. Churchill Livingstone, Edinburgh.

Abildgaard U (1995) Relative roles of tissue factor pathway inhibitor and antithrombin in the control of thrombogenesis. *Blood Coag. Fibrinol.* **6** (Suppl. 1): S45-S49.

Abumiya T, Yamaguchi T, Terasaki T, Kokawa T, Kario K, Kato H (1995) Decreased plasma tissue factor pathway inhibitor activity in ischaemic stroke patients. *Thromb. Haemost.* **74**: 1050–1054.

Aiach M, Gandrille S, Emmerich J (1995) A review of mutations causing deficiencies of antithrombin, protein C and protein S. *Thromb. Haemost.* **74**: 81–89.

Aillaud MF, Succo E, Alessi MC, Gandois JM, Gallian P, Morange P, Juhan-Vague I (1995) Resistance to activated protein C – diagnostic strategy in a laboratory of haemostasis. *Thromb. Haemost.* **74**: 1197–1207.

Al-Mondhiry HAB, Bilezikian SB, Nossel HL (1975) Fibrinogen 'New York' – an abnormal fibrinogen associated with thromboembolism: functional evaluation. *Blood* **45**: 607–619.

Alber T (1989) Mutational effects on protein stability. *Annu. Rev. Biochem.* **58**: 765–798.

Alessi MC, Aillaud MF, Paut O, Roquelaure B, Alhenc-Gelas M, Pellissier MC, Ghanen N, Juhan-Vague I (1996) Purpura fulminans in a patient homozygous for a mutation in the protein C gene – prenatal diagnosis in a subsequent pregnancy. *Thromb. Haemost.* **75**: 520–526.

Alexandre P, Larcan A, Briquel ME (1980) Recurring thromboembolic accidents caused by family-related deficiency of the fibrinolysis system. *Blut* **41**: 437–444.

Alhenc-Gelas M, Emmerich J, Gandrille S, Aubry ML, Benaily N, Fiessinger JN, Aiach M (1995) Protein C infusion in a patient with inherited protein C deficiency caused by two missense mutations: Arg 178 to Gln and Arg-1 to His. *Blood Coag. Fibrinol.* **6**: 35–41.

Allaart CF, Aronson DC, Ruys TH, Rosendaal FR, van Bockel JH, Bertina RM, Briët E (1990) Hereditary protein S deficiency in young adults with arterial occlusive disease. *Thromb. Haemost.* **64**: 206–210.

Allaart CF, Poort SR, Rosendaal FR, Reitsma PH, Bertina RM, Briët E (1993) Increased risk of venous thrombosis in carriers of hereditary protein C deficiency defect. *Lancet* **341**: 134–138.

Allaart CF, Rosendaal FR, Noteboom WMP, Vandenbroucke JP, Briët E (1995) Survival in families with hereditary protein C deficiency, 1820 to 1993. *Br. Med. J.* **311**: 910–913.

Allen DH, Tracy PB (1995) Human coagulation factor V is activated to the functional cofactor by elastase and cathepsin G expressed at the monocyte surface. *J. Biol. Chem.* **270**: 1408–1415.

Almer L-O, Öhlin H (1987) Elevated levels of the rapid inhibitor of plasminogen activator in acute myocardial infarction. *Thromb. Res.* **47**: 335–339.

Alpert JS, Dalen JE (1994) Epidemiology and natural history of venous thromboembolism. *Prog. Cardiovasc. Dis.* **36**: 417–422.

Amphlett GW, Kisiel W, Castellino FJ (1981) Interaction of calcium with bovine plasma protein C. *Biochemistry* **20**: 2156–2161.

Amundsen T, Ueland PM, Waage A (1995) Plasma homocysteine levels in patients with deep venous thrombosis. *Arterioscler. Thromb. Vasc. Biol.* **15**: 1321–1323.

Andersen BD, Lind B, Phillips M, Hanson AB, Ingerslev J, Thorsen S (1996) Two mutations in exon XII of the protein Sα gene in four thrombophilic families resulting in premature stop codons and depressed levels of mutation mRNA. *Thromb. Haemost.* **76**: 143–150.

Anderson FA, Wheeler HB, Goldber RJ (1991) A population-based perspective of the hospital incidence and case-fatality rates of deep vein thrombosis and pulmonary embolism. The Worcester DVT Study. *Arch. Intern. Med.* **151**: 933–938.

Anderson GM, Shaw AR, Shafer JA (1993) Functional characterization of promoter elements involved in regulation of human Bβ-fibrinogen expression. *J. Biol. Chem.* **268**: 22650–22655.

Andersson TR, Larsen ML, Abildgaard U (1987) Low heparin cofactor II associated with abnormal crossed immunoelectrophoresis pattern in two Norwegian families. *Thromb. Res*. **47**: 243–248.

Andersson A, Dahlbäck B, Hansen C, Hillarp A, Levan G, Szpirer J, Szpirer C (1990) Genes for C4b-binding protein α- and β-chains (*C4BPA* and *C4BPB*) are located on chromosome 1, band 1q32, in humans and on chromosome 13 in rats. *Somat. Cell Mol. Genet*. **16**: 493–500.

Andreasen PA, Riccio A, Welinder KG, Douglas R, Sartorio R, Nielsen LS, Oppenheimer C, Blasi F, Danø K (1986) Plasminogen activator inhibitor type-1: reactive center and amino-terminal heterogeneity determined by protein and cDNA sequencing. *FEBS Lett*. **209**: 213–218.

Andreasen PA, Georg B, Lund LR, Riccio A, Stacey SN (1990) Plasminogen activator inhibitors: hormonally regulated serpins. *Mol. Cell. Endocrinol*. **68**: 1–19.

Angeles-Cano E, Gris JC, Loyau S, Schved JF (1993) Familial association of high levels of histidine-rich glycoprotein and plasminogen activator inhibitor-1 with venous thromboembolism. *J. Clin. Lab. Med*. **121**: 646–653.

Antonarakis SE, Boehm CD, Giardina PJ, Kazazian HH (1982) Non-random association of polymorphic restriction sites in the β-globin gene cluster. *Proc. Natl Acad. Sci. USA* **79**: 137–141.

Antonarakis SE, Boehm CD, Serjeant GR, Theisen CE, Dover GJ, Kazazian HH (1984) Origin of the β^s-globin gene in Blacks: the contribution of recurrent mutation or gene conversion or both. *Proc. Natl Acad. Sci. USA* **81**: 853–856.

Aoki N, Moroi M, Sakata Y, Yoshida N, Matsuda M (1978) Abnormal plasminogen. A hereditary molecular abnormality found in a patient with recurrent thrombosis. *J. Clin. Invest*. **61**: 1186–1195.

Aoki N, Tateno K, Sakata Y (1984) Differences of frequency distribution of plasminogen phenotypes between Japanese and American populations: new methods for the detection of plasminogen variants. *Biochem. Genet*. **22**: 871–881.

Aparicio C, Dahlbäck B (1996) Molecular mechanisms of activated protein C resistance. *Biochem. J*. **313**: 467–472.

Aranda A, Paramo JA, Rocha E (1988) Fibrinolytic activity in plasma after gynecological and urolological surgery. *Haemostasis* **18**: 129–134.

Ardissino D, Peyvandi F, Merlini PA, Colombi E, Mannucci PM (1996) Factor V (Arg^{506}→Gln) mutation in young survivors of myocardial infarction. *Thromb. Haemost*. **75**: 701–702.

Arenzana N, Rodríguez de Córdoba S, Rey-Campos J (1995) Expression of the human gene coding for the α-chain of C4b-binding protein, C4BPA, is controlled by an HNF1-dependent hepatic-specific promoter. *Biochem. J*. **308**: 613–621.

Arima T, Motomura M, Nishiura Y, Tsujihata M, Okajima K, Abe H, Nagatake S (1992) Cerebral infarction in a heterozygote with variant antithrombin III. *Stroke* 23: 1822–1825.

Aritomi M, Watanabe N, Ohishi R, Gomi K, Kiyota T, Yamamoto S, Ishida T, Maruyama I (1993) Recombinant human soluble thrombomodulin delivers bounded thrombin to antithrombin III: thrombomodulin associates with free thrombin and is recycled to activate protein C. *Thromb. Haemost*. **70**: 418–422.

Arni RK, Padmanabhan K, Padmanabhan KP, Wu TP, Tulinsky A (1993) Structures of the non-covalent complexes of human and bovine prothrombin fragment 2 with human PPACK-thrombin. *Biochemistry* **32**: 4727–4737.

Arnljots B, Dahlbäck B (1995) Protein S as an *in vivo* cofactor to activated protein C in prevention of microarterial thrombosis in rabbits. *J. Clin. Invest*. **95**: 1987–1993.

Arnljots B, Bergqvist D, Dahlbäck B (1994) Inhibition of microarterial thrombosis by activated protein C in a rabbit model. *Thromb. Haemost*. **72**: 415–420.

Arnman V, Nilsson A, Stemme S, Risberg B, Rymo L (1994) Expression of plasminogen activator inhibitor-1 mRNA in healthy, atherosclerotic and thrombotic human arteries and veins. *Thromb. Res*. **76**: 487–499.

Arruda VR, Annichino-Bizzacchi JM, Costa FF, Reitsma PH (1995) Factor V Leiden (FVQ 506) is common in a Brazilian population. *Am. J. Hematol*. **49**: 242–243.

Atha DH, Lormean JC, Petiton M, Rosenberg RD, Choay J (1985) Contribution of monosaccharide residues in heparin binding to antithrombin III. *Biochemistry* **24**: 6723–6729.

Austin RC, Rachubinski RA, Ofosu FA, Blajchman MA (1991a) Antithrombin III Hamilton, Ala 382 to Thr: an antithrombin III variant that acts as a substrate but not an inhibitor of α-thrombin and factor Xa. *Blood* **77**: 2185–2189.

Austin RC, Rachubinski RA, Blajchman MA (1991b) Site-directed mutagenesis of alanine-382 of human antithrombin III. *FEBS Lett*. **280**: 254–258.

Awidi AS, Abu-Khalaf M, Herzallah U, Abu-Rajab A, Shannak MM, Abu-Obeid T, Al-Taher I, Anshasi B (1993) Hereditary thrombophilia among 217 consecutive patients with thromboembolic disease in Jordan. *Am. J. Hematol.* **44**: 95–100.

Aznar J, Dasi A, España F, Estellés A (1986) Fibrinolytic study in a homozygous protein C deficient patient. *Thromb. Res.* **42**: 313–322.

Azuma H, Uno Y, Shigekiyo T, Saito S (1993) Congenital plasminogen deficiency caused by a Ser572 to Pro mutation. *Blood* **82**: 475–480.

Bachmann F (1987) Fibrinolysis. In: Verstraete M, Vermylen J, Lijnen R, Lindhort J (eds) *Thrombosis and Haemostasis*, pp. 227–265. Leuven University Press, Leuven.

Bachmann F, Kruithof EKO (1984) Tissue plasminogen activator: chemical and physiological aspects. *Sem. Thromb. Haemost.* **10**: 6–17.

Bacon-Baguley T, Kudryk BJ, Walz DA (1987) Thrombospondin interaction with fibrinogen. Evidence for binding to the Aα and Bβ chains of fibrinogen. *J. Biol. Chem.* **262**: 1927–1931.

Bacon-Baguley T, Ogilvie ML, Gartner TK, Walz DA (1990) Thrombospondin binding to specific sequences within the Aα and Bβ chains of fibrinogen. *J. Biol. Chem.* **265**: 2317–2321.

Bajzar L, Nesheim M (1993) The effect of activated protein C on fibrinolysis in cell-free plasma can be attributed specifically to attenuation of prothrombin activation. *J. Biol. Chem.* **268**: 8608–8616.

Bajzar L, Fredenburgh JC, Nesheim ME (1990) The activated protein C-mediated enhancement of tissue-type plasminogen activator-induced fibrinolysis in a cell-free system. *J. Biol. Chem.* **265**: 16948–16954.

Bajzar L, Manuel R, Nesheim ME (1995) Purification and characterization of TAFI, a thrombin-activatable fibrinolysis inhibitor. *J. Biol. Chem.* **270**: 14477–14481.

Bajzar L, Morser J, Nesheim ME (1996a) TAFI, or plasma procarboxypeptidase B, couples the coagulation and fibrinolytic cascades through the thrombin–thrombomodulin complex. *J. Biol. Chem.* **271**: 16603–16607.

Bajzar L, Nesheim ME, Tracy PB (1996b) The profibrinolytic effect of activated protein C in clots formed from plasma is TAFI-dependent. *Blood* **88**: 2093–2100.

Bajzar L, Kalafatis M, Simioni P, Tracy PB (1996c) An antifibrinolytic mechanism describing the prothrombotic effect associated with factor V Leiden. *J. Biol. Chem.* **271**: 22949–22952.

Baker ME, French FS, Joseph DR (1987) Vitamin K-dependent protein S is similar to rat androgen-binding protein. *Biochem. J.* **243**: 293–296.

Bakker HM, Tans G, Janssen-Claessen T, Thomassen MCLGD, Hemker HC, Griffin JH, Rosing J (1992) The effect of phospholipids, calcium ions and protein S on rate constants of human factor Va inactivation by activated human protein C. *Eur. J. Biochem.* **208**: 171–178.

Bancroft JD, Schaefer LA, Degen SJF (1990) Characterization of the *Alu*-rich 5′-flanking region of the human prothrombin-encoding gene: identification of a positive *cis*-acting element that regulates liver-specific expression. *Gene* **95**: 253–260.

Bancroft JD, McDowell SA, Friezner Degen SJ (1992) The human prothrombin gene: transcriptional regulation in HepG2 cells. *Biochemistry* **31**: 12469–12476.

Banfield DK, MacGillivray RTA (1992) Partial characterization of vertebrate prothrombin cDNAs: amplification and sequence analysis of the B chain of thrombin from nine different species. *Proc. Natl Acad. Sci. USA* **89**: 2779–2783.

Banner DW, D'Arcy A, Chène C, Winkler FK, Guha A, Konigsberg WH, Nemerson Y, Kirchhofer D (1996) The crystal structure of the complex of blood coagulation factor VIIa with soluble tissue factor. *Nature* **380**: 41–46.

Bantia S, Mane SM, Bell WR, Dang CV (1990) Fibrinogen Baltimore I: polymerization defect associated with a γ292 Gly→Val (GGC→GTC) mutation. *Blood* **76**: 2279–2283.

Bar-Shavit R, Eldor A, Vlodavsky I (1989) Binding of thrombin to subendothelial extracellular matrix. *J. Clin. Invest.* **84**: 1096–1104.

Barnum SR, Dahlbäck B (1990) C4b-binding protein, a regulatory component of the classical pathway of complement, is an acute phase protein and is elevated in systemic lupus erythematosis. *Compl. Inflamm.* **7**: 71–77.

Barstad RM, Stormorken H, Örning L, Stephens RW, Petersen LB, Kierulf P, Sakariassen KS (1994) Reduced thrombus formation in native blood of homozygous factor VII-deficient patients at high arterial wall shear rate. *Blood* **84**: 3371–3377.

Bauer KA (1995) Management of patients with hereditary defects predisposing to thrombosis including pregnant women. *Thromb. Haemost.* **74**: 94–100.

Bauer KA, Rosenberg RD (1987) The pathophysiology of the prethrombotic state in humans: insights gained from studies using markers of hemostatic system activators. *Blood* **70**: 343–350.

Bauer KA, Rosenberg RD (1994) Activation markers of coagulation. *Baillière Clin. Haematol.* 7: 523–542.

Bauer KA, Kass BL, Beeler DL, Rosenberg RD (1984) The detection of protein C activation in humans. *J. Clin. Invest.* 74: 2033–2041.

Bauer KA, Weiss LM, Sparrow D, Vokonas PS, Rosenberg RD (1987) Aging-associated changes in indices of thrombin generation and protein C activation in humans. *J. Clin. Invest.* 80: 1527–1534.

Bauer KA, Broekmans AW, Bertina RM, Conard J, Horellou M-H, Samama MM, Rosenberg RD (1988) Hemostatic enzyme generation in the blood of patients with hereditary protein C deficiency. *Blood* 71: 1418–1426.

Bauer KA, Kass BL, ten Cate H, Bednarek MA, Hawiger JJ, Rosenberg RD (1989) Detection of factor X activation in humans. *Blood* 74: 2007–2015.

Baumann RE, Henschen AH (1994) Linkage disequilibrium relationships among four polymorphisms within the human fibrinogen gene cluster. *Hum. Genet.* 94: 165–170.

Bayston TA, Ireland H, Olds RJ, Thein SL, Lane DA (1994) A polymorphism in the human coagulation factor V gene. *Hum. Mol. Genet.* 3: 2085.

Beaglehole R (1990) International trends in coronary heart disease mortality, morbidity and risk factors. *Epidemiol. Rev.* 12: 1–15.

Beauchamp NJ, Daly ME, Hampton KK, Cooper PC, Preston FE, Peake IR (1994) High prevalence of a mutation in the factor V gene within the U.K. population: relationship to activated protein C resistance and familial thrombosis. *Br. J. Haematol.* 88: 219–222.

Beauchamp NJ, Daly ME, Cooper PC, Makris M, Preston FE, Peake IR (1996) Molecular basis of protein S deficiency in three families also showing independent inheritance of factor V Leiden. *Blood* 88: 1700–1707.

Beaumont V, Malinow MR, Sexton G, Wilson D, Lemort N, Upson B, Beaumont JL (1992) Hyperhomocyst(e)inemia, anti-estrogen antibodies and other risk factors for thrombosis in women on oral contraceptives. *Atherosclerosis* 94: 147–152.

Beck EA, Shainoff JR, Vogel A, Jackson DP (1971) Functional evaluation of an inherited abnormal fibrinogen: Fibrinogen Baltimore. *J. Clin. Invest.* 50: 1874–1879.

Beckmann RJ, Schmidt RJ, Santerre RF, Plutzky J, Crabtree GR, Long GL (1985) The structure and evolution of a 461 amino acid human protein C precursor and its messenger RNA, based upon the DNA sequence of cloned human liver cDNAs. *Nucleic Acids Res.* 13: 5233–5247.

Behn-Krappa A, Hölker I, Sandaradura de Silva U, Doerfler W (1991) Patterns of DNA methylation are indistinguishable in different individuals over a wide range of human DNA sequences. *Genomics* 11: 1–7.

Bell WR, Smith TL (1982) Current status of pulmonary thromboembolic disease: pathophysiology, diagnosis, prevention and treatment. *Am. Heart J.* 103: 289–262.

Bellissimo DB, Kirschbaum NE, Foster PA (1996) Improved method for Factor V Leiden typing by PCR-SSP. *Thromb. Haemost.* 75: 520–526.

Bennett WF, Paoni NF, Keyt BA, Botstein D, Jones ATS, Presta L, Wurm FM, Zoller MJ (1991) High resolution analysis of functional determinants on human tissue-type plasminogen activator. *J. Biol. Chem.* 266: 5191–5201.

Benzakour O, Formstone C, Rahman S, Kanthou C, Dennehy U, Scully MF, Kakkar VV, Cooper DN (1995) Evidence for a protein S receptor(s) on human vascular smooth muscle cells. *Biochem. J.* 308: 481–485.

Benzer S (1961) On the topography of the genetic fine structure. *Proc. Natl Acad. Sci. USA* 47: 403–417.

Beresford CH (1988) Antithrombin III deficiency. *Blood Rev.* 2: 239–250.

Beresford CH, Owen MC (1990) Antithrombin III. *Int. J. Biochem.* 22: 121–128.

Berg L-P, Grundy CB, Thomas F, Millar DS, Green PJ, Krawczak M, Reiss J, Kakkar VV, Cooper DN (1992) *De novo* splice site mutation in the antithrombin III (*AT3*) gene causing recurrent venous thrombosis: demonstration of exon skipping by ectopic transcript analysis. *Genomics* 13: 1359–1361.

Berg L-P, Scopes DA, Alhaq A, Kakkar VV, Cooper DN (1994) Disruption of a binding site for hepatocyte nuclear factor 1 in the protein C gene promoter is associated with hereditary thrombophilia. *Hum. Mol. Genet.* 3: 2147–2152.

Berg DT, Wiley MR, Grinnell BW (1996a) Enhanced protein C activation and inhibition of fibrinogen cleavage by a thrombin modulator. *Science* 273: 1389–1391.

Berg L-P, Soria JM, Formstone CJ, Morell M, Kakkar VV, Estivill X, Sala N, Cooper DN (1996b) Aberrant RNA splicing of the protein C and protein S genes in healthy individuals. *Blood Coag. Fibrinol.* 7: 625–631.

Bergqvist D (1983) Frequency of thromboembolic complications. In: Bergqvist D (ed.) *Post-operative Thromboembolism*, pp. 12–13. Springer, Berlin.

Berkner K, Busby S, Davie E, Hart C, Insley M, Kisiel W, Kumar A, Murray M, O'Hara P, Woodbury R, Hagen F (1986) Isolation and expression of cDNAs encoding human factor VII. *Cold Spring Harb. Symp. Quant. Biol.* **51**: 531–541.

Bernardi F, Marchetti G, Panicucci F, Tripodi M (1986) *Taq*I polymorphism at the human coagulation factor XII locus. *Nucleic Acids Res.* **14**: 5119.

Bernardi F, Marchetti G, Patracchini P, del Senno L, Tripodi M, Fantoni A, Bartolai S, Vannini F, Felloni L, Rossi L, Panicucci F, Conconi F (1987) Factor XII gene alteration in Hageman trait detected by *Taq*I restriction enzyme. *Blood* **69**: 1421–1424.

Bernardi F, Marchetti G, Volinia S, Patracchini P, Casonato A, Girolami A, Conconi F (1988) A frequent FXII gene mutation in Hageman trait. *Hum. Genet.* **80**: 149–151.

Bernardi F, Patracchini P, Gemmati D, Boninsegna S, Guerra S, Legnani C, Ballerini G, Marchetti G (1992) Rapid detection of a protein C gene mutation present in the asymptomatic and not in the thrombosis-prone lineage. *Br. J. Haematol.* **81**: 277–282.

Bernardi F, Legnani C, Micheletti F, Lunghi B, Ferraresi P, Palareti G, Biagi R, Marchetti G (1996) A heparin cofactor II mutation (HCII Rimini) combined with factor V Leiden or type I protein C deficiency in two unrelated thrombophilic subjects. *Thromb. Haemost.* **76**: 505–509.

Bernstein MJ (1986) Prevention of venous thrombosis and pulmonary embolism. Office of Medical Applications of Research. National Institute of Health. *J. Am. Med. Assoc.* **256**: 744.

Berrettini M, Lämmle B, Ciavarella G, Ciavarella N (1985) Functional and immunological studies of abnormal factor XII in a cross-reacting material positive (CRM+) factor XII deficiency. *Thromb. Haemost.* **54**: 120.

Bertina RM (1985) Hereditary protein S deficiency. *Haemostasis* **15**: 241–246.

Bertina RM (1988a) Molecular basis of thrombosis. In: Bertina RM (ed) *Protein C and Related Proteins; Biochemical and Clinical Aspects*, pp. 1–20. Churchill Livingstone, Edinburgh.

Bertina RM (1988b) Prevalence of hereditary thrombophilia and the identification of genetic risk factors. *Fibrinolysis* **2** (Suppl. 2): 7–13.

Bertina RM, Cupers R, van Wijngaardon A (1984) Factor IXa protects activated factor VIII against inactivation by activated protein C. *Biochem. Biophys. Res. Commun.* **125**: 177–183.

Bertina RM, van der Linden IK, Engesser L, Muller HP, Brommer EJP (1987) Hereditary heparin cofactor II deficiency and the risk of development of thrombosis. *Thromb. Haemost.* **57**: 196–200.

Bertina RM, Briët E, Engesser L, Reitsma PH (1988) Protein C deficiency and the risk of thrombosis. *New Engl. J. Med.* **318**: 931.

Bertina RM, Ploos van Amstel HK, van Wijngaarden A, Coenen J, Leemhuis MP, Deutz-Terlouw PP, van der Linden IK, Reitsma PH (1990) Heerlen polymorphism of protein S, an immunologic polymorphism due to dimorphism of residue 460. *Blood* **76**: 538–548.

Bertina RM, Koeleman BPC, Koster T, Rosendaal FR, Dirven RJ, de Ronde H, van der Velden PA, Reitsma PH (1994) Mutation in blood coagulation factor V associated with resistance to activated protein C. *Nature* **369**: 64–67.

Bertina RM, Reitsma PH, Rosendaal FR, Vandenbroucke JP (1995) Resistance to activated protein C and factor V Leiden as risk factors for venous thrombosis. *Thromb. Haemost.* **74**: 449–453.

Bhattacharya A, Shepard AR, Moser DR, Roberts LR, Holland LJ (1991) Molecular cloning of cDNA for the Bβ subunit of *Xenopus* fibrinogen, the product of a coordinately-regulated gene family. *Mol. Cell. Endocrinol.* **75**: 111–121.

Biasiutti FD, Merlo C, Furlan M, Sulzer I, Binder BR, Lämmle B (1995) No association of APC resistance with myocardial infarction. *Blood Coag. Fibrinol.* **6**: 456–459.

Biemond BJ, Levi M, Coronel R, Janse MJ, ten Cate JW, Pannekoek H (1995) Thrombolysis and reocclusion in experimental jugular vein and coronary artery thrombosis. *Circulation* **91**: 1175–1179.

Binnie CG, Lord ST (1991) A synthetic analog of fibrinogen α27–50 is an inhibitor of thrombin. *Thromb. Haemost.* **65**: 165–168.

Binnie CG, Lord ST (1993) The fibrinogen sequences that interact with thrombin. *Blood* **81**: 3186–3192.

Bird AP (1980) DNA methylation and the frequency of CpG in animal DNA. *Nucleic Acids Res.* **8**: 1499–1504.

Bird AP (1986) CpG-rich islands and the function of DNA methylation. *Nature* **321**: 209–213.

Bird AP (1995) Gene number, noise reduction and biological complexity. *Trends Genet.* **11**: 94–100.

Björk I, Jackson CM, Jornvall H, Lavine KK, Nordling K, Salsgiver WJ (1982) The active site of antithrombin. Release of the same proteolytically cleaved form of the inhibitor from complexes with factor IXa, factor Xa and thrombin. *J. Biol. Chem.* **257**: 2406–2411.

Björk I, Nordling K, Larsson I, Olson ST (1992) Kinetic characterization of the substrate reaction between a complex of antithrombin with a synthetic reactive-bond loop tetradecapeptide and four target proteinases of the inhibitor. *J. Biol. Chem.* **267**: 19047–19050.

Björk I, Nordling K, Olson ST (1993) Immunologic evidence for insertion of the reactive-bond loop of antithrombin into the Aβ-sheet of the inhibitor during trapping of target proteinases. *Biochemistry* **32**: 6501–6505.

Blackburn MN, Smith RL, Carson J, Sibley CC (1984) The heparin-binding site of antithrombin III. Identification of a critical tryptophan residue in the amino acid sequence. *J. Biol. Chem.* **259**: 939–941.

Blajchman MA, Fernandez-Rachubinski F, Sheffield WP, Austin RC, Schulman S (1992a) Antithrombin III Stockholm: a codon 392 (Gly→Asp) mutation with normal heparin binding and impaired serine protease reactivity. *Blood* **79**: 1428–1434.

Blajchman MA, Austin RC, Fernandez-Rachubinski F, Sheffield WP (1992b) Molecular basis of inherited human antithrombin deficiency. *Blood* **80**: 2159–2171.

Blake CCF, Harlos K, Holland SK (1987) Exon and domain evolution in the proenzymes of blood coagulation and fibrinolysis. *Cold Spring Harb. Symp. Quant. Biol.* **52**: 925–931.

Blanco A, Bonduel M, Peñalva L, Hepner M, Lazzari M (1994) Deep vein thrombosis in 13-yearold boy with hereditary protein S deficiency and a review of the pediatric literature. *Am. J. Hematol.* **45**: 330–334.

Blann AD, McCollum CN (1994) Von Willebrand factor, endothelial cell damage and atherosclerosis. *Eur. J. Vasc. Surg.* **8**: 10–15.

Blaszczyk R, Ritter R, Thiede C, Wehling C, Hintz G, Neubauer A, Riess H (1996) Simple and rapid detection of factor V Leiden by allele-specific PCR amplification. *Thromb. Haemost.* **75**: 757–759.

Blasi F, Riccio A, Sebastio G (1986) Human plasminogen activators; genes and protein structure. *Horiz. Biochem. Biophys.* **8**: 377–414.

Blinder MA, Tollefsen DM (1990) Site-directed mutagenesis of arginine 103 and lysine 185 in the proposed glycosaminoglycan-binding site of heparin cofactor II. *J. Biol. Chem.* **265**: 286–291.

Blinder MA, Marasa JC, Reynolds CH, Deaven LL, Tollefsen DM (1988) Heparin cofactor II: cDNA sequence, chromosome localization, restriction fragment length polymorphism and expression in *Escherichia coli*. *Biochemistry* **27**: 752–759.

Blinder MA, Andersson TR, Abildgaard U, Tollefsen DM (1989) Heparin cofactor II Oslo: mutation of Arg-189 to His decreases the affinity for dermatan sulfate. *J. Biol. Chem.* **264**: 5128–5133.

Bloemenkamp KWM, Rosendaal FR, Helmerhorst FM, Büller HR, Vandenbroucke JP (1995) Enhancement by factor V Leiden mutation of risk of deep-vein thrombosis associated with oral contraceptives containing a third-generation progestagen. *Lancet* **346**: 1593–1596.

Blombäck B, Gröndahl NJ, Hessel B, Iwanaga S, Wallén P (1973) Primary structure of human fibrinogen and fibrin. II. Structural studies on NH$_2$-terminal part of γ-chain. *J. Biol. Chem.* **248**: 5806–5820.

Blombäck B, Hessel B, Fields R, Procyk R (1988) Fibrinogen Aarhus: an abnormal fibrinogen with Aa19 Arg→Gly substitution. In: MW Mosesson, DL Amrani, KR Siebenlist, JP DiOrio (eds) *Fibrinogen 3. Biochemistry, Biological Functions, Gene Regulation and Expression* pp. 263–266. Elsevier, Amsterdam.

Bloom AL, Forbes CD, Thomas DP, Tuddenham EGD (eds) (1994) *Haemostasis and Thrombosis*, 3rd ed. Churchill Livingstone, Edinburgh.

Bock SC, Levitan DJ (1983) Characterization of an unusual DNA length polymorphism 5′ to the human antithrombin III gene. *Nucleic Acids Res.* **11**: 8569–8582.

Bock SC, Prochownik EV (1987) Molecular genetic survey of 16 kindreds with hereditary antithrombin III deficiency. *Blood* **70**: 1273–1278.

Bock SC, Radziejewska E (1991) A *Nhe*I RFLP in the human antithrombin III gene (1q23–q25). *Nucleic Acids Res.* **19**: 2519.

Bock SC, Wion KL, Vehar GA, Lawn RM (1982) Cloning and expression of the cDNA for human antithrombin III. *Nucleic Acids Res.* **10**: 8113–8125.

Bock SC, Harris JF, Balazs I, Trent JM (1985a) Assignment of the human antithrombin III structural gene to chromosome 1q23–25. *Cytogenet. Cell Genet.* **39**: 67–69.

Bock SC, Harris JF, Schwartz CE, Ward JH, Hershgold EJ, Skolnick MH (1985b) Hereditary thrombosis in a Utah kindred is caused by a dysfunctional antithrombin III gene. *Am. J. Hum. Genet.* **37**: 32–41.

Bock SC, Marrinan JA, Radziejewska E (1988) Antithrombin III Utah: proline-407 to leucine mutation in a highly conserved region near the inhibitor reactive site. *Biochemistry* **27**: 6171–6178.

Bock SC, Silbermann JA, Wikoff W, Abildgaard U, Hultin MB (1989) Identification of a threonine for alanine substitution of residue 404 of antithrombin III Oslo suggests integrity of the 404–407 region is important for maintaining normal plasma inhibitor levels. *Thromb. Haemost.* **62**: 494.

Bode W, Huber R (1992) Natural protein proteinase inhibitors and their interaction with proteinases. *Eur. J. Biochem.* **204**: 433–451.

Bode W, Mayr J, Baumann U, Huber R, Stone SR, Hofsteenge J (1989) The refined 1.9 Å crystal structure of human α-thrombin: interaction with D-Phe-Pro-Arg chloromethylketone and significance of the Tyr-Pro-Pro-Trp insertion segment. *EMBO J.* **8**: 3467–3475.

Bohmfalk JF, Fuller GM (1980) Plasminogen is synthesized by primary cultures of rat hepatocytes. *Science* **209**: 408–410.

Bohonus VL, Doolittle RF, Pontes M, Strong DD (1986) Complementary DNA sequence of lamprey fibrinogen β-chain. *Biochemistry* **25**: 6512–6516.

Boisclair MD, Ireland H, Lane DA (1990) Assessment of hypercoagulable states by measurement of activation fragments and peptides. *Blood Rev.* **4**: 25–40.

Bokarewa MI, Bremme K, Blombäck M (1996) Arg506-Gln mutation in factor V and risk of thrombosis during pregnancy. *Br. J. Haematol.* **92**: 473–478.

Bolyard MG, Lord ST (1988) High level expression of a functional human fibrinogen γ-chain in *Escherichia coli*. *Gene* **66**: 183–192.

Bolyard MG, Lord ST (1989) Expression in *Escherichia coli* of the human fibrinogen Bβ chain and its cleavage by thrombin. *Blood* **73**: 1202–1206.

Boomsma DI, Hennis BC, Van Wees AGM, Frants RR, Kluft C (1993) A parent–twin study of plasma levels of histidine-rich glycoprotein (HRG). *Thromb. Haemost.* **70**: 848–851.

Booth NA, Bennettt B (1994) Fibrinolysis and thrombosis. *Baillière Clin. Haematol.* **7**: 559–572.

Borg JY, Owen MC, Soria C, Soria J, Caen J, Carrell RW (1988) Proposed heparin binding site in antithrombin based on arginine 47. A new variant Rouen-II, 47 Arg to Ser. *J. Clin. Invest.* **81**: 1292–1296.

Borg JY, Brennan SO, Carrell RW, George P, Perry DJ, Shaw J (1990) Antithrombin Rouen IV 24 Arg→Cys. *FEBS Lett.* **266**: 163–166.

Borg JY, Perry D, Vasse M, Carrell RW (1991) Molecular characterization (Arg 189→ His) of a hereditary cofactor II (HCII) variant with low affinity for heparin and dermatan sulfate. *Thromb. Haemost.* **65**: 785.

Borgel D, Gandrille S, Gouault-Heilmann M, Aiach M (1994) First frameshift mutation in the active protein S gene associated with a quantitative hereditary deficiency. *Blood Coag. Fibrinol.* **5**: 593–600.

Borgel D, Jude B, Aiach M, Gandrille S (1996) First case of sporadic protein S deficiency due to a novel candidate mutation, Ala 484→Pro, in the protein S active gene (*PROS1*). *Thromb. Haemost.* **75**: 883–886.

Borrell M, Garí M, Coll I, Vallvé C, Tirado I, Soria JM, Sala N, Muñoz C, Oliver A, Garcia A, Fontcuberta J (1995) Abnormal polymerization and normal binding of plasminogen and t-PA in three new dysfibrinogenaemias: Barcelona III and IV (γArg275→His) and Villajoyosa (γArg275→Cys). *Blood Coag. Fibrinol.* **6**: 198–206.

Bosma PJ, van den Berg EA, Kooistra T, Siemieniak DR, Slightom JL (1988) Human plasminogen activator inhibitor-1 gene. *J. Biol. Chem.* **263**: 9129–9141.

Bosma PJ, Kooistra T, Siemieniak DR, Slightom JL (1991) Further characterization of the 5′-flanking DNA of the gene encoding human plasminogen activator inhibitor-1. *Gene* **100**: 261–266.

Bottema CDK, Ketterling RP, Yoon HS, Sommer SS (1990) The pattern of factor IX germ-line mutation in Asians is similar to that of Caucasians. *Am. J. Hum. Genet.* **47**: 835–841.

Bottema CDK, Ketterling RP, Setsuko L, Yoon H-S, Phillips JA, Sommer SS (1991) Missense mutations and evolutionary conservation of amino acids: evidence that many of the amino acids in factor IX function as 'spacer' elements. *Am. J. Hum. Genet.* **49**: 820–838.

Bovill EG, Bauer KA, Dickerman JD, Callas P, West B (1989) The clinical spectrum of heterozygous protein C deficiency in a large New England kindred. *Blood* **73**: 712–717.

Bovill EG, Tornczak JA, Grant B, Bhushan F, Pillemer E, Rainville IR, Longon GL (1992) Protein C Vermont: symptomatic type II protein C deficiency associated with two GLA domain mutations. *Blood* **79**: 1456–1465.

Boyer C, Wolf M, Vedrenne J, Meyer D, Larrieu MJ (1986) Homozygous variant of antithrombin III: ATIII Fontainebleau. *Thromb. Haemost.* **56**: 18–22.

Bradford HN, Annamalai A, Doshi K, Colman RW (1988) Factor V is activated and cleaved by platelet calpain: comparison with thrombin proteolysis. *Blood* **71**: 388–394.

Brandstetter H, Bauer M, Huber R, Lollar P, Bode W (1995) X-ray structure of clotting factor IXa: active site and molecular structure related to Xase activity and hemophilia B. *Proc. Natl Acad. Sci. USA* **92**: 9796–9800.

Brass EP, Forman WB, Edwards RV, Lindon O (1978) Fibrin formation: effect of calcium ions. *Blood* **52**: 654–658.

Brattström L, Tengborn L, Lagerstedt C, Israelsson B, Hultberg B (1991) Plasma homocysteine in venous thromboembolism. *Haemostasis* **21**: 51–57.

Braun A, Müller B, Rosche AA (1996) Population study of the G1691A mutation (R506Q, FV Leiden) in the human factor V gene that is associated with resistance to activated protein C. *Hum. Genet.* **97**: 263–264.

Brennan SO, George PM, Jordan RE (1987) Physiological variant of antithrombin III lacks carbohydrate sidechain at Asn 135. *FEBS Lett.* **219**: 431–436.

Brennan SO, Borg J-Y, George PM, Soria C, Soria J, Caen J, Carrell RW (1988) New carbohydrate site in mutant antithrombin (7 Ile→Asn) with decreased heparin affinity. *FEBS Lett.* **237**: 118–122.

Brenner S (1988) The molecular evolution of genes and proteins: a tale of two serines. *Nature* **334** : 528–530.

Brenner B, Zivelin A, Lanir N, Greengard JS, Griffin JH, Seligsohn U (1996) Venous thromboembolism associated with double heterozygosity for R506Q mutation of factor V and for T298M mutation of protein C in a large family of a previously described homozygous protein C-deficient newborn with massive thrombosis. *Blood* **88**: 877–880.

Brey RL, Coull BM (1996) Cerebral venous thrombosis. Role of activated protein C resistance and factor V gene mutation. *Stroke* **27**: 1719–1720.

Briët E, Broekmans AW, Engesser L (1988) Hereditary protein S deficiency. In: Bertina, RM (ed.). *Protein C and Related Proteins: Biochemical and Clinical Aspects*, pp. 203–212. Churchill Livingstone, Edinburgh.

Briginshaw GF, Shonberge JN (1974a) Identification of two distinct heparin cofactors in human plasma I. Separation and partial purification. *Arch. Biochem. Biophys.* **161**: 683–690.

Briginshaw GF, Shonberge JN (1974b) Identification of two distinct heparin cofactors in human plasma. II. Inhibition of thrombin and activated factor X. *Thromb. Res.* **4**: 463–477.

Bristol JA, Freedman SJ, Furie BC, Furie B (1994) Profactor IX: the propeptide inhibits binding to membrane surfaces and activation by factor XIa. *Biochemistry* **33**: 14136–14143.

British Committee for Standards in Haematology (1990) Guidelines on the investigation and management of thrombophilia. *J. Clin. Pathol.* **43**: 703–709.

Britten RJ (1986) Rates of DNA sequence evolution differ between taxonomic groups. *Science* **231**: 1393–1398.

Broekmans AW (1985) Hereditary protein C deficiency. *Haemostasis* **15**: 233–240.

Broekmans AW, Conard J (1988) Hereditary protein C deficiency. In: Bertina, RM (ed.) *Protein C and Related Proteins*, pp. 160–187. Churchill Livingstone, Edinburgh.

Broekmans AW, Veltkamp JJ, Bertina RM (1983) Congenital protein C deficiency and venous thrombembolism. *New Engl. J. Med.* **309**: 340–344.

Broekmans AW, Bertina RM, Reinalda-Poot J, Engesser L, Muller HP, Leeuw JA, Michiels JJ, Brommer EJP, Briët E (1985) Hereditary protein S deficiency and venous thromboembolism. *Thromb. Haemost.* **53**: 273–277.

Broekmans AW, Van der Linden IK, Jansen-Koeter Y, Bertina RM (1986) Prevalence of protein C (PC) and protein S (PS) deficiency in patients with thrombotic disease. *Thromb. Res.* **42** (Suppl. VI): 135.

Brook JD, McCurrach ME, Harley HG, Buckler AJ, Church D, Aburatani H, Hunter K, Stanton VP, Thirion JP, Hudson T, Sohn R, Zemelman B, Snell RG, Rundle SA, Crow S, Davies J, Shelbourne P, Buxton J, Harper PS, Shaw DJ, Housman DE (1992) Molecular basis of myotonic dystrophy: expansion of a trinucleotide (CTG) repeat at the 3′ end of a transcript encoding a protein kinase family member. *Cell* **68**: 799–808.

Brown CH, Crowe MF (1975) Defective α-polymerization in the conversion of fibrinogen Baltimore to fibrin. *J. Clin. Invest.* **55**: 1190–1194.

Brown WM, Dziegielewska KM, Foreman RC, Saunders NR (1989) Nucleotide and deduced amino acid sequence of a γ-subunit of bovine fibrinogen. *Nucleic Acids Res.* **17**: 6397.

Browne MJ, Tyrrell AWR, Chapman CG, Carey JE, Glover DM, Grosveld FG, Dodd I, Robinson JH (1985) Isolation of a human tissue-type plasminogen activator genomic clone and its expression in mouse L cells. *Gene* **33**: 279–284.

Browne MJ, Carey JE, Chapman CG, Tyrrell AWR, Entwisle C, Lawrence GMP, Reavy B, Dodd I, Esmail A, Robinson JH (1988) A tissue-type plasminogen activator mutant with prolonged clearance *in vivo*. *J. Biol. Chem.* **263**: 1599–1602.

Brox JH, Østerud B, Bjorklid E, Fenton JW (1984) Production and availability of thromboplastin in endothelial cells: the effect of thrombin, endotoxin and platelets. *Br. J. Haematol.* **57**: 239–246.

Broze GJ (1995a) Tissue factor pathway inhibitor and the current concept of blood coagulation. *Blood Coag. Fibrinol.* **6** (Suppl. 1): S7-S13.

Broze GJ (1995b) Tissue factor pathway inhibitor and the revised theory of coagulation. *Ann. Rev. Med.* **46**: 103–112.

Broze GJ, Likert K, Higuchi D (1993) Inhibition of factor VIIa/tissue factor by antithrombin III and tissue factor pathway inhibitor. *Blood* **82**: 1679–1680.

Bruce D, Perry DJ, Borg J-V, Carrell RW, Wardell MR (1994) Thromboembolic disease due to thermolabile conformational changes of antithrombin Rouen-VI (187 Asn→Asp). *J. Clin. Invest.* **94**: 2265–2274.

Brunel F, Duchange N, Fischer A-M, Cohen GN, Zakin MM (1987) Antithrombin III Alger: a new case of Arg→47Cys mutation. *Am. J. Hematol.* **25**: 223–224.

Brunner HG, Nillesen W, van Oost BA, Jansen G, Wieringa B, Ropers HH, Smeets HJM (1992) Presymptomatic diagnosis of myotonic dystrophy. *J. Med. Genet.* **29**: 780–784.

Budzynski AZ, Olexa SA, Pandya BV (1983) Fibrin polymerization sites in fibrinogen and fibrin fragments. *Ann. NY Acad. Sci.* **408**: 301–314.

Bugge TH, Flick MJ, Daugherty CC, Degen JL (1995) Plasminogen deficiency causes severe thrombosis but is compatible with development and reproduction. *Genes Devel.* **9**: 794–807.

Bugge TH, Kombrinck KW, Flick MJ, Daugherty CC, Danton MJS, Degen JL (1996) Loss of fibrinogen rescues mice from the pleiotropic effects of plasminogen deficiency. *Cell* **87**: 709–719.

Bulens F, Abañez-Tallon I, Van Acker P, De Vriese A, Nelles L, Belayew A, Collen D (1995) Retinoic acid induction of human tissue-type plasminogen activator gene expression via a direct repeat element (DR5) located at –7 kilobases. *J. Biol. Chem.* **270**: 7167–7175.

Bullock P, Champoux JJ, Botchan M (1985) Association of crossover points with topoisomerase I cleavage sites: a model for non-homologous recombination. *Science* **230**: 954–957.

Busby SJ, Mulvihill E, Rao D, Kumar AA, Lioubin P, Heipel M, Sprecher C, Halfpap L, Prunkard D, Gambee J, Foster DC (1991) Expression of recombinant human plasminogen in mammalian cells is augmented by suppression of plasmin activity. *J. Biol. Chem.* **266**: 15286–15292.

Butler HR, ten Cate JW (1989) Acquired antithrombin III deficiency. Laboratory diagnosis incidence, clinical implication and treatment with antithrombin III concentrate. *Am. J. Med.* **87** (Suppl. 3b): 445–485.

Byean I-JL, Kelley RF, Lhinas M (1991) Kringle-2 domain of the tissue-type plasminogen activator. *Eur. J. Biochem.* **197**: 155–165.

Byrne CD, Schwartz K, Meer K, Cheng J-F, Lawn RM (1994) The human apolipoprotein (a)/plasminogen gene cluster contains a novel homologue transcribed in liver. *Arterioscler. Thromb.* **14**: 534–541.

Cadroy Y, Sié P, Boneu B (1994) Frequency of a defective response to activated protein C in patients with a history of venous thrombosis. *Blood* **83**: 2008–2009.

Camani C, Bachmann F, Kruithof EKO (1994) The role of plasminogen activator inhibitor type 1 in the clearance of tissue-type plasminogen activator by rat hepatoma cells. *J. Biol. Chem.* **269**: 5770–5775.

Camire RM, Kalafatis M, Cushman M, Tracy RP, Mann KG, Tracy PB (1995) The mechanism of inactivation of human platelet factor Va from normal and activated protein C-resistant individuals. *J. Biol. Chem.* **270**: 20794–20800.

Candiani G, Cremonesi L, Bruno L, Tenan M, Ferrari M, Camerino G, Taramelli R (1989) A *Taq*I RFLP of the human plasminogen gene. *Nucleic Acids Res.* **17**: 8900.

Caput D, Beutler B, Hartog K, Thayer R, Brown-Shimer S, Cerami A (1986) Identification of a common nucleotide sequence in the 3′-untranslated region of mRNA molecules specifying inflammatory mediators. *Proc. Natl Acad. Sci. USA* **83**: 1670–1674.

Carmeliet P, Collen D (1994) Evaluation of the plasminogen/plasmin system in transgenic mice. *Fibrinolysis* **8** (Suppl. 1): 269–276.

Carmeliet P, Kiekens L, Schoonjans L, Ream B, van den Oord JJ, De Mol M, Mulligan RC, Collen D (1993a) Plasminogen activator inhibitor-1 gene-deficient mice. I. Generation by homologous recombination and characterization. *J. Clin. Invest.* **92**: 2746–2755.

Carmeliet P, Stassen JM, Schoonjans L, Ream B, van den Oord JJ, De Mol M, Mulligan RC, Collen D (1993b) Plasminogen activator inhibitor-1 gene-deficient mice. II. Effects on hemostasis, thrombosis and thrombolysis. *J. Clin. Invest.* **92**: 2756–2760.

Carmeliet P, Schoonjans L, Kleckens L, Ream B, Degan J, Bronson R, De Vos R, van den Oord JJ, Collen D, Mulligan RC (1994) Physiological consequences of loss of plasminogen activator gene function in mice. *Nature* **368**: 419–424.

Carrell N, McDonagh J (1982) Fibrinogen Chapel Hill II: defective in reactions with thrombin, factor XIIIa and plasmin. *Br. J. Haematol.* **52**: 35–47.

Carrell N, Gabriel DA, Blatt PM, Carr ME, McDonagh J (1983) Hereditary dysfibrinogenemia in a patient with thrombotic disease. *Blood* **62**: 439–447.

Carrell RW, Evans DL (1992) Serpins: mobile conformations in a family of proteinase inhibitors. *Curr. Opin. Struct. Biol.* **2**: 438–446.

Carrell RW, Travis J (1985) α1-antitrypsin and the serpins: variation and countervariation. *Trends Biochem. Sci.* **10**: 20–24.

Carrell RW, Owen M, Brennan S, Vaughan L (1979) Carboxy terminal fragment of human α1-antitrypsin from hydroxylamine cleavage: homology with antithrombin III. *Biochem. Biophys. Res. Commun.* **91**: 1032–1037.

Carrell RW, Christey PB, Boswell DR (1987) Serpins: antithrombin and other inhibitors of coagulation and fibrinolysis. Evidence from amino acid sequences. In: Verstraete M, Vermylen J, Lijnen R, Arnout J (eds) *Thrombosis and Haemostasis*, pp. 1–15. Leuven University Press, Leuven.

Carrell RW, Aulak KS, Owen MC (1989) The molecular pathology of the serpins. *Molec. Biol. Med.* **6**: 35–42.

Carrell RW, Evans DL, Stein PE (1991) Mobile reactive centre of serpins and the control of thrombosis. *Nature* **353**: 576–578.

Carrell RW, Stein PE, Fermi G, Wardell MR (1994) Biological implications of a 3Å structure of dimeric antithrombin. *Nature Structure* **2**: 257–270.

Carter PE, Duponchel C, Tosi M, Fothergill JE (1991) Complete nucleotide sequence of the gene for human C1 inhibitor with an unusually high density of *Alu* elements. *Eur. J. Biochem.* **197**: 301–308.

Carter CJ (1994a) The pathophysiology of venous thrombosis. *Prog. Cardiovasc. Dis.* **36**: 439–446.

Carter CJ (1994b) The natural history and epidemiology of venous thrombosis. *Prog. Cardiovasc. Dis.* **36**: 423–438.

Caskey CT, Pizzuti A, Fu YH, Fenwick RG, Nelson DL (1992) Triplet repeat mutations in human disease. *Science* **256**: 784–789.

Caso R, Lane DA, Thompson E, Zangouras D, Panico M, Morris H, Olds RJ, Thein SL, Girolami A (1990) Antithrombin Padua I: impaired heparin binding caused by an Arg 47 to His (CGT to CAT) substitution. *Thromb. Res.* **58**: 185–190.

Caso R, Lane DA, Thompson EA, Olds RJ, Thein SL, Panico M, Blench I, Morris HR, Freyssinet JM, Aiach M, Rodeghiero F, Finazzi G (1991) Antithrombin Vicenza, Ala 384 to Pro (GCA to CCA) mutation transforming the inhibitor into a substrate. *Br. J. Haematol.* **77**: 87–92.

Cassels-Brown A, Minford AMB, Chatfield SL, Bradbury JA (1993) Ophthalmic manifestations of neonatal protein C deficiency. *Br. J. Ophthalmol.* **78**: 486–487.

Castaman G, Ruggeri M, Burei F, Rodeghiero F (1993) High levels of histidine-rich glycoprotein and thrombotic diathesis. Report of two unrelated families. *Thromb. Res.* **69**: 297–305.

Castaman G, Ruggeri M, Tosetto A, Missiaglia E, Rodegheiro F (1996a) Thrombosis in patients with heterozygous and homozygous factor XII deficiency is not explained by the associated presence of factor V Leiden. *Thromb. Haemost.* **76**: 275–281.

Castaman G, Ruggeri M, Tosetto A, Bernardi F, Rodeghiero F (1996b) The Ser460Pro substitution of the protein S (PS) gene is rare in Italian patients with type IIa PS deficiency. *Blood* **88**: 3666–3667.

Castellino FJ (1984) Biochemistry of human plasminogen. *Sem. Thromb. Hemostas.* **10**: 18–23.

Castellino FJ, Ploplis VA, Powell JR (1981) The existence of independent domain structures in human Lys77 plasminogen. *J. Biol. Chem.* **256**: 4778–4782.

Casula L, Murru S, Pecorara M, Ristaldi MS, Restagno G, Mancuso G, Morfini M, DeBiasi R, Baudo F, Carbonara A, Mori PG, Cao A, Pirastu M (1990) Recurrent mutations and three novel rearrangements in the factor VIII gene of hemophilia A patients of Italian descent. *Blood* **75**: 662–670.

Catto AJ, Grant PJ (1995) Risk factors for cerebrovascular disease and the role of coagulation and fibrinolysis. *Blood Coag. Fibrinol.* **6**: 497–510.

Catto A, Carter A, Ireland H, Bayston TA, Philippou H, Barrett J, Lane DA, Grant PJ (1995) Factor V Leiden gene mutation and thrombin generation in relation to the development of acute stroke. *Arterioscl. Thromb. Vasc. Biol.* **15**: 783–785.

Cederholm-Williams SA (1984) Molecular biology of plasminogen activators and recombinant DNA progress. *BioEssays* **1**: 168–173.

Ceriello A, Pirisi M, Giacomello R, Stel G, Falleti E, Motz E, Lizzio S, Gonano F, Bartoli E (1994) Fibrinogen plasma levels as a marker of thrombin activation: new insights on the role of fibrinogen as a cardiovascular risk factor. *Thromb. Haemost.* **71**: 593–595.

Chafa O, Fischer AM, Meriane F, Chellali F, Rahal S, Sternberg C, Benabadji M (1989) A new case of 'type II' inherited protein S deficiency. *Br. J. Haematol.* **73**: 501–505.

Chaida A, Gialeraki C, Tsoukala C, Mandalaki T (1996) Prevalence of the FVQ506 mutation in the Hellenic population. *Thromb. Haemost.* **76**: 124.

Chan LC, Bourke C, Lam CK, Liu HW, Brookes S, Jenkins V, Pasi J (1996) Lack of activated protein C resistance in healthy Hong Kong Chinese blood donors – correlation with absence of Arg[506]–Gln mutation of factor V gene. *Thromb. Haemost.* **75**: 520–526.

Chandra T, Stackhouse R, Kidd VJ, Woo SLC (1983a) Isolation and sequence characterization of a cDNA clone of human antithrombin III. *Proc. Natl Acad. Sci. USA* **80**: 1845–1848.

Chandra T, Stackhouse R, Kidd VJ, Robson KJH, Woo SLC (1983b) Sequence homology between human α1-antichymotrypsin, α1-antitrypsin and antithrombin III. *Biochemistry* **22**: 5055–5060.

Chang J-Y (1989) Binding of heparin to human antithrombin III activates selective chemical modification at lysine 236. *J. Biol. Chem.* **264**: 3111–3115.

Chang JY, Tran TH (1986) Antithrombin III Basel. Identification of a Pro→Leu substitution in a hereditary abnormal antithrombin with impaired heparin cofactor activity. *J. Biol. Chem.* **261**: 1174–1176.

Chang GTG, Ploos van Amstel HK, Hessing M, Reitsma PH, Bertina RM, Bouma BN (1992) Expression and characterization of recombinant human protein S in heterologous cells – studies of the interaction of amino acid residues Leu-608 to Glu-612 with human C4b-binding protein. *Thromb. Haemost.* **67**: 526–532.

Chang GTG, Maas BHA, Ploos van Amstel IIK, Reitsma PH, Bertina RM, Bouma BN (1994a) Studies of the interaction between human protein S and human C4b-binding protein using deletion variants of recombinant human protein S. *Thromb. Haemost.* **71**: 461–467.

Chang GTG, Aaldering L, Hackeng TM, Reitsma PH, Bertina RM, Bouma BN (1994b) Construction and characterization of thrombin-resistant variants of recombinant human protein S. *Thromb. Haemost.* **72**: 693–697.

Cheng KC, Cahill DS, Kasai H, Nishimura S, Loeb LA (1992) 8-Hydroxyguanine, an abundant form of oxidative DNA damage, causes G→T and A→C substitutions. *J. Biol. Chem.* **267**: 166–172.

Cheresh DA, Berliner SA, Vicente V, Ruggeri ZM (1989) Recognition of distinct adhesive sites on fibrinogen by related integrins on platelets and endothelial cells. *Cell* **58**: 945–953.

Chodosh LA, Carthew RW, Morgan JG, Crabtree GR, Sharp PA (1987) The adenovirus major late transcription factor activates the rat γ-fibrinogen promoter. *Science* **238**: 684–688.

Chodosh LA, Baldwin AS, Carthew RW, Sharp PA (1988) Human CCAAT-binding proteins have heterologous subunits. *Cell* **53**: 11–24.

Chow TW, McIntyre LV, Petersen DM (1983) Importance of plasma fibronectin in determining PFP and PRP clot mechanical properties. *Thromb. Res.* **29**: 243–248.

Chow BK-C, Ting V, Tufaro F, MacGillivray RTA (1991) Characterization of a novel liver-specific enhancer in the human prothrombin gene. *J. Biol. Chem.* **266**: 18927–18933.

Chowdhury V, Olds RJ, Lane D, Conard J, Pabinger I, Ryan K, Bauer KA, Bhavani M, Abildgaard U, Finazzi G, Castaman G, Mannucci PM, Thein SL (1993) Identification of nine novel mutations in type I antithrombin deficiency by heteroduplex screening. *Br. J. Haematol.* **84**: 656–661.

Chowdhury V, Lane DA, Mille B, Auberger K, Gandenberger-Bachem S, Pabinger I, Olds RJ, Thein SL. (1994) Homozygous antithrombin deficiency: report of two new cases (99 Leu to Phe) associated with arterial and venous thrombosis. *Thromb. Haemost.* **72**: 198–202.

Chowdhury V, Mille B, Olds RJ, Lane DA, Watton J, Barrowcliffe TW, Pabinger I, Woodcock BE, Thein SL (1995) Antithrombins Southport (Leu 99 to Val) and Vienna (Gln 118 to Pro): two novel antithrombin variants with abnormal heparin binding. *Br. J. Haematol.* **89**: 602–609.

Christiansen WT, Castellino FJ (1994) Properties of recombinant chimeric human protein C and activated protein C containing the γ-carboxyglutamic acid and trailing helical stack domains of protein C replaced by those of human coagulation. factor IX. *Biochemistry* **33**: 5901–5911.

Christiansen WT, Tulinsky A, Castellino FJ (1994) Functions of individual γ-carboxyglutamic acid (Gla) residues of human protein C. Determination of functionally nonessential Gla residues and correlations with their mode of binding to calcium. *Biochemistry* 33: 14993–15000.

Chuansumrit A, Manco-Johnson MJ, Hathaway WE (1989) Heparin cofactor II in adults and infants with thrombosis and DIC. *Am. J. Hematol.* 31: 109–113.

Chung DW, Rixon MW, MacGillivray RTA, Davie EW (1981) Characterization of a cDNA clone coding for the β-chain of bovine fibrinogen. *Proc. Natl Acad. Sci. USA* 78: 1466–1470.

Chung DW, Que BG, Rixon MW, Mace M, Davie EW (1983a) Characterization of complementary deoxyribonucleic acid and genomic deoxyribonucleic acid for the β-chain of human fibrinogen. *Biochemistry* 22: 3244–3250.

Chung DW, Chan W-Y, Davie EW (1983b) Characterization of a complementary deoxyribonucleic acid coding for the γ chain of human fibrinogen. *Biochemistry* 22: 3250–3256.

Chung LP, Gagnon J, Reid KBM (1985a) Amino acid sequence studies of human C4b-binding protein: N-terminal sequence analysis and alignment of the fragments produced by limited proteolysis with chymotrypsin and the peptides produced by cyanogen bromide treatment. *Mol. Immunol.* 22: 427–435.

Chung LP, Bentley DR, Reid KBM (1985b) Molecular cloning and characterization of the cDNA coding for C4b-binding protein, a regulatory protein of the classical pathway of the human complement system. *Biochem. J.* 230: 133–141.

Church FC, Hoffman MR (1994) Heparin cofactor II and thrombin. Heparin-binding proteins linking hemostasis and inflammation. *Trends Cardiovasc. Med.* 4: 140–146.

Citarella F, Tripodi M, Fantoni A, Bernardi F, Romeo, G, Rocchi M (1988) Assignment of human coagulation factor XII (fXII) to chromosome 5 by cDNA hybridization to DNA from somatic cell hybrids. *Hum. Genet.* 80: 397–398.

Clarke BJ, Cote H, Cool DE, Clark-Lewis I, Saito H, Pixley RA, Colman RW, MacGillivray RTA (1989) Mapping of a putative surface-binding site of human coagulation factor XII. *J. Biol. Chem.* 264: 11497–11502.

Clarke JH, Light DR, Blasko E, Parkinson JF, Nagashima M, McLean K, Vilander L, Andrews WH, Morser J, Glaser CB (1993) The short loop between epidermal growth factor-like domains 4 and 5 is critical for human thrombomodulin function. *J. Biol. Chem.* 268: 6309–6315.

Clause LH, Comp RC (1986) The regulation of haemostasis: the protein C system. *New Engl. J. Med.* 314: 1298–1304.

Collen D, DeCock F (1975) A tanned red cell haemagglutination inhibition assay (TRCH II) for the quantitative estimation of thrombin–antithrombin III and plasma α2-antiplasmin complex in human plasma. *Thromb. Res.* 7: 235–239.

Collen D, Lijnen HR (1987) Fibrinolysis and the control of hemostasis. In Stamatoyanoupolous G, Nienhuis AW, Leder P, Majerus PW (eds) *The Molecular Basis of Blood Diseases*, pp. 662–688. WB Saunders, Philadelphia.

Collen D, Schetz J, DeCock F, Holmer E, Verstraete M (1977) Metabolism of antithrombin III (heparin cofactor) in man: effects of venous thrombosis and of heparin administration. *Eur. J. Clin. Invest.* 7: 27–35.

Collet JP, Soria J, Mirshahi M, Hirsch M, Dagonnet FB, Caen J, Soria C (1993) Dusart syndrome: a new concept of the relationship between fibrin clot architecture and fibrin clot degradability: hypofibrinolysis related to an abnormal clot structure. *Blood* 82: 2462–2469.

Collins DW, Jukes TH (1994) Rates of transition and transversion in coding sequences since the human–rodent divergence. *Genomics* 20: 386–396.

Colman RW, Hirsh J, Marder VJ, Salzman EW, Eds (1994) *Hemostasis and Thrombosis: Basic Principles and Clinical Practice*. JB Lippincott, Philadelphia.

Comp PC, Esmon CT (1981) Generation of fibrinolytic activity by infusion of activated protein C into dogs. *J. Clin. Invest.* 68: 1221–1228.

Comp PC, Esmon CT (1984) Recurrent venous thromboembolism in patients with a partial deficiency of protein S. *New Engl. J. Med.* 311: 1525–1528.

Comp PC, Nixon RR, Cooper MR, Esmon CT (1984) Familial protein S deficiency is associated with recurrent thrombosis. *J. Clin. Invest.* 74: 2082–2088.

Comp PC, Doray D, Patton D, Esmon CT (1986) An abnormal plasma distribution of protein S occurs in functional protein S deficiency. *Blood* 67: 504–508.

Comp PC, Forristall J, West CD, Trapp RG (1990) Free protein S levels are elevated in familial C4b-binding protein deficiency. *Blood* 76: 2527–2529.

Conard J, Horellou MH, Teger-Nilsson AC, Bertina RM, Samama M (1984) The fibrinolytic system in patients with congenital protein C deficiency. *Thromb. Res.* **36**: 363–367.

Conard J, Horellou MH, van Dreden P, Samama M, Reitsma PH, Poort S, Bertina RM (1992) Homozygous protein C deficiency with late onset and recurrent coumarin-induced skin necrosis. *Lancet* **339**: 743–744.

Conard J, Bauer KA, Gruber A, Griffin JH, Schwarz HP, Horellou M-H, Samama MM, Rosenberg RD (1993) Normalization of markers of coagulation activation with a purified protein C concentrate in adults with homozygous protein C deficiency. *Blood* **82**: 1159–1164.

Conlan MG, Folsom AR, Finch A, Davis CE, Sorlie P, Wu KK (1993) Correlation of plasma protein C levels with cardiovascular risk factors in middle-aged adults: the Atherosclerosis Risk in Commuinities (ARIC) study. *Thromb. Haemost.* **70**: 762–767.

Conlan MG, Folsom AR, Finch A, Davis CE, Marcucci G, Sorlie P, Wu KK (1994) Antithrombin III: associations with age, race, sex and cardiovascular disease risk factors. *Thromb. Haemost.* **72**: 551–556.

Conway EM, Nowakowski B, Steiner-Mosonyi M (1993) Thrombomodulin lacking the cytoplasmic domain efficiently internalizes thrombin via nonclathrin-coated, pit-mediated endocytosis. *J. Cell. Physiol.* **158**: 285–298.

Conway EM, Liu L, Nowakowski B, Steiner-Mosonyi M, Jackman RW (1994) Heat shock of vascular endothelial cells induces an up-regulatory transcriptional response of the thrombomodulin gene that is delayed in onset and does not attenuate. *J. Biol. Chem.* **269**: 22804–22810.

Cooke WL (1968) Acute myocardial infarction with decreased factor VII. *West Virginia Med. J.* **64**: 248.

Cool DE, MacGillivray RTA (1987) Characterization of the human blood coagulation factor XII gene. *J. Biol. Chem.* **262**: 13662–13673.

Cool DE, Edgell C-JS, Louie GV, Zoller MJ, Brayer GD, MacGillivray RTA (1985) Characterization of human blood coagulation factor XII cDNA. Prediction of factor XII and the tertiary structure of β-factor XIIa. *J. Biol. Chem.* **260**: 13666–13676.

Coon WW, Willis PW, Keller JB (1973) Venous thromboembolism and other venous disease in the Tecumseh community health study. *Circulation* **48**: 839–846.

Cooper DN, Krawczak M (1989) Cytosine methylation and the fate of CpG dinucleotides in vertebrate genomes. *Hum. Genet.* **83**: 181–189.

Cooper DN, Krawczak M (1990) The mutational spectrum of single base-pair substitutions causing human genetic disease: patterns and predictions. *Hum. Genet.* **85**: 55–74.

Cooper DN, Krawczak M (1991) Mechanisms of insertional mutagenesis in human genes causing genetic disease. *Hum. Genet.* **87**: 409–415.

Cooper DN, Krawczak M (1993) *Human Gene Mutation.* BIOS Scientific Publishers, Oxford.

Cooper DN, Youssoufian H (1988) The CpG dinucleotide and human genetic disease. *Hum. Genet.* **78**: 151–155.

Cooper DN, Krawczak M, Antonarakis SE (1995) The nature and mechanisms of human gene mutation. In: Scriver CR, Beaudet AL, Sly WS, Valle D (eds) *The Metabolic and Molecular Basis of Inherited Disease,* 7th edn, McGraw-Hill, New York.

Coronary Drug Project Research Group (1970) The coronary drug project: initial findings leading to modifications of its research protocol. *J. Am. Med. Assoc.* **214**: 1303–1307.

Corral J, Iniesta JA, Gonzalez-Conejero R, Vicente V (1996) Detection of factor V Leiden from a drop of blood by PCR-SSCP. *Thromb. Haemost.* **76**: 735–737.

Corrigan JJ, Sleeth JJ, Jeter M (1989) Newborn's fibrinolytic mechanism: components and plasmin generation. *Am. J. Hematol.* **32**: 273–278.

Cortellaro M, Cofrancesco E, Boscheti C, Mussoni L, Donati MB, Cardillo M, Catalano M, Gabrielli L, Lombardi B, Specchia G, Tavazzi L, Tremoli E, Pozzoli E, Turri M (1993) Increased fibrin turnover and high PAI-1 activity as predictors of ischemic events in aterosclerotic patients – a case–control study. *Arterioscler. Thromb.* **13**: 1412–1417.

Cosgriff TM, Bishop DT, Hershgold EJ, Skolnick MH, Martin BA, Baty BJ, Carlson KS (1983) Familial antithrombin III deficiency: its natural history, genetics, diagnosis and treatment. *Medicine* **62**: 209–220.

Coulondre C, Miller JH, Farabaugh PJ, Gilbert W (1978) Molecular basis of base substitution hotspots in *Escherichia coli. Nature* **274**: 775–780.

Courtois G, Morgan JG, Campbell LA, Fourel G, Crabtree GR (1987) Interaction of a liver-specific nuclear factor with the fibrinogen and α1-antitrypsin promoters. *Science* **238**: 688–692.

Courtois G, Baumhueter S, Crabtree GR (1988) Purified hepatocyte nuclear factor 1 interacts with a family of hepatocyte-specific promoters. *Proc. Natl Acad. Sci. USA* **85**: 7937–7941.

Crabtree GR, Kant JA (1981) Molecular cloning of cDNA for the α, β and γ chains of rat fibrinogen. *J. Biol. Chem.* **256**: 9718–9723.

Crabtree GR, Kant JA (1982a) Organization of the rat γ-fibrinogen gene: alternative mRNA splice patterns produce the γ A and γ B (γ) chains of fibrinogen. *Cell* **31**: 159–166.

Crabtree GR, Kant JA (1982b) Coordinate accumulation of the mRNAs for the α, β, and γ chains of rat fibrinogen following defibrination. *J. Biol. Chem.* **257**: 7277–7279.

Crabtree GR, Comeau CM, Fowlkes DM, Fornace AJ, Malley JD, Kant JA (1985) Evolution and structure of the fibrinogen genes; random insertion of introns or selective loss? *J. Molec. Biol.* **185**: 1–19.

Craig JM, Bickmore WA (1994) The distribution of CpG islands in mammalian chromosomes. *Nature Genetics.* **7**: 376–382.

Credo BR, Curtis CG, Lorand L (1981) α chain domain of fibrinogen controls generation of fibrinoligase (coagulation factor XIIIa): calcium ion regulatory aspects. *Biochemistry* **20**: 3770–3778.

Creighton TE, Darby NJ (1989) Functional evolutionary divergence of proteolytic enzymes and their inhibitors. *Trends Biochem. Sci.* **14**: 319–324.

Cripe LD, Moore KD, Kane WH (1992) Structure of the gene for human coagulation factor V. *Biochemistry* **31**: 3777–3785.

Crowther DC, Evans DLI, Carrell RW (1992) Serpins: implications of a mobile reactive centre. *Curr. Opin. Biotech.* **3**: 399–407.

Cucuianu MP, Cristea A, Roman S, Rus H, Missits I, Pechet L (1983) Comparative behavior of the components of the factor VIII complex in acute myocardial infarction. *Thromb. Res.* **30**: 487–497.

Cucuianu M, Blaga S, Pop S, Olinic D, Olinic N, Colhon D, Cristea A (1994) Homozygous or compound heterozygous qualitative antithrombin III deficiency. *Nouv. Rev. Franc. Hematol.* **36**: 335–337.

Cui J, O'Shea K, Purkayastha A, Saunders TL, Ginsburg D (1996) Fatal haemorrhage and incomplete block to embryogenesis in mice lacking coagulation factor V. *Nature* **384**: 66–70.

Dahlbäck B (1980) Human coagulation factor V purification and thrombin-catalysed activation. *J. Clin. Invest.* **66**: 583–591.

Dahlbäck B (1983) Purification of human vitamin K-dependent protein S and its limited proteolysis by thrombin. *Biochem. J.* **209**: 837–846.

Dahlbäck B (1984) Interaction between vitamin K-dependent protein S and the complement protein C, C4b-binding protein. *Sem. Thromb. Haemostas.* **10**: 139–148.

Dahlbäck B (1986) Inhibition of protein Ca cofactor function of human and bovine protein S by C4b-binding protein. *J. Biol. Chem.* **261**: 12022–12027.

Dahlbäck B (1991) Protein S and C4b-binding protein: components involved in the regulation of the protein C anticoagulant system. *Thromb. Haemost.* **66**: 49–61.

Dahlbäck B (1994) Physiological anticoagulation. Resistance to activated protein C and venous thromboembolism. *J. Clin. Invest.* **94**: 923–927.

Dahlbäck B (1995a) The protein C anticoagulant system: inherited defects as basis for venous thrombosis. *Thromb. Res.* **77**: 1–43.

Dahlbäck B (1995b) Inherited thrombophilia: resistance to activated protein C as a pathogenic factor of venous thromboembolism. *Blood* **85**: 607–614.

Dahlbäck B (1995c) Resistance to activated protein C, the Arg^{506} to Gln mutation in the factor V gene, and venous thrombosis. Functional tests and DNA-based assays, pros and cons. *Thromb. Haemost.* **73**: 739–742.

Dahlbäck B (1995d) New molecular insights into the genetics of thrombophilia. *Thromb. Haemost.* **74**: 139–148.

Dahlbäck B (1996) Are we ready for factor V Leiden screening? *Lancet* **347**: 1346–1347.

Dahlbäck B, Hildebrand B (1994) Inherited resistance to activated protein C is corrected by anticoagulant cofactor activity found to be a property of factor V. *Proc. Natl Acad. Sci. USA* **91**: 1396–1400.

Dahlbäck B, Stenflo J (1993) The protein C anticoagulant system. In: Stamatoyannopoulos G, Nienhuis AW, Majerus PW, Varmus H (eds) *The Molecular Basis of Blood Diseases*. WB Saunders, Philadelphia.

Dahlbäck B, Stenflo J (1994) A natural anticoagulant pathway: proteins C, S, C4b-binding protein and thrombomodulin. In: Bloom AL, Forbes CD, Thomas DP, Tuddenham EGD (eds) *Haemostasis and Thrombosis*, Vol. 1, pp. 671–698. Churchill Livingstone, Edinburgh.

Dahlbäck B, Smith CA, Müller-Eberhard HJ (1983) Visualization of human C4b-binding protein and its complexes with vitamin K-dependent protein S and complement protein C4b. *Proc. Natl Acad. Sci. USA* **80**: 346–350.

Dahlbäck B, Lundwall A, Stenflo J (1986a) Primary structure of bovine vitamin K-dependent protein S. *Proc. Natl Acad. Sci. USA* **83**: 4199–4203.

Dahlbäck B, Lundwall A, Stenflo J (1986b) Localization of thrombin cleavage sites in the amino-terminal region of bovine protein S. *J. Biol. Chem.* **261**: 5111–5115.

Dahlbäck B, Hansson C, Islam MQ, Szpirer J, Szpirer C, Lundwall A, Levan G (1988) Assignment of gene for coagulation factor V to chromosome 1 in man and to chromosome 13 in rat. *Somat. Cell Mol. Genet.* **14**: 509–514.

Dahlbäck B, Frohm B, Nelsesteuen G (1990a) High affinity interaction between C4b-binding protein and vitamin K-dependent protein S in the presence of calcium. *J. Biol. Chem.* **265**: 16082–16087.

Dahlbäck B, Hildebrand B, Malm J (1990b) Characterization of functionally important domains in human vitamin K-dependent protein S using monoclonal antibodies. *J. Biol. Chem.* **265**: 8127–8135.

Dahlbäck B, Hildebrand B, Linse S (1990c) Novel type of very high affinity calcium binding sites in β-hydroxyasparagine-containing epidermal growth factor-like domains in vitamin K-dependent protein S. *J. Biol. Chem.* **265**: 18481–18489.

Dahlbäck B, Wiedmer T, Sims PJ (1992) Binding of anticoagulant vitamin K-dependent protein S to platelet-derived microparticles. *Biochemistry* **31**: 12769–12777.

Dahlbäck B, Carlsson M, Svensson PJ (1993) Familial thrombophilia due to a previously unrecognized mechanism characterized by poor anticoagulant response to activated protein C. *Proc. Natl Acad. Sci. USA* **90**: 1004–1008.

Dalmon J, Laurent M, Courtois G (1993) The human β-fibrinogen promoter contains a hepatocyte nuclear factor 1-dependent interleukin-6-responsive element. *Mol. Cell. Biol.* **13**: 1183–1193.

Daly ME, Perry DJ (1990) *Dde*I polymorphism in intron 5 of the ATIII gene. *Nucleic Acids Res.* **18**: 5583.

Daly M, Bruce D, Perry DJ, Price J, Harper PL, O'Meara A, Carrell RW (1990) Antithrombin Dublin (-3 Val→Glu): an N-terminal variant which has an aberrant signal peptidase cleavage site. *FEBS Lett.* **273**: 87–90.

Daly M, Perry DJ, Harper PL, Daly HM, Rogues AWW, Carrell RW (1992) Insertions/ deletions in the antithrombin gene: 3 mutations associated with non-expression. *Thromb. Haemost.* **67**: 521–525.

Daly M, Perry DJ, Bruce DB, Harper PL, Tait RC, Walker ID, Mayne EE, Daly HM, Brown K, Carrell RW (1996a) Type I antithrombin deficiency: five novel mutations associated with thrombosis. *Blood Coag. Fibrinol.* **7**: 139–143.

Daly E, Vessey MP, Hawkins MM, Carson JL, Gough P, Marsh S (1996b) Risk of venous thromboembolism in users of hormone replacement therapy. *Lancet* **348**: 977–980.

Damus PS, Hicks M, Rosenberg RD (1973) Anticoagulant action of heparin. *Nature* **246**: 355–357.

Dang CV, Ebert RF, Bell WR (1985) Localization of a fibrinogen calcium binding site between γ subunit positions 311 and 336 by terbium fluorescence. *J. Biol. Chem.* **260**: 9713–9719.

Danø K, Andreasen PA, Grøndahl-Hansen J, Kristensen P, Nielsen LS, Skriver L (1985) Plasminogen activators, tissue degradation and cancer. *Adv. Cancer Res.* **44**: 139–266.

Davie EW (1995) Biochemical and molecular aspects of the coagulation cascade. *Thromb. Haemost.* **74**: 1–6.

Davie EW, Fujikawa K, Kisiel W (1991) The coagulation cascade: initiation, maitenance and regulation. *Biochemistry* **30**: 10363–10370.

Davies J, Yamagata H, Shelbourne P, Buxton J, Ogihara T, Nokelainen P, Nakagawa M, Williamson R, Johnson K, Miki T (1992) Comparison of the myotonic dystrophy associated CTG repeat in European and Japanese populations. *J. Med. Genet.* **29**: 766–769.

Davies MJ (1996) The contribution of thrombosis to the clinical expression of coronary atherosclerosis. *Thromb. Res.* **82**: 1–32.

Dawson S, Hamsten A, Wiman B, Henney A, Humphries S (1991) Genetic variation at the plasminogen activator inhibitor-1 locus is associated with altered levels of plasma plasminogen activator inhibitor-1 activity. *Arterioscl. Thromb.* **11**: 183–190.

Dawson SJ, Wiman B, Hamsten A, Green F, Humphries S, Henney AM (1993) The two allele sequences of a common polymorphism in the promoter of the plasminogen activator inhibitor-1 (PAI-1) gene respond differently to interleukin-1 in HepG2 cells. *J. Biol. Chem.* **268**: 10739–10745.

De Agostini A, Lijnen HR, Pixley RA, Colman RW, Schapira M (1984) Inactivation of factor XII active fragment in normal plasma. Predominant role of C1-inhibitor. *J. Clin. Invest.* **73**: 1542–1549.

De Bono D (1987) Clinical trials of new thrombolytic agents in acute myocardial infarction. In: Vertraete M, Vermylen J, Lijuen R, Arnont J (eds) *Thrombosis and Haemostasis*, pp. 267–280. Leuven University Press, Leuven.

De Cristofaro R, Furlan M, Landolfi R (1994) Fibrinogen Milano IV (Aα16 Arg-His): characterization of its abnormal interaction with human α-thrombin. *Biochem. J.* **302**: 623–624.

De Fouw NJ, van Hinsbergh VWM, de Jong YF, Haverkate F, Bertina RM (1987) The interaction of activated protein C and thrombin with the plasminogen activator inhibitor released from human endothelial cells. *Thromb. Haemost.* 57: 176–182.

De Fouw NJ, Bertina RM, Haverkate F (1988) Activated protein C and fibrinolysis. In: Bertina RM (ed.) *Protein C and Related Proteins; Biochemical and Clinical Aspects*, pp. 71–90. Churchill Livingstone, Edinburgh.

De Fouw NJ, Haverkate F, Bertina RM (1990) Protein C and fibrinolysis: a link between coagulation and fibrinolysis. In: Liu CY, Chien S (eds) *Fibrinogen, Thrombosis, Coagulation and Fibrinolysis*. Plenum Press, New York.

De Fouw NJ, van Tilburg NH, Haverkate F, Bertina RM (1993) Activated protein C accelerator clot lysis by virtue of its anticoagulant activity. *Blood Coag. Fibrinol.* 4: 201–210.

de Knijff P, Green F, Johansen LG, Grootendorst D, Temple A, Cruickshank JK, Humphries SE, Jespersen J, Kluft C (1994) New alleles in F7 VNTR. *Hum. Mol. Genet.* 3: 384.

de Maat MPM, Kluft C, Jespersen J, Gram J (1996) World distribution of factor V Leiden mutation. *Lancet* 347: 58.

de Roux N, Chadeuf G, Molho-Sabatier P, Plouin P-F, Aiach M (1990) Clinical and biochemical characterization of antithrombin III Franconville, a variant with Pro 41→Leu mutation. *Br. J. Haematol.* 75: 222–227.

DeSalle R, Gatesy J, Wheeler W, Grimaldi D (1992) DNA sequences from a fossil termite in Oligo-Miocene amber and their phylogenetic implications. *Science* 257: 1933–1936.

De Stefano V, Leone G (1991) Mortality related to thrombosis in congenital antithrombin III deficiency. *Lancet* 337: 847–848.

De Stefano V, Leone G, Micalizzi P, Teofili L, Falappa PG, Pollari G, Bizzi B (1991) Arterial thrombosis as clinical manifestation of congenital protein C deficiency. *Ann. Hematol.* 62: 180–183.

De Stefano V, Mastrangelo S, Schwarz HP (1993) Replacement therapy with a purified protein C concentrate during initiation of oral anticoagulation in severe protein C congenital deficiency. *Thromb. Haemost.* 69: 247–249.

De Stefano V, Finazzi G, Mannucci PM (1996) Inherited thrombophilia: pathogenesis, clinical syndromes and management. *Blood* 87: 3531–3544.

Degen SJF, Davie EW (1987) Nucleotide sequence of the gene for human prothrombin. *Biochemistry* 26: 6165–6177.

Degen SJF, MacGillivray RTA, Davie EW (1983) Characterization of the complementary deoxyribonucleic acid and gene coding for human prothrombin. *Biochemistry* 22: 2087–2097.

Degen SJF, Rajput B, Reich E (1986) The human tissue plasminogen activator gene. *J. Biol. Chem.* 261: 6972–6985.

Degen SJF, Bell SM, Schaefer LA, Elliot RW (1990) Characterization of the cDNA coding for mouse plasminogen and localization of the gene to mouse chromosome 17. *Genomics* 8: 49–61.

Deguchi K, Tsukada T, Iwasaki E, Wada H, Murashima S, Miyasaki M, Shirakawa S (1992) Late-onset homozygous protein C deficiency manifesting cerebral infarction as the first symptom at age 27. *Intern. Med.* 31: 922–925.

Demers C, Ginsberg JS, Hirsh J, Henderson P, Blajchman MA (1992) Thrombosis in antithrombin III-deficient persons. Report of a large kindred and literature review. *Ann. Int. Med.* 116: 754–761.

den Heijer M, Blom HJ, Gerrits WBJ, Koster T, Bos GMJ, Briët E, Reitsma PH, Vandenbroucke JP, Rosendaal FR (1995) Is hyperhomocysteinaemia a risk factor for recurrent venous thrombosis? *Lancet* 345: 882–885.

den Heijer M, Koster T, Blom HJ, Bos GMJ, Briët E, Reitsma PH, Vandenbroucke JP, Rosendaal FR (1996) Hyperhomocysteinemia as a risk factor for deep vein thrombosis. *New Engl. J. Med.* 334: 759–762.

Denault JB, Leduc R (1996) Furin/PACE/SPC1: a convertase involved in exocytic and endocytic processing of precursor proteins. *FEBS Lett.* 379: 113–116.

Derechin VM, Blinder MA, Tollefsen DM (1990) Substitution of arginine for Leu 444 in the reactive site of heparin cofactor II enhances the rate of thrombin inhibition. *J. Biol. Chem.* 265: 5623–5628.

Deschiens M-A, Conard J, Horellou MH, Ameri A, Preter M, Chedru F, Samama MM, Bousser M-G (1996) Coagulation studies, factor V Leiden and anticardiolipin antibodies in 40 cases of cerebral venous thrombosis. *Stroke* 27: 1724–1730.

Desmarais S, de Moerloose P, Reber G, Minazio P, Perrier A, Bounameaux H (1996) Resistance to activated protein C in an unselected population of patients with pulmonary embolism. *Lancet* 347: 1374–1375.

Devraj-Kizuk R, Chiu DHK, Prochownik EV, Carter CJ, Ofosu FA, Blajchman MA (1988) Antithrombin III-Hamilton: a gene with a point mutation (guanine to adenine) in codon 382 causing impaired serine protease reactivity. *Blood* **72**: 1518–1523.

Dhote R, Bachmeyer C, Horellou MH, Toulon P, Christoforov B (1995) Central retinal vein thrombosis associated with resistance to activated protein C. *Am. J. Ophthalmol.* **120**: 388–389.

Dichek D, Quertermous T (1989) Thrombin regulation of mRNA levels of tissue plasminogen activator and plasminogen activator inhibitor-1 in cultured human umbilical vein endothelial cells. *Blood* **74**: 222–228.

Dichek DA, Anderson J, Kelly AB, Hanson SR, Harker LA (1996) Enhanced *in vivo* antithrombotic effects of endothelial cells expressing recombinant plasminogen activators transduced with retroviral vectors. *Circulation* **93**: 301–309.

Diepstraten CM, Ploos van Amstel JK, Reitsma PH, Bertina RM (1991) A CCA/CCG neutral dimorphism in the codon for Pro 626 of the human protein S gene PSα (PROS1). *Nucleic Acids Res.* **19**: 5091.

Diquelou A, Dupouy D, Gaspin D, Constans J, Sie P, Boneu B, Sakariassen KS, Cadroy Y (1995) Relationship between endothelial tissue factor and thrombogenesis under blood flow conditions. *Thromb. Haemost.* **74**: 778–783.

DiScipio RG, Hermondson MA, Yates WG, Davie EW (1977) A comparison of human prothrombin, factor IX (Christmas factor), factor X (Stuart factor) and protein S. *Biochemistry* **16**: 698–706.

DiScipio RG, Davie EW (1979) Characterization of protein S, a γ-carboxyglutamic acid containing protein from bovine and human plasma. *Biochemistry* **18**: 899–904.

Dittman WA, Majcrus PW (1989) Sequence of a cDNA for mouse thrombomodulin and comparison of the predicted mouse and human amino acid sequences. *Nucleic Acids Res.* **17**: 802.

Dittman WA, Majerus PW (1990) Structure and function of thrombomodulin: a natural anticoagulant. *Blood* **75**: 329–336.

Dittman WA, Kumada T, Majerus PW (1989) Transcription of thrombomodulin mRNA in mouse hemangioma cells is increased by cycloheximide and thrombin. *Proc. Natl Acad. Sci. USA* **86**: 7179–7182.

Dittman WA, Nelson SC, Greer PK, Horton ET, Palomba ML, McCachren SS (1994) Characterization of thrombomodulin expression in response to retinoic acid and identification of a retinoic acid response element in the human thrombomodulin gene. *J. Biol. Chem.* **269**: 16925–16932.

Dobkin CS, Ding XH, Jenkins EC, Krawczun MS, Brown WT, Goonewardena P, Willner WT, Benson C, Heitz D, Rousseau F (1991) Prenatal diagnosis of fragile X syndrome. *Lancet* **338**: 957–958.

Doig RG, Begley CG, McGrath KM. (1994) Hereditary protein C deficiency associated with mutations in exon IX of the protein C gene. *Thromb. Haemost.* **72**: 203–208.

Dolan G, Preston FE (1988) Familial plasminogen deficiency and thromboembolism. *Fibrinolysis* **2** (Suppl. 2): 26–34.

Dolan G, Greaves M, Cooper P, Preston FE (1988) Thrombovascular disease and familial plasminogen deficiency: a report of three kindreds. *Br. J. Haematol.* **70**: 417–421.

Dolan G, Ball J, Preston FE (1989) Protein C and protein S. *Baillière's Clin. Haematol.* **2**: 999–1042.

Dolan G, Neal K, Cooper P, Brown P, Preston FE (1994) Protein C, antithrombin III and plasminogen: effect of age, sex and bood group. *Br. J. Haematol.* **86**: 798–803.

Donahue JP, Patel H, Anderson WF, Hawiger J (1994) Three-dimensional structure of the platelet integrin recognition segment of the fibrinogen γ chain obtained by carrier protein-driven crystallization. *Proc. Natl Acad. Sci. USA* **91**: 12178–12182.

Doolittle RF (1980) The evolution of vertebrate fibrinogen. In: Peeters H (ed.) *Protides of the Biologic Fluids*. Pergamon Press, New York.

Doolittle RF (1993) The evolution of vertebrate blood coagulation: a case of Yin and Yang. *Thromb. Haemostas.* **70**: 24–28.

Doolittle RF (1994) Fibrinogen and fibrin. In: Bloom AL, Forbes CD, Thomas DP, Tuddenham EGD (eds) *Haemostasis and Thrombosis*, Vol. 1. Churchill Livingstone, Edinburgh.

Doolittle RF, Feng DF (1987) Reconstructing the evolution of vertebrate blood coagulation from a consideration of the amino acid sequences of clotting proteins. *Cold Spring Harb. Symp. Quant. Biol.* **52**: 869–874.

Doolittle RF, Goldbaum DM, Doolittle LR (1978) Designation of sequences involved in the 'coiled coil' interdomainal connnection in fibrinogen: construction of an atomic scale model. *J. Mol. Biol.* **120**: 311–325.

Doolittle RF, Everse SJ, Spraggon G (1996) Human fibrinogen: anticipating a 3-dimensional structure. *FASEB J.* **10**: 1464–1470.

Drews R, Paleyanda RK, Lee TK, Chang RR, Rehemtulla A, Kaufman RJ, Drohan WN, Lubo H (1995) Proteolytic maturation of protein C upon engineering the mouse mammary gland to express furin. *Proc. Natl Acad. Sci USA* **92**: 10462–10466.

Dreyfus M, Magny JF, Bridey F, Schwarz HP, Planché C, Dehan M, Tchernia G (1991) Treatment of homozygous protein C deficiency and neonatal purpura fulminans with a purified protein C concentrate. *New Engl J. Med.* **325**: 1565–1568.

Dreyfus M, Masterson M, David M, Rivard GE, Müller F-M, Kreuz W, Beeg T, Minford A, Allgrove J, Cohen JD, Christoph J, Bergmann F, Mitchell VE, Haworth C, Nelson K, Schwarz HP (1995) Replacement therapy with a monoclonal antibody purified protein C concentrate in newborns with severe congenital protein C deficiency. *Sem. Thromb. Hemost.* **21**: 371–378.

Drill VA, Calhoun DM (1968) Oral contraceptives and thromboembolic disease. *J. Am. Med. Assoc.* **206**: 77–84.

Duchange N, Chasse J-F, Cohen GN, Zakin MM (1986) Antithrombin III Tours gene: identification of a point mutation leading to an arginine to cysteine replacement in a silent deficiency. *Nucleic Acids Res.* **14**: 2408.

Duchange N, Chasse J-F, Cohen GN, Zakin MM (1987) Molecular characterization of the antithrombin III Tours deficiency. *Thromb. Res.* **45**: 115–121.

Duchemin J, Gandrille S, Borgel D, Feurgard P, Alhenc-Gelas M, Matheron C, Dreyfus M, Dupuy E, Juhan-Vague I, Aiach M (1995) The Ser 460 to Pro substitution of the protein S a (PROS1) gene is a frequent mutation associated with free protein S (type IIa) deficiency. *Blood* **86**: 3436–3443.

Duchemin J, Borg J-Y, Borgel D, Vasse M, Leveque H, Alach M, Gandrille S (1996) Five novel mutations of the protein S active gene (PROS 1) in 8 Norman families. *Thromb. Haemost.* **75**: 437–444.

Dudani AK, Ganz PR (1996) Endothelial cell surface actin serves as a binding site for plasminogen, tissue plasminogen activator and lipoprotein (a). *Br. J. Haematol.* **95**: 168–178.

Dürr C, Hinney A, Luckenbach C, Kömpf J, Ritter H (1992) Genetic studies of antithrombin III with IEF and ASO hybridization. *Hum. Genet.* **90**: 457–459.

Eastman EM, Gilula NB (1989) Cloning and characterization of a cDNA for the Bβ chain of rat fibrinogen: evolutionary conservation of translated and 3'-untranslated sequences. *Gene* **79**: 151–158.

Eaton DL, Fless GM, Kohr WJ, McLean JW, Xu QT, Miller CG, Lawn RM, Scanu AM (1987) Partial amino acid sequence of apolipoprotein (a) shows that it is homologous to plasminogen. *Proc. Natl Acad. Sci. USA* **84**: 3224–3228.

Ebert RF (1991a) Dysfibrinogenemia: an overview of the field. *Thromb. Haemost.* **65**: 1317.

Ebert RF (1991b) *Index of Variant Human Fibrinogens.* CRC Press, Boca Raton, FL.

Edenbrandt C-M, Lundwall A, Wydro R, Stenflo J (1990) Molecular analysis of the gene for vitamin K-dependent protein S and its pseudogene. Cloning and partial gene organization. *Biochemistry* **29**: 7861–7868.

Edens RE, Fromm JR, Fromm SJ, Linhardt RJ, Weiler JM (1995) Two-dimensional affinity resolution electrophoresis demonstrates that three distinct heparin populations interact with antithrombin III. *Biochemistry* **34**: 2400–2407.

Edlund T, Ny T, Ranby M, Heden L-O, Palm G, Holmgren E, Josephson S (1983) Isolation of cDNA sequences coding for a part of human tissue plasminogen activator. *Proc. Natl Acad. Sci. USA* **80**: 349–352.

Egeberg O (1965) Inherited antithrombin deficiency causing thrombophilia. *Thromb. Diath. Haemorrh.* **13**: 516–530.

Ehrenforth S, Aygören-Pürsün E, Hach-Wunderle V, Scharrer I (1993) Prevalence of elevated histidine-rich glycoprotein in patients with thrombophilia – a study of 695 patients. *Thromb. Haemost.* **71**: 160–161.

Ehrlich HJ, Jaskunas SR, Grinnell BW, Yan SB, Bang NU (1989) Direct expression of recombinant activated human protein C, a serine protease. *J. Biol. Chem.* **264**: 14298–14304.

Ehrlich HJ, Grinnell BW, Jaskunas SR, Esmon CT, Yan SB, Bang NU (1990) Recombinant human protein C derivatives: altered response to calcium resulting in enhanced activation by thrombin. *EMBO J.* **9**: 2367–2373.

Ehrlich HJ, Keijer J, Preissner KT, Gebbink RK, Pannekoek H (1991) Functional interaction of plasminogen activator inhibitor type 1 (PAI-1) and heparin. *Biochemistry* **30**: 1021–1028.

Ehrlich HJ, Gebbink RK, Keijer J, Pannekoek H (1992) Elucidation of structural requirements on plasminogen activator inhibitor 1 for binding to heparin. *J. Biol. Chem.* **267**: 11606–11611.

Emmerich J, Chadeuf G, Alhenc-Gelas M, Gouault-Heilman M, Toulon P, Fiessinger JN, Aiach M. (1994a) Molecular basis of antithrombin type I deficiency: the first large in-frame deletion and two novel mutations in exon 6. *Thromb. Haemost.* **72**: 534–539.

Emmerich J, Chadeuf G, Coetzee MJ, Alhenc-Gelas M, Fiessinger J-N, Aiach M (1994b) A phenylalanine 402 to leucine mutation is responsible for a stable inactive conformation of antithrombin. *Thromb. Res.* **76**: 307–315.

Emmerich J, Poirier O, Evans A, Marques-Vidal P, Arveiler D, Luc G, Aiach M, Cambien F (1995) Myocardial infarction, Arg 506 to Gln factor V mutation, and activated protein C resistance. *Lancet* **345**: 321.

Engel H, Zwang L, van Vliet HHDM, Michiels JJ, Stibbe J, Lindemans J (1996) Phenotyping and genotyping of coagulation factor V Leiden. *Thromb. Haemost.* **75**: 267–269.

Engesser L, Kluft C, Briët E, Brommer EJ (1987a) Familial elevation of plasma histidine-rich glycoprotein in a family with thrombophilia. *Br. J. Haematol.* **67**: 355–358.

Engesser L, Broekmans AW, Briët E, Brommer EJP, Bertina RM (1987b) Hereditary protein S deficiency: clinical manifestations. *Ann. Intern. Med.* **106**: 677–682.

Engesser L, Kluft C, Juhan-Vague I, Briët E, Brommer EJP (1988a) Plasma histidine-rich glycoprotein and thrombophilia. *Fibrinolysis* **2**: 43–46.

Engesser L, Koopman J, De Munk G, Haverkate F, Novakova I, Verheijen JH, Briët E, Brommer EJP (1988b) Fibrinogen Nijmegen: congenital dysfibrinogenemia associated with impaired t-PA-mediated plasminogen activation and decreased binding of t-PA. *Thromb. Haemost.* **60**: 113–120.

Engesser L, Brommer EJP, Kluft C, Briët E (1989) Elevated plasminogen activator inhibitor (PAI), a cause of thrombophilia? A study in 203 patients with familial or sporadic venous thrombosis. *Thromb. Haemost.* **62**: 673–680.

Engh RA, Wright TH, Huber R (1990) Modeling the intact form of the α1-proteinase inhibitor. *Protein Eng.* **3**: 469–477.

Erdjument H, Lane DA, Ireland H, Panico M, Di Marzo V, Blench I, Morris HR (1987) Formation of a covalent disulphide-linked antithrombin complex by an antithrombin variant, antithrombin Northwick Park. *J. Biol. Chem.* **262**: 13381–13384.

Erdjument H, Lane DA, Panico M, Di Marzo V, Morris HR (1988a) Single amino acid substitutions in the reactive site of antithrombin leading to thrombosis. *J. Biol. Chem.* **263**: 5589–5593.

Erdjument H, Lane DA, Ireland H, Di Marzo V, Panico M, Morris HR, Tripodi A, Mannucci PM (1988b) Antithrombin Milano, single amino acid substitution at the reactive site, Arg 393 to Cys. *Thromb. Haemost.* **60**: 471–475.

Erdjument H, Lane DA, Panico M, Di Marzo V, Morris HR, Bauer K, Rosenberg RD (1989) Antithrombin Chicago, amino acid substitution of arginine 393 to histidine. *Thromb. Res.* **54**: 613–619.

Erickson LA, Fici GJ, Lund JE, Boyle TP, Polites HG, Marotti KR (1990) Development of venous occlusions in mice transgenic for the plasminogen activator inhibitor-1 gene. *Nature* **346**: 74–76.

Ernst E (1994) Fibrinogen: an important risk factor for atherothrombotic diseases. *Ann. Med.* **26**: 15–22.

Escoffre M, Zini JM, Schliamser L, Mazoyer E, Soria C, Tobelem G, Dupuy E (1995) Severe arterial thrombosis in a congenitally factor VII deficient patient. *Br. J. Haematol.* **91**: 739–741.

Esmon CT (1979) The subunit structure of thrombin-activated factor V. Isolation of activated factor V, separation of subunits and reconstitution of biological activity. *J. Biol. Chem.* **254**: 964–973.

Esmon CT (1987) The regulation of natural anticoagulant pathways. *Science* **235**: 1348–1352.

Esmon NL (1989a) Thrombomodulin. In: Coller BS (ed.) *Progress in Hemostasis and Thrombosis*, Vol. 9, pp. 29–56. WB Saunders, Philadelphia.

Esmon CT (1989b) The roles of protein C and thrombomodulin in the regulation of blood coagulation. *J. Biol. Chem.* **264**: 4743–4746.

Esmon CT (1993a) Cell mediated events that control blood coagulation and vascular injury. *Annu. Rev. Cell Biol.* **9**: 1–26.

Esmon CT (1993b) Molecular events that control the protein C anticoagulant pathway. *Thromb. Haemost.* **70**: 29–35.

Esmon CT (1995) Thrombomodulin as a model of molecular mechanisms that modulate protease specificity and function at the vessel surface. *FASEB J.* **9**: 946955.

Esmon CT, Owen WG (1980) Formation and functions of the anticoagulant protein, activated protein C. In: *Plasma and Cellular Modulatory Proteins*, pp. 203–214. Center for Blood Research Inc., Boston, MA.

Esmon CT, Owen WG (1981) Identification of an endothelial cell co-factor for thrombin-catalysed activation of protein C. *Proc. Natl Acad. Sci. USA* **78**: 2249–2252.

Espinosa R, Sadler JE, LeBeau MM (1989) Regional localization of the human thrombomodulin gene to 20p12–cen. *Genomics* 5: 649–650.

Estelles A, Aznar J, Gilabert J (1980) Dysfunctional plasminogen in a full-term newborn. *Pediatr. Res.* 14: 1180–1185.

Evans DL, Marshall CJ, Christey PB, Carrell RW (1992) Heparin binding site, conformational change, and activation of antithrombin. *Biochemistry* 31: 12629–12642.

Ezenagu LC, Brandt JT (1986) Laboratory determination of heparin cofactor II. *Arch. Pathol. Lab. Med.* 110: 1149–1153.

Faioni EM, Franchi F, Asti D, Sacchi E, Bernardi F, Mannucci PM (1993) Resistance to activatedprotein C in nine thrombophilic families: interference in a protein S functional assay. *Thromb. Haemost.* 70: 1067–1071.

Faioni EM, Franchi F, Asti D, Mannucci PM (1996) Resistance to activated protein C mimicking dysfunctional protein C: diagnostic approach. *Blood Coag. Fibrinol.* 7: 349–352.

Fair DS (1983) Quantitation of factor VII in the plasma of normal and warfarin-treated individuals by radioimmunoassay. *Blood* 62: 784–791.

Fair DS, Bahnak BR (1984) Human hepatoma cells secrete single chain factor X, prothrombin and antithrombin III. *Blood* 64: 194–204.

Fair DS, Marlar RA (1986) Biosynthesis and secretion of factor VIII, protein C, protein S and the protein C inhibitor from a human hepatoma cell line. *Blood* 67: 64–70.

Fair DS, Marlar RA, Levin EG (1986) Human endothelial cells synthesize protein S. *Blood* 67: 1168–1171.

Falcon CR, Cattaneo M, Panzeri D, Martinelli I, Mannucci PM (1994) High prevalence of hyperhomocysteinemia in patients with juvenile venous thrombosis. *Arterioscl. Thromb.* 14: 1080–1083.

Falkon L, Bari M, Montserrat I, Borell Fontcuberta J. (1992) Familial elevation of plasma histidine-rich glycoprotein: a case associated with recurrent venous thrombosis and high PAI-1 levels. *Thromb. Res.* 66: 265–270.

Fan B, Turko IV, Gettins PGW (1994) Antithrombin histidine variants ^{1}H NMR resonance assignments and functional properties. *FEBS Lett.* 354: 84–88.

Farrell DH, Mulvihill ER, Huang S, Chung DW, Davie EW (1991) Recombinant human fibrinogen and sulfation of the γ' chain. *Biochemistry* 30: 9414–9420.

Farrell DH, Huang S, Davie EW (1993) Processing of the carboxyl 15-amino acid extension in the α-chain of fibrinogen. *J. Biol. Chem.* 268: 10351–10355.

Fay PJ, Smudzin TM, Walker FJ (1991a) Activated protein C catalyzed inactivation of human factor VIII and factor VIIIa: identification of cleavage sites and correlation of proteolysis with cofactor activity. *J. Biol. Chem.* 266: 20139–20145.

Fay PJ, Coumans J-V, Walker FJ (1991b) von Willebrand factor mediates protection of factor VIII from activated protein C-catalyzed inactivation. *J. Biol. Chem.* 266: 2172–2177.

Fearon ER, Cho KR, Nigro JM, Kern SE, Simons JW, Ruppert JM, Hamilton SR, Preisinger AC, Thomas G, Kinzler KW, Vogelstein B (1990) Identification of a chromosome 18q gene that is altered in colorectal cancers. *Science* 247: 49–56.

Fenton JW, Olson TA, Zabinski MP, Wilner GD (1988) Anion binding exosite of human α-thrombin and fibrin(ogen) recognition. *Biochemistry* 27: 7106–7112.

Ferguson WS, Finlay TH (1983) Localization of the disulfide band in human antithrombin III required for heparin-accelerated thrombin inactivation. *Arch. Biochem. Biophys.* 221: 304–307.

Fermo I, D'Angelo SV, Paroni R, Mazzola G, Calori G, D'Angelo A (1995) Prevalence of moderate hyperhomocysteinemia in patients with early-onset venous and arterial occlusive disease. *Ann. Intern. Med.* 123: 747–753.

Fernandez JA, Griffin JH (1994) A protein S binding site on C4b-binding protein involves β chain residues 31–45. J. *Biol. Chem.* 269: 2535–2540.

Fernandez F, van Ryn J, Ofosu FA, Hirsh J, Buchanan MR (1986a) The haemorrhagic and antithrombotic effects of dermatan sulfate. *Br. J. Haematol.* 64: 309–317.

Fernandez J, Paramo JA, Cuesta B, Aranda B, Rocha E (1986b) Fibrinogen Pamplona II. A new congenital dysfibrinogenemia with abnormal fibrin-enhanced plasminogen activation and defective binding of thrombin to fibrin. In: Muller-Berghaus G, Scheefers-Borchel U, Selmayr E, Henschen A (eds) *Fibrinogen and its Derivatives. Biochemistry, Physiology and Pathophysiology*, pp. 25–30. Elsevier, Amsterdam,

Fernandez JA, Heeb MJ, Griffin JH (1993) Identification of residues 413–433 of plasma proteins as essential for binding to C4b-binding protein. *J. Biol. Chem.* 268: 16788–16794.

Fernandez JA, Villoutreix BO, Hackeng TM, Griffin JH, Bouma BN (1994) Analysis of protein S C4b-binding protein interactions by homology modeling and inhibitory antibodies. *Biochemistry* 33: 11073–11078.

Fernandez-Rachubinski F, Blajchman MA (1992) A useful restriction analysis for the determination of human antithrombin III variants with mutations from Ala 382 to Ala 384. *Thromb. Res.* 65: 117–120.

Fernandez-Rachubinski F, Rachubinski RA, Blajchman MA (1992) Partial deletion of an antithrombin III allele in a kindred with a type 1 deficiency. *Blood* 80: 1476–1485.

Fernlund P, Stenflo J (1983) β-Hydroxyaspartic acid in vitamin K-dependent proteins. *J. Biol. Chem.* 258: 12509–12512.

Finazzi G, Caccia R, Barbui T (1987) Different prevalence of thromboembolism in the subtypes of congenital antithrombin III deficiency: review of 404 cases. *Thromb. Haemost.* 58: 1094.

Fink LM, Eidt JF, Johnson K, Cook JM, Cook CD, Morser J, Marlar R, Collins CL, Schaefer R, Xie S-S, Hsu S-M, Hsu P-L (1993) Thrombomodulin activity and localization. *Int. J. Dev. Biol.* 37: 221–226.

Fish WW, Bjork I (1979) Release of the two chain form of antithrombin from the antithrombin–thrombin complex. *Eur. J. Biochem.* 101: 31–38.

Fisher R, Waller EK, Grossi G, Thompson D, Tizard R, Schleuning WD (1985) Isolation and characterization of the human tissue-type plasminogen activator structural gene including its 5' flanking region. *J. Biol. Chem.* 260: 11223–11230.

Fisher CL, Greengard JS, Griffin JH (1994) Models of the serine protease domain of the human antithrombotic plasma factor activated protein C and its zymogen. *Protein Science* 3: 588–599.

Fisher M, Fernandez JA, Ameriso SF, Xie D, Gruber A, Paganini-Hill A, Griffin JH (1996) Activated protein C resistance in ischemic strike not due to factor V arginine[506]→glutamine mutation. *Stroke* 27: 1163–1166.

Follo M, Ginsburg D (1989) Stucture and expression of the human gene encoding plasminogen activator inhibitor, PAI-1. *Gene* 84: 447–453.

Formstone CJ, Wacey AI, Berg L-P, Rahman S, Bevan D, Rowley M, Voke J, Bernardi F, Legnani C, Simioni P, Girolami A, Tuddenham EGD, Kakkar VV, Cooper DN (1995) Detection and characterization of seven novel protein S (*PROS*) gene lesions: evaluation of reverse transcript-polymerase chain reaction as a mutation screening strategy. *Blood* 86: 2623–2641.

Formstone CJ, Hallam PJ, Tuddenham EGD, Voke J, Layton M, Nicolaides K, Hann IM, Cooper DN (1996) Severe perinatal thrombosis in double and triple heterozygous offspring of a family segregating two independent protein S mutations and a protein C mutation. *Blood* 87: 3731–3737.

Fornace AJ, Cummings DE, Comeau CM, Kant JA, Crabtree GR (1984a) Structure of the human γ-fibrinogen gene; alternative mRNA splicing near the 3' end of the gene produces γ A and γB forms of γ-fibrinogen. *J. Biol. Chem.* 259: 12826–12830.

Fornace AJ, Cummings DE, Comeau CM, Kant JA, Crabtree GR (1984b) Single-copy inverted repeat associated with regional genetic duplications in γ-fibrinogen and imunoglobulin genes. *Science* 224: 161–164.

Forsgren M, Raden B, Israelsson M, Larsson K, Heden L-O (1987) Molecular cloning and characterization of a full-length cDNA clone for human plasminogen. *FEBS Lett.* 213: 254–260.

Forsyth PD, Dolan G (1995) Activated protein C resistance in cases of cerebral infarction. *Lancet* 345: 795.

Foster D, Davie EW (1984) Characterization of a cDNA coding for human protein C. *Proc. Natl Acad. Sci. USA* 81: 4766–4770.

Foster DC, Yoshitake S, Davie EW (1985) The nucleotide sequence of the gene for human protein C. *Proc. Natl Acad. Sci. USA* 82: 4673–4677.

Foster DC, Rudinski MS, Schach BG, Berkner KL, Kumar AA, Hagen FS, Sprecher CA, Insley MY, Davie EW (1987) Propeptide of human protein C is necessary for γ-carboxylation. *Biochemistry* 26: 7003–7011.

Foster DC, Sprecher CA, Holly RD, Gambee JE, Walker KM, Kumar A (1990) Endoproteolytic processing of the dibasic cleavage site in the human protein C precursor in transfected mammalian cells: effect of sequence alteration on efficiency of cleavage. *Biochemistry* 29: 347–354.

Foster DC, Holly RD, Sprecher CA, Walker KM, Kumar AA (1991) Endoproteolytic processing of the human protein C precursor by the yeast *Kex*2 endopeptidase coexpressed in mammalian cells. *Biochemistry* 30: 367–372.

Fowler WR, Erickson HP (1979) Trinodular structure of fibrinogen: confirmation by both shadowing and negative-stain electron microscopy. *J. Mol. Biol.* 134: 241–249.

Fowlkes OM, Mullis NT, Comeau CM, Crabtree GR (1984) Potential basis for regulation of the coordinately expressed fibrinogen genes: homology in the 5′ flanking regions. *Proc. Natl Acad. Sci. USA* **81**: 2313–2316.

Francis RB (1989) Clinical disorders of fibrinolysis: a critical review. *Blut* **59**: 1–14.

Frank SL, Klisak I, Sparkes RS, Mohandas T, Tomlinson JE, McLean JW, Lawn RM, Lusis AJ (1988) The apolipoprotein (a) gene resides on human chromosome 6q26–27 in close proximity to the homologous gene for plasminogen. *Hum. Genet.* **79**: 352–356.

Frank SL, Klisak I, Sparkes RS, Lusis AJ (1989) A gene homologous to plasminogen located on human chromosome 2q11–p11. *Genomics* **4**: 449–451.

Franzen L-E, Sverisson S, Lawn O (1980) Structural studies on the carbohydrate portion of human antithrombin III. *J. Biol. Chem.* **255**: 5090–5093.

Freedman SJ, Blostein MD, Baleja JD, Jacobs M, Furie BC, Furie B (1996) Identification of the phospholipid binding site in the vitamin K-dependent blood coagulation protein factor IX. *J. Biol. Chem.* **271**: 16227–16236.

Freyssinet JM, Wiesel ML, Grunebaum L, Pereillo J-M, Gauchy J, Schuhler S, Freund G, Cazenave J-P (1989) Activation of human protein C by blood coagulation factor Xa in the presence of anionic phospholipids. Enhancement by sulphated polysaccharides. *Biochem. J.* **261**: 341–348.

Friedrich M, Tautz D (1995) Ribosomal DNA phylogeny of the major extant arthropod classes and the evolution of myriapods. *Nature* **376**: 165–167.

Fritze LMS, Jackman RW, Rosenberg RW (1988) Activity of the thrombomodulin gene promoter in cells of mesodermal origin. *J. Cell Biol.* **107** (Suppl.): 309a.

Frosst P, Blom HJ, Milos R, Goyette P, Sheppard CA, Matthews RG, Boers GJH, den Heijer M, Kluijtmans LAJ, van den Heuvel LP, Rozen R (1995) A candidate genetic risk factor for vascular disease: a common mutation in methylenetetrahydrofolatereductase. *Nature Genetics* **10**: 111–113.

Fryer RH, Wilson BD, Gubler DB, Fitzgerald LA, Rodgers GM (1993) Homocysteine, a risk factor for premature vascular disease and thrombosis, induces tissue factor activity in endothelial cells. *Arterioscl. Thromb.* **13**: 1327–1333.

Fu Y, Weissbach L, Plant PW, Oddoux C, Cao Y, Liang TJ, Roy SN, Redman CM, Grieninger G (1992) Carboxy-terminal-extended variant of the human fibrinogen α subunit: a novel exon conferring marked homology to β and γ subunit. *Biochemistry* **31**: 11968–11972.

Fuchs G, Egbring R, Havemann K (1977) Fibrinogen Marburg. A new genetic variant of fibrinogen. *Blut* **34**: 107–118.

Fujii S, Lucore CL, Hopkins WE, Billadello JJ, Sobel BE (1990) Induction of synthesis of plasminogen activator inhibitor type-1 by tissue-type plasminogen activator in human hepatic and endothelial cells. *Thromb. Haemost.* **64**: 412–419.

Fujikawa K, McMullen BA (1983) Amino acid sequence of human β-factor XIIa. *J. Biol. Chem.* **258**: 10924–10933.

Fujikawa K, Saito H (1989) Contact activation. In: Scriver CR, Beaudet AL, Sly WS, Valle D (eds) *The Metabolic Basis of Inherited Disease*, 6th edn, pp. 2189–2206. McGraw-Hill, New York.

Fujikawa D, McMullen B, Heimark RL, Kurachi K, Davie EW (1980) The role of factor XII (Hageman factor) in blood coagulation and a partial amino acid sequence of human factor XII and its fragments. *Protides Biol. Fluids* **28**: 193–196.

Fujimura H, Kambayashi J, Monden M, Kato H, Miyata T (1995) Coagulation factor V Leiden mutation may have a racial background. *Thromb. Haemost.* **74**: 1381–1382.

Fujiwara J, Kimura T, Ayusawa D, Oishi M (1994) A novel regulatory sequence affecting the constitutive expression of tissue plasminogen activator (tPA) gene in human melanoma (Bowes) cells. *J. Biol. Chem.* **269**: 18558–18562.

Fukudome K, Esmon CT (1994) Identification, cloning and regulation of a novel endothelial cell protein C/activated protein C receptor. *J. Biol. Chem.* **269**: 26486–26491.

Fukudome K, Esmon CT (1995) Molecular cloning and expression of murine and bovine endothelial cell protein C/activated protein C receptor (EPCR). The structural and functional conservation in human, bovine and murine EPCR. *J. Biol. Chem.* **270**: 5571–5577.

Furie B, Furie BC (1988) The molecular basis of blood coagulation. *Cell* **53**: 505–518.

Furie B, Bing DH, Feldman RJ, Robinson DJ, Burnier JP, Furie BC (1982) Computer-generated models of blood coagulation factor Xa, factor IXa and thrombin based upon structural homology with other serine proteases. *J. Biol. Chem.* **257**: 3875–3882.

Furlan M, Steinmann C, Jungo M, Bögli C, Baudo F, Redaelli R, Fedeli F, Lämmle B (1994) A frameshift mutation in exon V of the Aα-chain gene leading to truncated Aα-chains in the homozygous dysfibrinogen Milano III. *J. Biol. Chem.* **269**: 33129–33134.

Furlan M, Stucki B, Steinmann C, Jungo M, Lämmle B (1996) Normal binding of calcium to five fibrinogen variants with mutations in the carboxy terminal part of the γ-chain. *Thromb. Haemost.* **76**: 377–383.

Fuster V, Badimon L, Badimon JJ, Chesebro JH (1992a) The pathogenesis of coronary artery disease and the acute coronary syndromes. Part 1. *New Engl. J. Med.* **326**: 242–250.

Fuster V, Badimon L, Badimon JJ, Chesebro JH (1992b) The pathogenesis of coronary artery disease and the acute coronary syndromes. Part 2. *New Engl. J. Med.* **326**: 310–318.

Galanakis DK (1993) Inherited dysfibrinogenemia: emerging abnormal structure associations with pathologic and nonpathologic dysfunctions. *Sem. Thromb. Hemost.* **19**: 386–395.

Galanakis DK, Henschen A, Schubach W, Lord S, Al-Mondhiry H (1990) Determination of abnormal structure and heterozygosity by amplification of genomic DNA using the polymerase chain reaction and by amino acid sequence analysis. In: Matsuda M, Iwanaga S, Takada A, Henschen A (eds) *Fibrinogen 4: Current Basic and Clinical Aspects*, pp. 173–178. Elsevier, Amsterdam.

Gan Z-R, Li Y, Chen Z, Lewis SD, Shafer JA (1994) Identification of basic amino acid residues in thrombin essential for heparin-catalyzed inactivation by antithrombin III. *J. Biol. Chem.* **269**: 1301–1305.

Gandrille S, Aiach M (1991) Polymorphism in the protein C gene detected by denaturing gradient gel electrophoresis. *Nucleic Acids Res.* **19**: 6982.

Gandrille S, Aiach M (1995) Identification of mutations in 90 of 121 consecutive symptomatic French patients with a type I protein C deficiency. *Blood* **86**: 2598–2605.

Gandrille S, Aiach M, Lane DA, Vidaud D, Molho-Sabatier P, Caso R, de Moerloose P, Fiessinger J-N, Clauser E (1990) Important role of arginine 129 in heparin-binding site of antithrombin III. *J. Biol. Chem.* **265**: 18997–19001.

Gandrille S, Vidaud D, Emmerich J, Clauser E, Sié P, Fiessinger JN, Alhenc-Gelas M (1991a) Molecular basis for hereditary antithrombin III quantitative deficiencies: a Stop codon in exon IIIa and a frameshift in exon VI. *Br. J. Haematol.* **78**: 414–420.

Gandrille S, Vidaud M, Aiach M, Alhenc-Gelas M, Fischer AM, Gouault-Heilman M, Toulon P, Goossens M (1991b) Six previously undescribed mutations in 9 families with protein C quantitative deficiency. *Thromb. Haemost.* **65**: 646.

Gandrille S, Vidaud M, Aiach M, Alhenc-Gelas M Fischer AM, Gouault-Heilman M, Toulon P, Fiessinger JN, Goossens M (1992) Two novel mutations responsible for hereditary type I protein C deficiency: characterization by denaturing gradient gel electrophoresis. *Hum. Mutat.* **1**: 491–500.

Gandrille S, Alhenc-Gelas M, Gaussem P, Aillaud M-F, Dupuy E, Juhan-Vague I, Aiach M (1993a) Five novel mutations located in exons III and IX of the protein C gene in patients presenting with defective protein C anticoagulant activity. *Blood* **82**: 159–168.

Gandrille S, Juda B, Alhenc-Gelas M, Millaire A, Aiach M (1993b) Compound heterozygosity in a family with protein C deficiency illustrating the complexity of the underlying molecular mechanism. *Thromb. Haemost.* **70**: 747–752.

Gandrille S, Goossens M, Aiach M (1994a) Scanning method to establish the molecular basis of protein C deficiencies. *Hum. Mutat.* **4**: 20–30.

Gandrille S, Jude B, Alhenc-Gelas M, Emmerich J, Aiach M. (1994b) First *de novo* mutations in the protein C gene of two patients with type I deficiency: a missense mutation and a splice site deletion. *Blood* **84**: 2566–2570.

Gandrille S, Borgel D, Eschwege-Gufflet V, Aillaud M, Dreyfus M, Matheron C, Gaussem P, Abgrall JF, Jude B, Sie P, Toulon P, Aiach M (1995a) Identification of 15 different candidate causal point mutations and three polymorphisms in 19 patients with protein S deficiency using a scanning method for the analysis of the protein S active gene. *Blood* **85**: 130–138.

Gandrille S, Greengard JS, Alhenc-Gelas M, Juhan-Vague I, Abgrall JF, Jude B, Griffin JH, Aiach M (1995b) Incidence of activated protein C resistance caused by the Arg 506 Gln mutation in factor V in 113 unrelated symptomatic protein C-deficient patients. *Blood* **86**: 219–224.

Gandrille S, Alhenc-Gelas M, Aiach M (1995c) A rapid screening method for the factor V Arg506→Gln mutation. *Blood Coag. Fibrinol.* **6**: 245–248.

Gandrille S, Borgel D, Ireland H, Lane DA, Simmonds R, Reitsma PH, Mannhalter C, Pabinger I, Saito H, Suzuki K, Formstone C, Cooper DN, Espinosa Y, Sala N, Bernardi F, Aiach M (1997) Protein S deficiency: a database of mutations. *Thromb. Haemost.*, in press.

Ganesan V, Kelsey H, Cookson J, Osborn A, Kirkham FJ (1996) Activated protein C resistance in childhood stroke. *Lancet* **347**: 260.

Garone L, Edmunds T, Hanson E, Bernasconi R, Huntington JA, Meagher JL, Fan B, Gettins PGW (1996) Antithrombin–heparin affinity reduced by fucosylation of carbohydrate at asparagine 155. *Biochemistry* 35: 8881–8889.

García de Frutos P, Alim RIM, Härdig Y, Zöller B, Dahlbäck B (1994) Differential regulation of α and β chains of C4b-binding protein during acute-phase response resulting in stable plasma levels of free anticoagulant protein S. *Blood* 84: 815–822.

Gardiner JE, McGann MA, Berridge CW, Fulcher CA, Zimmerman TS, Griffin JH (1984) Protein S as a cofactor for activated protein C in plasma and in the inactivation of purified factor VIII: C. *Circulation* 70: II: 205.

Gasic GP, Arenas CP, Gasic TB, Gasic GJ (1992) Coagulation factors X, Xa and protein S as potent mitogens of cultured aortic smooth muscle cells. *Proc. Natl Acad. Sci. USA* 89: 2317–2320.

Gaussem P, Gandrille S, Duchemin J, Emmerich J, Alhenc-Gelas M, Aillaud M-F, Aiach M (1994) Influence of six mutations of the protein C gene on the Gla domain conformation and calcium affinity. *Thromb. Haemost.* 71: 748–754.

Gebbink RK, Reynolds CH, Tollefsen DM, Mertens K, Pannekoek H (1993) Specific glycosaminoglycans support the inhibition of thrombin by plasminogen activator inhibitor 1. *Biochemistry* 32: 1675–1680.

Geerts WH, Code KI, Jay RM, Chen E, Szalai JP (1994) A prospective study of venous thromboembolism after major trauma. *New Engl. J. Med.* 331: 1601–1606.

Geng J-P, Castellino FJ (1996) The propeptides of human protein C, factor VII, and factor IX are exchangeable with regard to directing gamma-carboxylation of these proteins. *Thromb. Haemost.* 76: 205–207.

Geng J-P, Cheng C-H, Castelllino FJ (1996) Functional consequences of mutations in amino acid residues that stabilize calcium binding to the first epidermal growth factor homology domain of human protein C. *Thromb. Haemost.* 76: 720–728.

Gerard RD, Chien KR, Meidell RS (1986) Molecular biology of tissue plasminogen activator and endogenous inhibitors. *Mol. Biol. Med.* 3: 449–457.

Gerlitz B, Grinnell BW (1996) Mutation of protease domain residues Lys 37–39 in human protein C inhibits activation by the thrombomodulin-thrombin complex without affecting activation by free thrombin. *J. Biol. Chem.* 271: 22285–22288.

Gerlitz B, Hassell T, Vlahos CJ, Parkinson JF, Bang NU, Grinnell BW (1993) Identification of the predominant glycosaminoglycan-attachment site in soluble recombinant human thrombomodulin: potential regulation of functionality by glycosyltransferase competition for serine[474]. *Biochem J.* 295: 131–140.

Gerritsen ME, Bloor CM (1993) Endothelial cell gene expression in response to injury. *FASEB J.* 7: 523–532.

Gershagen S, Fernlund P, Lundwall A (1987) A cDNA coding for human sex hormone-binding globulin. Homology to vitamin K-dependent protein S. *FEBS Lett.* 220: 129–135.

Gershwin ME, Gude JK (1973) Deep vein thrombosis and pulmonary embolism in congenital factor VII deficiency. *New Engl. J. Med.* 288: 141–142.

Gettins P, Choay J, Crews BC, Zettlmeissl G (1992) Role of tryptophan 49 in the heparin cofactor activity of human antithrombin III. *J. Biol. Chem.* 267: 21946–21953.

Gettins P, Patston PA, Schapira M (1993a) The role of conformational change in serpin structure and function. *BioEssays* 15: 461–467.

Gettins PGW, Fan B, Crews BC, Turko IV, Olson ST, Streusand VJ (1993b) Transmission of conformational change from the heparin binding site to the reactive center of antithrombin. *Biochemistry* 32: 8385–8389.

Giannelli F, Green PM, Sommer SS, Poon MC, Ludwig M, Schwaab R, Reitsma PH, Goossens M, Yoshioka A, Brownlee GG (1996) Haemophilia B (sixth edition): a database of point mutations and short additions and deletions. *Nucleic Acids Res.* 24: 103–118.

Gibbs CS, Coutre SE, Tsiang M, Li W-X, Jain AK, Dunn KE, Law VS, Mao CT, Matsumara SY, Mejza SJ, Paborsky LR, Leung LLK (1995) Conversion of thrombin into an anticoagulant by protein engineering. *Nature* 378: 413–416.

Gibson JB, Carson NAJ, Neill DW (1964) Pathological findings in homocystinuria. *J. Clin. Pathol.* 17: 427–437.

Gierasch LM (1989) Signal sequences. *Biochemistry* 28: 923–930.

Ginsberg JS (1996) Management of venous thrombosis. *New Engl. J. Med.* 24: 1816–1828.

Ginsburg D, Zehab R, Yang AY, Rafferty UM, Andreasen PA, Nielsen L, Danÿi K, Lebo RV, Gelehrter TD (1986) cDNA cloning of human plasminogen activator inhibitor from endothelial cells. *J. Clin. Invest.* 78: 1673–1680.

Girolami A, Falezza G, Patrassi G, Stenico M, Vettore L (1977) Factor VII Verona coagulation disorder: double heterozygosis with an abnormal VII and heterozygous factor VII deficiency. *Blood* 50: 603–610.

Girolami A, Fabris F, Dal BO, Zanon LR, Ghiotto G, Burul A (1978) Factor VII Padua: a congenital coagulation disorder due to an abnormal factor VII with a peculiar activation pattern. *J. Lab. Clin. Med.* 91: 387–395.

Girolami A, Marafioti F, Rubertelli M, Cappellato MG (1986) Congenital heterozygous plasminogen deficiency associated with a severe thrombotic tendency. *Acta Haematol.* 75: 54–57.

Girolami A, Lazzaro AR, Simioni P (1988) The relationship between defective heparin cofactor activities and thrombotic phenomena in ATIII abnormalities. *Thromb. Haemost.* 59: 121.

Girolami A, Simioni P, Lazzaro AR, Cordiano I (1989) Severe arterial cerebral thrombosis in a patient with protein S deficiency (moderately reduced total and markedly reduced free protein S): a family study. *Thromb. Haemost.* 61: 144–147.

Girolami A, Simioni P, Girolami B, Marchiori A, Millar DS, Bignell P, Kakkar VV, Cooper DN (1993) A novel dysfunctional protein C (Protein C Padua 2) associated with a thrombotic tendency: substitution of Cys for Arg-1 results in a strongly reduced affinity for binding of Ca^{++}. *Br. J. Haematol.* 85: 521–527.

Girolami A, Sartori MT, Saggiorato G, Sgarabotto D, Patrassi GM (1994) Symptomatic versus asymptomatic patients in congenital hypoplasminogenemia: a statistical analysis. *Haematologia* 26: 59–65.

Girolami A, Simioni P, Girolami B, Radossi P (1996) Homozygous patients with APC resistance may remain paucisymptomatic or asymptomatic during oral contraception. *Blood Coag. Fibrinol.* 7: 590–594.

Gjores JE (1956) The incidence of venous thrombosis and its sequelae in certain districts in Sweden. *Acta Chir. Scand.* 111: 16–24.

Gladson CL, Scharrer I, Hach V, Beck KH, Griffin JH (1988) The frequency of type I heterozygous protein S and protein C deficiency in 141 unrelated young patients with venous thrombosis. *Thromb. Haemost.* 59: 18–22.

Godall HC, Madsen K, Nissen-Meyer R (1962) Thromboembolism in patient with total proconvertin (factor VII) deficiency. *Acta Med. Scand.* 171: 325–327.

Godowski PJ, Mark MR, Chen J, Sadick MD, Raab H, Hammonds RG (1995) Reevaluation of the roles of protein S and Gas6 as ligands for the receptor tyrosine kinase *Rse/Tyro3*. *Cell* 82: 355–358.

Gojobori T, Ikeo K (1994) Molecular evolution of serine protease and its inhibitor with special reference to domain evolution. *Phil. Trans. R. Soc. Lond. B* 344: 411–415.

Goldberg RJ, Seneff M, Gore JM (1987) Occult malignant neoplasms in patients with deep vein thrombosis. *Arch. Intern. Med.* 147: 251–253.

Gómez E, Ledford MR, Pegelow CH, Reitsma PH, Bertina RM (1994) Homozygous protein S deficiency due to a one base pair deletion that leads to a Stop codon in exon III of the protein S gene. *Thromb. Haemost.* 71: 723–726.

Gómez E, Poort SR, Bertina RM, Reitsma PH (1995) Identification of eight point mutations in protein S deficiency type I – analysis of 15 pedigrees. *Thromb. Haemost.* 73: 750–755.

Gomi K, Zushi M, Honda G (1990) Antithrombotic effect of recombinant human thrombomodulin on thrombin-induced thromboembolism in mice. *Blood* 75: 1396–1399.

Gonda Y, Hirata S, Saitoh K, Aoki Y, Mohri M, Gomi K, Sugihara T, Kiyota T, Yamamoto S, Ishida T (1993) Antithrombotic effect of recombinant human soluble thrombomodulin on endotoxin-induced disseminated intravascular coagulation in rats. *Thromb. Res.* 71: 325–335.

Gonzalez-Gronow M, Robbins KC (1984) *In vitro* biosynthesis of plasminogen in a cell-free system directed by mRNA fractions isolated from monkey liver. *Biochemistry* 23: 190–196.

Goodnough LT, Saito H, Ratnoff OD (1983) Thrombosis or myocardial infarction in congenital clotting factor abnormalities and chronic thrombocytopenias: a report of 21 patients and a review of 50 previously reported cases. *Medicine* 62: 248–255.

Goodwin CA, Melissari E, Kakkar VV, Scully MF (1993) Plasma levels of tissue factor pathway inhibitor in thrombophilic patients. *Thromb. Res.* 72: 363–366.

Gordon EM, Ratnoff OD, Jones PK (1982) The role of augmented Hageman factor (factor XII) titers in the cold-promoted activation of factor VII and spontaneous shortening of the prothrombin time in women using oral contraceptives. *J. Lab. Clin. Med.* 99: 363–369.

Gordon EM, Hellerstein HK, Ratnoff OD, Arafah BM, Yamashita TS (1987) Augmented Hageman factor and prolactin titers, enhanced cold activation of factor VII and spontaneous shortening of prothrombin time in survivors of myocardial infarction. *J. Lab. Clin. Med.* 109: 409–413.

Gordon EM, Johnson TR, Ramos LP, Schmeidler-Sapiro KT (1991) Enhanced expression of factor XII (Hageman factor) in isolated livers of estrogen- and prolactin-treated rats. *J. Lab. Clin. Med.* **117**: 353–358.

Grabowski EF, Lam FP (1995) Endothelial cell function, including tissue factor expression, under flow conditions. *Thromb. Haemost.* **74**: 123–128.

Grailhe P, Boyer-Neumann C, Haverkate F, Grimbergen J, Larrieu MJ, Anglés-Cano E (1993) The mutation in fibrinogen Bicêtre II (γ Asn308→Lys) does not affect the binding of t-PA and plasminogen to fibrin. *Blood Coag. Fibrinol.* **4**: 679–687.

Grantham R (1974) Amino acid difference formula to help explain protein evolution. *Science* **185**: 862–864.

Greaves M, Preston FE (1991) The hypercoagulable state in clinical practice. *Br. J. Haematol.* **79**: 148–151.

Green F, Humphries S (1989) Control of plasma fibrinogen levels. *Baillière's Clin. Haematol.* **2**: 945–959.

Green PM, Montandon AJ, Bentley DR, Ljung R, Nilsson IM, Giannelli F (1990) The incidence and distribution of CpG→TpG transitions in the coagulation factor IX gene. A fresh look at CpG mutational hotspots. *Nucleic Acids Res.* **18**: 3227–3231.

Green F, Kelleher C, Wilkes H, Temple A, Meade T, Humphries S (1991) A common genetic polymorphism associated with lower coagulation factor VII levels in healthy individuals. *Arteriosclerosis* **11**: 540–546.

Greenberg CS, Dobson JV, Miraglia CC (1985) Regulation of plasma factor XIII binding to fibrin *in vitro*. *Blood* **66**: 1028–1034.

Greenberg CS, Achyuthan KE, Fenton JW (1987) Factor XIIIa formation promoted by complexing of α-thrombin, fibrin and plasm factor XIII. *Blood* **69**: 867–871.

Greengard JS, Sun X, Xu X, Fernandez JA, Griffin JH, Evatt B (1994a) Activated protein C resistance caused by Arg506→Gln mutation in factor Va. *Lancet* **343**: 1361–1362.

Greengard JS, Eichinger S, Griffin JH, Bauer KA (1994b) Variability of thrombosis among homozygous siblings with resistance to activated protein C due to an Arg→Gln mutation in the gene for factor V. *New Engl. J. Med.* **331**: 1559–1562.

Greengard JS, Fisher CL, Villoutreix B, Griffin JH (1994c) Structural basis for type I and type II deficiencies of antithrombotic plasma protein C: patterns revealed by three-dimensional molecular modelling of mutations of the protease domain. *Prot. Struct. Func. Genet.* **18**: 367–380.

Greengard JS, Griffin JH, Fisher CL (1994d) Possible structural implications of 20 mutations in the protein C protease domain. *Thromb. Haemost.* **72**: 869–873.

Greengard JS, Fernandez JA, Radtke K-P, Griffin JH (1995a) Identification of candidate residues for interaction of protein S with C4b binding protein and activated protein C. *Biochem. J.* **305**: 397–403.

Greengard JS, Xu X, Gandrille S, Griffin JH (1995b) Alternative PCR method for diagnosis of mutation causing activated protein C resistant Gln506-factor V. *Thromb. Res.* **80**: 441–443.

Greer G (1990) Comparative modelling methods: application to the family of the mammalian serine proteases. *Prot. Struct. Func. Genet.* **7**: 317–334.

Greffe BS, Manco-Johnson MJ, Marlar RA (1989) Molecular forms of human protein C: comparison and distribution in human adult plasma. *Thromb. Haemost.* **62**: 902.

Grey ST, Hau H, Salem HH, Hancock WW (1993) Selective effects of protein C on activation of human monocytes by lipopolysaccharide, interferon-γ, or PMA: modulation of effects on CD11b and CD14 but not CD25 or CD54 induction. *Transplant. Proc.* **25**: 2913–2914.

Grey ST, Tsuchida A, Hau H, Orthner CL, Salem HH, Hancock WW (1994) Selective inhibitory effects of the anticoagulant activated protein C on the responses of human mononuclear phagocytes to LPS, IFN-γ, or phorbol ester. *J. Immunol.* **153**: 3664–3672.

Griffin JH, Evatt B, Zimmerman TS, Kleiss AJ, Wideman C (1981) Deficiency of protein C in congenital thrombotic disease. *J. Clin. Invest.* **68**: 1370–1373.

Griffin JH, Gruber A, Fernandez JA (1992) Reevaluation of total, free, and bound protein S and C4b-binding protein levels in plasma anticoagulated with citrate or hirudin. *Blood* **79**: 3203–3208.

Griffin JH, Evatt B, Wideman C, Fernandez JA (1993) Anticoagulant protein C pathway defective in majority of thrombophilic patients. *Blood* **82**: 1989–1993.

Griffin JH, Heeb MJ, Kojima Y, Fernandez JA, Kojima K, Hackeng TM, Greengard JS (1995) Activated protein C resistance: molecular mechanisms. *Thromb. Haemost.* **74**: 444–448.

Griffith MJ (1982) The heparin-enhanced antithrombin III/thrombin reaction is saturable with respect to both thrombin and antithrombin III. *J. Biol. Chem.* **257**: 13899–13902.

Griffiths MJ, Noyes CM, Church FC (1985) Reactive site peptide structural similarities between heparin cofactor II and antithrombin III. *J. Biol. Chem.* **260**: 2218–2225.

Grimaudo V, Bachmann F, Hauert J, Christie MA, Kruitof E (1992) Hypofibrinolysis in patients with a history of idiopathic deep vein thrombosis and/or pulmonary embolism. *Thromb. Haemost.* **67**: 397–401.

Grinnell BW, Berg DT, Walls J, Yan SB (1987) Trans-activated expression of fully γ-carboxylated recombinant human protein C, an antithrombotic factor. *Biotechnology* **5**: 1189–1192.

Grinnell BW, Walls JD, Gerlitz B (1991) Glycosylation of human protein C affects its secretion, processing, functional activities and activation by thrombin. *J. Biol. Chem.* **266**: 9778–9785.

Grinnell BW, Hermann RB, Yan SB (1994) Human protein C inhibits selectin-mediated cell adhesion: role of unique fucosylated oligosaccharide. *Glycobiology* **4**: 221–226.

Gritzmacher CA (1989) Molecular aspects of heavy-chain class switching. *Crit. Rev. Immunol.* **9**: 173–200.

Gruber A, Hanson SR, Kelly AB (1990) Inhibition of thrombus formation by activated recombinant protein C in a primate model of arterial thrombosis. *Circulation* **82**: 578–585.

Gruber A, Mori E, del Zoppo GJ, Waxman L, Griffin JH. (1994) Alteration of fibrin network by activated protein C. *Blood* **83**: 2541–2548.

Grundy C, Chitolie A, Talbot S, Bevan D, Kakkar V, Cooper DN (1989) Protein C London 1: recurrent mutation at Arg 169 (CGG→TGG) in the protein C gene causing thrombosis. *Nucleic Acids Res.* **17**: 10513.

Grundy C, Plendl H, Grote W, Zoll B, Kakkar VV, Cooper DN (1991a) A single base-pair deletion in the protein C gene causing recurrent thromboembolism. *Thromb. Res.* **61**: 335–340.

Grundy CB, Thomas F, Millar DS, Krawczak M, Melissari E, Lindo V, Moffat E, Kakkar VV, Cooper DN (1991b) Recurrent deletion in the human antithrombin III gene. *Blood* **78**: 1027–1032.

Grundy CB, Melissari E, Lindo V, Scully MF, Kakkar VV, Cooper DN (1991c) The molecular basis of late-onset homozygous protein C deficiency. *Lancet* **338**: 575–576.

Grundy CB, Schulman S, Tengborn L, Kakkar VV, Cooper DN (1992a) Two different missense mutations at Arg 178 of the protein C (*PROC*) gene causing recurrent venous thrombosis. *Hum. Genet.* **89**: 685–686.

Grundy CB, Holding S, Millar DS, Kakkar VV, Cooper DN (1992b) A novel missense mutation in the antithrombin III gene (Ser 349→Pro) causing recurrent venous thrombosis. *Hum. Genet.* **88**: 707–708.

Grundy CB, Chisholm M, Kakkar VV, Cooper DN (1992c) A novel homozygous missense mutation in the protein C (*PROC*) gene causing recurrent venous thrombosis. *Hum. Genet.* **89**: 683–684.

Grundy CB, Schulman S, Kobosko J, Kakkar VV, Cooper DN (1992d) Protein C deficiency and thromboembolism: recurrent mutation at Arg 306 in the protein C gene. *Hum. Genet.* **88**: 586–588.

Guillerm C, Lellouche F, Darnige L, Schandelong A, Geffroy F, Dorval I (1996) Rapid detection of the G161A mutation of coagulation factor V by PCR-mediated site-directed mutagenesis. *Clin. Chem.* **42**: 329.

Guinto ER, Esmon CT (1984) Loss of prothrombin and of factor Xa-factor Va interactions upon inactivation of factor Va by activated protein C. *J. Biol. Chem.* **259**: 13986–13992.

Guinto ER, Esmon CT, Mann KG, MacGillivray RTA (1992) The complete cDNA sequence of bovine coagulation factor V. J. Biol. Chem. *J. Biol. Chem.* **267**: 2971–2978.

Gurgey A, Mesci L, Renda Y, Olcay L, Kocak N, Erdem G (1996) Factor V Q506 mutation in children with thrombosis. *Am. J. Hematol.* **53**: 37–39.

Hackeng TM, Hessing M, van't Veer C, Meijer-Huizinga F, Meijers JCM, de Groot PG, van Mourik JA, Bouma BN (1993) Protein S binding to human endothelial cells is required for expression of cofactor activity for activated protein C. *J. Biol. Chem.* **268**: 3993–4000.

Hackeng TM, van't Veer C, Meijers JCM, Bouma BN (1994) Human protein S inhibits prothrombinase complex activity on endothelial cells and platelets via direct interactions with factors Va and Xa. *J. Biol. Chem.* **269**: 21051–21058.

Hackeng TM, Tans G, Koppelman SJ, De Groot PG, Rosing J, Bouma BN (1996) Protein C activation on endothelial cells by prothrombin activation products generated *in situ*: meizothrombin is a better protein C activator than α-thrombin. *Biochem. J.* **319**: 399–405.

Hagan KW, Ruiz-Echevarria MJ, Quan Y, Peltz SW (1995) Characterization of *cis*-acting sequences and decay intermediates involved in nonsense-mediated mRNA turnover. *Mol. Cell. Biol.* **15**: 809–823.

Hagen FS, Gray CL, O'Hara P, Grant FJ, Saari GC, Woodbury RG, Hart CE, Insley M, Kisiel W, Kurachi K, Davie EW (1986) Characterization of a cDNA coding for human factor VII. *Proc. Natl Acad. Sci. USA* **83**: 2412–2416.

Haidaris PJ, Courtney MA (1992) Liver-specific RNA processing of the ubiquitously transcribed rat fibrinogen γ-chain gene. *Blood* **79**: 1218–1224.

Hajjar KA (1991) The endothelial cell tissue plasminogen activator receptor. *J. Biol. Chem.* **266**: 21962–21970.

Hajjar KA, Jacovina AT, Chacko J (1994) An endothelial cell receptor for plasminogen/ tissue plasminogen activator. I. Identity with annexin II. *J. Biol. Chem.* **269**: 21191–21197.

Hakkert BC, Rentenaar JM, van Mourik JA (1990) Monocytes enhance the bidirectional release of type I plasminogen activator inhibitor by endothelial cells. *Blood* **76**: 2272–2276.

Halbmeyer W-M, Mannhalter C, Feichtinger C, Rubi K, Fischer M (1992) The prevalence of factor XII deficiency in 103 orally anticoagulated outpatients suffering from recurrent venous and/or arterial thromboembolism. *Thromb. Haemost.* **68**: 285–290.

Halbmeyer W-M, Haushofer A, Schön R, Mannhalter C, Strohmer E, Baumgarten K, Fischer M (1994a) The prevalence of moderate and severe FXII (Hageman factor) deficiency among the normal population: evaluation of the incidence of FXII deficiency among 300 healthy blood donors. *Thromb. Haemost.* **71**: 68–72.

Halbmeyer W.-M., Haushofer A, Schön R, Fischer M (1994b) The prevalence of poor anticoagulant response to activated protein C (APC resistance) among patients suffering from stroke or venous thrombosis and among healthy subjects. *Blood Coag. Fibrinol.* **5**: 51–57.

Haley PE, Doyle MF, Mann KG. (1989) The activation of bovine protein C by factor Xa. *J. Biol. Chem.* **264**: 16303–16310.

Hall CA, Rapaport SI, Ames SB (1964) A clinical and family study of hereditary proconvertin (factor VII) deficiency. *Am. J. Med.* **37**: 172–181.

Hallam PJ, Millar DS, Krawczak M, Kakkar VV, Cooper DN (1995a) Population differences in the frequency of the factor V Leiden variant among individuals with clinically symptomatic protein C deficiency. *J. Med. Genet.* **32**: 543–545.

Hallam PJ, Wacey AI, Mannucci PM, Legnani C, Kühnau W, Krawczak M, Kakkar VV, Cooper DN (1995b) A novel missense mutation (Thr176→Ile) at the putative hinge of the neo N-terminus of activated protein C. *Hum. Genet.* **95**: 447–450.

Hamaguchi M, Morishita Y, Takahashi I, Ogura M, Takamatsu J, Saito H (1991) FDP-D-dimer induces the secretion of interleukin-1, urokinase-type plasminogen activator and plasminogen activator inhibitor-2 in a human promonocytic leukemic cell line. *Blood* **77**: 94–100.

Hamsten A, Wiman B, De Faire U, Blombäck M (1985) Increased plasma levels of a rapid inhibitor of tissue plasminogen activator in young survivors of myocardial infarction. *New Engl. J. Med.* **313**: 1557–1563.

Hamsten A, Blombäck M, Wiman B, Svensson J, Szamosi A, De Faire U, Mettinger L (1986) Haemostatic function in myocardial infarction. *Br. Heart J.* **55**: 58–66.

Hamsten A, De Fairell, Walldius G, Dahlen G, Szamosi A, Landou C, Blombäck M, Wiman B (1987) Plasminogen activator inhibitor in plasma: risk factor for recurrent myocardial infarction. *Lancet* **ii**: 3–9.

Hanawalt PC (1990) Selective DNA repair in active genes. *Acta Biol. Hung.* **41**: 77–91.

Handford PA, Mayhew M, Baron M, Winship PR, Campbell ID, Brownlee GG (1991) Key residues involved in calcium-binding motifs in EGF-like domains. *Nature* **351**: 164–166.

Handt S, Jerome WG, Tietze L, Hantgan RR (1996) Plasminogen activator inhibitor-1 secretion by endothelial cells increases fibrinolytic resistance of an *in vitro* fibrin clot: evidence for a key role of endothelial cells in thrombolytic resistance. *Blood* **87**: 4204–4213.

Hanna LS, Scheraga HA, Francis CW, Marder VJ (1984) Comparison of stuctures of various fibrinogens and a derivative thereof by a study of the kinetics of release of fibrinopeptide. *Biochemistry* **23**: 40681–40687.

Hansen MS, Clemmensen I, Winther D (1980) Fibrinogen Copenhagen; an abnormal fibrinogen with defective polymerization and release of fibrinopeptide A, but normal adsorption of plasminogen. *Scand. J. Clin. Lab. Invest.* **40**: 221–226.

Hanson SR, Griffin JH, Harker LA, Kelly AB, Esmon CT, Gruber A (1993) Antithrombotic effects of thrombin-induced activation of endogenous protein C in primates. *J. Clin. Invest.* **92**: 2003–2012.

Hantgan RR, Fowler RW, Erickson HP, Hermans J (1980) Fibrin assembly: a comparison of electron microscopic and light scattering results. *Thromb. Haemost.* **44**: 119–124.

Hantgan RR, Francis CW, Scheraga HA, Marder VJ (1987) Fibrinogen structure and physiology. In: Colman RW, Hirsch J, Marder VJ, Saltman EW (eds) *Hemostasis and Thrombosis*, 2nd edn, pp. 269–288. JB Lippincott, Philadelphia.

Hantgan RR, Francis CW, Marder VJ (1994) Fibrinogen structure and physiology. In: Colman RW, Hirsch J, Marder VJ, Salzman EW (eds) *Hemostasis and Thrombosis*, 3rd edn, pp. 277–300. J.B. Lippincott, Philadelphia.

Harbourne T, O'Brien D, Nicolaides AN (1991) Fibrinolytic activity in patients with idiopathic and secondary deep venous thrombosis. *Thromb. Res.* **64**: 543–550.

Härdig Y, Rezaie A, Dahlbäck B (1993) High affinity binding of human vitamin K-dependent protein S to a truncated recombinant β chain of C4b-binding protein expressed in *Escherichia coli. J. Biol. Chem.* **268**: 3033–3038.

Harker LA, Schlichter SJ (1972) Platelet and fibrinogen consumption in man. *New Engl. J. Med.* **287**: 999–1005.

Harley HG, Brook JD, Rundle SA, Crow S, Reardon W, Buckler AJ, Harper PS, Housman DE, Shaw DJ (1992) Expansion of an unstable DNA region and phenotypic variation in myotonic dystrophy. *Nature* **355**: 545–546.

Harper PL, Carrell RW (1994) The serpins. In: Bloom AL, Forbes CD, Thomas DP, Tuddenham EGD (eds) *Thrombosis and Haemostasis*, 2nd edn, pp. 641–653. Churchill Livingstone, Edinburgh.

Harper PL, Luddington RJ, Daly M, Bruce D, Williamson D, Edgar PF, Perry DJ, Carrell RW (1991) The incidence of dysfunctional antithrombin variants: four cases in 210 patients with thromboembolic disease. *Br. J. Haematol.* **77**: 360–364.

Harris KW, Esmon CT (1985) Protein S is required for bovine platelets to support activated protein C binding and activity. *J. Biol. Chem.* **260**: 2007–2010.

Harris MG, Exner T, Rickard KA, Kronenberg H (1985) Multiple cerebral thrombosis in Fletcher factor (prekallikrein) deficiency: a case report. *Am. J. Hematol.* **19**: 387–393.

Harris TJR, Patel T, Marston FAO, Little S, Emtage JS, Opdenakker G, Volckaert G, Rombauts W, Billiau A, De Somer P (1986) Cloning of cDNA coding for human tissue-type plasminogen activator and its expression in *Escherichia coli. Molec. Biol. Med.* **3**: 279–292.

Hasegawa N, Sasaki S (1990) Location of the binding site 'b' for lateral polymerization of fibrin. *Thromb. Res.* **57**: 183–195.

Hasegawa DK, Tyler BJ, Edson JR (1982) Thrombotic disease in three families with inherited plasminogen deficiency. *Blood* **60**: 213a.

Hassan HJ, Leonardi A, Chelucci C, Mattia G, Macioce G, Guerriero R, Russo G, Mannucci PM, Peschle C (1990) Blood coagulation factors in human embryonic-fetal development: preferential expression of the FVII/tissue factor pathway. *Blood* **76**: 1158–1164.

Hassan JH, Chelucci C, Peschle C, Sorrentino V (1992) Transforming growth factor β inhibits expression of fibrinogen and factor VII in a hepatoma cell line. *Thromb. Haemost.* **67**: 478–483.

Hathaway WE (1991) Clinical aspects of antithrombin III deficiency. *Sem. Hematol.* **28**: 19–23.

Hathaway WE, Belhasen LP, Hathaway HS (1965) Evidence for a new plasma thromboplastin factor. I. Case report, coagulation studies and physicochemical properties. *Blood* **26**: 521–532.

Haverkate F, Timan G (1977) Protective effect of calcium on the plasmin degradation of fibrinogen and fibrin fragments D. *Thromb. Res.* **10**: 803–812.

Haverkate F, Samama M (1995) Familial dysfibrinogenemia and thrombophilia. Report on a study of the SSC subcommittee on fibrinogen. *Thromb. Haemost.* **73**: 151–161.

Haverkate F, Koopman J, Kluft C, D'Angelo A, Cattaneo M, Mannucci PM (1986) Fibrinogen Milano II: a congenital dysfibrinogenaemia associated with juvenile arterial and venous thrombosis. *Thromb. Haemost.* **55**: 131–135.

Hayashi T, Zushi M, Yamamoto S, Suzuki K (1990) Further localization of binding sites for thrombin and protein C in human thrombomodulin. *J. Biol. Chem.* **265**: 20156–20159.

Hayashi T, Nishioka J, Shigekiyo T, Saito S, Suzuki K (1994) Protein S Tokushima: abnormal molecule with a substitution of Glu for Lys-155 in the second epidermal growth factor-like domain of protein S. *Blood* **83**: 683–690.

Hayashi T, Nishioka J, Suzuki K (1995) Molecular mechanism of the dysfunction of protein S (Lys155→Glu) for the regulation of the blood coagulation system. *Biochim. Biophys. Acta* **1272**: 159–167

Hayward CPM, Furmaniak-Kazmierczak E, Cieutat A-M, Moore JC, Bainton DF, Nesheim ME, Kelton JG, Côté G (1995) Factor V is complexed with multimerin in resting platelet lysates and colocalizes with multimerin in platelet α-granules. *J. Biol. Chem.* **270**: 19217–19224.

He X, Dahlbäck B (1993) Molecular cloning, expression and functional characterization of rabbit anticoagulant vitamin K-dependent protein S. *Eur. J. Biochem.* **217**: 857–865.

He X, Dahlbäck B (1994) Rabbit plasma, unlike its human counterpart, contains no complex between protein S and C4b-binding protein. *Thromb. Haemost.* **71**: 446–451.

Healy AM, Rayburn HB, Rosenberg RD, Weiler H (1995) Absence of the blood-clotting regulator thrombomodulin causes embryonic lethality in mice before development of a functional cardiovascular system. *Proc. Natl Acad. Sci. USA* **92**: 850–854.

Heeb MJ, Griffin JH (1988) The biochemistry of protein S. In: Bertina RM (ed.) *Protein C and Related Proteins*. Churchill Livingstone, Edinburgh.

Heeb MJ, Espana F, Griffin JH (1989) Inhibition and complexation of activated protein C by two major inhibitors in plasma. *Blood* **73**: 446–451.

Heeb MJ, Gruber A, Griffin JH (1991) Identification of divalent metal ion-dependent inhibition of activated protein C by α2-macroglobulin and α2-antiplasmin in blood and comparisons to inhibition of factor Xa, thrombin, and plasmin. *J. Biol. Chem.* **266**: 17606–17612.

Heeb MJ, Mesters RM, Tans G, Rosing J, Griffin JH (1993) Binding of protein S to factor Va associated with inhibition of prothrombinase that is independent of activated protein C. *J. Biol. Chem.* **268**: 2872–2877.

Heeb MJ, Rosing J, Bakker HM, Fernandez JA, Tans G, Griffin JH (1994) Protein S binds to and inhibits factor Xa. *Proc. Natl Acad. Sci. USA* **91**: 2728–2732.

Heeb MJ, Kojima Y, Greengard JS, Griffin JH (1995) Activated protein C resistance: molecular mechanisms based on studies using purified Gln^{506}-factor V. *Blood* **85**: 3405–3411.

Heijboer H, Brandjes DPM, Büller HR, Sturk A, ten Cate JW (1990) Deficiencies of coagulation-inhibiting and fibrinolytic proteins in outpatients with deep vein thrombosis. *New Engl. J. Med.* **323**: 1512–1516.

Heinrich J, Balleisen L, Schulte H, Assmann G, van de Loo J (1994) Fibrinogen and factor VII in the prediction of coronary risk. Results from the PROCAM study in healthy men. *Arterioscler. Thromb.* **14**: 54–59.

Heistinger M, Rumpl E, Illiasch H, Türck H, Kyrle PA, Lechner K, Pabinger I (1992) Cerebral sinus thrombosis in a patient with hereditary protein S deficiency: case report and review of the literature. *Ann. Hematol.* **64**: 105–109.

Hekman CM, Loskutoff DJ (1985) Endothelial cells produce a latent inhibitor of plasminogen activators that can be activated by denaturants. *J. Biol. Chem.* **260**: 11581–11587.

Hekman CM, Loskutoff DJ (1988) Kinetic analysis of the interactions between plasminogen activator inhibitor 1 and both urokinase and tissue plasminogen activator. *Biochem. Biophys.* **262**: 199–210.

Hellgren M, Svensson PJ, Dahlbäck B (1995) Resistance to activated protein C as a basis for venous thromboembolism associated with pregnancy and oral contraceptives. *Am. J. Obstet. Gynecol.* **173**: 210–214.

Henkens CMA, van der Meer J, Hillege JL, van der Schaaf W, Bom VJJ, Halie MR (1993) The clinical expression of hereditary protein C and protein S deficiency: a relation to clinical thrombotic risk factors and to levels of protein C and protein S? *Blood Coag. Fibrinol.* **4**: 555–562.

Henkens CMA, Bom VJJ, van der Meer J (1995a) Lowered APC-sensitivity ratio related to increased factor VIII-clotting activity. *Thromb. Haemost.* **74**: 1198–1199.

Henkens CMA, Bom VJJ, van der Schaaf W, Pelsma PM, Smit Sibinga C Th, de Kam PJ, van der Meer J (1995b) Plasma levels of protein S, protein C, and factor X: effects of sex, hormonal state and age. *Thromb. Haemost.* **74**: 1271–1275.

Hennekens CH, Buring JE (1987) *Epidemiology in Medicine*. Little, Brown and Company, Boston.

Hennis BC, Kluft C (1991) *Kpn* I RFLP in the human histidine-rich glycoprotein gene. *Nucleic Acids Res.* **19**: 4311.

Hennis BC, Havelaar AC, Kluft C (1992) PCR detection of a dinucleotide repeat in the human histidine-rich glycoprotein (HRG) gene. *Hum. Mol. Genet.* **1**: 78.

Hennis BC, Frants RR, Bakker E, Vossen RHAM, Van der Poort EW, Blonden LA, Cox S, Meera Khan P, Spurr NK, Kluft C (1994) Evidence for the absence of intron H of the histidine-rich glycoprotein (HRG) gene: genetic mapping and *in situ* hybridization of HRG to chromosome 3q28–29. *Genomics* **19**: 195–197.

Hennis BC, van Boheemen PA, Wakabayashi S, Koide T, Hoffmann JJML, Kievit P, Dooijewaard G, Jansen JG, Kluft C (1995a) Identification and genetic analysis of a common molecular variant of histidine-rich glycoprotein with a difference of 2kD in apparent molecular weight. *Thromb. Haemost.* **74**: 1491–1496.

Hennis BC, Boomsma DI, van Boheemen PA, Engesser L, Kievit P, Dooihewaard G, Kluft C (1995b) An amino acid polymorphism in histidine-rich glycoprotein (HRG) explains 59% of the variance in plasma HRG levels. *Thromb. Haemost.* **74**: 1497–1500.

Hennis BC, Van Boheemen PA, Koeleman BPC, Boomsma DI, Engesser L, Van Wees AGM, Novakova I, Brommer EJP, Kluft C (1995c) A specific allele of the histidine-rich glycoprotein (HRG) locus is linked with elevated plasma levels of HRG in a Dutch family with thrombosis. *Brit. J. Haematol.* **89**: 845–852.

Henschen AH (1993) Human fibrinogen – structural variants and functional sites. *Thromb. Haemost.* **70**: 42–47.

Henschen A, Southan C, Soria J, Soria C, Samama M (1981) Structure abnormality of fibrinogen Metz and its relationship to the clotting defect. *Thromb. Haemost.* **46**: 103A.

Henschen A, Lottspeich F, Kohl M, Southan C (1983) Covalent structure of fibrinogen. *Ann. NY Acad. Sci.* **408**: 28–43.

Henschen AH, Theodor I, Pirkle H (1991) Hydroxyproline, a posttranslational modification of proline, is a constituent of human fibrinogen. *Thromb. Haemost.* **65**: 821.

Henthorn PS, Mager DL, Huisman THJ, Smithies O (1986) A gene deletion ending within a complex array of repeated sequences 3′ to the human β-globin gene cluster. *Proc. Natl Acad. Sci. USA* **83**: 5194–5198.

Henthorn PS, Smithies O, Mager DL (1990) Molecular analysis of deletions in the human β-globin gene cluster: deletion junctions and locations of breakpoints. *Genomics* **6**: 226–237.

Herbert JM, Lamarche I, Prabonnaud V, Dol F, Gauthier T (1994) Tissue-type plasminogen activator is a potent mitogen for human aortic smooth muscle cells. *J. Biol. Chem.* **269**: 3076–3080.

Hernandez A, Uhrberg M, Enczmann J, Witt I, Reitsma PH, Wernet P (1995) Rapid identification of gene defects in protein C deficiency by temperature gradient gel electrophoresis. *Blood Coag. Fibrinol.* **6**: 23–30.

Herzog R, Lutz S, Blin N, Marasa JC, Blinder MA, Tollefsen DM (1991) Complete nucleotide sequence of the gene for human heparin cofactor II and mapping to chromosomal band 22q11. *Biochemistry* **30**: 1350–1357.

Hessel B, Stenbjerg S, Dyr J, Kudryk B, Therkildsen L, Blombäck B (1986) Fibrinogen Aarhus – a new case of dysfibrinogenemia. *Thromb. Res.* **42**: 21–26.

Hessel B, Silveira AM, Carlsson K, Oksa H, Rasi V, Vahtera E, Procyk R, Blombäck B (1995) Fibrinogenemia Tampere – a dysfibrinogenemia with defective gelation and thromboembolic disease. *Thromb. Res.* **78**: 323–339

Hesselvik JF, Malm J, Dahlbäck B, Blombäck M (1991) Protein S and C4b-binding protein in severe infection and septic shock. *Thromb. Haemost.* **65**: 126.

Heywood DM, Ossei-Gerning NO, Grant PJ (1996) Association of factor VIII: C levels with environmental and genetic factors in patients with ischaemic heart disease and coronary atheroma characterised by angiography. *Thromb. Haemost.* **76**: 161–165.

Higgins DL, Lewis SD, Shafer JA (1983) Steady state kinetic parameters for the thrombin-catalysed conversion of human fibrinogen to fibrin. *J. Biol. Chem.* **258**: 9276–9282.

Hill RE, Hastie ND (1987) Accelerated evolution in the reactive centre regions of serine protease inhibitors. *Nature* **326**: 96–99.

Hillarp A, Dahlback B (1989) Cloning of cDNA coding for the β-chain of human complement component C4b-binding protein: sequence homology with the α-chain. *Proc. Natl Acad. Sci. USA* **87**: 1183–1187.

Hillarp A, Hessing A, Dahlbäck B (1989) Protein S binding in relation to the subunit composition of human C4b-binding protein. *FEBS Lett.* **259**: 53–56.

Hillarp A, Pardo-Manuel F, Ruiz RR, Rodriguez de Cordoba S, Dahlbäck B (1993) The human C4b-binding protein β-chain gene. *J. Biol. Chem.* **268**: 15017–15023.

Hillarp A, Dahlbäck B, Zöller B (1995) Activated protein C resistance: from phenotype to genotype and clinical practice. *Blood Rev.* **9**: 201–212.

Hirokawa K, Aoki N (1991) Regulatory mechanisms for thrombomodulin expression in human umbilical vein endothelial cells *in vitro*. *J. Cell. Physiol.* **147**: 157–165.

Hirsh J, Piovella F, Pina M (1989) Congenital antithrombin III deficiency. Incidence and clinical features. *Am. J. Med.* **87** (Suppl. 3B): 345–385.

Hobart MJ (1979) Genetic polymorphism of human plasminogen. *Ann. Hum. Genet.* **42**: 419–423.

Hochschwender SM, Laursen RA (1981) The lysine binding sites of human plasminogen: evidence for a critical tryptophan in the binding site of Kringle 4. *J. Biol. Chem.* **256**: 11172–11176.

Hofferbert S, Müller J, Kösterling H, von Ohlen W-D, Schloesser M (1996) A novel 5′-upstream mutation in the factor XII gene is associated with a *Taq*I restriction site in an *Alu* repeat in factor XII-deficient patients. *Hum. Genet.* **97**: 838–841.

Hoffman JJML, Hennis BC, Kluft C, Vijgen M (1993) Hereditary increase of plasma histidine-rich glycoprotein associated with abnormal heparin binding (HRG Eindhoven). *Thromb. Haemost.* **70**: 894–899.

Hofsteenge J, Stone SR (1987) The effect of thrombomodulin on the cleavage of fibrinogen and fibrinogen fragments by thrombin. *Eur. J. Biochem.* **168**: 49–56.

Hofsteenge J, Taguchi H, Stone SR (1986) Effect of thrombomodulin on the kinetics of the interaction of thrombin with substrates and inhibitors. *Biochem. J.* **237**: 243–251.

Hogg PJ, Jackson CM (1989) Fibrin monomer protects thrombin from inactivation by heparin-antithrombin III: implications for heparin efficacy. *Proc. Natl Acad. Sci. USA* **86**: 3619–3623.

Hogg PJ, Öhlin A-K, Stenflo J (1992) Identification of structural domains in protein C involved in its interaction with thrombin–thrombomodulin on the surface of endothelial cells. *J. Biol. Chem.* **267**: 703–706.

Holm J, Zöller B, Svensson PJ, Berntorp E, Erhardt L, Dahlbäck B (1994) Myocardial infarction associated with homozygous resistance to activated protein C. *Lancet* **344**: 952–953.

Hoogendoorn H, Toh CH, Nesheim ME, Giles AR (1991) α2-Macroglobulin binds and inhibits activated protein C. *Blood* **78**: 2283–2288.

Hooper WC, Phillips DJ, Ribeiro MJA, Benson JM, George VG, Ades EW, Evatt BL (1994) Tumor necrosis factor-α downregulates protein S secretion in human microvascular and umbilical vein endothelial cells but not in the HepG-2 hepatoma cell line. *Blood* **84**: 483–489.

Hopkins PCR, Carrell RW, Stone SR (1993) Effects of mutations in the hinge region of serpins. *Biochemistry* **32**: 7650–7657.

Horrevoets AJG, Smilde A, de Vries C, Pannekoek H (1994) The specific roles of finger and kringle 2 domains of tissue-type plasminogen activator during *in vitro* fibrinolysis. *J. Biol. Chem.* **269**: 12639–12644.

Hortin G, Tollefsen DM, Strauss AW (1986) Identification of two sites of sulfation of human heparin cofactor II. *J. Biol. Chem.* **261**: 15827–15836.

Hortin GL, Tollefsen DM, Benitto BM (1989) Antithrombin activity of a peptide corresponding to residues 54–75 of heparin cofactor II. *J. Biol. Chem.* **264**: 13979–13982.

Horvat R, Palade GE (1993) Thrombomodulin and thrombin localization on the vascular endothelium: their internalization and transcytosis by plasmalemmal vesicles. *Eur. J. Cell Biol.* **61**: 299–313.

Hoskins J, Norman DK, Beckmann RJ, Long GL (1987) Cloning and characterization of human liver cDNA encoding a protein S precursor. *Proc. Natl Acad. Sci. USA* **84**: 349–353.

Howell WH, Holt E (1918) Two new factors in blood coagulation: heparin and pro-antithrombin. *Am. J. Physiol.* **47**: 328–341.

Hoylaerts MD, Rijken DC, Lijnen HR, Collen D (1982) Kinetics of the activation of plasminogen by human tissue plasminogen activator. *J. Biol. Chem.* **257**: 2912–2919.

Hoylaerts M, Owen WG, Collen D (1984) Involvement of heparin chain length in the heparin-catalyzed inhibition of thrombin by antithrombin III. *J. Biol. Chem.* **259**: 5670–5677.

Hu X, Worton RG (1992) Partial gene duplication as a cause of human disease. *Hum. Mutat.* **1**: 3–12.

Hu X, Ray PN, Worton RG (1991) Mechanisms of tandem duplication in the Duchenne muscular dystrophy gene include both homologous and nonhomologous intrachromosomal recombination. *EMBO J.* **10**: 2471–2477.

Hu C-H, Harris JE, Davie EW, Chung DW (1995) Characterization of the 5′-flanking region of the gene for the α chain of human fibrinogen. *J. Biol. Chem.* **270**: 47: 28342–28349.

Huang LS, Ripps ME, Korman SH, Deckelbaum RJ, Breslow JL (1989) Hypobetalipoproteinaemia due to an apolipoprotein B gene exon 21 deletion derived by *Alu–Alu* recombination. *J. Biol. Chem.* **264**: 11394–11400.

Huang S, Cao Z, Chung DW, Davie EW (1996) The role of βγ and αγ complexes in the assembly of human fibrinogen. *J. Biol. Chem.* **271**: 27942–27947.

Huber R, Carrell RW (1989) Implications of the three-dimensional structure of α1-antitrypsin for structure and function of serpins. *Biochemistry* **28**: 8951–8966.

Huber P, Dalmon J, Courtois G, Laurent M, Assouline Z, Marguerie G (1987) Characterization of the 5′-flanking region for the human fibrinogen β-gene. *Nucleic Acids Res.* **15**: 1615–1625.

Huber P, Laurent M, Dalmon J (1990) Human β-fibrinogen gene expression. *J. Biol. Chem.* **265**: 5695–5701.

Hull R, Hirsh J, Sackett DL (1981) Replacement of venography in suspected venous thrombosis by impedence plethysmography and I^{125} fibrinogen scanning: a less invasive approach. *Ann. Intern. Med.* **94**: 12–15.

Hull RD, Feldstein W, Pineo GF, Raskob GE (1995) Cost effectiveness of diagnosis of deep vein thrombosis in symptomatic patients. *Thromb. Haemost.* **74**: 189–196.

Hultin MB, McKay J, Abildgaard U (1988) Antithrombin Oslo: type Ib classification of the first reported antithrombin-deficient family with a review of hereditary antithrombin variants. *Thromb. Haemost.* **59**: 468–473.

Humphries JE, De La Cadena RA, Atkins TL, Colman RW, Gonias SL (1994) Interaction of high molecular weight kininogen with plasminogen inhibits binding of plasminogen to cell surfaces. *Fibrinolysis* **8**: 245–254.

Humphries S (1994) The genetic contribution to the risk of thrombosis and cardiovascular disease. *Trends Cardiovasc. Med.* **4**: 8–17.

Humphries SE, Imam AMA, Robbins TP, Cook M, Carritt B, Ingle C, Williamson R (1984) The identification of a DNA polymorphism of the α-fibrinogen gene and the regional assignment of the human fibrinogen genes to 4q 26–qter. *Hum. Genet.* **68**: 148–153.

Humphries SE, Cook M, Dubowitz M, Stirling Y, Meade TW (1987) Role of genetic variation at the fibrinogen locus in determination of plasma fibrinogen concentrations. *Lancet* **i**: 1452–1455.

Hunt LT, Dayhoff MO (1980) A surprising new protein superfamily containing ovalbumin, antithrombin III and α1-proteinase inhibitor. *Biochem. Biophys. Res. Commun.* **95**: 864–871.

Hwu HR, Roberts JW, Davidson EH, Britten RJ (1986) Insertion and/or deletion of many repeated DNA sequences in human and higher ape evolution. *Proc. Natl Acad. Sci. USA* **83**: 3875–3879.

Hyers TM, Hull RD, Weg JG (1995) Antithrombotic therapy for venous thromboembolic disease. *Chest* **108**: 335S–351S.

Ichinose A (1992) Multiple members of the plasminogen-apolipoprotein (a) gene family associated with thrombosis. *Biochemistry* **31**: 3113–3118.

Ichinose A, Espling ES, Takamatsu J, Saito H, Shinmyozu K, Maruyama I, Petersen TE, Davie EW (1991) Two types of abnormal genes for plasminogen in families with a predisposition for thrombosis. *Proc. Natl Acad. Sci. USA* **88**: 115–119.

Ido M, Ohiwa M, Hayashi T, Nishioka J, Hatada T, Watanabe Y, Wada H, Shirakawa S, Suzuki K (1993) A compound heterozygous protein C deficiency with a single nucleotide G deletion encoding Gly-381 and an amino acid substitution of Lys for Gla-26. *Thromb. Haemost.* **70**: 636–641.

Ido M, Hayashi T, Nishioka J, Itoh M, Minoura H, Toyoda N, Hirayama M, Kawasaki H, Sakurai M, Suzuki K (1996) Prenatal diagnosis of compound heterozygous deficiency of protein C by direct detection of the mutation sites. *Thromb. Haemost.* **76**: 275–281.

Ieko M, Sawada K, Sakurama S, Yamagishi I, Isogawa S, Nakagawa S, Satoh M, Yasukouchi T, Matsuda M (1991) Fibrinogen Date: congenital hypodysfibrinogenemia associated with decreased binding of tissue-plasminogen activator. *Am. J. Hematol.* **37**: 228–233.

Imada S, Yamaguchi H, Nagumo M, Katayanagi S, Iwaraki H, Imada M (1990) Identification of fetomodulin, a surface marker protein of fetal development, as thrombomodulin by gene cloning and functional assays. *Devel. Biol.* **140**: 113–122.

Imam AMA, Eaton MAW, Williamson R, Humphries S (1983) Isolation and characterization of cDNA clones for the Aα- and β-chains of human fibrinogen. *Nucleic Acids Res.* **11**: 7427–7434.

Inhorn RC, Tollefsen DM (1986) Isolation and characterization of a partial cDNA clone for heparin cofactor II. *Biochem. Biophys. Res. Commun.* **137**: 431–436.

Ireland H, Lane DA, Thompson E, Walker ID, Blench I, Morris HR, Freyssinet JM, Grunebaum L, Olds R, Thein SL (1991a) Antithrombin Glasgow II: alanine 382 to threonine mutation in the serpin P12 position resulting in a substrate reaction with thrombin. *Br. J. Haematol.* **79**: 70–74.

Ireland H, Lane DA, Thompson E, Olds R, Thein SL, Hach-Wunderle V, Scharrer I (1991b) Antithrombin Frankfurt I: arginine to cysteine substitution at the reactive site and formation of a variant antithrombin albumin covalent complex. *Thromb. Haemost.* **65**: 913.

Ireland H, Bayston T, Thompson E, Adami A, Gonçalves C, Lane DA, Finazzi G, Barbui T (1995) Apparent heterozygous type II protein C deficiency caused by the factor V 506 Arg to Gln mutation. *Thromb. Haemost.* **73**: 731–732.

Ireland HA, Boisclair MD, Taylor J, Thompson E, Thein SL, Girolami A, De Caterina M, Scopacasa F, De Stefano V, Leone G, Finazzi G, Cohen H, Lane DA (1996a) Two novel [R(-11)C; T394D] and two repeat missense mutations in the protein C gene associated with venous thrombosis in six kindreds *Hum. Mutat.* **7**: 176–179.

Ireland H, Thompson E, Lane DA and the Protein C Study Group (1996b) Gene mutations in 21 unrelated cases of phenotypic heterozygous protein C deficiency and thrombosis. *Thromb. Haemost.* **76**: 867–873.

Ishii H, Majerus PW (1985) Thrombomodulin is present in human plasma and urine. *J. Clin. Invest.* **76**: 2178–2181.

Ishii H, Uchiyama H, Kazama M (1991) Soluble thrombomodulin antigen in conditioned medium is increased by damage of endothelial cells. *Thromb. Haemost.* **65**: 618–622.

Iso H, Folsom AR, Winkelmann JC, Koike K, Harada S, Greenberg B, Sato S, Shimamoto T, Iida M, Komachi Y (1995) Polymorphisms of the beta fibrinogen gene and plasma fibrinogen concentration in Caucasian and Japanese population samples. *Thromb. Haemost.* **73**: 106–111.

Jackman RW, Beeler DL, VanderWater L, Rosenberg RD (1986) Characterization of a thrombomodulin cDNA reveals structural similarity to the low density lipoprotein receptor. *Proc. Natl Acad. Sci. USA* **83**: 8834–8838.

Jackman RW, Beeler DL, Fritze L, Soff G, Rosenberg RD (1987) Human thrombomodulin gene is intron depleted: nucleic acid sequence of the cDNA and gene predict protein structure and suggest sites of regulatory control. *Proc. Natl Acad. Sci. USA* **84**: 6425–6429.

Jackson DE, Mitchell CA, Salem HH (1994) Protein C is responsible for the rapid inactivation of factor Va following blood clotting *in vitro*. *Thromb. Haemost.* **72**: 70–73.

Jagd S, Vibe-Pedersen K, Magnusson S (1985) Location of two of the introns in the antithrombin III gene. *FEBS Lett.* **193**: 213–216.

Jamison CS, McDowell SA, Marlar RA, Friezner Degen SJ (1995) Developmental expression of protein C and protein S in the rat. *Thromb. Res.* **78**: 407–419.

Jane SM, Hau L, Salem HH (1992) Protein S negates the activated protein C inhibitory activity of plasma. *Blood Coag. Fibrinol.* **3**: 257–261.

Jansson JH, Nilsson TK, Olofsson BO (1991) Tissue plasminogen activator and other risk factors as predictors of cardiovascular events in patients with severe angina pectoris. *Eur. Heart J.* **12**: 157–161.

Jansson JH, Olofsson BO, Nilsson TK (1993) Predictive value of tissue plasminogen activator mass concentration on long-term mortality in patients with coronary artery disease. A 7-year follow-up. *Circulation* **88**: 2030–2034.

Janus TJ, Lewis SD, Lorand L, Shafer JA (1983) Promotion of thrombin-catalyzed activation of factor XIII by fibrinogen. *Biochemistry* **26**: 6269–6272.

Jeffcoate TNA, Miller J, Roos RF (1968) Puerperal thromboembolism in relation to the inhibition of lactation by oestrogen therapy. *Br. Med. J.* **4**: 19–22.

Jenny RJ, Mann KG (1989) Factor V: a prototype procofactor vitamin K-dependent enzyme complexes in blood clotting. *Baillière's Clin. Haematol.* **2**: 919–944.

Jenny RJ, Pittman DD, Toole JJ, Kriz RW, Aldape RA, Hewick RM, Kaufman RJ, Mann KG (1987) Complete cDNA and derived amino acid sequence of human factor V. *Proc. Natl Acad. Sci. USA* **84**: 4846–4850.

Jesty J, Lorenz A, Rodriguez J, Wun T-C (1996) Initiation of the tissue factor pathway of coagulation in the presence of heparin: control by antithrombin III and tissue factor pathway inhibitor. *Blood* **87**: 2301–2307.

Jhingan A, Zhang L, Christiansen WT, Castellino FJ (1994) The activities of recombinant γ-carboxyglutamic-acid-deficient mutants of activated human protein C toward human coagulation factor Va and factor VIII in purified systems and in plasma. *Biochemistry* **33**: 1869–1875.

Jick H, Derby LE, Myers MW, Vasilakis C, Newton KM (1996) Risk of hospital admission for idiopathic venous thromboembolism among users of postmenopausal oestrogens. *Lancet* **348**: 981–983.

Jobin F, Vu L, Lessard M (1991) Two cases of inherited triple deficiency in a large kindred with thrombotic diathesis and deficiencies of antithrombin III, heparin cofactor II, protein C and protein S. *Thromb. Haemost.* **66**: 295–299.

Jochmans K, Lissens W, Vervoort R, Peeters S, De Waele M, Liebaers I (1994a) Antithrombin Gly424→Arg: a novel point mutation responsible for type I antithrombin deficiency and neonatal thrombosis. *Blood* **83**: 146–151.

Jochmans K, Lissens W, Yin T, Michiels JJ, van der Luit L, Peerlinck K, De Waele M, Liebaers I (1994b) Molecular basis for type 1 antithrombin deficiency: identification of two novel point mutations and evidence for a *de novo* splice site mutation. *Blood* **84**: 3742–3748.

Johansson L, Hedner U, Nilsson IM (1978) A family with thromboembolic disease associated with deficient fibrinolytic activity in vessel wall. *Acta Med. Scand.* **203**: 477–480.

Johnson EJ, Prentice CRM, Parapia LA (1990) Premature arterial disease associated with familial antithrombin III deficiency. *Thromb. Haemost.* **63**: 13–15.

Jordan RE, Beeler D, Rosenberg R (1979) Fractionation of low-molecular weight heparin species and their interaction with antithrombin. *J. Biol. Chem.* **254**: 2902–2912.

Jørgensen M, Bonnevie-Nielsen V (1987) Increased concentration of the fast-acting plasminogen activator inhibitor in plasma associated with familial venous thrombosis. *Br. J. Haematol.* **65**: 175–180.

Jørgensen M, Mortensen JZ, Madson AG, Thorsen S, Jacobsen B (1982) A family with reduced plasminogen activator activity in blood associated with recurrent venous thrombosis. *Scand. J. Haematol.* **29**: 217–223.

Joseph DR, Baker ME (1992) Sex hormone-binding globulin, and androgen-binding protein, and vitamin K-dependent protein S are homologous to laminin A, merosin, and *Drosophila* crumbs protein. *FASEB J.* **6**: 2477–2481.

Joslin G, Fallon RJ, Bullock J, Adams SP, Perlmutter DH (1991) The SEC receptor recognizes a pentapeptide neodomain of α_1-antitrypsin-protease complexes. *J. Biol. Chem.* **266**: 11282–11288.

Juhan-Vague I, Collen D (1992) On the role of coagulation and fibrinolysis in atherosclerosis. *Ann. Epidemiol.* **2**: 427–438.

Juhan-Vague I, Moerman B, De Cock F, Aillaud MF, Collen D (1984) Plasma levels of a specific inhibitor of tissue-type plasminogen activator (and urokinase) in normal and pathological conditions. *Thromb. Res.* **33**: 523–530.

Juhan-Vague I, Valadier J, Alessi MC, Aillaud MF, Ansaldi J, Philip-Joet C, Holvoet P, Serradimigni A, Collen D (1987) Deficient t-PA release and elevated PA inhibitor levels in patients with spontaneous or recurrent deep venous thrombosis. *Thromb. Haemost.* **57**: 67–72.

Juhan-Vague I, Alessi M-C, Declerck PJ (1995) Pathophysiology of fibrinolysis. *Bailliére's Clin. Haematol.* **8**: 329–343.

Kaczmarek E, Lee MH, McDonagh J (1993) Initial interaction betwen fibrin and tissue plasminogen activator (t-PA). *J. Biol. Chem.* **268**: 2474–2479.

Kalafatis M, Mann KG (1993) Role of the membrane in the inactivation of factor Va by activated protein C. *J. Biol. Chem.* **268**: 27246–27257.

Kalafatis M, Jenny RJ, Mann KG (1990) Identification and characterization of a phospholipid-binding site of bovine factor Va. *J. Biol. Chem.* **265**: 21580–21589.

Kalafatis M, Rand MD, Jenny RJ, Ehrlich YH, Mann KG (1993) Phosphorylation of factor Va and factor VIIIa by activated platelets. *Blood* **81**: 704–719.

Kalafatis M, Rand MD, Mann KG (1994) The mechanism of inactivation of human factor V and human factor Va by activated protein C. *J. Biol. Chem.* **269**: 31869–31880.

Kalafatis M, Bertina RM, Rand MD, Mann KG (1995a) Characterization of the molecular defect in factor V^{R506Q}. *J. Biol. Chem.* **270**: 4053–4057.

Kalafatis M, Lu D, Bertina RM, Long GL, Mann KG (1995b) Biochemical prototype for familial thrombosis: a study combining a functional protein C mutation and factor V Leiden. *Arterioscler. Thromb. Vasc. Biol.* **15**: 2181–2187.

Kambayashi J, Fujimura H, Kawasaki T, Sakon M, Monden M, Sehisa E, Amino N (1995) The incidence of activated protein C resistance among patients with deep vein thrombosis and healthy subjects in Osaka. *Thromb. Res.* **79**: 227–229.

Kaminski M, McDonagh J (1987) Inhibited thrombins. Interactions with fibrinogen and fibrin. *Biochem. J.* **242**: 881–887.

Kaminski M, Siebenlist KR, Mosesson MW (1991) Evidence for thrombin enhancement of fibrin polymerization that is independent of its catalytic activity. *J. Lab. Clin. Med.* **117**: 218–225.

Kamiya T, Sugihara T, Ogata K, Saito H, Suzuki K, Nishioka J, Hashimoto S, Yamagata K (1986) Inherited deficiency of protein S in a Japanese family with recurrent venous thrombosis: a study of three generations. *Blood* **67**: 406–410.

Kane WH, Davie EW (1986) Cloning of a cDNA coding for human factor V, a blood coagulation factor homologous to factor VIII and ceruloplasmin. *Proc. Natl Acad. Sci. USA* **83**: 6800–6804.

Kane WH, Davie EW (1988) Blood coagulation factors V and VIII: structural and functional similarities and their relationship to hemorrhagic and thrombotic disorders. *Blood* **71**: 539–555.

Kane WH, Ichinose A, Hagan FS, Davie EW (1987) Cloning of cDNAs coding for the heavy chain region and connecting region of human factor V, a blood coagulation factor with four types of internal repeats. *Biochemistry* **26**: 6508–6514.

Kane WH, Devore-Cater D, Ortel TL (1990) Expression and characterization of recombinant human factor V and a mutant lacking a major protein of the connecting region. *Biochemistry* **29**: 6762–6768.

Kant JA, Crabtree GR (1983) The rat fibrinogen genes; linkage of the Aa and γ chain genes. *J. Biol. Chem.* **258**: 4666–4667.

Kant JA, Lord ST, Crabtree GR (1983) Partial mRNA sequences for human Aα, Bβ and γ fibrinogen chains: evolutionary and functional implications. *Proc. Natl Acad. Sci. USA* **80**: 3953–3957.

Kant JA, Fornace AJ, Saxe D, Simon MI, McBride OW, Crabtree GR (1985) Evolution and organization of the fibrinogen locus on chromosome 4: gene duplication accompanied by transposition and inversion. *Proc. Natl Acad. Sci. USA* **82**: 2344–2348.

Kao FT, Morse HG, Law ML, Lidsky A, Chandra T, Woo SLC (1984a) Genetic mapping of the structural gene for antithrombin III to human chromosome 1. *Hum. Genet.* **67**: 34–36.

Kao FT, Morse HG, Law ML, Lidsky A, Chandra T, Woo SLC (1984b) Molecular genetic mapping of the structural gene for human antithrombin III to chromosome 1, region 1p31.3-qter. *Cytogenet. Cell Genet.* **37**: 505.

Kario K, Matsuo T, Tai S, Sakamoto S, Yamada T, Miki T, Matsuo M (1992) Congenital protein C deficiency and myocardial infarction: concomitant factor VII hyperactivity may play a role in the onset of arterial thrombosis. *Thromb. Res.* **67**: 95–103.

Kariya Y, Kato K, Hayashizaki Y, Himeno S, Tarui S, Matsubara K (1987) Revision of consensus sequence of human *Alu* repeat – a review. *Gene* **53**: 1–10.

Kato A, Miura O, Sumi Y, Aoki N (1988) Assignment of the human protein C gene to chromosome region 2q14→q21 by *in situ* hybridization. *Cytogenet. Cell Genet.* **47**: 46–47.

Katsumi A, Senda T, Yamashita Y, Yamazaki T, Hamaguchi M, Kojima T, Kobayashi S, Saito H (1996) Protein C Nagoya, an elongated mutant of protein C, is retained within the endoplasmic reticulum and is associated with GRP78 and GRP94. *Blood* **89**: 10: 4164–4175.

Kawamura S, Saitou N, Ueda S (1992) Concerted evolution of the primate immunoglobulin α-gene through gene conversion. *J. Biol. Chem.* **267**: 7359–7367.

Kawasaki T, Kambayashi J, Sakon M (1995) Hyperlipidemia: a novel etiologic factor in deep vein thrombosis. *Thromb. Res.* **79**: 147–151.

Kazama M, Tahara C, Suzuki Z, Gohchi K, Abe T (1981) Abnormal plasminogen, a case of recurrent thrombosis. *Thromb. Res.* **21**: 517–522.

Kazazian HH, Wong C, Youssoufian H, Scott AF, Phillips DG, Antonarakis SE (1988) Haemophilia A resulting from *de novo* insertion of L1 sequences represents a novel mechanism for mutation in man. *Nature* **332**: 164–166.

Kazui S, Kuriyama Y, Sakata T, Hiroki M, Miyashita K, Sawada T (1993) Accelerated brain infarction in hypertension comlicated by hereditary heterozygous protein C deficiency. *Stroke* **24**: 2097–2103.

Kehl M, Henschen A, Tavory S, Rimon A, Soria J, Soria C (1984) The structural error in fibrinogen Haifa, a genetically abnormal fibrinogen with a polymerisation defect. *Proceedings, VIII Int. Cong. Thromb.*, Istanbul: 82.

Keller FG, Ortel TL, Quinn-Allen M, Kane WH (1995) Thrombin-catalyzed activation of recombinant human factor V. *Biochemistry* **34**: 4118–4124.

Keyt BA, Paoni NF, Refino CJ, Berleau L, Nguyen H, Chow A, Lai J, Pena L, Pater C, Ogez J, Etcheverry T, Botstein D, Bennett WF (1994) A faster acting and more potent form of tissue plasminogen activator. *Proc. Natl Acad. Sci. USA* **91**: 3670–3674.

Kierkegaard A (1983) Incidence and diagnosis of deep vein thrombosis associated with pregnancy. *Acta Obstet. Gynecol. Scand.* **62**: 239–243.

Kierkegaard A (1985) Deep vein thrombosis and the oestrogen content in oral contraceptives – an epidemiological analysis. *Contraception* **31**: 29–41.

Kimura S, Aoki N (1986) Cross linking site in fibrinogen for α2-antiplasmin inhibitor. *J. Biol. Chem.* **261**: 15591–15595.

Kirschbaum NE, Foster PA (1995) The polymerase chain reaction with sequence specific primers for the detection of the factor V mutation associated with activated protein C resistance. *Thromb. Haemost.* **74**: 874–878.

Kisiel W (1979) Human plasma protein C. Isolation, characterization and mechanism of activation by α-thrombin. *J. Clin. Invest.* **64**: 761–769.

Kitchens CS (1985) Concept of hypercoagulability: a review of its development, clinical application and recent progress. *Sem. Thromb. Hemost.* **11**: 293–315.

Klinger KW, Winqvist R, Riccio A, AndreasenPA, Sartorio R, Nielsen LS, Stuart N, Stanislovitis P, Watkins P, Douglas R, Grzeschik K-H, Alitalo K, Blasi F, Danø K (1987) Plasminogen activator inhibitor type 1 gene is located at region q21.3–q22 of chromosome 7 and genetically linked with cystic fibrosis. *Proc. Natl Acad. Sci. USA* **84**: 8548–8552.

Kluft C, Verheijen JH, Jie AFH, Rijken DL, Preston FH, Sue-ling HM, Jesperson J, Arsen AO (1985) The post-operative fibrinolytic shutdown: a rapidly reverting acute phase pattern for the fast-acting inhibitor of tissue-type plasminogen activator after trauma. *Scand. J. Clin. Lab. Invest.* **45**: 605–610.

Kluft C, Jie AFH, Lowe GDO, Blamey SL, Forbes CD (1986) Association between post-operative hyper-response in t-PA inhibition and deep vein thrombosis. *Thromb. Haemost.* **56**: 107.

Kluijtmans LAJ, van den Heuvel LPWJ, Boers GHJ, Frosst P, Stevens EMB, van Oost BA, den Heijer M, Trijbels FJM, Rozen R, Blom HJ (1996) Molecular genetic analysis in mild hyperhomocysteinemia: a common mutation in the methylenetetrahydrofolate reductase gene is a genetic risk factor for cardiovascular disease. *Am. J. Hum. Genet.* **58**: 35–41.

Ko Y-L, Hsu T-S, Wu S-M, Ko Y-S, Chang C-J, Wang S-M, Chen W-J, Cheng N-J, Kuo C-T, Chiang C-W, Lee Y-S (1996) The G1691A mutation of the coagulation factor V gene (factor V Leiden) is rare in Chinese: an analysis of 618 individuals. *Hum. Genet.* **98**: 176–177.

Koeberl DD, Bottema CDK, Ketterling RP, Bridge PJ, Lillicrap DP, Sommer SS (1990) Mutations causing hemophilia B: direct estimate of the underlying rates of spontaneous germ-line transitions, transversions and deletions in a human gene. *Am. J. Hum. Genet.* **47**: 202–217.

Koeleman BPC, Reitsma PH, Allaart CF, Bertina RM (1994) Activated protein C resistance as an additional risk factor for thrombosis in protein C-deficient families. *Blood* **84**: 1031–1035.

Koeleman BPC, van Rumpt D, Hamulyak K, Reitsma PH, Bertina RM (1995) Factor V Leiden: an additional risk factor for thrombosis in protein S deficient families? *Thromb. Haemost.* **74**: 580–583.

Koelman JH, Bakker CM, Pladsoen WC, Peeters FL, Barth PG (1992) Hereditary protein S deficiency presenting with cerebral sinus thrombosis in an adolescent girl. *J. Neurol.* **239**: 105–106.

Koenhen E, Bertina RM, Reitsma PH (1989) *Msp*I RFLP in intron 8 of the human protein C gene. *Nucleic Acids Res.* **17**: 8401.

Koide T (1988) Human histidine-rich glycoprotein gene: evidence for evolutionary relatedness to cystatin supergene family. *Thromb. Res.* (Suppl. VIII): 91–97.

Koide T, Odani S, Takahashi K, Ono T, Sakuragawa N (1984) Antithrombin III Toyama: replacement of arginine-47 by cysteine in hereditary abnormal antithrombin III that lacks heparin-binding ability. *Proc. Natl Acad. Sci. USA* **81**: 289–293.

Koide T, Foster D, Yoshitake S, Davie EW (1986) Amino acid sequence of human histidine-rich glycoprotein derived from the nucleotide sequence of its cDNA. *Biochemistry* **25**: 2220–2225.

Komiyama Y, Pedersen AH, Kisiel W (1990) Proteolytic activation of human factors IX and X by recombinant human factor VIIa: effects of calcium, phospholipids and tissue factor. *Biochemistry* **29**: 9418–9425.

Kondo S, Tokunaga F, Kario K, Matsuo T, Koide T (1996) Molecular and cellular basis for type I heparin cofactor II deficiency (Heparin Cofactor II Awaji). *Blood* **87**: 1006–1012.

Kondrashov AS, Crow JF (1993) A molecular approach to estimating the human deleterious mutation rate. *Hum. Mutat.* **2**: 229–234.

Kontula K, Ylikorkala A, Mietinen H, Vuorio A, Kauppinen-Mäkelin, Hämäläinen L, Palomäki II, Kaste M (1995) Arg506Gln Factor V mutation (Factor V Leiden) in patients with ischaemic cerebrovascular disease and survivors of myocardial infarction. *Thromb. Haemost.* **73**: 558–560.

Koopman J, Haverkate F (1994) Hereditary variants of human fibrinogens. In: Bloom AL, Forbes CD, Thomas DP, Tuddenham EGD (eds) *Haemostasis and Thrombosis*, Vol. 1, pp. 515–529. Churchill Livingstone, Edinburgh.

Koopman J, Engesser L, Nieveen M, Haverkate F, Brommer EJP (1986) Fibrin polymerization associated with tissue-type plasminogen activator (t-PA) induced glu-plasminogen activation. In: Müller-Berghaus G, Scheefers-Borchel U, Selmayr E, Henschen A (eds) *Fibrinogen and its Derivatives*, pp. 315–318. Excerpta Medica, Amsterdam.

Koopman J, Haverkate F, Briët E, Lord ST (1991) A congenitally abnormal fibrinogen (Vlissingen) with a 6-base deletion in the γ-chain gene causing defective calcium binding and impaired fibrin polymerization. *J. Biol. Chem.* **266**: 13456–13461.

Koopman J, Haverkate F, Lord ST, Grimbergen J, Mannucci PM (1992a) Molecular basis of fibrinogen Naples associated with defective thrombin binding and thrombophilia: homozygous substitution of Bβ 68 Ala→Thr. *J. Clin. Invest.* **90**: 238–244.

Koopman J, Haverkate F, Grimbergen J, Egbring R, Lord ST (1992b) Fibrinogen Marburg: a homozygous case of dysfibrinogenemia lacking amino acids Aα 461–610 (Lys 461 AAA→Stop TAA). *Blood* **80**: 1972–1979.

Koopman J, Haverkate F, Grimbergen J, Engesser L, Novakova I, Kerst AFJA, Lord ST (1992c) Abnormal fibrinogens IJmuiden (Bβ Arg→14Cys) and Nijmegen (Bβ Arg→ Cys) form disulfide-linked fibrinogen–albumin complexes. *Proc. Natl Acad. Sci. USA* **89**: 3478–3482.

Koopman J, Haverkate F, Grimbergen J, Lord ST, Mosesson MW, DiOrio JP, Siebenlist KS, Legrand C, Soria J, Soria C, Caen J (1993) Molecular basis for fibrinogen Dusart (Aα 554 Arg→Cys) and its association with abnormal fibrin polymerization and thrombophilia. *J. Clin. Invest.* **91**: 1637–1643.

Koppelman SJ, Hackeng TM, Sixma JJ, Bouma BN (1995a) Inhibition of the intrinsic factor X activating complex by protein S: evidence for a specific binding of protein S to factor VIII. *Blood* **86**: 1062–1071.

Koppelman SJ, van't Veer C, Sixma JJ, Bouma BN (1995b) Synergistic inhibition of the intrinsic factor X activation by protein S and C4b-binding protein. *Blood* **86**: 2653–2660.

Korninger C, Lechner K, Niessner H, Gossinger H, Kundi M (1984) Impaired fibrinolytic capacity predisposes for recurrence of venous thrombosis. *Thromb. Haemost.* **52**: 127–130.

Kornreich R, Bishop DF, Desnick RJ (1990) α-Galactosidase A gene rearrangements causing Fabry disease. *J. Biol. Chem.* **265**: 9319–9326.

Koster T, Rosendaal FR, de Ronde H, Briët E, Vandenbroucke JP, Bertina RM (1993) Venous thrombosis due to poor anticoagulant response to activated protein C: Leiden Thrombophilia Study. *Lancet* **342**: 1503–1506.

Koster T, Rosendaal FR, Briët E, Vandenbroucke JP (1994a) John Hageman's factor and deep-vein thrombosis: Leiden Thrombophilia Study. *Br. J. Haematol.* **87**: 422–424.

Koster T, Rosendaal FR, Reitsma PH, van der Velden PA, Briët E, Vandenbroucke JP (1994b) Factor VII and fibrinogen levels as risk factors for venous thrombosis. *Thromb. Haemost.* **71**: 719–722.

Koster T, Blann AD, Briët E, Vandenbroucke JP, Rosendaal FR (1995a) Role of clotting factor VIII in effect of von Willebrand factor on occurrence of deep-vein thrombosis. *Lancet* **345**: 152–155.

Koster T, Rosendaal FR, Briët E, van der Meer FJM, Colly LP, Trienekens PH, Poort SR, Reitsma PH, Vandenbroucke JP (1995b) Protein C deficiency in a controlled series of unselected outpatients: an infrequent but clear risk factor for venous thrombosis (Leiden Thrombophilia Study). *Blood* **85**: 2756–2761.

Kourteva Y, Schapira M, Patston PA (1995) The effect of sex and age on antithrombin biosynthesis in the rat. *Thromb. Res.* **78**: 521–529.

Koyama I, Parkinson JF, Aoki N, Bang NU, Muller-Berhaus G, Preissner KT (1991) Realationship between post-translational glycosylation and anticoagulant function of secretable recombinant mutants of human thrombomodulin. *Br. J. Haematol.* **78**: 515–522.

Koyama T, Hirosawa S, Kawamata N, Tohda S, Aoki N (1994) All-*trans* retinoic acid upregulates throbomodulin and downregulates tissue factor expression in acute promyelocytic leukemia cells: distinct expression of thrombomodulin and tissue factor in human leukemic cells. *Blood* **84**: 3001–3009.

Kozich V, Kraus E, de Franchis R, Fowler B, Boers GHJ, Graham I, Kraus JP (1995) Hyperhomocysteinemia in premature arterial disease: examination of cystathionine β-synthase alleles at the molecular level. *Hum. Mol. Genet.* **4**: 623–629.

Krainer AR, Maniatis T (1985) Multiple factors including the small nuclear ribonucleoproteins U1 and U2 are necessary for pre-mRNA splicing *in vitro*. *Cell* **42**: 725–736.

Krakow W, Endres GF, Siegel BM, Scheraga HA (1972) An electron microscopic investigation of the polymerization of bovine fibrin monomer. *J. Mol. Biol.* **71**: 95–103.

Kraus M, Zander N, Fickenscher K (1995) Coagulation assay with improved specificity to factor V mutants insensitive to activated protein C. *Thromb. Res.* **80**: 255–264.

Krawczak M, Cooper DN (1991) Gene deletions causing human genetic disease: mechanisms of mutagenesis and the role of the local DNA sequence environment. *Hum. Genet.* **86**: 425–441.

Krawczak M, Cooper DN (1996) Single base-pair substitutions in pathology and evolution: two sides to the same coin. *Hum. Mutat.* **8**: 23–31.

Krawczak M, Reiss J, Cooper DN (1992) The mutational spectrum of single base-pair substitutions in mRNA splice junctions of human genes: causes and consequences. *Hum. Genet.* **90**: 41–54.

Krawczak M, Reitsma PH, Cooper DN (1995) The mutational demography of protein C deficiency. *Hum. Genet.* **96**: 142–146.

Krawczak M, Wacey A, Cooper DN (1996) Molecular reconstruction and homology modelling of the catalytic domain of the common ancestor of the haemostatic vitamin K-dependent serine proteinases. *Hum. Genet.* **98**: 351–370.

Krishnaswamy S, Williams EB, Mann KG (1986) The binding of activated protein C to factors V and Va. *J. Biol. Chem.* **261**: 9684–9693.

Kruithof E, Tran-Thang C, Ransijn A, Bachmann F (1984) Demonstration of a fast acting inhibitor of plasminogen activators in human plasma. *Blood* **64**: 907–913.

Kruithof EKO, Tran-Thang C, Bachmann F (1986) Studies on the release of a plasminogen activator inhibitor by human platelets. *Thromb. Haemost.* **55**: 201–205.

Kudryk BJ, Collen D, Woods KR, Blombäck B (1974) Evidence for localization of polymerization sites in fibrinogen. *J. Biol. Chem.* **249**: 3322–3325.

Kugler W, Wagner U, Ryffel GU (1988) Tissue-specificity of liver gene expression: a common liver-specific promoter element. *Nucleic Acids Res.* **16**: 3165–3174.

Kuhn LA, Griffin JH, Fisher CL, Greengard JS, Bouma BN, Espana F, Tainer JA (1990) Elucidating the structural chemistry of glycosaminoglycan recognition of protein C inhibitor. *Proc. Natl Acad. Sci. USA* **87**: 8506–8510.

Kumada T, Dittman WA, Majerus PW (1988) A role for thrombomodulin in the pathogenesis of thrombin-induced thromboembolism in mice. *Blood* **71**: 728–733.

Kumar A, Blumenthal DK, Fair DS (1991) Identification of molecular sites on factor VII which mediate its assembly and function in the extrinsic pathway activation complex. *J. Biol. Chem.* **266**: 915–921.

Kunkel TA (1984) Mutational specificity of depurination. *Proc. Natl Acad. Sci. USA* **81**: 1494–1498.

Kunkel TA (1985) The mutational specificity of DNA polymerase β during *in vitro* DNA synthesis. *J. Biol. Chem.* **260**: 5787–5796.

Kunkel TA, Soni A (1988) Mutagenesis by transient misalignment. *J. Biol. Chem.* **264**: 14784–14789.

Kurosawa J, Stearns DJ, Jackson KW, Esmon CT (1988) A 10 kDa cyanogen bromide fragment from the epidermal growth factor homology domain of rabbit thrombomodulin contains the primary thrombin binding site. *J. Biol. Chem.* **263**: 5993–5996.

Kuryavyi VV, Jovin TM (1995) Triad-DNA: a model for trinucleotide repeats. *Nature Genetics* **9**: 339–341.

Lämmle B, Wuillemin WA, Huber I, Krauskopf M, Zürcher C, Pflugshaupt R, Furlan M (1991) Thromboembolism and bleeding tendency in congenital factor XII deficiency – a study on 74 subjects from 14 Swiss families. *Thromb. Haemost.* **65**: 117–121.

Lane DA, Caso R (1989) Antithrombin: structure, genomic organization, function and inherited deficiency. *Baillière's Clin. Haematol.* **2**: 961–998.

Lane DA, Erdjument H, Flynn A, di Marzo V, Panico M, Morris HR, Greaves M, Dolan G, Preston FE (1989a) Antithrombin Sheffield: amino acid substitution at the reactive site (Arg 393 to His) causing thrombosis. *Br. J. Haematol.* **71**: 91–96.

Lane DA, Erdjument H, Thompson E, Panico M, Di Marzo V, Morris HW, Leone G, De Stefano V, Thein SL (1989b) A novel amino acid substitution in the reactive site of a congenital variant antithrombin. *J. Biol. Chem.* **264**: 10200–10204.

Lane DA, Ireland H, Olds RJ, Thein SL, Perry DJ, Aiach M (1991) Antithrombin III: a database of mutations. *Thromb. Haemost.* **66**: 657–661.

Lane DA, Olds RR, Thein S-L (1992a) Antithrombin and its deficiency states. *Blood Coag. Fibrinol.* **3**: 315–341.

Lane DA, Olds RJ, Conard J, Boisclair M, Bock SC, Hultin M, Abildgaard U, Ireland H, Thompson E, Sas G, Horellou MH, Tamponi G, Thein S-L (1992b) Pleiotropic effects of antithrombin strand 1C substitution mutations. *J. Clin. Invest.* **90**: 2422–2433.

Lane DA, Olds RJ, Boisclair M, Chowdhury V, Thein SL, Cooper DN, Blajchman M, Perry D, Emmerich J, Aiach M (1993) Antithrombin III mutation database: first update. *Thromb. Haemost.* **70**: 361–369.

Lane DA, Olds RJ, Thein SL (1994) Antithrombin III: summary of first database update. *Nucleic Acids Res.* **22**: 3556–3559.

Lane DA, Kunz G, Olds RJ, Thein SL (1996a) Molecular genetics of antithrombin deficiency. *Blood Rev.* **10**: 59–74.

Lane DA, Auberger K, Ireland H, Roscher AA, Thein SL (1996b) Prenatal diagnosis in combined antithrombin and factor V gene mutation. *Br. J. Haematol.* **94**: 753–755.

Lane DA, Mannucci PM, Bauer KA, Bertina RM, Bochkov NP, Boulyjenkov V, Chandy M, Dahlbäck B, Ginter EK, Miletich JP, Rosendaal FR, Seligsohn U (1996c) Inherited thrombophilia: Part 1. *Thromb. Haemost.* **76**: 651–662.

Lane DA, Mannucci PM, Bauer KA, Bertina RM, Bochkov NP, Boulyjenkov V, Chandy M, Dahlbäck B, Ginter EK, Miletich JP, Rosendaal FR, Seligsohn U (1996d) Inherited thrombophilia: Part 2. *Thromb. Haemost.* **76**: 824–834.

Lane D, Bayston T, Olds RJ, Fitches AC, Cooper DN, Millar DS, Jochmans K, Perry DJ, Okajima K, Thein SL, Emmerich J (1997) Antithrombin mutation database: 2nd (1997) update. *Thromb. Haemost.* **77**: 197–211.

Lau Y-F, Kuzma G, Wei C-M, Livinston DJ, Hsiung N (1987) A modified human tissue plasminogen activator with extended half-life *in vivo. Biotechnology* **5**: 953–958.

Laudano AP, Doolittle RF (1978) Studies on synthetic peptides that bind to fibrinogen and present fibrin polymerization. *Proc. Natl Acad. Sci. USA* **75**: 3085–3089.

Laudano AP, Doolittle RF (1980) Studies on synthetic peptides that bind to fibrinogen and prevent polymerization: structural requirements and species differences. *Biochemistry* **19**: 1013–1019.

Lawrence DA, Olson ST, Palaniappan S, Ginsburg D (1994) Engineering plasminogen activator inhibitor 1 mutants with increased functional stability. *Biochemistry* **33**: 3643–3648.

Lawrence DA, Ginsburg D, Day DE, Berkenpas MB, Verhamme IM, Kvassman J-O, Shore JD (1995) Serpin–protease complexes are trapped as stable acyl-enzyme intermediates. *J. Biol. Chem.* **270**: 25309–25310.

Lawrence DA, Strandberg L, Ericson J, Ny T (1990) Structure–function studies of the serpin plasminogen activator inhibitor type 1. *J. Biol. Chem.* **265**: 20293–20301.

Lawson JH, Butenas S, Ribarik N, Mann KG (1993) Complex-dependent inhibition of factor VIIa by antithrombin III and heparin. *J. Biol. Chem.* **266**: 767–770.

Le DT, Griffin JH, Greengard JS, Mujumdar V, Rapaport SI (1995) Use of a generally applicable tissue factor-dependent factor V assay to detect activated protein C-resistant factor Va in patients receiving warfarin and in patients with a lupus anticoagulant. *Blood* **85**: 1704–1711.

Le Bonniec F, MacGillivray RTA, Esmon CT (1991a) Thrombin Glu39 restricts the P3′ specificity to non-acidic residues. *J. Biol. Chem.* **266**: 13796–13803.

Le Bonniec BF, Esmon CT (1991b) Glu192 to Gln substitution in thrombin mimics the catalytic switch induced by thrombomodulin. *Proc. Natl Acad. Sci. USA* **88**: 7371–7375.

Le Bonniec BF, Guinto ER, Stone SR (1995) Identification of thrombin residues that modulate its interactions with antithrombin III and α1-antitrypsin. *Biochemistry* **34**: 12241–12248.

Leclerc D, Campeau E, Goyette P, Adjalla CE, Christensen B, Ross M, Eydoux P, Rosenblatt DS, Rozen R, Gravel RA (1996) Human methionine synthase: cDNA cloning and identification of mutations in patients of the *cblG* complementation group of folate/cobalamin disorders. *Hum. Mol. Genet.* **5**: 1867–1874.

Lee CD, Mann KG (1989) Activation/inactivation of human factor V by plasmin. *Blood* **73**: 185–190.

Lee LH, Jennings I, Luddington R, Baglin T (1994) Markers of thrombin and plasmin generation in patients with inherited thrombophilia. *J. Clin. Pathol.* **47**: 631–634.

Legnani C, Palaretti G, Biagi R, Coccheri S, Bernardi F, Rosendaal FR, Reitsma PH, De Ronde H, Bertina RM (1996) Activated protein C resistance: a comparison between two clotting assays and their relationship to the presence of the factor V Leiden mutation. *Br. J. Haematol.* **93**: 694–699.

Lensen RPM, Rosendaal FR, Koster T, Allaart CF, de Ronde H, Vandenbroucke JP, Reitsma PH, Bertina RM (1996) Apparent different thrombotic tendency in patients with factor V Leiden and protein C deficiency. *Blood* **88**: 4205–4208.

Lentz SR, Chen Y, Sadler JE (1993) Sequences required for thrombomodulin cofactor activity within the fourth epidermal growth factor-like domain of human thrombomodulin. *J. Biol. Chem.* **268**: 15312–15317.

Leone G, Stefano V, Di Donfrancesco A, Ferrelli R, Traisci G, Bizzi B (1987) Antithrombin III Pescara: a defective ATIII variant with no alterations of plasma crossed immunoelectrophoresis but with an abnormal crossed immunoelectrofocusing pattern. *Br. J. Haematol.* **65**: 187–191.

Lerch PG, Rickli EE, Lergier W, Gillessen D (1980) Localization of individual lysine-binding regions in human plasminogen and investigations on their complex-forming properties. *Eur. J. Biochem.* **107**: 7–13.

Leroy-Matheron C, Levent M, Pignon J-M, Mendonça C, Gouault-Heilmann M (1996) The 1691 G→A mutation in the factor V gene: relationship to activated protein C (APC) resistance and thrombosis in 65 patients. *Thromb. Haemost.* **75**: 4–10.

Leung LLK (1986) Interaction of histidine-rich glycoprotein with fibrinogen and fibrin. *J. Clin. Invest.* **77**: 1305–1311.

Leung LLK, Harpel PC, Nachman RL, Rabellino EM (1983) Histidine-rich glycoprotein is present in human platelets and is released following thrombin stimulation. *Blood* **62**: 1016–1021.

Leung LLK, Nachman RL, Harpel PC (1984) Complex formation of platelet thrombospondin with histidine-rich glycoprotein. *J. Clin. Invest.* **73**: 5–12.

Leven RM, Schick PK, Budzynski AZ (1985) Fibrinogen biosynthesis in isolated guinea pig megakaryocytes. *Blood* **65**: 501–504.

Levi M, Hack CE, de Boer JP, Brandjes DPM, Büller HR, ten Cate JW (1991) Reduction of contact activation related fibrinolytic activity in factor XII deficient patients. *J. Clin. Invest.* **88**: 1155–1160.

Levi M, Biemond BJ, van Zonnefeld AJ, ten Cate JW, Pannakoek H (1992) Inhibition of plasminogen activator inhibitor-1 activity results in promotion of endogenous thrombolysis and inhibition of thrombus extension in models of experimental thrombosis. *Circulation* **85**: 305–311.

Levoir D, Emmerich J, Alhenc-Gelas M, Dumontier I, Petite J-P, Fiessinger J-N, Aiach M (1995) Portal vein thrombosis and factor V Arg 506 to Gln mutation. *Thromb. Haemost.* **73**: 550–551.

Li L, Kikuchi S, Arinami T, Kobayashi K, Tsuchiya S, Hamaguchi H (1994) Plasminogen with type-I mutation in the Chinese Han population. *Clin. Genet.* **45**: 285–287.

Liebhaber SA, Goossens M, Kan YW (1981) Homology and concerted evolution at the α1 and α2 loci of human α-globin. *Nature* **290**: 26–29.

Lijnen HR, Hoylaerts M, Collen D (1980) Isolation and characterization of a human plasma protein with affinity for the lysine-binding sites in plasminogen. Role in the regulation of fibrinolysis and identification as histidine-rich glycoprotein. *J. Biol. Chem.* **255**: 10214–10222.

Lijnen R, Hoylaerts M, Collen D (1983) Heparin binding properties of human histidine-rich glycoprotein. Mechanism and role in the neutralization of heparin in plasma. *J. Biol. Chem.* **258**: 3803–3808.

Lijnen R, van Hoef B, Collen D (1984a) Histidine-rich glycoprotein modulates the anticoagulant activity of heparin in human plasma. *Thromb. Haemost.* **57**: 266–268.

Lijnen HR, Soria J, Soria C, Collen D, Caen JP (1984b) Dysfibrinogenemia (fibrinogen Dusard) associated with impaired fibrin-enhanced plasminogen activation. *Thromb. Haemost.* **51**: 108–109.

Lim WA, Faruggio DC, Sauer RT (1992) Structural and energetic consequences of disruptive mutations in a protein core. *Biochemistry* **31**: 4324–4333.

Lin J-H, Wang M, Andrews WH, Wydro R, Morser J (1994) Expression efficiency of the human thrombomodulin-encoding gene in various vector and host systems. *Gene* **147**: 287–292.

Lind B, van Solinge WW, Schwartz M, Thorsen S (1993) Splice site mutation in the human protein C gene associated with venous thrombosis: demonstration of exon skipping by ectopic transcript analysis. *Blood* **82**: 2423–2432.

Lind B, Schwartz M, Thorsen S (1995) Six different point mutations in seven Danish families with symptomatic protein C deficiency. *Thromb. Haemost.* **73**: 186–193.

Lindahl AK, Sandset PM, Thune-Wiiger M, Nordfang O, Sakariassen KS (1994) Tissue factor pathway inhibitor prevents thrombus formation on procoagulant subendothelial matrix. *Blood Coag. Fibrinol.* **5**: 755–760.

Lindblad B, Svensson PJ, Dahlbäck B (1994) Arterial and venous thromboembolism with fatal outcome in a young man with inherited resistance to activated protein C. *Lancet* **343**: 917.

Lindo VS, Kakkar VV, Melissari E (1995) Cleaved antithrombin (Atc): a new marker for thrombin generation and activation of the coagulation system. *Br. J. Haematol.* **89**: 157–162.

Lintel-Hekkert W te, Bertina RM, Reitsma PH (1988) Two RFLPs approximately 7kb 5' of the human protein C gene. *Nucleic Acids Res.* **16**: 11849.

Liu CS, Chang JY (1987) The heparin binding site of human antithrombin III. Selective chemical modification at Lys114, Lys125 and Lys287 impairs its heparin cofactor activity. *J. Biol. Chem.* **262**: 17356–17361.

Liu CY, Koehn JA, Morgan FJ (1985) Characterization of fibrinogen New York 1. A dysfunctional fibrinogen with a deletion of Bβ (9–72) corresponding exactly to exon 2 of the gene. *J. Biol. Chem.* **260**: 4390–4396.

Liu CY, Wallén P, Handley DA (1986) Fibrinogen New York I: The structural, functional and genetic defects and a hypothesis of the role of fibrin in the regulation of coagulation and fibrinolysis. In: Lane DA, Henschen A, Jasani MK (eds) *Fibrinogen, Fibrin Formation and Fibrinolysis*, pp. 79–90. De Gruyter, Berlin.

Liu L, Dewar L, Song Y, Kulczycky M, Blajchman MA, Fenton JW, Andrew M, Delorme M, Ginsberg J, Preissner KT, Ofosu FA (1995) Inhibition of thrombin by antithrombin III and heparin cofactor II *in vivo*. *Thromb. Haemost.* **73**: 405–412.

Liu Y, Lyons RM, McDonagh J (1988) Plasminogen San Antonio: an abnormal plasminogen with a more cathodic migration, decreased activation and associated thrombosis. *Thromb. Haemost.* **59**: 49–53.

Liu LW, Rezaie AR, Carson CW, Esmon NL, Esmon CT (1994) Occupancy of anion binding exosite 2 on thrombin determines Ca²⁺ dependence of protein C activation. *J. Biol. Chem.* **269**: 11807–11812.

Liu Z, Fuller GM (1995) Detection of a novel transcription factor for the Aα fibrinogen gene in response to interleukin-6. *J. Biol. Chem.* **270**: 7580–7586.

Lodish H (1988) Transport of secretory and membrane glycoprotein from the rough endoplasmic reticulum to the Golgi. A rate-limiting step in protein maturation and secretion. *J. Biol. Chem.* **263**: 2107–2110.

Loeb LA, Kunkel TA (1982) Fidelity of DNA synthesis. *Annu. Rev. Biochem.* **52**: 429–457.

Loeb LA, Preston BD (1986) Mutagenesis by apurinic/apyrimidinic sites. *Annu. Rev. Genet.* **20**: 201–230.

Loebermann H, Tokuoka R, Diesenhofer J, Huber R (1984) Human α1-proteinase inhibitor: crystal structure analysis of two crystal modifications, molecular model and preliminary analysis of the implications for function. *J. Mol. Biol.* **177**: 531–556.

Long GL, Belagaje RM, MacGillivray RT (1984) Cloning and sequencing of liver cDNA coding for bovine protein C. *Proc. Natl Acad. Sci. USA* **81**: 5653–5656.

Long GL, Marshall A, Gardner JC, Naylor SL (1988) Genes for human vitamin K-dependent plasma proteins C and S are located on chromosomes 2 and 3 respectively. *Somat. Cell Mol. Genet.* **14**: 93–98.

Long GL, Tomczak JA, Rainville IR, Dreyfus M, Schramm W, Schwarz HP. (1994) Homozygous type I protein C deficiency in two unrelated families exhibiting thrombophilia related to Ala136→Pro or Arg286→His mutations. *Thromb. Haemost.* **72**: 526–533.

Longbin L, Dewar L, Song Y, Kulczycky M, Blajchman MA, Fenton JW, Andrew M, Delorme M, Ginsberg J, Preissner KT, Ofosu FA (1995) Inhibition of thrombin by antithrombin III and heparin cofactor II *in vivo*. *Thromb. Haemost.* **73**: 405–412.

Lord ST (1985) Expression of a cloned human fibrinogen cDNA in *Escherichia coli*: synthesis of an Aα polypeptide. *DNA* **4**: 33–38.

Loskutoff DJ (1991) Regulation of PAI-1 gene expression. *Fibrinolysis* **5**: 197–206.

Loskutoff DJ, Linders M, Keijer J, Veerman H, van Heerikulzen H, Pannekoek H (1987) Structure of the human plasminogen activator inhibitor 1 gene: nonrandom distribution of introns. *Biochemistry* **26**: 3763–3768.

Loskutoff DJ, Sawdey M, Mimuro J (1988) Type 1 plasminogen activator inhibitor. *Prog. Hemost. Thromb.* **9**: 87–115.

Lottenberg R, Dolly FR, Kitchens CS (1985) Recurring thromboembolic disease and pulmonary hypertension associated with severe hypoplasminogenemia. *Am. J. Hematol.* **19**: 181–193.

Lozano M, Escolar G, Schwarz HP, Hernandez R, Bozzo J, Ordinas A (1996) Activated protein C inhibits thrombus formation in a system with flowing blood. *Br. J. Haematol.* **95**: 179–183.

Lu D, Bovill EG, Long GL (1994a) Molecular mechanism for familial protein C deficiency and thrombosis in Protein C$_{VERMONT}$ (Glu^{20}→Ala and Val^{34}→Met). *J. Biol. Chem.* **269**: 29032–29038.

Lu D, Kalafatis M, Mann KG, Long GL (1994b) Loss of membrane-dependent factor Va cleavage: a mechanistic interpretation of the pathology of protein C$_{VERMONT}$. *Blood* **84**: 687–690.

Lucas MA, Fretto LJ, McKee PA (1983) The binding of human plasminogen to fibrin and fibrinogen. *J. Biol. Chem.* **258**: 4249–4256.

Lucore CL, Sobel BE (1988) Interactions of tissue-type plasminogen activator with plasma inhibitors and their pharmacologic implications. *Circulation* **78**: 660–669.

Lucore CL, Fujii S, Wun T-C, Sobel BE, Billadello JJ (1988) Regulation of the expression of type 1 plasminogen activator inhibitor in HepG2 cells by epidermal growth factor. *J. Biol. Chem.* **263**: 15845–15848.

Lundwall A, Dackowski W, Cohen E, Shaffer M, Mahr A, Dahlbäck B, Stenflo J, Wydro R (1986) Isolation and sequence of the cDNA for human protein S, a regulator of blood coagulation. *Proc. Natl Acad. Sci. USA* **83**: 6716–6720.

Lunghi B, Iacoviello L, Gemmati D, Dilasio MG, Castoldi E, Pinotti M, Castaman G, Redaelli R, Mariani G, Marchetti G, Bernardi F (1996) Detection of new polymorphic markers in the factor V gene: association with factor V levels in plasma. *Thromb. Haemost.* **75**: 45–48.

Lüscher TF (1994) The endothelium and cardiovascular disease – a complex relation. *New Engl. J. Med.* **330**: 1081–1083.

Lüscher TF, Tanner FC, Tschudi MR, Noll G (1993) Endothelial dysfunction in coronary artery disease. *Annu. Rev. Med.* **44**: 395–418.

Ma DD, Aboud MR, Williams BG, Isbister JP (1995) Activated protein C resistance and inherited factor V missense mutation in patients with venous and arterial thrombosis in a haematology clinic. *Austr. N.Z. J. Med.* **25**: 151–154.

MacCluer JW, Kammerer CM (1991) Dissecting the genetic contribution to coronary heart disease. *Am. J. Hum. Genet.* **49**: 1139–1144.

MacFarlane AS, Todd D, Cromwell S (1964) Fibrinogen catabolism in humans. *Clin. Sci.* **26**: 415–420.

Madison EL (1994) Probing structure-function relationships of tissue-type plasminogen activator by site-specific mutagenesis. *Fibrinolysis* **8** (Suppl. 1): 221–236.

Madison EL, Goldsmith EJ, Gerard RD, Gething M-J, Sambrook JF (1989) Serpin-resistant mutants of human tissue-type plasminogen activator. *Nature* **339**: 721–724.

Madison EL, Goldsmith EJ, Gerard RD, Gething M-JH, Sambrook JF, Bassel-Duby RS (1990) Amino acid residues that affect interaction of tissue-type plasminogen activator with plasminogen activator inhibitor 1. *Proc. Natl Acad. Sci. USA* **87**: 3530–3533.

Mager DL, Goodchild NL (1989) Homologous recombination between the LTRs of a human retrovirus-like element causes a 5-kb deletion in two siblings. *Am. J. Hum. Genet.* **45**: 848–854.

Magnaghi P, Mihalich A, Taramelli R (1994) Several liver specific DNase hypersensitive sites are present in the intergenic region separating human plasminogen and apolipoprotein (a) gene. *Biochem. Biophys. Res. Commun.* **205**: 930–935.

Mahasandana C, Suvatte V, Marlar RA, Manco-Johnson MJ, Jacobson LJ, Hathaway WE (1990a) Neonatal *purpura fulminans* associated with homozygous protein S deficiency. *Lancet* **335**: 61–62.

Mahasandana C, Suvatte V, Chuansumrit A, Marlar RA, Manco-Johnson MJ, Jacobson LJ, Hathaway WE (1990b) Homozygous protein S deficiency in an infant with purpura fulminans. *J. Pediatr.* **117**: 750–753.

Mahasandra C, Veerakul G, Tanphaichitr VS, Suvatte V, Opartkiattikul N, Hathaway WE (1996) Homozygous protein S deficiency: 7-year follow-up. *Thromb Haemost.* **76**: 1118 1122.

Malcolm BA, Wilson KP, Matthews BW, Kirsch JF, Wilson AC (1990) Ancestral lysozymes reconstructed, neutrality tested, and thermostability linked to hydrocarbon packing. *Nature* **345**: 86–89.

Malek AM, Jackman R, Rosenberg RD, Izumo S (1994) Endothelial expression of thrombomodulin is reversibly regulated by fluid shear stress. *Circulation Res.* **74**: 852–860.

Malgaretti U, Bruno L, Pontoglio M, Candiani G, Meroni G, Ottolenghi S, Taramelli R (1990) Definition of the transcription initiation site of human plasminogen gene in liver and non-hepatic cell lines. *Biochem. Biophys. Res. Commun.* **173**: 1013–1018.

Malinov MR (1994) Homocyst(e)ine and arterial occlusive diseases. *J. Intern. Med.* **236**: 603–608.

Malinowski DP, Sadler JE, Davie EW (1984) Characterization of a complementary deoxyribonucleic acid coding for human and bovine plasminogen. *Biochemistry* **23**: 4243–4250.

Malm J, Laurell M, Dahlbäck B (1988) Changes in the plasma levels of vitamin K-dependent proteins C and S and of C4b-binding protein during pregnancy and oral contraception. *Br. J. Haematol.* **68**: 437–443.

Manabe S-I, Matsuda M (1985) Homozygous protein C deficiency combined with heterozygous dysplasminogenemia found in a 21 year old thrombophilic male. *Thromb. Res.* **39**: 333–341.

Manco-Johnson MJ, Marlar RA, Jacobson LJ, Hays T, Waraday BA (1988) Severe protein C deficiency in newborn infants. *J. Pediatr.* **113**: 359–366.

Mandel H, Brenner B, Berant M, Rosenberg N, Lanir N, Jakobs C, Fowler B, Seligsohn U (1996) Coexistence of hereditary homocystinuria and factor V Leiden – effect on thrombosis. *New Engl. J. Med.* **334**: 763–768.

Manfioletti G, Brancolini C, Avazi G, Schneider C (1993) The protein encoded by a growth arrest-specific gene (*gas*6) is a new member of the vitamin K-dependent proteins related to protein S, a negative coregulator in the blood coagulation cascade. *Mol. Cell. Biol.* **13**: 4976–4985.

Mann KG, Lawler CM, Vehar GA, Church WR (1984) Coagulation factor V contains copper ion. *J. Biol. Chem.* **259**: 12949–12951.

Mann KG, Jenny RJ, Krishnaswamy S (1988) Cofactor proteins in the assembly and expression of blood clotting enzyme complexes. *Annu. Rev. Biochem.* **57**: 915–956.

Mann KG, Nesheim ME, Church WR, Haley P, Krishnaswamy S (1990) Surface dependent reactions of the vitamin K-dependent enzyme complexes. *Blood* **76**: 1–16.

Mannhalter C, Fischer M, Hopmeier P, Deutsch E (1987) Factor XII activity and antigen concentrations in patients suffering from recurrent thrombosis. *Fibrinolysis* **1**: 259–263.

Mannucci PM, Giangrande PLF (1992) Detection of the prethrombotic state due to procoagulant imbalance. *Eur. J. Haematol.* **48**: 65–69.

Mannucci PM, Tripodi A (1987) Laboratory screening of inherited thrombotic syndromes. *Thromb. Haemost.* **57**: 247–251.

Mannucci PM, Vigano S, Bottasso B, Candotti G, Bozzetti P, Rossi E, Pardi G (1984) Protein C antigen during pregnancy, delivery and puerperium. *Thromb. Haemost.* **52**: 217.

Mannucci PM, Kluft C, Traas DW, Seveso P, D'Angelo A (1986a) Congenital plasminogen deficiency associated with venous thromboembolism: therapeutic trial with Stanozolol. *Br. J. Haematol.* **63**: 753–759.

Mannucci PM, Tripodi A, Bertina RM (1986b) Protein S deficiency associated with 'juvenile' arterial and venous thrombosis. *Thromb. Haemost.* **55**: 440.

Mannucci PM, Valsecchi C, Krachmalnicoff A, Faioni EM, Tripodi A (1989) Familial dysfunction of protein S. *Thromb. Haemost.* **62**: 763–766.

Manotti C, Quintavalla R, Pini M, Jeran M, Paolicelli M, Dettori AG (1989) Thromboembolic manifestations and congenital factor V deficiency: a family study. *Haemostasis* **19**: 331–334.

Manten B, Westendorp RGJ, Koster T, Reitsma PH, Rosendaal FR (1996) Risk factor profiles in patients with different clinical manifestations of venous thromboembolism: a focus on the factor V Leiden mutation. *Thromb. Haemost.* **76**: 510–515.

Marchetti G, Legnani C, Patracchini P, Gemmati D, Ferrati M, Palareti G, Coccheri S, Bernardi F (1993) Study of a protein S gene polymorphism at DNA and mRNA level in a family with symptomatic protein S deficiency. *Br. J. Haematol.* **85**: 173–175.

Marciniak E, Wilson HD, Marlar RA (1985) Neonatal purpura fulminans: a genetic disorder related to the absence of protein C in blood. *Blood* **65**: 15–21.

Marder VJ (1995) Thrombolytic therapy: overview of results in major vascular occlusions. *Thromb. Haematol.* **74**: 101–105.

Margaglione M, D'Andrea G, Cappucci G, Grandone E, Giuliani N, Colaizzo D, Vecchione G, Di Minno G (1996) Detection of the factor V Leiden using SSCP. *Thromb. Haemost.* **76**: 813–821.

Margalit H, Fischer N, Ben-Sasson SA (1993) Comparative analysis of structurally defined heparin binding sequences reveals a distinct spatial distribution of basic residues. *J. Biol. Chem.* **268**: 19228–19231.

Marguerie GA, Plow EF, Edgington TS (1979) Human platelets possess an inducible and saturable receptor specific for fibrinogen. *J. Biol. Chem.* **254**: 5357–5363.

Marino MW, Fuller GM, Elder FFB (1986) Chromosomal localization of human and rat Aα, Bβ and γ fibrinogen genes by *in situ* hybridization. *Cytogenet. Cell Genet.* **42**: 36–41.

Markland W, Pollock D, Livingston DJ (1989) Structure–function analysis of tissue-type plasminogen activator by linker insertion, point and deletion mutagenesis. *Prot. Eng.* **3**: 117–125.

Marlar RA, Griffin JH (1980) Deficiency of protein C inhibitor in combined factor V/VIII deficiency disease. *J. Clin. Invest.* **66**: 1186–1189.

Marlar RA, Neumann A (1990) Neonatal purpura fulminans due to homozygous protein C or protein S deficiency. *Sem. Thromb. Haemost.* **16**: 299–309.

Marlar RA, Montgomery RR, Broekmans AW (1989) Diagnosis and treatment of homozygous protein C deficiency. *J. Pediatr.* **114**: 528–534.

Marlar RA, Sills RH, Groncy PK, Montgomery RR, Madden RM (1992) Protein C survival during replacement therapy in homozygous protein C deficiency. *Am. J. Hematol.* **41**: 24–31.

Marquette KA, Pittman DD, Kaufman RJ (1995) The factor V B-domain provides two functions to facilitate thrombin cleavage and release of the light chain. *Blood* **86**: 3026–3034.

Martin PD, Robertson W, Turk D, Bode W, Edwards BFP (1992) The structure of residues 7–16 of the Aα-chain of human fibrinogen bound to bovine thrombin at 2.3 Å resolution. *J. Biol. Chem.* **267**: 7911–7920.

Martinelli I, Landi G, Merati G, Cella R, Tosetto A, Mannucci PM (1996a) Factor V gene mutation is a risk factor for cerebral venous thrombosis. *Thromb. Haemost.* **75**: 393–394.

Martinelli I, Magatelli R, Cattaneo M, Mannucci PM (1996b) Prevalence of mutant factor V in Italian patients with hereditary deficiencies of antithrombin, protein C or protein S. *Thromb. Haemost.* **75**: 694–695.

Maruyama I, Majerus PW (1985) The turnover of thrombin–thrombomodulin complex in cultured human umbilical vein endothelial cells and A549 lung cancer cells. *J. Biol. Chem.* **260**: 15432–15438.

Maruyama I, Salem HH, Majerus PW (1984) Coagulation factor Va binds to human umbilical vein endothelial cells and accelerates protein C activation. *J. Clin. Invest.* **74**: 224–228.

März W, Seydewitz H, Winkelmann B, Chen M, Nauck M, Witt I (1995) Mutation in coagulation factor V associated with resistance to activated protein C in patients with coronary artery disease. *Lancet* **345**: 526–527.

Mast AE, Enghild JJ, Pizzo SV, Salvesen G (1991) Analysis of the plasma elimination kinetics and conformational stabilities of native, proteinase-complexed, and reactive site cleaved serpins:

comparison of α1-proteinase inhibitor, α1-antichymotrysin, antithrombin III, α2-antiplasmin and ovalbumin. *Biochemistry* **30**: 1723–1730.

Matsuda M, Terukina S, Miyata T, Iwanaga S (1986) Defective release of fibrinopeptide A due to substitution of Aα arginine-16 by cysteine in two abnormal fibrinogens Kawaguchi and Osaka. In: Muller-Berghaus G, Scheefers-Borschel U, Selmayr E, Henschen A (eds) *Fibrinogen and its Derivatives; Biochemistry, Physiology and Pathophysiology,* pp. 37–40. Elsevier, Amsterdam.

Matsuda M, Sugo T, Sakata Y, Murayama H, Mimuro J, Tanabe S, Yoshitake S (1988) A thrombotic state due to an abnormal protein C. *New Engl. J. Med.* **319**: 1265–1268.

Matsuka YV, Medved LV, Migliorini MM, Ingham KC (1996) Factor XIIIa-catalyzed cross-linking of recombinant αC fragments of human fibrinogen. *Biochemistry* **35**: 5810–5816.

Matsuo T, Kario K, Sakamoto S, Yamada T, Miki T, Hirase T, Kobayashi H (1992) Hereditary heparin cofactor II deficiency and coronary artery disease. *Thromb. Res.* **65**: 495–501.

Matsushita K, Kuriyama Y, Sawada T, Uchida K (1992) Cerebral infarction associated with protein C deficiency. *Stroke* **23**: 108–111.

Mayo M, Halil T, Browse NL (1971) The incidence of deep vein thrombosis after prostatectomy. *Br. J. Urol.* **43**: 738–742.

Mazzorana M, Baffet G, Kneip B, Launois B, Guguen-Guillouzo C (1991) Expression of coagulation factor V gene by normal adult human hepatocytes in primary culture. *Br. J. Haematol.* **78**: 229–235.

McAlpine PJ, Dixon M, Guinto ER, MacGillivray RTA (1990) A *Pst*I polymorphism in the human coagulation factor V (*F5*) gene. *Nucleic Acids Res.* **18**: 7471.

McCance SG, Menhart N, Castellino FJ (1994) Amino acid residues of the kringle-4 and kringle-5 domains of human plasminogen that stabilize their interactions with ω-amino acid ligands. *J. Biol. Chem.* **269**: 32405–32410.

McClure DB, Walls JD, Grinnell BW (1992) Post-translational processing events in the secretion pathway of human protein C, a complex vitamin K-dependent antithrombotic factor. *J. Biol. Chem.* **267**: 19710–19717.

McDonagh J, Carrell N, Lee MH (1994) Dysfibrinogenemia and other disorders of fibrinogen structure and function. In: Colman RW, Hirsh J, Marder VJ, Salzman EW (eds) *Hemostasis and Thrombosis: Basic Principles and Clinical Practice,* pp. 314–334. JB Lippincott, Philadelphia.

McLean JW, Tomlinson JE, Kuang WJ, Eaton DL, Chen EY, Fless GM, Scanu AM, Lawn RM (1987) cDNA sequence of human apolipoprotein (a) is homologous to plasminogen. *Nature* **330**: 132–137.

McMullen BA, Fujikawa K (1985) Amino acid sequence of the heavy chain of human α-factor XIIa (activated Hageman factor). *J. Biol. Chem.* **260**: 5328–5341.

Meade TW (1983) Factor VII and ischaemic heart disease: epidemiologic evidence. *Haemostasis* **13**: 178–185.

Meade TW, Stirling Y, Wilkes H, Mannucci PM (1985) Effects of oral contraceptives and obesity on protein C antigen. *Thromb. Haemost.* **53**: 198–199.

Meade TW, Mellows S, Brozovic M, Miller GJ, Chakrabarti RR, North WRS, Haines AP, Stirling Y, Imeson JD, Thompson SG (1986) Haemostatic function and ischaemic heart disease: principal results of the Northwick Park Heart Study. *Lancet* **ii**: 533–537.

Meade TW, Dyer S, Howarth DJ, Imeson JD, Stirling Y (1990) Antithrombin III and procoagulant activity: sex differences and effects of the menopause. *Br. J. Haematol.* **74**: 77–81.

Meade TW, Cooper J, Miller GJ, Howarth DJ, Stirling Y (1991) Antithrombin III and arterial disease. *Lancet* **338**: 850–851.

Medcalf RL, Rueeg M, Scleuning W-D (1990) A DNA motif related to the cAMP-responsive element and an exon-located activator protein-2 binding site in the human tissue-type plasminogen activator gene promoter cooperate in basal expression and convey activation by phorbol ester and cAMP. *J. Biol. Chem.* **265**: 14618–14626.

Medved LV, Litvinovitch SV, Ugarova TP, Lukinova NI, Kalikhevich VN, Ardemasova ZA (1993) Localization of a fibrin polymerization site complementary to Gly-His-Arg sequence. *FEBS Lett.* **320**: 239–242.

Meh DA, Siebenlist KR, Mosesson MW (1996) Identification and characterization of the thrombin binding sites on fibrin. *J. Biol. Chem.* **271**: 23121–23125.

Mehta J, Mehta P, Lawson D, Saldeen T (1987) Plasma tissue plasminogen activator inhibitor levels in coronary artery disease: correlation with age and serum triglyceride concentrations. *J. Am. Coll. Cardiol.* **9**: 263–268.

Meijers JCM, Chung DW (1991) Organization of the gene coding for human protein C inhibitor (plasminogen activator inhibitor-3). *J. Biol. Chem.* **266**: 15028–15034.

Meijers JC, Vlooswijk RA, Kanters DH, Hessing M, Bouma BN (1988) Identification of monoclonal antibodies that inhibit the function of protein C inhibitor. Evidence for heparin independent inhibition of activated protein C in plasma. *Blood* 72: 1401–1406.

Melissari E, Monte G, Lindo VS, Pemberton KD, Wilson NV, Edmondson R, Das S, Kakkar VV (1992) Congenital thrombophilia among patients with venous thromboembolism. *Blood Coag. Fibrinol.* 3: 749–758.

Mellbring G, Dahlgren S, Wiman B, Sunnegardh O (1985) Relationship between preoperative status of the fibrinolytic system and occurrence of deep vein thrombosis after major abdominal surgery. *Thromb. Res.* 39: 157–163.

Menache D, O'Malley JP, Schorr JB, Wagner B, Williams C, Alving BM, Ballard JO, Goodnight SH, Hathaway WE, Hultin MB, Kitchens CS, Lessner HE, Makary AZ, Manco-Johnson M, McGeheeWG, Penner JA, Sanders JE (1990) Evaluation of the safety, recovery, half-life, and clinical efficacy of antithrombin III (human) in patients with hereditary antithrombin III deficiency. *Blood* 75: 33–39.

Menhart N, Hoover GJ, McCance SG, Castellino FJ (1995) Roles of individual kringle domains in the functioning of positive and negative effectors of human plasminogen activation. *Biochemistry* 34: 1482–1488.

Mesters RM, Houghten RA, Griffin JH (1991) Identification of a sequence of human activated protein C (residues 390–404) essential for its anticoagulant activity. *J. Biol. Chem.* 266: 24514–24519.

Mesters RM, Heeb MJ, Griffin JH (1993a) A novel exosite in the light chain of human activated protein C essential for interaction with blood coagulation factor Va. *Biochemistry* 32: 12656–12663.

Mesters RM, Heeb MJ, Griffin JH (1993b) Interactions and inhibition of blood coagulation factor Va involving residues 311–325 of activated protein C. *Protein Sci.* 2: 1482–1489.

Mey DA, Siebenlist KR, Galanakis DK, Bergtrom G, Mosesson MW (1995) The dimeric Aα chain composition of dysfibrinogenemic molecules with mutations at Aα16. *Thromb. Res.* 78: 531–539.

Miao CH, Leytus SP, Chung DW, Davie EW (1992) Liver-specific expression of the gene coding for human factor X, a blood coagulation factor. *J. Biol. Chem.* 267: 7395–7401.

Miao CH, Ho W-T, Greenberg DL, Davie EW (1996) Transcriptional regulation of the gene coding for human protein C. *J. Biol. Chem.* 271: 9587–9594.

Michiels JJ, van der Luit L, van Vliet HHDM, Jochmans K, Lissens W (1995) Nonsense mutation Arg197Stop in a Dutch family with type 1 hereditary antithrombin (AT) deficiency causing thrombophilia. *Thromb. Res.* 78: 251–254.

Miles LA, Dahlberg CM, Plescia J, Felez J, Kato K, Plow EF (1991) Role of cell-surface lysines in plasminogen binding to cells: identification of α-enolase as a candidate plasminogen receptor. *Biochemistry* 30: 1682–1691.

Miletich JP, Broze GJ (1990) β-Protein C is not glycosylated at asparagine 329. *J. Biol. Chem.* 265: 11397–11404.

Miletich JP, Leykam FJ, Broze GJ (1983) Detection of a single chain protein C in human plasma. *Blood* 62 (Suppl.): 306a.

Miletich J, Sherman L, Broze G (1987) Absence of thrombosis in subjects with heterozygous protein C deficiency. *New Engl. J. Med.* 317: 991–996.

Miletich JP, Prescott SM, White R, Majerus PW, Bovill EG (1993) Inherited predisposition to thrombosis. *Cell* 72: 477–480.

Millar DS, Grundy CB, Bignell P, Mitchell DC, Corden D, Woods P, Kakkar VV, Cooper DN (1993a) A novel nonsense mutation in the protein C (*PROC*) gene (Trp29→Term) causing recurrent venous thrombosis. *Hum. Genet.* 91: 196.

Millar DS, Grundy CB, Bignell P, Moffat EH, Martin R, Kakkar VV, Cooper DN (1993b) A Gla domain mutation (Arg15→Trp) in the protein C gene causing type 2 protein C deficiency and recurrent venous thrombosis. *Blood Coag. Fibrinol.* 4: 345–347.

Millar DS, Lopez A, White D, Abraham G, Laursen B, Holding S, Reverter JC, Reynaud J, Martinowitz U, Hayes JPLA, Kakkar VV, Cooper DN (1993c) Screening for mutations in the antithrombin III (*AT3*) gene causing recurrent venous thrombosis by single strand conformation polymorphism analysis. *Hum. Mutat.* 2: 324–326.

Millar DS, Wacey AI, Ribando J, Melissari E, Laursen B, Woods P, Kakkar VV, Cooper DN (1994a) Three novel missense mutations in the antithrombin III (*AT3*) gene causing recurrent venous thrombosis. *Hum. Genet.* 94: 509–512.

Millar DS, Allgrove J, Rodeck C, Kakkar VV, Cooper DN (1994b) A homozygous deletion/insertion mutation in the protein C (*PROC*) gene causing neonatal purpura fulminans: prenatal diagnosis in an at-risk pregnancy. *Blood Coag. Fibrinol.* 5: 647–649.

Millar DS, Bevan D, Chitolie A, Reynaud J, Chisholm M, Kakkar VV, Cooper DN (1995) Three novel mutations in the protein C (*PROC*) gene causing venous thrombosis. *Blood Coag. Fibrinol.* **6**: 138–140.

Miller SP (1965) Coagulation dynamics in factor V deficiency: a family study with a note on the occurrence of thrombophlebitis. *Thromb. Diath. Haemorr.* **13**: 500–515.

Mima N, Azuma H, Shigekiyo T, Saito S (1996) A novel missense mutation in two families with congenital plasminogen deficiency: identification of an Ala675 to Thr675 substitution. *Thromb. Haemost.* **75**: 96–100.

Mimuro J, Loskutoff DL (1989) Purification of a protein from bovine plasma that binds to type 1 plasminogen activator inhibitor and prevents its interaction with extracellular matrix; evidence that the protein is vitronectin. *J. Biol. Chem.* **264**: 936–938.

Mimuro J, Sawdey M, Hattori M, Loskutoff DJ (1989) cDNA for bovine type 1 plasminogen activator inhibitor (PAI-1). *Nucleic Acids Res.* **17**: 8872.

Mimuro J, Matsumara S, Kaneko M, Yoshitake S, Iijima K, Nakamura K, Sakata M (1993) An abnormal protein C (protein C Yonago) with an amino acid substitution of Gly for Arg-15 caused by a single base mutation of C to G in codon 57 (CGG→GGG): deteriorated calcium-dependent conformation of the γ-carboxyglutamic acid domain relative to a thrombotic tendency. *Int. J. Haematol.* **57**: 9–14.

Minford AMB, Parapia LA, Stainforth C, Lee D (1996) Treatment of homozygous protein C deficiency with subcutaneous protein C concentrate. *Br. J. Haematol.* **93**: 215–216.

Mirshahi M, Soria J, Soria C, Faivre R, Lu H, Courtney M, Roitsch C, Tripier D, Caen JP (1989) Evaluation of the inhibition by heparin and hirudin of coagulation activation during r-tPA-induced thrombolysis. *Blood* **74**: 1025–1030.

Mitchell CA, Hau L, Salem HH (1986) Control of thrombin mediated cleavage of protein S. *Thromb. Haemost.* **56**: 151–154.

Miura O, Sugahara Y, Aoki N (1993) Intracellular transport-deficient mutants causing hereditary deficiencies of factors involved in coagulation and fibrinolysis. *Thromb. Haemost.* **69**: 296–297.

Miyata T, Iwanaga S, Sakata Y, Aoki N (1982) Plasminogen Tochigi: inactive plasmin resulting from replacement of alanine-600 by threonine in the active site. *Proc. Natl Acad. Sci. USA* **79**: 6132–6136.

Miyata T, Iwanaga S, Sakata Y, Aoki N, Takamatsu J, Kamiya T (1984) Plasminogens Tochigi II and Nagoya: two additional molecular defects with Ala→600Thr replacement found in plasmin light chain variants. *J. Biochem.* **96**: 277–287.

Miyata T, Terukina S, Matsuda M, Kasamatsu A, Takeda Y, Murukami T, Iwanaga S (1987) Fibrinogens Kawaguchi and Osaka: an amino acid substitution of Aα arginine-16 to cysteine which forms an extra interchain disulfide bridge between the two Aα chains. *J. Biochem.* **102**: 93–101.

Miyata T, Kawabata S-I, Iwanaga S, Takahashi I, Alving B, Saito H (1989) Coagulation factor XII (Hageman factor) Washington DC: inactive factor XIIa results from Cys571→Ser substitution. *Proc. Natl Acad. Sci. USA* **86**: 8319–8322.

Miyata T, Zheng Y-Z, Sakata T, Tsushima N, Kato H (1994) Three missense mutations in the protein C heavy chain causing type I and type II protein C deficiency. *Thromb. Haemost.* **71**: 32–37.

Miyata T, Zheng Y-Z, Sakata T, Kato H (1995) Protein C Osaka 10 with aberrant propeptide processing: loss of anticoagulant activity due to an amino acid substitution in the protein C precursor. *Thromb. Haemost.* **74**: 1003–1008.

Miyata T, Sakata T, Zheng Y-Z, Tsukamoto H, Umeyama H, Uchiyama S, Ikusaka M, Yoshioka A, Imanaka Y, Fujimura H, Kambayashi J, Kato H (1996) Genetic characterization of protein C deficiency in Japanese subjects using a rapid and nonradioactive method for single-strand conformational polymorphism analysis and a model building. *Thromb. Haemost.* **76**: 302–311.

Mizuguchi J, Hu C-H, Cao Z, Loeb KR, Chung DW, Davie EW (1995) Characterization of the 5′-flanking region of the gene for the γ chain of human fibrinogen. *J. Biol. Chem.* **270**: 28350–28356.

Modrich P (1987) DNA mismatch correction. *Annu. Rev. Biochem.* **56**: 435–466.

Mohri M, Oka M, Aoki Y, Gonda Y, Hirata S, Gomi K, Kiyota T, Sugihara T, Yamamoto S, Ishida T (1994) Intravenous extended infusion of recombinant human soluble thrombomodulin prevented tissue factor-induced disseminated intravascular coagulation in rats. *Am. J. Hematol.* **45**: 298–303.

Molho-Sabatier P, Aiach M, Gaillard I, Fiessinger J-N, Fischer A-M, Chadeuf G, Clauser E (1989) Molecular characterization of antithrombin III variants using polymerase chain reaction. *J. Clin. Invest.* **84**: 1236–1242.

Monkovic DD, Tracy PB (1990) Activation of human factor V by factor Xa and thrombin. *Biochemistry* **29**: 1118–1128.

Monnat RJ, Hackmann AFM, Chiaverotti TA (1992) Nucleotide sequence analysis of human hypoxanthine phosphoribosyltransferase (*HPRT*) gene deletions. *Genomics* **13**: 777–787.

Montaruli B, Voorberg J, Tamponi G, Borchiellini A, Muleo G, Pannocchia A, van Mourik JA, Schinco P (1996) Arterial and venous thrombosis in two Italian families with the factor V Arg^{506}→Gln mutation. *Eur. J. Haematol.* **57**: 96–100.

Morgan WT (1985) The histidine-rich glycoprotein of serum has a domain rich in histidine, proline and glycine that binds heme and metals. *Biochemistry* **24**: 1496–1501.

Morgan JG, Holbrook NJ, Crabtree GR (1987) Nucleotide sequence of γ chain gene of rat fibrinogen: conserved intron sequences. *Nucleic Acids Res.* **15**: 2774–2776.

Morgan JG, Courtois G, Fourel G, Chodosh LA, Campbell L, Evans E, Crabtree GR (1988) Sp1, a CAAT-binding factor and the adenovirus major late promoter transcription factor interact with functional regions of the γ-fibrinogen promoter. *Molec. Cell. Biol.* **8**: 2628–2637.

Morrissey JH (1995) Tissue factor interactions with factor VII: measurement and clinical significance of factor VIIa in plasma. *Blood Coag. Fibrinol.* **6** Suppl.1: S14-S19.

Morita T, Mizuguchi J, Kawabata S, Iwanaga S (1986) Proteolytic cleavage of vitamin K-dependent bovine plasma protein S by α-thrombin and plasma serine protease. *J. Biol. Chem.* **99**: 561–568.

Mosesson MW (1990) Fibrin polymerization and its regulatory role in hemostasis. *J. Lab. Clin. Med.* **116**: 8–17.

Mosesson MW (1993) Thrombin interactions with fibrinogen and fibrin. *Sem. Thromb. Hemost.* **19**: 361–367.

Mosesson MW, Beck EA (1969) Chromatographic, ultracentrifugal and related studies of fibrinogen Baltimore. *J. Clin. Invest.* **48**: 1656–1660.

Mosesson MW, Doolittle RF (eds) (1983) Molecular biology of fibrinogen and fibrin. *Ann. NY Acad. Sci.* **408**: 1–671.

Mosesson MW, Siebenlist KR, Hainfeld JF, Wall JS (1995a) The covalent structure of factor XIIIa-cross-linked fibrinogen fibrils. *J. Struct. Biol.* **115**: 88–101.

Mosesson MW, Siebenlist KR, DiOrio JP, Matsuda M, Hainfeld JF, Wall JS (1995b) The role of fibrinogen D domain intermolecular association sites in the polymerization of fibrin and fibrinogen Tokyo II (γ275Arg→Cys). *J. Clin. Invest.* **96**: 1053–1058.

Mosesson MW, Siebenlist KR, Hainfeld JF, Wall JS, Soria J, Soria C, Caen JP (1996) The relationship between the fibrinogen D domain self-association/cross-linking site (γXL) and the fibrinogen Dusard abnormality (Aα R554C-albumin). *J. Clin. Invest.* **97**: 2342–2350.

Mosher DF (1976) Action of fibrin-stabilizing factor on cold-insoluble globulin and α2-macroglobulin in clotting plasma. *J. Biol. Chem.* **251**: 1639–1645.

Mottonen J, Strand A, Symersky J, Sweet RM, Danley DE, Geoghegan KF, Gerard RD, Goldsmith EJ (1992) Structural basis of latency in plasminogen activator inhibitor-1. *Nature* **355**: 270–273.

Mourey L, Samama J-P, Delarue M, Petitou M, Choay J, Moras D (1993) Crystal structure of cleaved bovine antithrombin III at 3.2Å resolution. *J. Mol. Biol.* **232**: 223–241.

Mudd SH, Levy HL, Skovby F (1989) Disorders of transsulfuration. In: Scriver CR, Beaudet AL, Sly WS, Valle D (eds) *The Metabolic Basis of Inherited Disease*, 6th edn, pp. 693–734. McGraw-Hill, New York.

Mulichak AM, Tulinsky A, Ravichandran KG (1991) Crystal and molecular structure of human plasminogen kringle 4 refined at 1.9-Å resolution. *Biochemistry* **30**: 10576–10588.

Munnich A, Saudubray J-M, Dautzenberg M-D, Parvy P, Ogier H, Girot R, Manigne P, Frèzal J (1983) Diet-responsive proconvertin (factor VII) deficiency in homocystinuria. *J. Pediatr.* **102**: 730–734.

Murakawa M, Okamura T, Kamura T, Shibuya T, Harada M, Niho Y (1993) Diversity of primary structures of the carboxy-terminal regions of mammalian fibrinogen Aα-chains. *Thromb. Haemost.* **69**: 351–360.

Murakawa M, Okamura T, Kamura T, Kuroiwa M, Harada M (1994) A comparative study of partial primary structures of the catalytic region of mammalian protein C. *Br. J. Haematol.* **86**: 590–600.

Murray JC (1991) Coagulation and cancer. *Br. J. Cancer* **64**: 422–424.

Murray JC, Buetow K, Chung D, Aschbacher A (1986) Linkage disequilibrium of RFLPs at the β-fibrinogen and γ-fibrinogen loci on chromosome 4. *Cytogenet. Cell Genet.* **40**: 707–708.

Murray JC, Buetow KH, Donovan M, Hornung S, Motulsky AG, Disteche C, Dyer K, Swisshelm K, Anderson J, Giblett E, Sadler E, Eddy R, Shows TB (1987) Linkage disequilibrium of plasminogen polymorphisms and assignment of the gene to human chromosome 6q26–6q27. *Am. J. Hum. Genet.* **40**: 338–350.

Musci G, Berliner LJ, Esmon CT (1988) Evidence for multiple conformational changes in the active centre of thrombin induced by complex formation with thrombomodulin: an analysis employing nitroxide spin labels. *Biochemistry* **27**: 769–773.

Mustafa S, Pabinger I, Mannhalter C (1995) Protein S deficiency type I: identification of point mutations in 9 of 10 families. *Blood* **86**: 3444–3451.

Mustafa S, Pabinger I, Mannhalter C (1996) Two new frequent dimorphisms in the protein S (PROS1) gene. *Thromb. Haemost.* **76**: 393–396.

Myerowitz R, Hogikyan ND (1987) A deletion involving *Alu* sequences in the β-hexosaminidase α-chain gene of French Canadians with Tay-Sachs disease. *J. Biol. Chem.* **262**: 15396–15399.

Nachman RL (1992) Thrombosis and atherogenesis: molecular connections. *Blood* **79**: 1879–1906.

Nagashima M, Lundh E, Leonard JC, Morser J, Parkinson JF (1993) Alanine-scanning mutagenesis of the epidermal growth factor-like domains of human thrombomodulin identifies critical residues for its cofactor activity. *J. Biol. Chem.* **268**: 2888–2892.

Najjam S, Chadeuf G, Gandrille S, Aiach M (1994) Arg 129 plays a specific role in the conformation of antithrombin and in the enhancement of factor Xa inhibition by the pentasaccharide sequence of heparin. *Biochim. Biophys. Acta* **1225**: 135–143.

Nakagawa M, Tanaka S, Tsuji H, Takada O, Uno M, Hashimoto-Gotoh T, Wagatsuma M (1991) Congenital antithrombin III deficiency (ATIII Kyoto): identification of a point mutation altering arginine 406 to methionine behind the reactive site. *Thromb. Res.* **64**: 101–108.

Nakamura K (1991) A new hereditary abnormal protein C (Protein C Yonago) with a dysfunctional Gla-domain. *Thromb. Haemost.* **65**: 1197.

Nakagawa K, Tsuji H, Masuda H, Kitamura H, Nakahara Y, Ogasahara Y, Okajima Y, Sawada S, Nakagawa M (1994) Protein C deficiency found in a patient with acute myocardial infarction: a single base mutation 157 Arg (CGA) to stop codon (TGA). *Int. J. Hematol.* **60**: 273–280.

Nawa K, Sakano K, Fujiwara H, Sato Y, Sugiyama N, Teruuchi T, Iwamoto M, Marumoto Y (1990) Presence and function of chondroitin-4-sulfate on recombinant human soluble thrombomodulin. *Biochem. Biophys. Res. Commun.* **171**: 729–734.

Nawroth P, Kisiel W, Stern D (1985) The role of the endothelium in the homeostatic balance of haemostasis. *Clin. Haematol.* **14**: 531–546.

Nei M (1987) *Molecular Evolutionary Genetics*. Columbia University Press, New York.

Nelson RM, Long GL (1992) Binding of protein S to C4b-binding protein. *J. Biol. Chem.* **267**: 8140–8145.

Nelson RM, van Dusen WJ, Friedman PA, Long GL (1991) β-Hydroxyaspartic acid and β-hydroxyasparagine residues in recombinant human protein S are not required for anticoagulation cofactor activity or for binding to C4b-binding protein. *J. Biol. Chem.* **266**: 20586–20589.

Nemerson Y (1988) Tissue factor and hemostasis. *Blood* **71**: 1–8.

Nesbitt JE, Fuller GM (1991) Transcription and translation are required for fibrinogen mRNA degradation in hepatocytes. *Biochim. Biophys. Acta* **1089**: 88–94.

Nesheim ME, Myrmel KH, Hibbard L, Mann KG (1979a) Isolation and characterization of single chain bovine factor V. *J. Biol. Chem.* **254**: 508–517.

Nesheim ME, Taswell JB, Mann KG (1979b) The contribution of bovine factor V and factor Va to the activity of prothrombinase. *J. Biol. Chem.* **254**: 10952–10962.

Nesheim ME, Canfield WM, Kisiel W, Mann KG (1982) Studies of the capacity of factor X to protect factor Va from inactivation by activated protein C. *J. Biol. Chem.* **257**: 1443–1447.

Neurath H (1984) Evolution of proteolytic enzymes. *Science* **224**: 350–357.

Ni F, Konishi Y, Frazier RB, Scheraga HA, Lord ST (1988a) High resolution NMR studies of fibrinogen-like peptides in solution. Interaction of thrombin with residues 1–23 of the Aα chain of human fibrinogen. *Biochemistry* **28**: 3082–3094.

Ni F, Meinwald YS, Vasquez M, Scheroga HA (1988b) High resolution NMR studies of fibrinogen-like peptides in solution. Structure of a thrombin-bound peptide corresponding to residues 7–16 of the Aα chain of human fibrinogen. *Biochemistry* **28**: 3094–3105.

Ni H, Waye JS, Austin RC, Calverley D, Eng B, Sheffield WP, Blajchman MA (1994) Characterization of a highly polymorphic trinucleotide short tandem repeat within the human antithrombin gene. *Thromb. Res.* **74**: 303–307.

Nicholls RD, Fischel-Ghodsian N, Higgs DR (1987) Recombination at the human α-globin gene cluster: sequence features and topological constraints. *Cell* **49**: 369–378.

Nichols WC, Amano K, Cacheris PM, Figueiredo MS, Michaelides K, Schwaab R, Hoyer L, Kaufman RJ, Ginsburg D (1996) Moderation of hemophilia A phenotype by the factor V R506Q mutation. *Blood* **88**: 1183–1187.

Nicolaes GAF, Tans G, Thomassen CLGD, Hemker HC, Pabinger I, Varadi K, Schwarz HP, Rosing J (1995) Peptide bond cleavages and loss of functional activity during inactivation of factor Va and factor Va^{R506Q} by activated protein C. *J. Biol. Chem.* **270**: 21158–21166.

Nicolaides AN, Field ES, Kakkar VV (1972) Prostatectomy and deep-vein thrombosis. *Br. J. Surg.* **59**: 487–488.

Niessen RWLM, Sturk A, Hordijk PL, Michiels F, Peters M (1992) Sequence characterization of a sheep cDNA for antithrombin III. *Biochim. Biophys. Acta* **1171**: 207–210.

Niessen RWLM, Lamping RJ, Sturk A, Peters M (1996) DNA length polymorphism present in the antithrombin III promoter region does not influence plasma antithrombin III activity levels. *Thromb. Haemost.* **75**: 372–377.

Nilsson IM, Ljungner H, Tengborn L (1985) Two different mechanisms in patients with venous thrombosis and defective fibrinolysis: low concentration of plasminogen activator or increased concentration of plasminogen activator inhibitor. *Br. Med. J.* **290**: 1453–1455.

Nishimura DY, Murray JC (1992) Binding of protein S to C4b-binding protein. *J. Biol. Chem.* **267**: 8140–8145.

Nishioka J, Suzuki K (1990) Inhibition of cofactor activity of protein S by a complex of protein S and C4b-binding protein. *J. Biol. Chem.* **265**: 9072–9076.

Nishioka J, Taneda H, Suzuki K (1993) Estimation of the possible recognition sites for thrombomodulin, procoagulant, and anticoagulant proteins around the active center of α-thrombin. *J. Biochem.* **114**: 148–155.

Nishioka J, Ido M, Hayashi T, Suzuki K (1996) The Gla26 residue of protein C is required for the binding of protein C to thrombomodulin and endothelial cell protein C receptor, but not to protein S and factor Va. *Thromb. Haemost.* **75**: 275–282.

Nordstrom M, Lindblad B, Berqvist D (1992) A prospective study of the incidence of deep-vein thrombosis within a defined urban population. *J. Intern. Med.* **232**: 155–160.

Nothwehr SF, Gordon JI (1990) Targetin of proteins into the eukaryotic secretory pathway: signal peptide structure/function relationships. *BioEssays* **12**: 479–484.

Novacek MJ (1992) Mammalian phylogeny: shaking the tree. *Nature* **356**: 121–125.

Novotny WF, Brown SG, Miletich JP, Rader DJ, Broze GJ (1991) Plasma antigen levels of the lipoprotein-associated coagulation inhibitor in patient samples. *Blood* **78**: 387–393.

Nowak-Göttl U, Koch HG, Aschka I, Kohlhase B, Vielhaber H, Kurlemann G, Oleszcuk-Raschke K, Kehl HG, Jürgens H, Schneppenheim R (1996) Resistance to activated protein C (APCR) in children with venous or arterial thromboembolism. *Br. J. Haematol.* **92**: 992–998.

Ny T, Elgh F, Lund B (1984) The structure of the human tissue-type plasminogen activator gene: correlation of intron and exon structures to functional and structural domains. *Proc. Natl Acad. Sci. USA* **81**: 5355–5359.

Ny T, Sawdey M, Lawrence D, Millan JL, Loskutoff DJ (1986) Cloning and sequence of a cDNA coding for the human β-migrating endothelial-cell-type plasminogen activator inhibitor. *Proc. Natl Acad. Sci. USA* **83**: 6776–6780.

O'Brien DP, Gale KM, Anderson JS, McVey JH, Miller GJ, Meade TW, Tuddenham EGD (1991) Purification and characterization of factor VII 304-Gln: a variant molecule with reduced activity isolated from a clinically unaffected male. *Blood* **78**: 132–140.

O'Hara PJ, Grant FJ (1988) The human factor VII gene is polymorphic due to variation in repeat copy number in a minisatellite. *Gene* **66**: 147–158.

O'Hara PJ, Grant FJ, Haldeman BA, Gray CL, Insley MY, Hagen FS, Murray MJ (1987) Nucleotide sequence of the gene coding for human factor VII, a vitamin K-dependent protein participating in blood coagulation. *Proc. Natl Acad. Sci. USA* **84**: 5158–5162.

Oberlé, I, Rousseau F, Heitz D, Kretz C, Devys D, Hanauer A, Boué J, Bertheas MF, Mandel JL (1991) Instability of a 550 base pair DNA segment and abnormal methylation in fragile X syndrome. *Science* **252**: 1097–1102.

Ochoa A, Brunel F, Medelzon D, Cohen GN, Zakin MM (1989) Different liver nuclear proteins bind to similar DNA sequences in the 5' flanking regions of three hepatic genes. *Nucleic Acids Res.* **17**: 119–133.

Odegaard B, Mann KG (1987) Proteolysis of factor Va by factor Xa and activated protein C. *J. Biol. Chem.* **262**: 11233–11238.

Ofosu FA, Modi GJ, Smith LM, Cerskus AL, Hirsch J, Blajchman MA (1984) Heparan sulphate and dermatan sulphate inhibit the generation of thrombin activity in plasma by complementary pathways. *Blood* **64**: 742–747.

Ofosu FA, Liu L, Freedman J (1996) Control mechanisms in thrombin generation. *Sem. Thromb. Hemost.* **22**: 303–308.

Ogston D (1987) *Venous Thrombosis. Causation and Prediction.* John Wiley & Sons, Chichester.

Ogura M, Tanabe N, Nishioka J, Suzuki K, Saito H (1987) Biosynthesis and secretion of functional protein S by a human megakaryoblastic cell line (MEG-01). *Blood* **70**: 301–306.

Ohishi R, Watanabe N, Aritomi M, Gomi K, Kiyota T, Yamamoto S, Ishida T, Maruyama I (1993) Evidence that the protein C activation pathway amplifies the inhibition of thrombin generation by recombinant human thrombomodulin in plasma. *Thromb. Haemost.* **70**: 423–426.

Öhlin A-K, Stenflo J (1987) Calcium-dependent interaction between the EGF region of human protein C and a monoclonal antibody. *J. Biol. Chem.* **262**: 13798–13804.

Öhlin A-K, Marlar RA (1995) The first mutation identified in the thrombomodulin gene in a 45-year-old man presenting with thromboembolic disease. *Blood* 85: 330–336.

Öhlin A-K, Linse S, Stenflo J (1988a) Calcium binding to the epidermal growth factor homology region of bovine protein C. *J. Biol. Chem.* **263**: 7411–7417.

Öhlin A-K, Landes G, Bourdon P, Oppenheimer C, Wydro R, Stenflo J (1988b) Beta-hydroxy aspartic acid in the first epidermal growth factor-like domain of protein C: its role in Ca^{2+} binding and biological activity. *J. Biol. Chem.* **263**: 19240–19248.

Öhlin A-K, Bjork I, Stenflo J (1990) Proteolytic formation and properties of a fragment of protein C containing the γ-carboxyglutamic acid-rich domain and the EGF-like region. *Biochemistry* 29: 644–651.

Ohno S (1970) *Evolution by Gene Duplication.* George Allen & Unwin, London.

Ohta T (1991a) Role of the diversifying selection and gene conversion in evolution of major histocompatibility loci. *Proc. Natl Acad. Sci. USA* 88: 6716–6720.

Ohta T (1991b) Multigene families and the evolution of complexity. *J. Mol. Evol.* 33: 34–41.

Ohta T (1994) On hypervariability at the reactive center of proteolytic enzymes and their inhibitors. *J. Mol. Evol.* 39: 614–619.

Ohwada A, Takahashi H, Uchida K, Nukiwa T, Kira S (1992) Gene analysis of heterozygous protein C deficiency in a patient with pulmonary arterial thromboembolism. *Am. Rev. Respir. Dis.* 145: 1491–1494.

Okafuji T, Maekawa K, Nawa K, Marumoto Y (1992) The cDNA cloning and mRNA expression of rat protein C. *Biochim. Biophys. Acta* **1131**: 329–332.

Okajima K, Koga S, Kaji M, Inoue M, Nakagaki T, Funatsu A, Okabe H, Takatsuki K, Aoki N (1990) Effect of protein C and activated protein C on coagulation and fibrinolysis in normal human subjects. *Thromb. Haemost.* **63**: 48–52.

Okajima K, Abe H, Maeda S, Motomura M, Tsujihata M, Nagataki S, Okabe H, Takatsuki K. (1993) Antithrombin III Nagasaki (Ser116→Pro): a heterozygous variant with defective heparin binding associated with thrombosis. *Blood* 81: 1300–1305.

Okajima K, Abe H, Wagatsuma M, Okabe H, Takatsuki K (1995) Antithrombin III Kumamoto II: a single mutation at Arg393–His increased the affinity of antithrombin III for heparin. *Am. J. Hematol.* **48**: 12–18.

Okamoto Y, Yamazaki T, Katsumi A, Kojima T, Takamatsu J, Nishida M, Saito H (1996) A novel nonsense mutation associated with an exon skipping in a patient with hereditary protein S deficiency type 1. *Thromb. Haemost.* **75**: 877–882.

Oldenberg M, Wijnen JT, Berg EAV, Le Clercq E, Kluft C, Koide T, Meera Khan P (1989) Assignment of histidine-rich glycoprotein to chromosome 3. *Cytogenet. Cell Genet.* **51**: 1055.

Olds RJ, Lane D, Caso R, Tripodi A, Mannucci PM, Thein S-L (1989) Antithrombin III Milano 2: a single base substitution in the thrombin binding domain detected with PCR and direct genomic sequencing. *Nucleic Acids Res.* **17**: 10511.

Olds RJ, Lane DA, Caso R, Girolami A, Thein S-L (1990a) Antithrombin III Padua 2 : a single base substitution in exon 2 detected with PCR and direct genomic sequencing. *Nucleic Acids Res.* **18**: 1926.

Olds RJ, Lane DA, Finazzi G, Barbui T, Thein S-L (1990b) A frameshift mutation leading to type I antithrombin deficiency and thrombosis. *Blood* 76: 2182–2186.

Olds RJ, Lane DA, Ireland H, Leone G, De Stefano V, Wiesel ML, Cazenave J-P, Thein SL (1991a) Novel point mutations leading to type I antithrombin deficiency and thrombosis. *Br. J. Haematol.* 78: 408–413.

Olds RJ, Lane DA, Ireland H, Finazzi G, Barbui T, Abildgaard U, Girolami A, Thein S-L (1991b) A common point mutation producing type 1A antithrombin III deficiency AT129 CGA to TGA (Arg→Stop). *Thromb. Res.* 64: 621–625.

Olds RJ, Lane DA, Caso R, Panico M, Morris HR, Sas G, Dawes J, Thein S-L (1992a) Antithrombin III Budapest: a single amino acid substitution (429Pro to Leu) in a region highly conserved in the serpin family. *Blood* 79: 1206–1212.

Olds RJ, Lane DA, Boisclair M, Sas G, Bock SC, Thein S-L (1992b) Antithrombin Budapest 3. An antithrombin variant with reduced heparin affinity resulting from the substitution L99F. *FEBS Lett.* **300**: 241–246.

Olds RJ, Lane DA, Chowdhury V, DeStefano V, Leone G, Thein SL (1993a) Complete nucleotide sequence of the antithrombin gene: evidence for homologous recombination causing thrombophilia. *Biochemistry* **33**: 4216–4224.

Olds RJ, Lane DA, Beresford CH, Abildgaard U, Hughes PM, Thein SL (1993b) A recurrent deletion in the antithrombin gene, AT106–108 (-6 bp), identified by DNA heteroduplex detection. *Genomics* **16**: 298–299.

Olds RJ, Lane DA, Thein SL (1994a) The molecular genetics of antithrombin deficiency. *Br. J. Haematol.* **87**: 221–226.

Olds RJ, Lane DA, Chowdhury V, Sas G, Pabinger I, Auberger K, Thein SL (1994b) (ATT) trinucleotide repeats in the antithrombin gene and their use in determining the origin of repeated mutations. *Hum. Mutat.* **4**: 31–41.

Olds RJ, Lane DA, Mille B, Chowdhury V, Thein SL (1995) Antithrombin: the major inhibitor of thrombin. *Sem. Thromb. Hemost.* **20**: 353–372.

Olexa SA, Budzynski AZ (1980) Evidence for four different polymerization sites involved in human fibrin formation. *Proc. Natl Acad. Sci. USA* **77**: 1374–1378.

Olivieri A, Friso S, Manzato F, Guella A, Bernardi F, Lunghi B, Girelli D, Azzini M, Brocco G, Russo C, Corrocher R (1995) Resistance to activated protein C in healthy women taking oral contraceptives. *Br. J. Haematol.* **91**: 465–470.

Olsen PH, Esmon NL, Esmon CT, Laue TM (1992) Ca^{2+} dependence of the interactions between protein C, thrombin and the elastase fragment of thrombomodulin. Analysis by ultracentrifugation. *Biochemistry* **31**: 746–754.

Olson ST (1988) Transient kinetics of heparin-catalyzed protease inactivation by antithrombin III. *J. Biol. Chem.* **263**: 1698–1708.

Olson ST, Shore JD (1981) Binding of high affinity heparin to antithrombin III. Characterization of the protein fluorescence enhancement. *J. Biol. Chem.* **256**: 11065–11072.

Olson ST, Shore JD (1982) Demonstration of a two-step reaction mechanism for inhibition of α-thrombin by antithrombin III and identification of the step affected by heparin. *J. Biol. Chem.* **257**: 14891–14895.

Olson ST, Björk I (1991) Predominant contribution of surface approximation to the mechanism of heparin acceleration of the antithrombin–thrombin reaction. *J. Biol. Chem.* **266**: 6353–6364.

Olson ST, Björk I (1992) Regulation of thrombin by antithrombin and heparin cofactor II. In: Berliner LJ (ed.) *Thrombin: Structure and Function.* Plenum Press, New York.

Olson ST, Srinivasan KR, Bjork I, Shore JD (1981) Binding of high affinity heparin to antithrombin III. Stopped flow kinetic studies of the binding interaction. *J. Biol. Chem.* **256**: 11073–11079.

Olson ST, Halvorson HR, Björk I (1991a) Quantitative characterization of the thrombin-heparin interaction. Discrimination between specific and non-specific binding models. *J. Biol. Chem.* **266**: 6342–6352.

Olson ST, Sheffer R, Stephens AW, Hirs CHW (1991b) Molecular basis of the reduced activity of antithrombin Denver with thrombin and factor Xa. Role of the P1' residue. *Thromb. Haemost.* **65**: 670.

Olson ST, Björk I, Sheffer R, Craig PA, Shore JD, Choay J (1992) Role of the antithrombin-binding pentasaccharide in heparin acceleration of antithrombin-proteinase reactions. Resolution of the antithrombin conformational change contribution to heparin rate enhancement. *J. Biol. Chem.* **267**: 12528–12538.

Ono M, Nawa K, Marumoto Y (1994) Antithrombotic effects of recombinant human soluble thrombomodulin in a rat model of vascular shunt thrombosis. *Thromb. Haemost.* **72**: 421–425.

Ortel TL, Devore-Carter D, Quinn-Allen MA, Kane WH (1992) Deletion analysis of recombinant human factor V. *J. Biol. Chem.* **267**: 4189–4198.

Orth K, Willnow T, Herz J, Gething MJ, Sambrook J (1994) Low density lipoprotein receptor-related protein is necessary for the internalization of both tissue-type plasminogen activator-inhibitor complexes and free tissue-type plasminogen activator. *J. Biol. Chem.* **269**: 21117–21122.

Orthner CL, Madurawe RD, Velander WH, Drohan WN, Battey FD, Strickland DK (1989) Conformational changes in an epitope localized to the NH_2-terminal region of protein C. *J. Biol. Chem.* **264**: 18781–18788.

Owen MC, Borg JY, Soria C, Soria J, Cain J, Carrell RW (1987) Heparin binding defect in a new antithrombin III variant: Rouen, 47 Arg→His. *Blood* **65**: 1275–1279.

Owen MC, Beresford CH, Carrell RW (1988) Antithrombin Glasgow, 393 Arg to His: a P1 reactive site variant with increased heparin affinity but no thrombin inhibitory activity. *FEBS Lett.* **231**: 317–320.

Owen MC, Shaw GJ, Grau E, Fontcuberta J, Carrell RW, Boswell DR (1989) Molecular characterization of antithrombin Barcelona 2: 47 Arginine to Cysteine. *Thromb. Res.* **55**: 451–457.

Owen MC, George PM, Lane DA, Boswell DR (1991) P1 variant antithrombins Glasgow (393 Arg to His) and Pescara (393 Arg to Pro) have increased heparin affinity and are resistant to catalytic cleavage by elastase. *FEBS Lett.* **280**: 216–220.

Pabinger I, Schneider B (1996) Thrombotic risk in hereditary antithrombin III, protein C or protein S deficiency. A Cooperative, retrospective study. *Arterioscler. Thromb. Vasc. Biol.* **16**: 742–748.

Pabinger I, Allaart CF, Hermans J, Briët E, Bertina RM (1992a) Hereditary protein C deficiency: laboratory values in transmitters and guidelines for the diagnostic procedure report on a study of the SSC subcommittee on protein C and protein S. *Thromb. Haemost.* **68**: 470–474.

Pabinger I, Brücker S, Kyrle PA, Schneider B, Korninger HC, Niessner H, Lechner K (1992b) Hereditary deficiency of antithrombin III, protein C and protein S: prevalence in patients with a history of venous thrombosis and criteria for rational patient screening. *Blood Coag. Fibrinol.* **3**: 547–553.

Pabinger I, Schneider B, Scharrer I, Hach-Wunderle V, Lechner K, Eichinger S, Kyrle PA, Vinazzer H, Lämmle B, Demarmels-Biasiutti F, Tilsner V, Marx G, Seifried E, Gabelman A, Aspöck, Fischer M, Halbmayer WM (1994a) Thrombotic risk of women with hereditary antithrombin III-, protein C- and protein S-deficiency taking oral contraceptive medication. *Thromb. Haemost.* **71**: 548–552.

Pabinger I, Kyrle PA, Heistinger M, Eichinger S, Wittmannn E, Lechner K (1994b) The risk of thromboembolism in asymptomatic patients with protein C and protein S deficiency: a prospective cohort study. *Thromb. Haemost.* **71**: 441–445.

Padro-Manuel F, Rey-Campos J, Hillarp A, Dahlbäck B, Rodriguez de Cordoba S (1990) Human genes for the α- and β-chains of complement C4b-binding protein are closely linked in a head-to-tail arrangement. *Proc. Natl Acad. Sci. USA* **87**: 4529–4532.

Pakula AA, Sauer RT (1989) Genetic analysis of protein stability and function. *Annu. Rev. Genet.* **23**: 289–310.

Pan Y, Doolittle RF (1992) cDNA sequence of a second fibrinogen a chain in lamprey: an archetypal version alignable with full-length β and γ chains. *Proc. Natl Acad. Sci. USA* **89**: 2066–2070.

Pandya BV, Gabriel JL, O'Brien J, Budzynski AZ (1991) Polymerization site in the β chain of fibrin: mapping of the Bβ 1–55 sequence. *Biochemistry* **30**: 162–168.

Pannekoek H, Veerman H, Lambers H, Diergaarde P, Verweij CL, van Zonneveld A-J, van Mourik JA (1986) Endothelial plasminogen activator inhibitor (PAI): a new member of the serpin gene family. *EMBO J.* **5**: 2539–2544.

Paramo JA, Colucci M, Collen D (1985a) Plasminogen activator inhibitor in the blood of patients with coronary heart disease. *Br. Med. J.* **291**: 573–574.

Paramo JA, Alfaro MJ, Rocha E (1985b) Postoperative changes in the plasmatic levels of tissue-type plasminogen activator and its fast-acting inhibitor-relationship to deep vein thrombosis and influence of prophylaxis. *Thromb. Haemost.* **54**: 713–716.

Park S, Jeong HYP, Kim M, Kim YS, Yoo OJ (1993) Antithrombin III Seoul: a new variant of hereditary dysfunctional antithrombin III with replacement of Phe 368 by Ser. *Thromb. Haemost.* **69**: 1258.

Parkinson JF, Grinnell BW, Moore RE, Hoskins J, Vlahos CJ, Bang NU (1990) Stable expression of a secretable deletion mutant of recombinant human soluble thrombomodulin. *J. Biol. Chem.* **265**: 12602–12606.

Parkinson JF, Vrahos CJ, Yan SC, Bang NU (1992a) Recombinant human thrombomodulin. Regulation of cofactor activity and anticoagulant function by a glycosaminoglycan side chain. *Biochem. J.* **283**: 151–156.

Parkinson JF, Nagashima M, Kuhn I, Leonard J, Morser J (1992b) Structure–function studies of the epidermal growth factor domains of human thrombomodulin. *Biochem. Biophys. Res. Commun.* **185**: 567–571.

Parkinson JF, Bang NU, Garcia JGN (1993) Recombinant human thrombomodulin attenuates human endothelial cell activation by human thrombin. *Arterioscl. Thromb.* **13**: 1119–1123.

Patracchini P, Aiello V, Palazzi P, Calzolari E, Bernardi F (1989) Sublocalization of the human protein C gene on chromosome 2q13–q14. *Hum. Genet.* **81**: 191–192.

Patrassi GM, Sartori MT, Saggiorato G, Boeri G, Girolami A (1992) Familial thrombophilia associated with high levels of plasminogen activator inhibitor. *Fibrinolysis* **6**: 99–103.

Patthy L (1985) Evolution of the proteases of blood coagulation and fibrinolysis by assembly from modules. *Cell* **41**: 657–663.

Patthy L (1988) Detecting distant homologies of mosaic proteins: analysis of the sequences of thrombomodulin, thrombospondin, complement components C9, C8a and C8b, vitronectin and plasma cell membrane glycoprotein PC-1. *J. Mol. Biol.* **202**: 689–696.

Patthy L (1990) Evolutionary assembly of blood coagulation proteins. *Sem. Thromb. Hemostas.* **16**: 245–259.

Pattinson JK, Millar DS, Grundy CB, Wieland K, Mibashan RS, Martinowitz U, McVey J, Tan-Un K, Vidaud M, Goossens M, Sampietro M, Krawczak M, Reiss J, Zoll B, Whitmore D, Bradshaw A, Wensley R, Ajani A, Mitchell V, Rizza C, Maia R, Winter P, Mayne EE, Schwartz M, Green PJ, Kakkar VV, Tuddenham EGD, Cooper DN (1990) The molecular genetic analysis of hemophilia A; a directed-search strategy for the detection of point mutations in the human factor VIII gene. *Blood* **76**: 2242–2248.

Pearson JD (1993) The control of production and release of haemostatic factors in the endothelial cell. *Baillière's Clin. Haematol.* **6**: 629–651.

Pegelow CH, Ledford M, Young J, Zilleruelo G (1992) Severe protein S deficiency in a newborn. *Pediatrics* **89**: 674–677.

Pennica D, Holmes WE, Kohr WJ, Harkins RN, Vehar GA, Ward CA, Bennett WF, Yelverton E, Seeburg PH, Heyneker HL, Goeddel DV, Collen D (1983) Cloning and expression of human tissue-type plasminogen activator cDNA in *E. coli. Nature* **301**: 214–221.

Perry DJ (1993) Trinucleotide repeat polymorphism in the human antithrombin III (*AT3*) gene. *Hum. Mol. Genet.* **2**: 618.

Perry DJ (1994) Antithrombin and its inherited deficiencies. *Blood Rev.* **8**: 37–55.

Perry DJ (1995) Ectopic transcript analysis in human antithrombin deficiency. *Blood Coag. Fibrinol.* **6**: 531–536.

Perry DJ, Carrell RW (1989) CpG dinucleotides are hotspots for mutation in the antithrombin III gene. *Mol. Biol. Med.* **6**: 239–243.

Perry DJ, Carrell RW (1996) Molecular genetics of human antithrombin deficiency. *Hum. Mutat.* **7**: 7–22.

Perry DJ, Harper PL, Fairham S, Daly M, Carrell RW (1989) Antithrombin Cambridge, 384 Ala to Pro: a new variant identified using the polymerase chain reaction. *FEBS Lett.* **254**: 174–176.

Perry DJ, Daly ME, Borg J-Y, Harper PL, Carrell RW (1991) Identification and significance of 6 novel variants of antithrombin. *Thromb. Haemost.* **65**: 838.

Perry DJ, Marshall C, Borg J-Y, Tait RC, Daly ME, Walker ID, Carrell RW (1995a) Two novel antithrombin variants, Asn187Asp and Asn187Lys, indicate a functional role for asparagine 187. *Blood Coag. Fibrinol.* **6**: 51–54.

Perry DJ, Daly ME, Colvin BT, Brown K, Carrell RW (1995b) Two antithrombin mutations in a compound heterozygote: Met20Thr and Tyr166Cys. *Am. J. Hematol.* **50**: 215–216.

Petäjä J, Rasi V, Vahtera E, Myllylä G (1991) Familial clustering of defective release of t-PA. *Br. J. Haematol.* **79**: 291–295.

Petäjä J, Rasi V, Myllylä G, Vahtera E, Hallman H (1996) Familial hypofibrinolysis and venous thrombosis. *Br. J. Haematol.* **71**: 393–398.

Petersen CB, Blackburn MN (1985) Isolation and characterization of an antithrombin III variant with reduced carbohydrate content and enhanced heparin binding. *J. Biol. Chem.* **260**: 610–615.

Petersen CB, Noyes CM, Pecon JM, Church FC, Blackburn MN (1987) Identification of a lysyl residue in antithrombin which is essential for heparin binding. *J. Biol. Chem.* **262**: 8061–8065.

Petersen EE, Dudeck-Wojciechowska G, Sottrup-Jensen L, Magnussen S (1979) The primary structure of antithrombin III (heparin cofactor): partial homology between α1-antitrypsin and antithrombin III. In: Collen D, Wiman B, Vertstraete M (eds) *The Physiological Inhibitors of Coagulation and Fibrinolysis*, p. 43. Elsevier/North Holland, Amsterdam.

Petersen TE, Martzen MR, Ichinose A, Davie EW (1990) Characterization of the gene for human plasminogen, a key enzyme in the fibrinolytic system. *J. Biol. Chem.* **265**: 6104–6111.

Pewarchuk WJ, Fernandez-Rachubinski F, Rachubinski RA, Blajchman MA (1990) Antithrombin III Sudbury: an Ala384→Pro mutation with abnormal thrombin-binding activity and thrombin diathesis. *Thromb. Res.* **59**: 793–797.

Pfeffer SR, Rothman JE (1987) Biosynthetic protein transport and sorting by the endoplasmic reticulum and Golgi. *Annu. Rev. Biochem.* **56**: 829–852.

Picado-Leonard J, Miller WL (1988) Homologous sequences insteroidogenic enzymes, steroid receptors and a steroid binding protein suggest a consensus steroid-binding sequence. *Mol. Endocrinol.* **2**: 1145–1151.

Pitney WR (1981) *Venous and Arterial Thrombosis. Evaluation, Prevention and Management.* Churchill Livingstone, Edinburgh.

Pittman DD, Marquette KA, Kaufman RJ (1994) Role of the B domain for factor VIII and factor V expression and function. *Blood* **84**: 4214–4225.

Pixley RA, Schapira M, Colman RW (1985) The regulation of human factor XIIa by plasma proteinase inhibitors. *J. Biol. Chem.* **260**: 1723–1729.

Pizzo SV, Petruska DB, Doman KA, Soong S-J, Fuchs HE (1985) Releasable vascular plasminogen activator and thrombotic strokes. *Am. J. Med.* **79**: 407–411.

Pizzo SV, Fuchs HE, Doman KA, Petruska DB, Berger H (1986) Release of tissue plasminogen activator and its fast-acting inhibitor in defective fibrinolysis. *Arch. Intern. Med.* **146**: 188–191.

Pizzuti A, Pieretti M, Fenwick RG, Gibbs RA, Caskey CT (1992) A transposon-like element in the deletion-prone region of the dystrophin gene. *Genomics* **13**: 594–600.

Plaisancié H, Alexandre Y, Uzan G, Besmond C, Benarous R, Frain M, Trepat JS, Dreyfus J-C, Kahn A (1984) Immunological screening of standard cDNA libraries in pBR322 vectors: detection of human fibrinogen and prothrombin cDNA clones. *Analyt. Biochem.* **142**: 271–276.

Pletcher CH, Nelsestuen GL (1983) Two substrate reaction model for the heparin-catalyzed bovine antithrombin/protease reaction. *J. Biol. Chem.* **258**: 1086–1091.

Ploos van Amstel JK, van der Zanden AL, Bakker E, Reitsma PH, Bertina RM (1987a) Independent isolation of human protein S cDNA and the assignment of the gene to chromosome 3. *Thromb. Haemost.* **58**: 497.

Ploos van Amstel JK, van der Zanden AL, Bakker E, Reitsma PH, Bertina RM (1987b) Two genes homologous with protein S cDNA are located on chromosome 3. *Thromb. Haemost.* **58**: 982–987.

Ploos van Amstel HK, van der Zanden L, Reitsma PH, Bertina RM (1987c) Human protein S cDNA encodes Phe-16 and Tyr 222 in consensus sequences for the post-translational processing. *FEBS Lett.* **222**: 186–190.

Ploos van Amstel HK, Reitsma PH, Bertina RM (1988) The human protein S locus: identification of the PSα gene as a site of liver protein S messenger RNA synthesis. *Biochem. Biophys. Res. Commun.* **157**: 1033–1038.

Ploos van Amstel HK, Huisman MV, Reitsma PH, ten Cate JW, Bertina RM (1989a) Partial protein S gene deletion in a family with hereditary thrombophilia. *Blood* **73**: 479–483.

Ploos van Amstel HK, Reitsma PH, Hamulyk K, Die-Smulders CEM de, Mannucci PM, Bertina RM (1989b) A mutation in the protein S pseudogene is linked to protein S deficiency in a thrombophilic family. *Thromb. Haemost.* **62**: 897–901.

Ploos van Amstel HK, Reitsma PH, van der Logt CPE, Bertina RM (1990) Intron-exon organization of the active human protein S gene PSα and its pseudogene PSβ: duplication and silencing during primate evolution. *Biochemistry* **29**: 7853–7861.

Plow EF, Srouji AH, Meyer D, Marguerie G, Ginsberg MH (1984) Evidence that three adhesive proteins interact with a common recognition site on activated platelets. *J. Biol. Chem.* **259**: 5388–5391.

Plutzky J, Hoskins JA, Long GL, Crabtree GR (1986) Evolution and organization of the human protein C gene. *Proc. Natl Acad. Sci. USA* **83**: 546–550.

Ponting CP, Marshall JM, Cederholm-Williams SA (1992) Plasminogen: a structural review. *Blood Coag. Fibrinol.* **3**: 605–614.

Poort SR, Pabinger-Fasching I, Mannhalter C, Reitsma PH, Bertina RM (1993) Twelve novel and two recurrent mutations in 14 Austrian families with hereditary protein C deficiency. *Blood Coag. Fibrinol.* **4**: 273–280.

Poort SR, Rosendaal FR, Reitsma PH, Bertina RM (1996) A common genetic variation in the 3'-untranslated region of the prothrombin gene is associated with elevated plasma prothrombin levels and an increase in venous thrombosis. *Blood* **68**: 3698–3703.

Porter JB (1982) Oral contraceptives and non-fatal vascular disease – recent experience. *Obstet. Gynecol.* **59**: 299–302.

Potempa J, Korzus E, Travis J (1994) The serpin superfamily of proteinase inhibitors: structure, function and regulation. *J. Biol. Chem.* **269**: 15957–15960.

Pottinger P, Sigurdsson F, Berliner N (1995) Detection of the factor V Leiden mutation in an unselected black population. *Blood* **86** (Suppl.): 203a.

Powers PA, Smithies O (1986) Short gene conversion in the human fetal globin gene region: a by-product of chromosome pairing during during meiosis? *Genetics* **112**: 343–358.

Prandoni P, Lensing AWA, Buller HR (1992) Deep-vein thrombosis and the incidence of subsequent symptomatic cancer. *New Engl. J. Med.* **327**: 1128–1133.

Prandoni P, Lensing AW, Cogo A, Cuppini S, Villalta S, Carta M, Cattelan AM, Polistena P, Bernardi E, Prins MH (1996) The long-term clinical course of acute venous thrombosis. *Ann. Intern. Med.* **125**: 1–7.

Pratt CW, Church FC (1993) General features of the heparin-binding serpins antithrombin, heparin cofactor II and protein C inhibitor. *Blood Coag. Fibrinol.* **4**: 479–490.

Pratt CW, Whinna HC, Church FC (1992) A comparison of three heparin-binding serine proteinase inhibitors. *J. Biol. Chem.* **267**: 8795–8801.

Preissner KT (1990) Biological relevance of the protein C system and laboratory diagnosis of protein C and protein S deficiencies. *Clin. Sci.* **78**: 351–364.

Preissner KT, Delvos U, Muller-Berghaus G (1987) Binding of thrombin to thrombomodulin accelerates inhibition of the enzyme by antithrombin III. Evidence for a heparin-independent mechanism. *Biochemistry* **26**: 2521–2528.

Preissner KT, Grulich-Henn J, Ehrlich HS, Declerde P, Justus C, Collen D, Pannekoek H, Muller-Berghaus G (1990a) Structural requirement for the extracellular interaction of plasminogen activator/inhibitor 1 with endothelial cell matrix-associated vitronectin. *J. Biol. Chem.* **265**: 18490–18498.

Preissner KT, Koyami T, Muller D, Tschopp J, Muller-Berghaus G (1990b) Domain structure of the endothelial cell receptor thrombomodulin as deduced from modulation of its anticoagulant functions. Evidence for a glycosaminoglycan-dependent secondary binding site for thrombin. *J. Biol. Chem.* **265**: 4915–4922.

Press RD, Liu X-Y, Beamer N, Coull BM (1996) Ischemic stroke in the elderly. Role of the common factor V mutation causing resistance to activated protein C. *Stroke* **27**: 44–48.

Preston FE, Rosendaal FR, Walker ID, Briët E, Berntorp E, Conard J, Fontcuberta J, Makris M, Mariani G, Noteboom W, Pabinger I, Legnani C, Scharrer I, Schulman S, van der Meer FJM (1996) Increased fetal loss in women with heritable thrombophilia. *Lancet* **348**: 913–916.

Prochownik EV (1985) Relationship between an enhancer element in the human antithrombin III gene and an immunoglobulin light-chain gene enhancer. *Nature* **316**: 845–848.

Prochownik EV, Orkin SH (1984) *In vivo* transcription of a human antithrombin III 'minigene'. *J. Biol. Chem.* **259**: 15386–15392.

Prochownik EV, Markham AF, Orkin SH (1983a) Isolation of a cDNA clone for human antithrombin III. *J. Biol. Chem.* **258**: 8389–8394.

Prochownik EV, Antonarakis S, Bauer KA, Rosenberg RD, Fearon ER, Orkin SH (1983b) Molecular heterogeneity of inherited antithrombin III deficiency. *New Engl. J. Med.* **308**: 1549–1552.

Prochownik EV, Bock SC, Orkin SH (1985) Intron structure of the human antithrombin III gene differs from that of other members of the serine protease inhibitor superfamily. *J. Biol. Chem.* **260**: 9608–9612.

Prochownik EV, Smith MJ, Markham A (1987) Two regions downstream of AATAAA in the human antithrombin III gene are important for cleavage-polyadenylation. *J. Biol. Chem.* **262**: 9004–9010.

Prowse CV, Cash JD (1984) Physiologic and pharmacologic enhancement of fibrinolysis. *Sem. Thromb. Haemost.* **10**: 51–60.

Pulak R, Anderson P (1993) mRNA surveillance by the *Caenorhabditis elegans* smg genes. *Genes Dev.* **7**: 1885–1897.

Quattrone A, Colucci M, Donati MB (1979) Cerebral thrombosis in two siblings with dysfibrinogenemia. *Neurosci. Letts.* 3 (Suppl. 1): 54–62.

Que BG, Davie EW (1986) Characterization of a cDNA coding for human factor XII (Hageman factor) *Biochemistry* **25**: 1525–1528.

Quehenberger P, Loner U, Kapiotis S, Handler S, Schneider B, Huber J, Speiser W (1996) Increased levels of activated factor VII and decreased plasma protein S activity and circulating thrombomodulin during use of oral contraceptives. *Thromb. Haemost.* **76**: 729–734.

Rabes JP, Trossaert M, Conard J, Samama M, Giraudet P, Boileau C (1995) Single point mutation of Arg[506] of factor V associated with APC resistance and venous thromboembolism: improved detection by PCR-mediated site-directed mutagenesis. *Thromb. Haemost.* **74**: 1379–1387.

Ragg H (1986) A new member of the plasma protease inhibitor gene family. *Nucleic Acids Res.* **14**: 1073–1088.

Ragg H, Preibisch G (1988) Structure and expression of the gene coding for the human serpin hLS2. *J. Biol. Chem.* **263**: 12129–12134.

Ragg H, Ulshoefer T, Gerewitz J (1990) On the activation of human leuserpin-2, a thrombin inhibitor by glycosaminoglycans. *J. Biol. Chem.* **265**: 5211–5218.

Raghunath PN, Tomaszewski JE, Brady ST, Caron RJ, Okada SS, Barnathan ES (1995) Plasminogen activator system in human coronary atherosclerosis. *Arterioscl. Thromb. Vasc. Biol.* **15**: 1432–1437.

Rai R, Regan L, Hadley E, Dave M, Cohen (1996) Second trimester pregnancy loss is associated with activated protein C resistance. *Br. J. Haematol.* **92**: 489–493.

Rallidis LS, Megalou AA, Papageoergakis NH, Trikas AG, Chatzidimitrou GI, Tsitouris GK (1996) Plasminogen activator inhibitor 1 is elevated in the children of men with premature myocardial infarction. *Thromb. Haemost.* **76**: 417–421.

Rånby M (1982) Studies on the kinetics of plasminogen activation by tissue plasminogen activator. *Biochim. Biophys. Acta* **704**: 461–469.

Rånby M, Norman IB, Wallen P (1982) A sensitive assay for tissue plasminogen activator. *Thromb. Res.* **27**: 743–749.

Rand MD, Hanson SR, Mann KG (1995) Factor V turnover in a primate model. *Blood* **86**: 2616–2623.

Rao LVM, Rapaport SI (1988) Activation of factor VII bound to tissue factor: a key early step in the tissue factor pathway of blood coagulation. *Proc. Natl Acad. Sci. USA* **84**: 6687–6691.

Rao LVM, Rapaport SI, Hoang AD (1993) Binding of factor VIIa to tissue factor permits rapid antithrombin III/heparin inhibition of factor VIIa. *Blood* **81**: 2600–2607.

Rao LVM, Nordfang O, Hoang AD, Pendurthi UR (1995) Mechanism of antithrombin III inhibition of factor VIIa/tissue factor activity on cell surfaces. Comparison with tissue factor pathway inhibitor/factor Xa-induced inhibition of factor VIIa/tissue factor activity. *Blood* **85**: 121–129.

Rapaport SI, Rao LVM (1995) The tissue factor pathway: how it has become a 'prima ballerina'. *Thromb. Haemost.* **74**: 7–17.

Ratnoff OD (1968) Activation of Hageman factor by L-homocystine. *Science* **162**: 1007–1009.

Ratnoff OD, Colopy JE (1955) A familial hemorrhage trait associated with deficiency of a clot-promoting fraction of plasma. *J. Clin. Invest.* **34**: 603–613.

Ratnoff OD, Saito H (1979) Surface-mediated reactions. *Curr. Topics Haematol.* **2**: 1–57.

Ratnoff OD, Buss RJ, Sheon RP (1968) The demise of John Hageman. *New Engl. J. Med.* **279**: 760–761.

Raum D, Marcus D, Alper CA, Levey R, Taylor PD, Starzl TE (1980a) Synthesis of human plasminogen by the liver. *Science* **208**: 1036–1037.

Raum D, Marcus D, Alper CA (1980b) Genetic polymorphism of human plasminogen. *Am. J. Hum. Genet.* **32**: 681–689.

Reber P, Furlan M, Beck EA, Finazzi G, Buelli M, Barbui T (1985) Fibrinogen Bergamo 1 (Aα 16 Arg→Cys): susceptibility towards thrombin following aminoethylation, methylation, or carboxymethylation of cysteine residues. *Thromb. Haemost.* **54**: 390–393.

Reber P, Furlan M, Henschen A, Kaudewitz H, Barbui T, Hilgard P, Nenci G, Berrettini M, Beck EA (1986) Three abnormal fibrinogen variants with the same amino acid substitution (γ275 Arg→His): fibrinogens Bergamo II, Essen and Perugia. *Thromb. Haemost.* **56**: 401–406.

Reddigar SR, Shibayama Y, Brunée T, Kaplan AP (1993) Human Hageman factor (factor XII) and high molecular weight kininogen compete for the same binding site on human umbilical vein endothelial cells. *J. Biol. Chem.* **268**: 11982–11987.

Redlitz A, Plow EF (1995) Receptors for plasminogen and t-PA: an update. *Baillière's Clin. Haematol.* **8**: 313–321.

Redlitz A, Tan AK, Eaton DL, Plow EF (1995) Plasma carboxypeptidases as regulators of the plasminogen system. *J. Clin. Invest.* **96**: 2534–2538.

Rees MM, Rodgers GM (1993) Homocysteinemia: association of a metabolic disorder with vascular disease and thrombosis. *Thromb. Res.* **71**: 337–359.

Rees DC, Cox M, Clegg JB (1995) World distribution of factor V Leiden. *Lancet* **346**: 1133–1134.

Rees DC, Cox M, Clegg JB (1996) Detection of the factor V Leiden mutation using whole blood PCR. *Thromb. Haemost.* **75**: 520–521.

Regan LM, Lamphear BJ, Walker FJ, Fay PJ (1994) Factor IXa protects factor VIIIa from activated protein C: factor IXa inhibits activated protein C-catalyzed cleavage of factor VIIIa at Arg562. *J. Biol. Chem.* **269**: 9445–9452.

Reich NE, Hoffman GC, de Wolfe VG (1976) Recurrent thrombophlebitis and pulmonary emboli in congenital factor V deficiency. *Chest* **69**: 113–114.

Reiner AP, Davie EW (1995) Introduction to hemostasis and the vitamin K-dependent coagulation factors. In: Scriver CR, Beaudet AL, Sly WS, Valle D (eds) *The Metabolic and Molecular Bases of Inherited Disease*, pp. 3181–3221. McGraw-Hill, New York.

Reitsma PH (1996) Protein C deficiency: summary of the 1995 database update. *Nucleic Acids Res.* **24**: 157–159.

Reitsma PH, Lintel Hekkert W, Koenhen E, van der Velden PA, Allaart CF, Deutz-Terlouw PP, Poort SR, Bertina RM (1990) Application of two neutral *Msp*I DNA polymorphisms in the analysis of hereditary protein C deficiency. *Thromb. Haemost.* **64**: 239–244.

Reitsma PH, Poort SR, Allaart CF, Briët E, Bertina RM (1991) The spectrum of genetic defects in a panel of 40 Dutch families with symptomatic protein C deficiency type I: heterogeneity and founder effects. *Blood* **78**: 890–894.

Reitsma PH, Poort SR, Bernardi F, Gandrille S, Long GL, Sala N, Cooper DN (1993) Protein C deficiency: a database of mutations. *Thromb. Haemost.* **69**: 77–84.

Reitsma PH, Ploos van Amstel HK, Bertina RM (1994) Three novel mutations in five unrelated subjects with hereditary protein S deficiency type I. *J. Clin. Invest.* **93**: 486–492.

Reitsma PH, Bernardi F, Doig RG, Gandrille S, Greengard S, Ireland H, Krawczak M, Lind B, Long GL, Poort SR, Saito H, Sala N, Witt I, Cooper DN (1995) Protein C deficiency: a database of mutations, 1995 update. *Thromb. Haemost.* **73**: 876–889.

Reitsma PH, van der Velden PA, Vogels E, van Strijp D, Tacken N, Adriaansen H, van Gemen B (1996) Use of the direct RNA amplification technique NASBA to detect factor V Leiden, a point mutation associated with APC resistance. *Blood Coag. Fibrinol.* **7**: 659–663.

Rejante M, Llinas M (1994) ^{1}H-NMR assignments and secondary structure of human plasminogen kringle 1. *Eur. J. Biochem.* **221**: 927–937.

Repke K, Gemmell CH, Guha A, Turitto VT, Broze GJ, Nemerson Y (1990) Hemophilia as a defect of the tissue factor pathway of blood coagulation. effect of factors VIII and IX on factor X activation in a continuous flow reactor. *Proc. Natl Acad. Sci. USA* **87**: 7623–7627.

Rezaie AR (1996) Role of residue 99 at the S2 subsite of factor Xa and activated protein C in enzyme specificity. *J. Biol. Chem.* **271**: 23807–23814.

Rezaie AR, Esmon CT (1992) The function of calcium in protein C activation by thrombin and the thrombin–thrombomodulin complex can be distinguished by mutational analysis of protein C derivatives. *J. Biol. Chem.* **267**: 26104–26109.

Rezaie AR, Esmon CT (1994a) Calcium inhibition of the activation of protein C by thrombin. Role of the P3 and P3′ residues. *Eur. J. Biochem.* **223**: 575–579.

Rezaie AR, Esmon CT (1994b) Proline at the P2 position in protein C is important for calcium-mediated regulation of protein C activation and secretion. *Blood* **83**: 2526–2531.

Rezaie AR, Esmon NL, Esmon CT (1992) The high affinity calcium-binding site involved in protein C activation is outside the first epidermal growth factor homology domain. *J. Biol. Chem.* **267**: 11701–11704.

Rezaie AR, Mather T, Sussman F, Esmon CT (1994) Mutation of Glu80→Lys results in a protein C mutant that no longer requires Ca^{2+} for rapid activation by the thrombin–thrombomodulin complex. *J. Biol. Chem.* **269**: 3151–3154.

Rezaie AR, Cooper ST, Church FC, Esmon CT (1995) Protein C inhibitor is a potent inhibitor of the thrombin–thrombomodulin complex. *J. Biol. Chem.* **270**: 25336–25339.

Richards RI, Holman K, Friend K, Kremer E, Hillen D, Staples A, Brown WT, Goonewardena P, Tarleton J, Schwartz C, Sutherland GR (1992) Evidence of founder chromosomes in fragile X syndrome. *Nature Genetics* **1**: 257–260.

Richards RI, Sutherland GR (1994) Simple repeat DNA is not replicated simply. *Nature Genetics* **6**: 114–116.

Richardson MA, Gerlitz B, Grinnell BW (1992) Enhancing protein C interaction with thrombin results in a clot-activated anticoagulant. *Nature* **360**: 261–264.

Rideout WM, Coetzee GA, Olumi AF, Jones PA (1990) 5-Methylcytosine as an endogenous mutagen in the human LDL receptor and p53 genes. *Science* **249**: 1288–1290.

Ridker PM, Vaughan PE, Stampfer MJ, Manson JE, Hennehens CH (1993) Endogenous tissue-type plasminogen activator and risk of myocardial infarction. *Lancet* **341**: 1165–1168.

Ridker PM, Hennekens CH, Lindpainter K, Stampfer MJ, Eisenberg PR, Miletich JP (1995a) Mutation in the gene coding for coagulation factor V and the risk of myocardial infarction, stroke, and venous thrombosis in apparently healthy men. *New Engl. J. Med.* **332**: 912–917.

Ridker PM, Miletich JP, Stampfer MJ, Goldhaber SZ, Lindpainter K, Hennekens CH (1995b) Factor V Leiden and risks of recurrent idiopathic venous thromboembolism. *Circulation* **92**: 2800–2802.

Riess H, Bisack T, Hiller E (1985) Protein C antigen in prothrombin complex concentrates: content, recovery and half life. *Blut* **50**: 303–307.

Rijken DC, Hoylaerts M, Collen D (1982) Fibrinolytic properties of one-chain and two-chain human extrinsic (tissue-type) plasminogen activator. *J. Biol. Chem.* **257**: 2920–2925.

Rintelen C, Pabinger I, Knöbl P, Lechner K, Mannhalter C (1996a) Probability of recurrence of thrombosis in patients with and without factor V Leiden. *Thromb. Haemost.* **75**: 229–232.

Rintelen C, Mannhalter C, Ireland H, Lane DA, Knöbl P, Lechner K, Pabinger I (1996b) Oral contraceptives enhance the risk of clinical manifestation of venous thrombosis at a young age in females homozygous for factor V Leiden. *Br. J. Haematol.* **93**: 487–490.

Ritchie DG, Levy BA, Adams MA, Fuller GM (1982) Regulation of fibrinogen synthesis by plasmin-derived fragments of fibrinogen and fibrin: an indirect feedback pathway. *Proc. Natl Acad. Sci. USA* **79**: 1530–1534.

Rixon MW, Chan W-Y, Davie EW, Chung DW (1983) Characterization of a complementary deoxyribonucleic acid coding for the α-chain of human fibrinogen. *Biochemistry* **22**: 3237–3244.

Rixon MW, Chung DW, Davie EW (1985) Nucleotide sequence of the gene for the γ-chain of human fibrinogen. *Biochemistry* **24**: 2077–2086.

Robbie LA, Booth NA, Croll AM, Bennett B (1993) The roles of α2-antiplasmin and plasminogen activator inhibitor 1 (PAI-1) in the inhibition of clot lysis. *Thromb. Haemost.* **70**: 301–304.

Robbins KC (1987) The plasminogen-plasmin enzyme system. In: Colman RW, Hirsh J, Marder VJ, Salzman EW (eds) *Hemostasis and Thrombosis*, pp. 340–357. J.B. Lippincott, Philadelphia.

Robbins KC (1988) Dysplasminogenemias. *Enzyme* **40**: 70–78.

Robbins KC (1990) Classification of abnormal plasminogens: dysplasminogenemias. *Sem. Thromb. Hemost.* **16**: 217–220.

Robbins KC, Boreisha IG, Hach-Wunderle V (1991a) Congenital plasminogen deficiency with an abnormal plasminogen: Frankfurt II, dysplasminogenemia-hypoplasminogenemia. *Fibrinolysis* 5: 145–153.

Robbins KC, Boreisha I, Godwin JE (1991b) Abnormal plasminogen Maywood I. *Thromb. Haemost.* **66**: 575–580.

Robbins KC (1992) Dysplasminogenemias. *Prog. Cardiovasc. Dis.* **34**: 295–308.

Roberts LR, Nichols LA, Holland LJ (1995) cDNA and amino-acid sequences and organization of the gene encoding the Bβ subunit of fibrinogen from *Xenopus laevis*. *Gene* **160**: 223–228.

Rocchi M, Roncuzzi L, Santamaria R, Archidiacono N, Dente L, Romeo G (1986) Mapping through somatic cell hybrids and cDNA probes of protein C to chromosome 2, factor X to chromosome 13 and α1-acid glycoprotein to chromosome 9. *Hum. Genet.* **74**: 30–33.

Rodeghiero F, Tosetto A (1993) Some remarks on the epidemiology of thrombotic disorders. *Thromb. Haemost.* **69**: 527–528.

Rodeghiero F, Tosetto A (1996) The VITA project: population–based distributions of protein C, antithrombin III, heparin-cofactor II and plasminogen – relationship with physiological variables and establishment of reference ranges. *Thromb. Haemost.* **76**: 226–233.

Rodeghiero F, Castaman G, Ruggeri M, Tosetto A (1992) Thrombosis in subjects with homozygous and heterozygous factor XII deficiency. *Thromb. Haemost.* **67**: 590.

Rodgers GM (1988) Hemostatic properties of normal and perturbed vascular cells. *FASEB J.* **2**: 116–123.

Rodgers GM, Conn MT (1990) Homocysteine, an atherogenic stimulus, reduces protein C activation by arterial and venous endothelial cells. *Blood* **75**: 895–901.

Rodriguez de Cordoba S, Sanchez-Corral P, Rey-Campos J (1991) Structure of the gene coding for the α polypeptide chain of the human complement component C4b-binding protein. *J. Exp. Med.* **173**: 1073–1082.

Roelse JC, Koopman MMW, Büller HR, ten Cate JW, Montaruli B, van Mourik JA, Voorberg J (1996) Absence of mutations at the activated protein C cleavage sites of factor VIII in 125 patients with venous thrombosis. *Br. J. Haematol.* **92**: 740–743.

Rogers J (1985) Exon shuffling and intron insertion in serine protease genes. *Nature* **315**: 458–459.

Romeo G, Hassan HJ, Staempfli S, Roncuzzi L, Cianetti L, Leonardi A, Vicente V, Mannucci PM, Bertina R, Peschle C, Cortese R (1987) Hereditary thrombophilia: identification of nonsense and missense mutations in the protein C gene. *Proc. Natl Acad. Sci. USA* **84**: 2829–2932.

Rosenberg RD (1975) Actions and interactions of antithrombin and heparin. *New Engl. J. Med.* **292**: 146–151.

Rosenberg RD (1987) Regulation of the hemostatic mechanism. In: Stamatoyannopoulos G, Nienhuis AW, Leder P, Majerus PW (eds). *The Molecular Basis of Blood Diseases*, pp. 534–574. WB Saunders Co., Philadelphia.

Rosenberg RD (1995) The absence of the blood clotting regulator thrombomodulin causes embryonic lethality in mice before development of a functional cardiovascular system. *Thromb. Haemost.* **74**: 52–57.

Rosenberg RD, Damus PS (1973) The purification and mechanism of action of human antithrombin-heparin co-factor. *J. Biol. Chem.* **248**: 6490–6505.

Rosendaal FR, Heijboer H (1991) Mortality related to thrombosis in congenital antithrombin III deficiency. *Lancet* **337**: 1545.

Rosendaal FR, Varekamp I, Smit C, Bröcker-Vriends AH, van Dijck H, Vandenbroucke JP, Hermans J, Suurmeijer TP, Briët E (1989) Mortality and causes of death in Dutch haemophiliacs 1973–1986. *Br. J. Haematol.* **71**: 71–76.

Rosendaal FR, Heijboer H, Briët E, Büller HR, Brandjes DPM, De Bruin K, Hommes DW, Vandenbroucke JP (1991) Mortality in hereditary antithrombin III deficiency – 1830–1989. *Lancet* **337**: 260–262.

Rosendaal FR, Bertina RM, Reitsma PH (1994) Evaluation of activated protein C resistance in stored plasma. *Lancet* **343**: 1289–1290.

Rosendaal FR, Koster T, Vandenbroucke JP, Reitsma PH (1995) High risk of thrombosis in patients homozygous for factor V Leiden (activated protein C resistance). *Blood* **85**: 1504–1508.

Rosing J, Tans G, Govers-Riemslaj JWP, Zwaal IRFA, Hemker HC (1980) The role of phospholipids and factor Va in the prothrombinase complex. *J. Biol. Chem.* **255**: 274–283.

Ross R (1993) The pathogenesis of atherosclerosis: a perspective for the 1990s. *Nature* **362**: 801–809.

Rousseau F, Heitz D, Mandel JL (1992) The unstable and methylatable mutations causing the fragile X syndrome. *Hum. Mutat.* **1**: 91–96.

Roussel B, Dieval J, Delobel J, Fernandez-Rachubinski F, Eng B, Rachubinski RA, Blajchman MA (1991) Antithrombin III Amiens: a new family with an Arg 47→Cys inherited variant of antithrombin III with impaired heparin cofactor activity. *Am. J. Hematol.* **36**: 25–29.

Roy SN, Mukhopadhyay G, Redman CM (1990) Regulation of fibrinogen assembly. *J. Biol. Chem.* **265**: 6389–6393.

Roy SN, Procyk R, Kudryk BJ, Redman CM (1991) Assembly and secretion of recombinant human fibrinogen. *J. Biol. Chem.* **266**: 4758–4763.

Roy S, Overton O, Redman C (1994) Overexpression of any fibrinogen chain by HepG2 cells specifically elevates the expression of the other two chains. *J. Biol. Chem.* **269**: 691–695.

Roy SN, Kudryk B, Redman CM (1995) Secretion of biologically active recombinant fibrinogen by yeast. *J. Biol. Chem.* **270**: 23761–23767.

Royal College of General Practitioner's Oral Contraception Study (1978) Oral contraceptives, venous thrombosis, and varicose veins. *J. R. Coll. Gen. Pract.* **28**: 393–399.

Royle NJ, Irwin DM, Koschinsky ML, MacGillivray RTA, Hamerton JL (1987) Human genes encoding prothrombin and ceruloplasmin map to 11p11–q12 and 3q21–24 respectively. *Somat. Cell. Mol. Genet.* **13**: 285: 292.

Royle NJ, Nigli M, Cool D, MacGillivray RT, Hamerton JL (1988) Structural gene encoding human factor XII is located at 5q33–qter. *Somat. Cell Mol. Genet.* **14**: 217–221.

Rubinstein I, Murray D, Hoffsten V (1988) Fatal pulmonary emboli in hospitalized patients. *Arch. Int. Med.* **145**: 1425–1426.

Ryan MP, Kutz SM, Higgins PJ (1996) Complex regulation of plasminogen activator inhibitor type-1 (PAI-1) gene expression by serum and substrate adhesion. *Biochem. J.* **314**: 1041–1046.

Sadler LA, Blanton SH, Daiger SP (1991) Dinucleotide repeat polymorphism at the human tissue plasminogen activator gene (*PLAT*). *Nucleic Acids Res.* **19**: 6058.

Sadler JE, Lentz SR, Sheehan JP, Tsiang M, Wu Q (1993) Structure–function relationships of the thrombin–thrombomodulin interaction. *Haemostasis* **23**: 183–193.

Saito H (1981) Hageman factor (factor XII) in health and disease. *Acta Haemal. Jap.* **44**: 216–221.

Saito H (1987) Contact factors in health and disease. *Sem. Thromb. Hemostas.* **13**: 36–49.

Saito H, Scialla SJ (1981) Isolation and properties of an abnormal Hageman factor (factor XII) molecule in a cross-reacting material positive Hageman trait plasma. *J. Clin. Invest.* **68**: 1028–1035.

Saito H, Scott JG, Movat HZ, Sciallo SJ (1979) Molecular heterogeneity of Hageman trait (factor XII deficiency). Evidence that two of 49 subjects are cross-reacting material positive (CRM+). *J. Lab. Clin. Med.* **94**: 256–265.

Saito H, Hamilton SM, Tavill AS, Louis L, Ratnoff OD (1980) Production and release of plasminogen by isolated perfused rat liver. *Proc. Natl Acad. Sci. USA* **77**: 6837–6840.

Sakai T, Lund-Hansen T, Thim L, Kisiel W (1990) The γ-carboxyglutamic acid domain of human factor VIIa is essential for its interaction with cell surface tissue factor. *J. Biol. Chem.* **265**: 1890–1894.

Sakamoto T, Ogawa H, Yasue H, Oda Y, Kitajima S, Tsumoto K, Mizokami H (1994) Prevention of arterial reocclusion after thrombolysis with activated protein C. *Circulation* **90**: 427–432.

Sakata Y, Aoki N (1980) Molecular abnormality of plasminogen. *J. Biol. Chem.* **255**: 5442–5447.

Sakata Y, Loskutoff DJ, Gladson CL, Hekman CM, Griffin JH (1986) Mechanism of protein C-dependent clot lysis: role of plasminogen activator inhibitor. *Blood* **68**: 1218–1223.

Sakata Y, Griffin JH, Loskutott DJ (1988) Effect of activated protein C on the fibrinolytic components released by cultured bovine endothelial cells. *Fibrinolysis* **2**: 7–9.

Sakuragawa N, Takahashi K, Kondo S, Koide T (1983) Antithrombin III Toyama: a hereditary abnormal antithrombin III of a patient with recurrent thrombophlebitis. *Thromb. Res.* **31**: 305–317.

Sala N, Poort SR, Bertina RM, Soria JM, Fontcuberta J, Reitsma PH (1991) Identification of two deletions and four point mutations in the protein C gene of 6 unrelated Spanish patients with hereditary protein C deficiency. *Thromb. Haemost.* **65**: 1197.

Salem HH, Broze GJ, Miletich JP, Majerus PW (1983a) Human coagulation factor Va is a cofactor for the activation of protein C. *Proc. Natl Acad. Sci. USA* **80**: 1584–1588.

Salem HH, Broze GJ, Miletich JP, Majerus PW (1983b) The light chain of factor Va contains the activity of factor Va that accelerates protein C activation by thrombin. *J. Biol. Chem.* **258**: 8531–8534.

Salem HH, Esmon NL, Esmon CT, Majerus PW (1984) Effects of thrombomodulin and coagulation factor Va light chain on protein C activation *in vitro*. *J. Clin. Invest.* **73**: 968–972.

Saleun S, De Moerloose P, Bura A, Aiach M, Emmerich J (1996) A novel nonsense mutation in the antithrombin III gene (Cys-4→Stop) causing recurrent venous thrombosis. *Blood Coag. Fibrinol.* **7**: 578–579.

Samama MM, Simmoneau G, Wainstein JP (1993) Sirius Study: epidemiology of risk factors for deep vein thrombosis (DVT) of the lower limbs in community practice. *Thromb. Haemost.* **69**: 797A.

Samama MM, Trossaërt M, Horellou MH, Elalamy I, Conard J, Deschamps A (1995) Risk of thrombosis in patients homozygous for factor V Leiden. *Blood* **85**: 4700–4702.

Samani NJ, Lodwick D, Martin D, Kimber P (1994) Resistance to activated protein C and risk of premature myocardial infarction. *Lancet* **344**: 1709–1710.

Samsioe G (1994) Coagulation and anticoagulation effects of contraceptive steroids. *Am. J. Obstet. Gynecol.* **170**: 1523–1527.

Sandler MA, Zhang J-N, Westerhausen DR, Billadello JJ (1994) A novel protein interacts with the major transforming growth factor-β responsive element in the plasminogen activator inhibitor type-1 gene. *J. Biol. Chem.* **269**: 21500–21504.

Sanson B-J, Friederich PW, Simioni P, Zanardi S, Hulsman MV, Girolami A, Cate J-W ten, Prins MH (1996) The risk of abortion and stillbirth in antithrombin-, protein C-, and protein S-deficient women. *Thromb. Haemost.* **75**: 387–388.

Sarkar G, Koeberl DD, Sommer SS (1990) Direct sequencing of the activation peptide and the catalytic domain of the factor IX gene in six species. *Genomics* **6**: 133–143.

Sartori MT, Patrassi GM, Theodoridis P, Perin A, Pietrogrande F, Girolami A (1994) Heterozygous type I plasminogen deficiency is associated with an increased risk for thrombosis: a statistical analysis in 20 kindreds. *Blood Coag. Fibrinol.* **5**: 889–893.

Sas G, Blaskó G, Petrö I, Griffin JH (1985) A protein S deficient family with portal vein thrombosis. *Thromb. Haemost.* **54**: 724.

Sawa H, Sobel BE, Fujii S (1994) Inhibition of type-1 plasminogen activator inhibitor production by antisense oligonucleotides in human vascular endothelial and smooth muscle cells. *J. Biol. Chem.* **269**: 14149–14152.

Sawada K, Yamamoto H, Matsumoto K, Yago H, Suehiro S, Tahara C, Ishii H, Kazama M, Abe T (1992) Changes in thrombomodulin levels in plasma of endotoxin-infused rabbits. *Thromb. Res.* **65**: 199–204.

Scarabin PY, Plu-Bureau G, Zitoun D, Bara L, Guize L, Samama MM (1995) Changes in haemostatic variables induced by oral contraceptives containing 50μg or 30μg oestrogen: absence of dose-dependent effect of PAI-1 activity. *Thromb. Haemost.* **74**: 928–932.

Schaaper RM, Danforth BN, Glickman BW (1986) Mechanisms of spontaneous mutagenesis: an analysis of the spectrum of spontaneous mutation in the *Escherichia coli lacI* gene. *J. Mol. Biol.* **189**: 273–284.

Schafer AI (1994) Hypercoagulable states: molecular genetics to clinical practice. *Lancet* **344**: 1739–1742.

Schafer AI, Kroll MH (1993) Nonatheromatous arterial thrombosis. *Annu. Rev. Med.* **44**: 155–170.

Scharrer IM, Wohl RC, Hach V, Sinio L, Boreisha I, Robbins KC (1986) Investigation of a congenital abnormal plasminogen, Frankfurt I, and its relationship to thrombosis. *Thromb. Haemost.* **55**: 396–401.

Scharrer I, Hach-Wunderle V, Wohl RC, Sinio L, Boreisha I, Robbins KC (1988a) Congenital abnormal plasminogen, Frankfurt I, a cause for recurrent venous thrombosis. *Haemostasis* **18** (Suppl. 1): 77–86.

Scharrer I, Hach-Wunderle V, Aygoren E, Lottenberg R, Robbins KC (1988b) Frequency and clinical manifestations of congenital plasminogen and fibrinogen deficiency compared to ATIII, PC and PS deficiency in patients suffering from thromboses. *Fibrinolysis* **2** (Suppl. 2): 16–17.

Scheraga HA (1986) Chemical basis of thrombin interaction with fibrinogen. *Ann. NY Acad. Sci.* **485**: 124–133.

Schmeidler-Sapiro KT, Ratnoff OD, Gordon EM (1991) Mitogenic effects of coagulation factor XII and factor XIIa on HepG2 cells. *Proc. Natl Acad. Sci. USA* **88**: 4382–4385.

Schmidel DK, Tatro AV, Phelps LG, Tomczak JA, Long GL (1990) Organization of the human protein S genes. *Biochemistry* **29**: 7845–7852.

Schmidel DK, Nelson RM, Broxson EH, Comp PC, Marlar RA, Long GL (1991) A 5.3kb deletion including exon XIII of the protein Sα gene occurs in two protein S-deficient families. *Blood* **77**: 551–559.

Schneider DJ, Sobel BE (1991) Augmentation of synthesis of plasminogen activator inhibitor type 1 by insulin and insulin-like growth factor type I: implications for vascular disease in hyperinsulinemic states. *Proc. Natl Acad. Sci. USA* **88**: 9959–9963.

Schneiderman J, Sawdey MS, Keeton MR, Bordin GM, Bernstein EF, Dilley RB, Loskutoff DJ (1992) Increased type I plasminogen activator inhibitor gene expression in atherosclerotic human arteries. *Proc. Natl Acad. Sci. USA* **89**: 6998–7003.

Schreuder HA, de Boer B, Dijkema R, Mulders J, Theunissen HJM, Grootenhuis PDJ, Hol WGJ (1994) The intact and cleaved human antithrombin III complex as a model for serpin-proteinase interactions. *Nature Struct. Biol.* **1**: 48–54.

Schröder W, Koesling M, Wulff K, Wehnert M, Herrmann FH (1996) Large-scale screening for factor V Leiden mutation in a North-Eastern German population. *Haemostasis* **26**: 233–236.

Schulze AJ, Huber R, Bode W, Engh RA (1994) Structural aspects of serpin inhibition. *FEBS Lett.* **344**: 117–124.

Schved JF, Gris JC, Martinez P, Sarlat C, Sanchez N, Arnaud A (1991) Familial thrombophilia with familial elevation of plasma histidine-rich glycoprotein and type I plasminogen activator inhibitor. *Thromb. Haemost.* **65**: 1044.

Schwalbe R, Dahlbäck B, Hillarp A, Nelsestuen G (1990) Assembly of protein S and C4b-binding protein on membranes. *J. Biol. Chem.* **265**: 16074–16081.

Schwarz HP, Fischer M, Hopmeier P, Batard MA, Griffin JH (1984) Plasma protein S deficiency in familial thrombotic disease. *Blood* **64**: 1297–1300.

Schwarz HP, Heeb MJ, Wencel-Drake JD, Griffin JH (1985) Identification and quantitation of protein S in human platelets. *Blood* **66**: 1452–1455.

Schwarz HP, Heeb MJ, Lämmle B, Berrettini, Griffin JH (1986) Quantitative immunoblotting of plasma and platelet protein S. *Thromb. Haemost.* **56**: 382–386.

Schwarz HP, Heeb MJ, Lottenberg R, Roberts H, Griffin JH (1989) Familial protein S deficiency with a variant protein S molecule in plasma and platelets. *Blood* **74**: 213–221.

Scopes D, Berg L-P, Krawczak M, Kakkar VV, Cooper DN (1995) Polymorphic variation in the human protein C (*PROC*) gene promoter can influence transcriptional effiency *in vitro*. *Blood Coag. Fibrinol.* **6**: 317–321.

Scott CF, Schapira M, James HL, Cohen AB, Colman RW (1982) Inactivation of factor Xia by plasma protease inhibitors: predominant role of α1-protease inhibitor and protective effect of high molecular weight kininogen. *J. Clin. Invest.* **69**: 844–852.

Seigneur M, Dufourcq P, Conri C, Constans J, Mercié P, Pruvost A, Amiral J, Midy D, Baste J-C, Boisseau MR (1993) Levels of plasma thrombomodulin are increased in atheromatous arterial disease. *Thromb. Res.* **71**: 423–428.

Selander-Sunnerhagen M, Ullner M, Persson E, Teleman O, Stenflo J, Drakenberg T (1992) How an epidermal growth factor (EGF)-like domain binds calcium. *J. Biol. Chem.* **267**: 19642–19647.

Seligsohn U, Griffin JH (1995) Contact activation and factor XI. In: Scriver CR, Beaudet AL, Sly WS, Valle D (eds) *The Metabolic and Molecular Bases of Inherited Disease*, 7th edn, pp. 3289–3311. McGraw-Hill, New York.

Seligsohn U, Berger A, Abend M, Rubin L, Attias D, Zivelin A, Rapaport SI (1984) Homozygous protein C deficiency manifested by massive thrombosis in the newborn. *New Engl. J. Med.* **310**: 559–562.

Seydewitz HH, Witt I (1985) Increased phosphorylation of human fibrinopeptide A under acute phase conditions. *Thromb. Res.* **40**: 29–39.

Sheehan JP, Sadler JE (1994) Molecular mapping of the heparin-binding exosite of thrombin. *Proc. Natl Acad. Sci. USA* **91**: 5518–5522.

Sheehan JP, Tollefsen DM, Sadler JE (1994) Heparin cofactor II is regulated allosterically and not primarily by template effects. *J. Biol. Chem.* **269**: 32747–32751.

Sheffield WP, Blajchman MA (1994) Site-directed mutagenesis of the P2 residue of human antithrombin. *FEBS Lett.* **339**: 147–150.

Sheffield WP, Blajchman MA (1995) Deletion mutagenesis of heparin cofactor II: defining the minimum size of a thrombin inhibiting serpin. *FEBS Lett.* **365**: 189–192.

Sheffield WP, Brothers AB, Wells MJ, Hatton MWC, Clarke BJ, Blajchman MA (1992) Molecular cloning and expression of rabbit antithrombin III. *Blood* **79**: 2330–2339.

Sheffield WP, Schuyler PD, Blajchman MA (1994) Molecular cloning and expression of rabbit heparin cofactor II: a plasma thrombin inhibitor highly conserved between species. *Thromb. Haemost.* **71**: 778–782.

Shen L, Dahlbäck B (1994) Factor V and protein S as synergistic cofactors to activated protein C in degradation of factor VIIIa. *J. Biol. Chem.* **269**: 18735–18738.

Shen S, Slightom JL, Smithies O (1981) A history of the human fetal globin gene duplication. *Cell* **26**: 191–203.

Shifter T, Machtey I, Creter D (1984) Thromboembolism in congenital factor VII deficiency. *Acta Haematol.* **71**: 60–62.

Shigekiyo T, Tomonari A, Uno Y, Azuma H, Shunto R, Saito S, Kaneko H, Satoh K, Ueda S. (1991) Familial elevation of plasma histidine-rich glycoprotein in a Japanese family. *Jap. J. Thromb. Hemost.* **2**: 530–536.

Shigekiyo T, Uno Y, Tomonari A, Satoh K, Hondo H, Ueda S, Saito S (1992) Type 1 congenital plasminogen deficiency is not a risk factor for thrombosis. *Thromb. Haemost.* **67**: 189–192.

Shigekiyo T, Ohshima T, Oka H, Tomonari A, Azuma H, Saito S (1993) Congenital histidine-rich glycoprotein deficiency. *Thromb. Haemost.* **70**: 263–265.

Shigekiyo T, Kanakuza M, Azuma H, Ohshima T, Kusaka K, Saito S (1995) Congenital deficiency of histidine-rich glycoprotein: failure to identify abnormalities in routine laboratory assays of hemostatic function, immunologic function and trace elements. *J. Lab. Clin. Med.* **125**: 719–723.

Shih CC, Stoye JP, Coffin JM (1988) Highly preferred targets for retrovirus integration. *Cell* **53**: 531–537.

Shih P, Malcolm BA, Rosenberg S, Kirsch JF, Wilson AC (1993) Reconstruction and testing of ancestral proteins. *Meth. Enzymol.* **224**: 576–591.

Shimizu A, Nagel GM, Doolittle RF (1992) Photoaffinity labeling of the primary fibrin polymerization site: isolation and characterization of a labeled cyanogen bromide fragment corresponding to γ-chain residues 337–379. *Proc. Natl Acad. Sci. USA* **89**: 2888–2892.

Shirai T, Shiojiri S, Ito H, Yamamoto S, Kusumoto H, Deyashika Y, Maruyama I, Suzuki K (1988) Gene structure of human thrombomodulin, a cofactor for thrombin-catalysed activation of protein C. *J. Biochem.* **103**: 281–285.

Shirk RA, Church FC, Wagner WD (1996) Arterial smooth muscle cell heparan sulfate proteoglycans accelerate thrombin inhibition by heparin cofactor II. *Arterioscl. Thromb. Vasc. Biol.* **16**: 1138–1146.

Shubeita HE, Cottey TL, Frank AE, Gerard RD (1990) Mutational and immunochemical analysis of plasminogen activator inhibitor 1. *J. Biol. Chem.* **265**: 18379–18385.

Sié P, Dupouy D, Pichon J, Boneu B (1985) Constitutional heparin cofactor II deficiency associated with recurrent thrombosis. *Lancet* **ii**: 414–416.

Siebenlist KR, Mosesson MW, DiOrio JP, Tavori S, Tatarsky I, Rimon A (1989) The polymerization of fibrin prepared from Fibrinogen Haifa (γ275Arg→His). *Thromb. Haemost.* **62**: 875–879.

Siebenlist KR, DiOrio JP, Budzynski AZ, Mosesson MW (1990) The polymerization and thrombin-binding properties of Des-(Bβ1–42)-fibrin. *J. Biol. Chem.* **265**: 18650–18656.

Siebenlist KR, Mosesson MW, DiOrio JP, Soria J, Soria C, Caen JP (1993) The polymerization of fibrinogen Dusart (Aα 554 Arg→Cys) after removal of carboxy terminal regions of the Aα chains. *Blood Coag. Fibrinol.* **4**: 61–65.

Sills RH, Marlar RA, Montgomery RR, Deshpande GN, Humbert JR (1984) Severe homozygous protein C deficiency. *J. Pediatr.* **105**: 409–414.

Simioni P, Lazzaro AR, Corer E, Salmistraro G, Girolami A (1990) Hereditary heparin cofactor II deficiency and thrombosis: report of six patients belonging to two separate kindreds *Blood Coag. Fibrinol.* **1**: 351–356.

Simioni P, de Ronde H, Prandoni P, Saladini M, Bertina RM, Girolami A (1995) Ischaemic stroke in young patients with activated protein C resistance. *Stroke* **2**: 885–890.

Simioni P, Scarano L, Gavasso S, Sardella C, Girolami B, Scudeller A, Girolami A (1996a) Prothrombin fragment 1+2 and thrombin-antithrombin complex levels in patients with inherited APC resistance due to factor V Leiden mutation. *Br. J. Haematol.* **92**: 435–441.

Simioni P, Scudeller A, Radossi P, Gavasso S, Girolan B, Tormene D, Girolami A (1996b) 'Pseudo homozygous' activated protein C resistance due to double heterozygous factor V defects (Factor V Leiden mutation and type I quantitative factor V defect) associated with thrombosis: report of two cases belonging to two unrelated kindreds *Thromb. Haemost.* **75**: 422–426.

Simioni P, Kalafatis M, Millar DS, Henderson SC, Luni S, Cooper DN, Girolami A (1996c) Compound heterozygous protein C deficiency resulting in the presence of only the β-form of protein C in plasma. *Blood* **88**: 2101–2108.

Simioni P, Prandoni P, Burlina A, Tormene D, Sardella C, Ferrari V, Benedetti L, Girolami A (1996d) Hyperhomocysteinemia and deep-vein thrombosis. *Thromb. Haemost.* **76**: 883–886.

Simioni P, Prandoni P, Lensing AWA, Sendeller A, Sardella C, Prins MH, Villalta S, Dazzi F, Girolami A (1997) The risk of recurrent venous thromboembolism in patients with an Arg506→Gln mutation in the gene for factor V (factor V Leiden). *New Engl. J. Med.* **336**: 399–403.

Simmonds RE, Ireland H, Kunz G, Lane DA and the Protein S Study Group (1996) Identification of 19 protein S gene mutations in patients with phenotypic protein S deficiency and thrombosis. *Blood* **88**: 4195–4204.

Skoda U, Goldmann SF, Händler C, Hummel K, Lechler E, Lübcke I, Manff G, Meyer-Börnecke D, Pesch S, Pulverer G (1988) Plasminogen hemizygosity. Detection of a silent allele in 7 members of a family by determination of plasminogen phenotypes, antigenic levels, and functional activity. *Vox Sang.* **54**: 210–214.

Skriver K, Wikoff WR, Patston PA, Tausk F, Schapira M, Kaplan AP, Bock SC (1991) Substrate properties of C1 inhibitor Ma (A434E) – genetic and structural evidence suggesting that the 'P12 region' contains critical determinants of serpin inhibitor/substrate status. *J. Biol. Chem.* **266**: 9216–9221.

Sloan IG, Firkin BG (1989) Impaired fibrinolysis in patients with thrombotic or haemostatic defects. *Thromb. Res.* **55**: 559–567.

Smith A, Alam J, Tatum F, Morgan WT (1989) Histidine-rich glycoprotein is synthesized by the liver and is a negative acute-phase response protein. *J. Cell Biol.* **107**: 584.

Smith JW, Deg N, Knauer DJ (1990) Heparin binding domain of antithrombin III: characterization using a synthetic peptide directed polyclonal antibody. *Biochemistry* **29**: 8950–8957.

Snow TR, Deal MT, Dickey DT, Esmon CT (1991) Protein C activation following coronary artery occlusion in the *in situ* porcine heart. *Circulation* **84**: 293–299.

Sobel BE, Collen D, Grossbard EB (Eds) (1987) *Tissue Plasminogen Activator in Thrombolytic Therapy*. Marcel Dekker, New York.

Sobel JH, Gawinowicz MA (1996) Identification of the α chain lysine donor sites involved in factor XIIIa fibrin cross-linking. *J. Biol. Chem.* **271**: 19288–19297.

Sobel JH, Trakht I, Wu HQ, Rudchenko S, Egbring R (1995) α-Chain cross-linking in fibrin(ogen) Marburg. *Blood* **86**: 989–1000.

Soff GA, Jackman RW, Rosenberg RD (1991) Expression of thrombomodulin by smooth muscle cells in culture: different effects of tumor necrosis factor and cyclic adenosine monophosphate on thrombomodulin expression by endothelial cells and smooth muscle cells in culture. *Blood* **77**: 515–518.

Solanki DL, Corn M (1980) Thromboembolism in patients with hereditary deficiency of coagulation disorders. *South. Med. J.* **73**: 944–946.

Solis MM, Cook C, Cook J, Glaser C (1991) Intravenous recombinant soluble human thrombomodulin prevents venous thrombosis in a rat model. *J. Vasc. Surg.* **14**: 599–604.

Solis MM, Vitti M, Cook J, Young D, Glaser C, Light D, Morser J, Wydro R, Yu S, Fink L (1994) Recombinant soluble human thrombomodulin: a randomized, blinded assessment of prevention of venous thrombosis and effects on hemostatic parameters in a rat model. *Thromb. Res.* **73**: 385–394.

Solymoss S, Tucker MM, Tracy PB (1988) Kinetics of inactivation of membrane bound factor Va by activated protein C. Protein S modulates factor Xa protection. *J. Biol. Chem.* **263**: 14884–14890.

Sommer SS, Bowie EJW, Ketterling RP, Bottema CDK (1992) Missense mutations and the magnitude of functional deficit: the example of factor IX. *Hum. Genet.* **89**: 295–297.

Sørensen JV, Lassen MR, Borris LC, Jørgensen PS, Schøtt P, Weber S, Murphy R, Walenga J (1990) Postoperative deep vein thrombosis and plasma levels of tissue plasminogen activator inhibitor. *Thromb. Res.* **60**: 247–251.

Soria J, Soria C, Bertrand O, Dunn F, Drouet L, Caen JP (1983a) Plasminogen Paris I: congenital abnormal plasminogen and its incidence in thrombosis. *Thromb. Res.* 32: 229–238.

Soria J, Soria C, Caen JP (1983b) A new type of congenital dysfibrinogenemia with defective fibrin lysis-Dusard syndrome: possible relation to thrombosis. *Br. J. Haematol.* 53: 575–586.

Soria J, Soria C, Hedner U, Nilsson IM, Bergqvist D, Samama M (1985) Episodes of increased fibronectin level observed in a patient suffereing from recurrent thrombosis related to congenital hypodysfibrinogenemia (fibrinogen Malmoe). *Br. J. Haematol.* 61: 727–738.

Soria J, Soria C, Samama M, Tabori S, Kehl M, Henschen A, Nieuwenhuizen W, Rimon A, Tatarski I (1987) Fibrinogen Haifa: fibrinogen variant with absence of protective effect of calcium on plasmin degradation of γ-chains. *Thromb. Haemost.* 57: 310–313.

Soria JM, Fontcuberta J, Borrell M, Estivill X, Sala N (1992) Protein C deficiency: identification of a novel two-base pair insertion and two point mutations in exon 7 of the protein C gene in Spanish families. *Hum. Mutat.* 1: 428–431.

Soria JM, Fontcuberta J, Chillón M, Borrell M, Estivill X, Sala N (1993) Acceptor slice site mutation in the invariant AG of intron 5 of the protein C gene, causing type I protein C deficiency. *Hum. Genet.* 92: 506–508.

Soria JM, Fontcuberta J, Borrell M, Estivill X, Sala N (1994a) Two novel mutations in exon 5 of the protein C gene in two Spanish families with thrombophilia due to protein C deficiency. *Hum. Mol. Genet.* 3: 1205–1206.

Soria JM, Brito D, Barceló J, Fontcuberta J, Botero L, Maldonado J, Estivill X, Sala N (1994b) Severe homozygous protein C deficiency: identification of a splice site missense mutation (184, Q→H) in exon 7 of the protein C gene. *Thromb. Haemost.* 72: 65–69.

Soria JM, Morell M, Estivill X, Sala N (1995a) A novel polymorphism (6376 G/T) in intron 7 of the human protein C gene. *Hum. Genet.* 96: 243–244.

Soria JM, Morell M, Jiménez-Astorga C, Estivill X, Sala N (1995b) Severe type I protein C deficiency in a compound heterozygote for Y124C and Q132X mutations in exon 6 of the *PROC* gene. *Thromb. Haemost.* 74: 1215–1220.

Soria JM, Morell M, Nicolau I, Estivill X, Sala N (1996a) Homozygosity for R87H missense mutation and for a rare intron 7 DNA variant (7054G→A) in the *PROC* genes of three siblings initially classified as heterozygotes for protein C deficiency. *Blood Coag. Fibrinol.* 7: 15–23.

Soria JM, Berg L-P, Fontcuberta J, Kakkar VV, Estivill X, Cooper DN, Sala N (1996b) Ectopic transcript analysis indicates that allelic exclusion is an important cause of type 1 protein C deficiency in patients with nonsense and frameshift mutations in the *PROC* gene. *Thromb. Haemost.* 75: 870–876

Soria JM, Morell M, Estivill X, Sala N (1996c) Recurrence of the *PROC* gene mutation R178Q: independent origins in Spanish protein C deficiency patients. *Hum. Mutat.* 8: 71–73

Sorice M, Arcieri P, Griggi T, Circella A, Misasi R, Lenti L, Di Nucci GD, Mariani G (1996) Inhibition of protein S by autoantibodies in patients with acquired protein S deficiency. *Thromb. Haemost.* 75: 555–559.

Sottrup-Jensen L, Claeys H, Zajdel M, Petersen TE, Magnusson S (1978) The primary structure of human plasminogen. In: Davidson JF, Rowan RM, Samama MM, Desnoyers PC (eds) *Progress in Chemical Fibrinolysis and Thrombolysis*, vol. 4. Raven Press, New York.

Southan C, Kehl M, Henschen A, Lane DA (1983) Fibrinogen Manchester: identification of an abnormal fibrinopeptide A with a C-terminal arginine→histidine substitution. *Br. J. Haematol.* 54: 143–151.

Southan C, Lane DA, Bode W, Henschen A (1985a) Thrombin-induced fibrinopeptide release from a fibrinogen variant (fibrinogen Sydney I) with an Aα Arg16→His substitution. *Eur. J. Biochem.* 147: 593–600.

Southan D, Thompson E, Panico M, Etieme T, Morris HR, Lane DA (1985b) Characterization of the peptides cleaved by plasmin from the C-terminal polymerization domain of human fibrinogen. *J. Biol. Chem.* 260: 13095–13101.

Souto JC, Gari M, Falkon L, Fontcuberta J (1996) A new case of hereditary histidine-rich glycoprtein deficiency with familial thrombophilia. *Thromb. Haemost.* 75: 374–375.

Spek CA, Poort SR, Bertina RM, Reitsma PH (1994) Determination of the allelic and haplotype frequencies of three polymorphisms in the promoter region of the human protein C gene. *Blood Coag. Fibrinol.* 5: 309–311.

Spek CA, Koster T, Rosendaal FR, Bertina RM, Reitsma PH (1995a) Genotypic variation in the promoter region of the protein C gene is associated with plasma protein C levels and thrombotic risk. *Arterioscler. Thromb. Vasc. Biol.* 15: 214–218.

Spek CA, Greengard JS, Griffin JH, Bertina RM, Reitsma PH (1995b) Two mutations in the promoter region of the human protein C gene both cause type I protein C deficiency by disruption of two HNF-3 binding sites. *J. Biol. Chem.* **270**: 24216–24221.

Sprengers ED, Kluft C (1987) Plasminogen activator inhibitors. *Blood* **69**: 381–387.

Stackhouse R, Chandra T, Robson KJH, Woo SLC (1983) Purification of antithrombin III mRNA and cloning of its cDNA. *J. Biol. Chem.* **258**: 703–706.

Stalder M, Hauert J, Kruithof EKO, Bachmann F (1985) Release of vascular plasminogen activator (v-PA) after venous stasis: electrophoretic zymographic analysis of free and complexed v-PA. *Br. J. Haematol.* **61**: 169–176.

Stead NW, Bauer KA, Kinney TR, Lewis JG, Campbell EE, Shifman MA, Rosenberg RD, Pizzo SV (1983) Venous thrombosis in a family with defective release of vascular plasminogen activator and elevated plasma factor VIII/von Willebrand factor. *Am. J. Med.* **74**: 33–39.

Stearns DJ, Kurosawa S, Sims PJ, Esmon NL, Esmon CT (1988) The interaction of a Ca^{2+} dependent monoclonal antibody with the protein C activation peptide region: evidence for obligatory Ca^{2+} binding to both antigen and antibody. *J. Biol. Chem.* **263**: 826–832.

Stearns DJ, Kurosawa S, Esmon CT (1989) Microthrombomodulin. Residues 310–486 from the epidermal growth factor precursor homology domain of thrombomodulin will accelerate protein C activation. *J. Biol. Chem.* **264**: 3352–3356.

Stearns-Kurosawa DJ, Kurosawa S, Mollica JS, Ferrell GL, Esmon CT (1996) The endothelial cell protein C receptor augments protein C activation by the thrombin–thrombomodulin complex. *Proc. Natl Acad. Sci. USA* **93**: 10212–10216.

Stefansson S, Lawrence DA (1996) The serpin PAI-1 inhibits cell migration by blocking integrin $\alpha_v\beta_3$ binding to vitronectin. *Nature* **383**: 441–443.

Stein PE, Carrell RW (1995) What do dysfunctional serpins tell us about molecular mobility and disease? *Nature Struct. Biol.* **2**: 96–113.

Stein PE, Leslie AGW, Finch JT, Turnell WG, McLaughlin PJ, Carrell RW (1990) Crystal structure of ovalbumin as a model for the reactive centre of serpins. *Nature* **347**: 99–102.

Stenbjerg S, Hessel B, Blombäck B, Larsson U, Therkildsen L, Rigler R (1983) Studies on the activation, polymerization, and gelation in abnormal fibrinogen: Fibrinogen Aarhus. *Thromb. Haemost.* **50**: 337–341.

Stenflo J (1976) A new vitamin K-dependent protein. *J. Biol. Chem.* **251**: 355–363.

Stenflo J, Lundwall A, Dahlbäck B (1987) β-Hydroxyasparagine in domains homologous to the epidermal growth factor precursor in vitamin K-dependent protein S. *Proc. Natl Acad. Sci. USA* **84**: 368–372.

Stenflo J, Ohlin AK, Owen WS, Schneider WJ (1988) β-Hydroxyaspartic acid or β-hydroxyasparagine in bovine low density lipoprotein receptor and in bovine thrombomodulin. *J. Biol. Chem.* **263**: 21–24.

Stephens AW, Thalley BS, Hirs CHW (1987) Antithrombin III Denver, a reactive site variant. *J. Biol. Chem.* **262**: 1044–1048.

Stephens AW, Siddiqui A, Hirs CHW (1988) Site-directed mutagenesis of the reactive center (serine 394) of antithrombin III. *J. Biol. Chem.* **263**: 15849–15852.

Stern A, Mattes R, Buckel P, Weidle UH (1989) Functional topology of human tissue-type plasminogen activator: characterization of two deletion derivatives and of a duplication derivative. *Gene* **79**: 333–344.

Stern DM, Nawroth PP, Harris K, Esmon CT (1986) Cultured bovine aortic endothelial cells promote activated protein C-protein S-mediated inactivation of factor Va. *J. Biol. Chem.* **261**: 713–718.

Stitt TN, Conn G, Gore M, Lai C, Bruno J, Radziejewski C, Mattsson K, Fisher J, Gies DR, Jones PF, Masiakowski P, Ryan TE, Tobkes NJ, Chen DH, DiStefano PS, Long GL, Basilico C, Goldfarb MP, Lemke G, Glass DJ, Yancopoulos GD (1995) The anticoagulation factor protein S and its relative, Gas6, are ligands for the Tyro 3/Axl family of receptor tyrosine kinases. *Cell* **80**: 661–670.

Stoeckert CJ, Collins FS, Weissman S (1984) Human fetal globin DNA sequences suggest novel conversion event. *Nucleic Acids Res.* **12**: 4469–4479.

Stolley PD, Tonascia JA, Tockman MS (1975) Thrombosis with low-estrogen oral contraceptives. *Am. J. Epidemiol.* **102**: 197–208.

Stoppa-Lyonnet D, Duponchel C, Meo T, Laurent J, Carter PE, Arala-Chaves M, Cohen JHM, Dewald G, Goetz J, Hauptmann G, Lagrue G, Lesavre P, Lopez-Trascasa M, Misiano G, Moraine C, Sobel A, Spth PJ, Tosi M (1991) Recombinational biases in the rearranged C1-inhibitor genes of hereditary angioedema patients. *Am. J. Hum. Genet.* **49**: 1055–1062.

Strandberg L, Lawrence D, Ny T (1988) The organization of the human plasminogen activator inhibitor-1 gene. Implications on the evolution of the serine protease inhibitor family. *Eur. J. Biochem.* **176**: 609–616.

Stringer HAR, Swieten PV, Heijnen HFG, Sixma JJ, Pannekoek H (1994) Plasminogen activator inhibitor-1 released from activated platelets plays a role in thrombolysis resistance. *Arterioscl. Thromb.* **14**: 1452–1457.

Strong DD, Moore M, Cottrell BA, Bohonus V$_L$, Pontes M, Evans B, Riley M, Doolittle RF (1985) Lamprey fibrinogen γ chain: cloning, cDNA sequencing and general characterization. *Biochemistry* **24**: 92–101.

Stubbs MT, Bode W (1993a) A player of many parts: the spotlight falls on thrombin's structure. *Thromb. Res.* **69**: 1–58.

Stubbs MT, Bode W (1993b) A model for the specificity of thrombin cleavage by thrombin. *Sem. Thromb. Hemost.* **19**: 344–351.

Stubbs JD, Lekutis C, Singer KL, Bui A, Yuzuki D, Srinivasan U, Parry G (1990) cDNA cloning of a mouse mammary epithelial cell surface protein reveals the existence of epidermal growth factor-like domains linked to factor VIII-like sequences. *Proc. Natl Acad. Sci. USA* **87**: 8417–8421.

Stubbs MT, Oschkinat H, Mayr I, Huber R, Angliker H, Stone SR, Bode W (1992) The interaction of thrombin with fibrinogen – a structural basis for its specificity. *Eur. J. Biochem.* **206**: 187–195.

Sue-Ling HM, Johnston D, Verheijen JH, Kluft C, Phillips PR, Davies JA (1987) Indicators of depressed fibrinolytic activity in pre-operative prediction of deep vein thrombosis. *Br. J. Surg.* **74**: 275–278.

Suenson E, Petersen LC (1986) Fibrin and plasminogen structures essential to stimulation of plasmin formation by tissue-type plasminogen activator. *Biochim. Biophys. Acta* **870**: 510–519.

Sugahara Y, Miura O, Yuen P, Aoki N (1992) Protein C Hong Kong: hereditary protein C deficiency caused by two mutant alleles, a five nucleotide deletion and a missense mutation. *Blood* **80**: 126–133.

Sugahara Y, Miura O, Hirosawa S, Aoki N (1994) Compound heterozygous protein C deficiency caused by two mutations, Arg-178 to Gln and Cys-331 to Arg, leading to impaired secretion of mutant protein C. *Thromb. Haemost.* **72**: 814–818.

Sun X-J, Chang J-Y (1990) Evidence that arginine-129 and arginine-145 are located within the heparin binding site of human antithrombin III. *Biochemistry* **29**: 8957–8962.

Sun X, Evatt B, Griffin JH (1994) Blood coagulation factor Va abnormality associated with resistance to activated protein C in venous thrombophilia. *Blood* **83**: 3120–3125.

Sutherland GR, Richards RI (1995) Simple tandem DNA repeats and human genetic disease. *Proc. Natl Acad. Sci. USA* **92**: 3636–3641.

Suttie JW (1993) Synthesis of vitamin K-dependent proteins. *FASEB J.* **7**: 445–452.

Suttie JW, Hoskins JA, Engelke J, Hopfgartner A, Ehrlich H, Bang NU, Belagaje RM, Schoner B, Long GL (1987) Vitamin K-dependent carboxylase: possible role of the substrate 'propeptide' as an intracellular recognition site. *Proc. Natl Acad. Sci. USA* **84**: 634–637.

Suzuki K, Dahlbäck B, Stenflo J (1982) Thrombin-catalyzed activation of human coagulation factor V. *J. Biol. Chem.* **257**: 6556–6564.

Suzuki K, Nishioka J, Hashimoto S (1983a) Regulation of activated protein C by thrombin-modified protein S. *J. Biol. Chem.* **266**: 699–705.

Suzuki K, Nishioka J, Hashimoto S (1983b) Protein C inhibitor: purification from human plasma and characterization. *J. Biol. Chem.* **258**: 163–168.

Suzuki K, Stenflo J, Dahlbäck B, Teodorsson B (1983c) Inactivation of human factor V by activated protein C. *J. Biol. Chem.* **258**: 1914–1920.

Suzuki K, Nishioka J, Matsuda M, Murayama H, Hashimoto S (1984) Protein S is essential for the activated protein C-catalysed inactivation of platelet associated factor Va. *J. Biochem.* **96**: 455–460.

Suzuki K, Deyashiki Y, Nishioka J, Kurachi K, Akira M, Yamamoto S, Hashimoto S (1987a) Characterization of a cDNA for human protein C inhibitor. *J. Biol. Chem.* **262**: 611–616.

Suzuki K, Kusumoto H, Deyashiki Y, Nishioka J, Maruyama I, Zushi M, Kawahara S, Honda G, Yamamoto S, Horiguchi S (1987b) Structure and expression of human thrombomodulin: a thrombin receptor on endothelium acting as cofactor for protein C activation. *EMBO J.* **6**: 1891–1897.

Suzuki K, Deyashiki Y, Nishioka J, Toma K (1989) Protein C inhibitor: structure and function. *Thromb. Haemost.* **61**: 337–342.

Svensson PJ, Dahlbäck B (1994) Resistance to activated protein C as a basis for venous thrombosis. *New Engl. J. Med.* **330**: 517–522.

Swindells MB, Thornton JM (1991) Modelling by homology. *Curr. Opin. Struct. Biol.* **1**: 219–223.

Tabernero MD, Thomas JF, Alberca I, Orfao A, Borrasca AL, Vincente V (1991) Incidence and clinical characteristics of hereditary disorders associated with venous thrombosis. *Am. J. Hematol.* **36**: 249–253.

Tada N, Sato M, Tsujimura A, Iwase R, Hashimoto-Gotoh T (1992) Isolation and characterization of a mouse protein C cDNA. *J. Biochem.* **111**: 491–495.

Tait RC, Walker ID, Islam SIA, Mitchell R, Davidson JF (1991a) Plasminogen levels and putative prevalence of deficiency in 4500 blood donors (abstr.). *Br. J. Haematol.* **77** (Suppl. 1): 10.

Tait RC, Walker ID, Perry DJ, Carrell RW, Islam SIA, McCall F, Mitchell R, Davidson JF (1991b) Prevalence of antithrombin III deficiency subtypes in 4000 healthy blood donors. *Thromb. Haemost.* **65**: 839.

Tait RC, Walker ID, Islam SIAM, McCall F, Conkie JA, Mitchell R, Davidson JF (1993a) Influence of demographic factors on antithrombin III activity in a healthy population. *Br. J. Haematol.* **84**: 476–480.

Tait RC, Walker ID, Islam SIAM, McCall F, Conkie JA, Wight M, Mitchell R, Davidson JF (1993b) Protein C activity in healthy volunteers: influence of age, sex, smoking and oral contraceptives. *Thromb. Haemost.* **70**: 281–285.

Tait RC, Walker ID, Perry DJ, Islam SIAM, Daly ME, McCall F, Conkie JA, Carrell RW (1994) Prevalence of antithrombin deficiency in the healthy population. *Br. J. Haematol.* **87**: 106–112.

Tait RC, Walker ID, Reitsma PH, Islam SIAM, McCall F, Poort SR, Conkie JA, Bertina RM (1995) Prevalence of protein C deficiency in the healthy population. *Thromb. Haemost.* **73**: 87–93.

Tait RC, Walker ID, Conkie JA, Islam SIAM, McCall F (1996) Isolated familial plasminogen deficiency may not be a risk factor for thrombosis. *Thromb. Haemost.* **76**: 1004–1008.

Takahashi I, Saito H (1988) A rapid purification with high recovery of factor XII (Hageman factor) on immunoaffinity column: application to an abnormal clotting factor XII (factor XII Toronto). *J. Biochem.* **103**: 641–643.

Takahashi H, Ito S, Hanano M, Wada K, Niwano K, Seki Y, Shibata A (1992) Circulating thrombomodulin as a novel endothelial cell marker: comparison of its behaviour with von Willebrand factor and tissue type plasminogen activator. *Am. J. Hematol.* **41**: 32–38.

Takamiya O, Ishida F, Kodaira H, Kitano K (1995) APC-resistance and *Mnl*I genotype (Gln 506) of coagulation factor V are rare in Japanese population. *Thromb. Haemost.* **74**: 990–997.

Takano S, Kimura S, Ohdama S, Aoki N (1990) Plasma thrombomodulin in health and diseases. *Blood* **76**: 2024–2028.

Takeshita M, Chang CN, Johnson F, Will S, Grollman A (1987) Oligodeoxynucleotides containing synthetic abasic sites. *J. Biol. Chem.* **262**: 10171–10179.

Tamaki T, Aoki N (1982) Cross-linking of α2-plasmin inhibitor to fibrin catalyzed by activated fibrin-stabilizing factor. *J. Biol. Chem.* **257**: 14767–14771.

Tanabe S, Sugo T, Matsuda M (1991) Synthesis of protein C in human umbilical vein endothelial cells. *J. Biochem.* **109**: 924–928.

Tanimoto M, Matsushita T, Sugiura I, Hamaguchi M, Takamatsu, J, Kamiya T, Saito H (1989) Structural gene analysis of a family with protein S deficiency by use of its cDNA probe. *Thromb. Haemost.* **62**: 273.

Tans G, Rosing J, Griffin JH (1983) Sulfatide-dependent auto-activation of human blood coagulation factor XII (Hageman factor). *J. Biol. Chem.* **258**: 8215–8222.

Tans G, Rosing J, Christella M, Thomassen LGD, Heeb MJ, Zwaal RFA, Griffin JH (1991) Comparison of anticoagulant and procoagulant activities of stimulated platelets and platelet-derived microparticles. *Blood* **77**: 2641–2648.

Tans G, Nicolaes GAF, Thomassen MCLGD, Hemker HC, van Zonnveld A-J, Pannekoek H, Rosing J (1994) Activation of human factor V by meizothrombin. *J. Biol. Chem.* **269**: 15969–15972.

Taylor FB (1992) Protein S, C4b binding protein, and the hypercoagulable state. *J. Lab. Clin. Med.* **119**: 596–597.

Taylor FB, Chang A, Esmon CT, D'Angelo A, Vigano-D'Angelo S, Blick KE (1987) Protein C prevents the coagulopathic and lethal effects of *Escherichia coli* infusion in the baboon. *J. Clin. Invest.* **79**: 918–923.

Tazawa R, Yamamoto K, Suzuki K, Hirokawa T, Hirosawa S, Aoki N (1994) Presence of functional cyclic AMP responsive element in the 3' untranslated region of the thrombomodulin gene. *Biochem. Biophys. Res. Commun.* **200**: 1391–1397.

Tejada ML, Deeley RG (1995) Cloning of an avian antithrombin: developmental and hormonal regulation of expression. *Thromb. Haemost.* **73**: 654–661.

ten Cate JW, Peters M, Buller H (1983) Isolated plasminogen deficiency in a patient with recurrent thromboembolic complications. *Thromb. Haemost.* **50**: 59.

ten Cate JW, Koopman MMW, Prins MH, Büller HR (1995) Treatment of venous thromboembolism. *Thromb. Haemost.* **74**: 197–203.

Thaler E, Lechner K (1981) Antithrombin III deficiency and thromboembolism. *Clin. Haematol.* **10**: 369–390.

Thein SL, Lane DA (1988) Use of synthetic oligonucleotides in the characterization of antithrombin III Northwick Park (393 CGT→TGT) and antithrombin III Glasgow (393 CGT→CAT). *Blood* **72**: 1817–1821.

Theunissen HJM, Dijkema R, Grootenhuis PDJ, Swinkels JC, de Poorter TL, Carati P, Visser A (1993) Dissociation of heparin-dependent thrombin and factor Xa inhibitory activities of antithrombin III by mutations in the reactive site. *J. Biol. Chem.* **268**: 9035–9040.

Thomas W, Drayna D (1992) A polymorphic dinucleotide repeat in intron 1 of the human tissue plasminogen activator gene. *Hum. Mol. Genet.* **1**: 138.

Thommen D, Buhrfeind E, Felix R, Sulzer I, Furlan M, Lämmle B (1989) Hämostase-parameter bei 55 Patienten mit venösen und/oder arteriellen Thromboembolien. *Schweiz. med. Wschr.* **119**: 493–499.

Thompson EA, Salem HH (1986) Inhibition by human thrombomodulin of factor Xa-mediated cleavage of prothrombin. *J. Clin. Invest.* **78**: 13–18.

Thompson MD, Dave JR, Nakhasi HL (1985) Molecular cloning of mouse mammary gland kappa-casein: comparison with rat κ-casein and rat and human γ-fibrinogen. *DNA* **4**: 263–271.

Thompson SG, Kienast J, Pyke SDM, Haverkate F, Van de Loo JCW (1995) Hemostatic factors and the risk of myocardial infarction or sudden death in patients with angina pectoris. *New Engl. J. Med.* **332**: 635–641.

Thorogood M, Mann J, Murphy M, Vessey M (1992) Risk factors for fatal venous thromboembolism in young women: a case–control study. *Int. J. Epidemiol.* **21**: 48–52.

Thunberg L, Backstrom G, Lindahl U (1982) Further characterization of the antithrombin-binding sequence in heparin. *Carbohydrate Res.* **100**: 393–410.

Tipping PG, Davenport P, Gallicchio M, Filonzi EL, Apostopoulos J, Wojta J (1993) Atheromatous plaque macrophages produce plasminogen activator inhibitor type-1 and stimulate its production by endothelial cells and vascular smooth muscle cells. *Am. J. Pathol.* **143**: 875–882.

Tishkoff SA, Ruano G, Kidd JR, Kidd KK (1996) Distribution and frequency of a polymorphic *Alu* insertion at the plasminogen activator locus in humans. *Hum. Genet.* **97**: 759–764.

Tokunaga F, Wakabayashi S, Sato H, Arakawa M, Tawaraya H, Koide T (1992) Identification of one base deletion in exon IX of the protein C gene that causes a type I deficiency. *Thromb. Res.* **68**: 417–423.

Tokunaga F, Tsukamoto T, Koide T (1996) Cellular basis for protein C deficiency caused by a single amino acid substitution at Arg15 in the γ-carboxyglutamic acid domain. *J. Biochem.* **120**: 360–368.

Tollefsen DM (1995) Insight into the mechanism of action of heparin cofactor II. *Thromb. Haemost.* **74**: 1209–1214.

Tollefsen DM, Blank MK (1981) Detection of a new heparin-dependent inhibitor of thrombin in human plasma. *J. Clin. Invest.* **68**: 589–596.

Tollefsen DM, Pestka CA, Monafo WJ (1983) Activation of heparin cofactor II by dermatan sulfate. *J. Biol. Chem.* **258**: 6713–6716.

Tomczak JA, Ando RA, Sobel HG, Bovill EG, Long GL. (1994) Genetic analysis of a large kindred exhibiting type I protein C deficiency and associated thrombosis. *Thromb. Res.* **74**: 243–254.

Tomonari A, Iwahana H, Yoshimoto K, Shigekiyo T, Saito S, Itakura M (1992) Two new nonsense mutations in type Ia antithrombin III deficiency at Leu 140 and Arg 197. *Thromb. Haemost.* **68**: 455–459.

Toossi Z, Sedor JR, Mettler MA, Everson B, Young T, Ratnoff OD (1992) Induction of expression of monocyte interleukin-1 by Hageman factor (factor XII). *Proc. Natl Acad. Sci. USA* **89**: 11969–11972.

Töpfer-Petersen E, Lottspeich F, Henschen A (1976) Carbohydrate linkage site in the β-chain of human fibrin. *Hoppe-Seyler Z. Physiol. Chem.* **357**: 1509–1513.

Toulon P, Gandrille S, Mathiot C, Sultan Y, Aiach M (1991) Hereditary heparin cofactor II variant Arg 189→His (HCII-Paris). *Blood* **78** (Suppl. 1): 74a.

Tracy PB, Mann KG (1983) Prothrombinase complex assembly on the platelet surface is mediated through the 74,000 dalton component of factor Va. *Proc. Natl Acad. Sci. USA* **80**: 2380–2384.

Tracy PB, Eide LL, Bowie EJ, Mann KG (1982) Radioimmunoassay of factor V in human plasma and platelets. *Blood* **60**: 59–63.

Tran TH, Marbert GA, Duckert F (1985) Association of hereditary heparin cofactor II deficiency with thrombosis. *Lancet* **ii**: 413–414.

Tremp GL, Duchange N, Branellec D, Cereghini S, Tailleux A, Berthou L, Fievet C, Touchet N, Schombert B, Fruchart J-C, Zakin MM, Denèfle P (1995) A 700-bp fragment of the human antithrombin III promoter is sufficient to confer high, tissue-specific expression on human apolipoprotein A-II in transgenic mice. *Gene* **156**: 199–205.

Trexler M, Vali Z, Patthy L (1982) Structure of the γ amino-carboxylic acid binding sites of human plasminogen: arginine 70 and aspartic acid 56 are essential for binding of ligand to kringle 4. *J. Biol. Chem.* **257**: 7401–7406.

Triplett DA, Brandt JT, Batard MA, Dixon JS, Fair DS (1985) Hereditary factor VII deficiency: heterogeneity defined by combined functional and immunochemical analysis. *Blood* **66**: 1284–1287.

Tripodi M, Citarella F, Guida S, Galeffi P, Gallo E, Ferrazza P, Amicone L, Mariani R, Longobardi C, Fantoni A (1986a) Molecular studies on DNA sequences coding for factor VII and factor XII of human coagulation. *Ital. J. Biochem.* **35**: 328–332.

Tripodi M, Citarella F, Guida S, Galeffi P, Fantoni A, Cortese R (1986b) cDNA sequence coding for human coagulation factor XII (Hageman). *Nucleic Acids Res.* **14**: 3146.

Tripodi A, Franchi F, Krachmalnicoff A, Mannucci PM (1990) Asymptomatic homozygous protein C deficiency. *Acta Haematol.* **83**: 152–155.

Tsai J-C, Perrella MA, Yoshizumi M, Hsieh C-M, Haber E, Scleigel R, Lee M-E (1994) Promotion of vascular smooth muscle cell growth by homocysteine: a link to atherosclerosis. *Proc. Natl Acad. Sci. USA* **91**: 6369–6373.

Tsay W, Greengard JS, Montgomery R, Griffin JH (1991) Five previously undescribed mutations in protein C that identify elements critical for gene and protein activity. *Blood* **78** (Suppl.1): 184a.

Tsay W, Greengard JS, Montgomery RR, McPherson RA, Fucci JC, Koerper MA, Coughlin J, Griffin JH (1993) Genetic mutations in ten unrelated American patients with symptomatic type I protein C deficiency. *Blood Coag. Fibrinol.* **4**: 791–796.

Tsay W, Greengard JS, Griffin JH. (1994) Exonic polymorphisms in the protein C gene: interethnic comparison between Caucasians and Asians. *Hum. Genet.* **94**: 177–178.

Tsiang M, Lentz SR, Dittman WA, Wen D, Scarpati EM, Sadler JE (1990) Equilibrium binding of thrombin to recombinant human thrombomodulin: effect of hirudin, fibrinogen, factor Va and peptide analogues. *Biochemistry* **29**: 10602–10612.

Tsiang M, Lentz SR, Sadler JE (1992) Functional domains of membrane-bound human thrombomodulin. EGF-like domains four to six and the serine/threonine-rich domain are required for cofactor activity. *J. Biol. Chem.* **267**: 6164–6170.

Tsiang M, Jain AK, Dunn KE, Rojas ME, Leung LLK, Gibbs CS (1995) Functional mapping of the surface residues of human thrombin. *J. Biol. Chem.* **270**: 16854–16863.

Tsuda S, Reitsma P, Miletich J (1991) Molecular defects causing heterozygous protein C deficiency in three asymptomatic kindreds. *Thromb. Haemost.* **65**: 647.

Tsuji H, Hashimoto-Gotoh T, Takado O, Uno M, Tanaka S, Nakagawa M (1991) Antithrombin III Kyoto: identification of an arginine 406 to methionine point mutation near protease reactive site. *Thromb. Haemost.* **65**: 913.

Tsutsumi S, Saito T, Sakata T, Miyata T, Ichinose A (1996) Genetic diagnosis of dysplasminogenemia: detection of an Ala601-Thr mutation in 118 out of 125 families and identification of a new Asp676–Asn mutation. *Thromb. Haemost.* **76**: 135–138.

Tuddenham EGD, Cooper DN (1994) *The Molecular Genetics of Haemostasis and its Inherited Disorders*. Oxford University Press, Oxford.

Tuddenham EGD, Takase T, Thomas AE, Awidi AS, Madanat FF, Abu Hajir MM, Kernoff PBA, Hoffbrand AV (1989) Homozygous protein C deficiency with delayed onset of symptoms at 7 to 10 months. *Thromb. Res.* **53**: 475–484.

Turko IV, Fan B, Gettins PGW (1993) Carbohydrate isoforms of antithrombin variant N135Q with different heparin affinities. *FEBS Lett.* **335**: 9–12.

Turner J, Grundy CB, Kakkar VV, Cooper DN (1990) *Msp* I RFLP in the human heparin cofactor II (*HCF2*) gene. *Nucleic Acids Res.* **18**: 1664.

Ueyama H, Hashimoto Y, Uchino M, Sasaki Y, Ueyama E, Okajima K, Araki S (1989) Progressing ischemic stroke in a homozygote with variant antithrombin III. *Stroke* **20**: 815–818.

Ueyama H, Murakami T, Nishiguchi S, Maeda S, Hashimoto Y, Okajima K, Shimada K, Araki S (1990) Antithrombin III Kumamoto: identification of a point mutation and genotype analysis of the family. *Thromb. Haemost.* **63**: 231–234.

Umlauf SW, Cox MM (1988) The functional significance of DNA sequence structure in a site-specific genetic recombination reaction. *EMBO J.* **7**: 1845–1852.

Uzan G, Courtois G, Besmond C, Frain M, Sala-Trepat J, Kahn A, Marguerie G (1984) Analysis of fibrinogen genes in patients with congenital afibrinogenemia. *Biochem. Biophys. Res. Commun.* **120**: 376–383.

Vali Z, Scheraga HA (1988) Localization of the binding site on fibrin for the secondary binding site of thrombin. *Biochemistry* **27**: 1956–1963.

Vandenbroucke JP, Koster T, Briët E, Reitsma PH, Rosendaal FR (1994) Increased risk of venous thrombosis in oral-contraceptive users who are carriers of factor V Leiden mutation. *Lancet* **344**: 1453–1457.

Vanhoutte PM, Scott-Burden T (1994) The endothelium in health and disease. *Texas Heart Inst. J.* **21**: 62–67.

van Bockxmeer FM, Baker RI, Taylor RR (1995) Premature ischaemic heart disease and the gene for coagulation factor V. *Nature Medicine* **1**: 185.

van Boeckel CAA, Grootenhuis PDJ, Visser A (1994) A mechanism for heparin-induced potentiation of antithrombin III. *Nature Struct. Biol.* **1**: 423–425.

van Boven HH, Olds RJ, Thein S-L, Reitsma PH, Lane DA, Briët E, Vandenbroucke JP, Rosendaal FR (1994) Hereditary antithrombin deficiency: heterogeneity of the molecular basis and mortality in Dutch families. *Blood* **84**: 4209–4213.

van Boven HH, Reitsma PH, Rosendaal FR, Briët E, Vandenbroucke JP, Bayston TA, Chowdhury V, Bauer K, Scharrer I, Lane DA (1995) Interaction of factor V Leiden with inherited antithrombin deficiency. *Blood Coag. Fibrinol.* **6**: 153–154.

van Boven HH, Reitsma PH, Rosendaal FR, Bayston TA, Chowdhury V, Bauer KA, Scharrer I, Conard J, Lane DA (1996) Factor V Leiden (FV R506Q) in families with inherited antithrombin deficiency. *Thromb. Haemost.* **75**: 417–421.

van Meijer M, Gebbink RK, Preissner KT, Pannekoek H (1994) Determination of the vitronectin binding site on plasminogen activator inhibitor 1 (PAI-1). *FEBS Lett.* **352**: 342–346.

van Mourik JA, Lawrence DA, Loskutoff DJ (1984) Purification of an inhibitor of plasminogen activator (antiactivator) synthesized by endothelial cells. *J. Biol. Chem.* **259**: 14914–14921.

van Wijnen M, Stam JG, van't Veer C, Meijers JCM, Reitsma PH, Bertina RM, Bouma BN (1996) The interaction of protein S with the phospholipid surface is essential for the activated protein C-independent activity of protein S. *Thromb. Haemost.* **76**: 397–403.

van Zonnefeld A-J, Chang GTG, Berg van den J, Kooistra T, Verheijen JH, Pannekoek H, Kluft C (1986) Quantification of tissue-type plasminogen activator (t-PA) mRNA in human endothelial cell cultures by hybridization with a t-PA cDNA probe. *Biochem. J.* **235**: 385–390.

van Zonnefeld A-J, Curriden SC, Loskutoff DJ (1988) Type I plasminogen activator inhibitor gene: functional analysis and glucocorticoid regulation of its promoter. *Proc. Natl Acad. Sci. USA* **85**: 5525–5529.

van de Locht LTF, Kuypers AWHM, Verbruggen BW, Linssen PCM, Novakova IRO, Mensink EJBM (1995) Semi-automated detection of the factor V mutation by allele specific amplification and capillary electrophoresis. *Thromb. Haemost.* **74**: 1276–1279.

van den Berg EA, LeClercq E, Kluft C, Koide T, Van der Zee A, Oldenburg M, Wijnen JT, Meera Khan P (1990b) Assignment of the human gene for histidine-rich glycoprotein to chromosome 3. *Genomics* **7**: 276–279.

van den Eijnden-Schrauwen Y, Lakenberg N, Emeis JJ, de Knijff P (1995) *Alu*-repeat polymorphism in the tissue-type plasminogen activator (tPA) gene does not affect basal endothelial tPA synthesis. *Thromb. Haemost.* **74**: 1197–1207.

van der Bom JG, Bots ML, van Vliet HHDM, Pols HAP, Hofman A, Grobbee DE (1996) Antithrombin and atherosclerosis in the Rotterdam study. *Arterioscl. Thromb. Vasc. Biol.* **16**: 864–867.

van der Meer FJ, van Tilburg N, van Wijngaarden A, van der Linden IK, Briët E, Bertina RM (1989) A second plasma inhibitor of activated protein C: α1-antitrypsin. *Thromb. Haemost.* **62**: 756–757.

van der Velden PA, Krommenhoek-Van Es T, Allaart CF, Bertina RM, Reitsma PH (1991) A frequent thrombomodulin amino acid dimorphism is not associated with thrombophilia. *Thromb. Haemost.* **65**: 511–513.

Varadi A, Patthy L (1984) β(Leu 121-Lys 122) Segment of fibrinogen is in a region essential for plasminogen binding by fibrin fragment E. *Biochemistry* **23**: 2108–2112.

Varadi A, Scheraga HA (1986) Localization of segments essential for polymerization and for calcium binding in the γ-chain of human fibrinogen. *Biochemistry* **25**: 519–528.

Varadi K, Philapitsch A, Santa T, Schwarz HP (1994) Activation and inactivation of human protein C by plasmin. *Thromb. Haemost.* **71**: 615–621.

Varadi K, Moritz B, Lang H, Bauer K, Preston E, Peake I, Rivard GE, Keil B, Schwarz HP (1995) A chromogenic assay for activated protein C resistance. *Br. J. Haematol.* **90**: 884–891.

Varadi K, Rosing J, Tans G, Pabinger I, Keil B, Schwarz HP (1996) Factor V enhances the cofactor function of protein S in the APC-mediated inactivation of factor VIII: influence of the factor V[R506Q] mutation. *Thromb. Haemost.* **76**: 208–214.

Vassalli J-D, Sappino A-P, Belin D (1991) The plasminogen activator/plasmin system. *J. Clin. Invest.* **88**: 1067–1072.

Vasse M, Borg JY, Monconduit M (1989) Protein C Rouen, a new hereditary protein C abnormality with low anticoagulant but normal amidolytic activities. *Thromb. Res.* **56**: 387–398.

Vehar GA, Keyte B, Eaton D, Rodriguez H, O'Brien DP, Rotblat F, Opermann H, Keck R, Wood W, Harkins R, Tuddenham EGD, Lawn R, Capon D (1984) Structure of human factor VIII. *Nature* **312**: 337–342.

Velander WH, Johnson JL, Page RL, Russell CG, Subramanian A, Wilkins TD, Gwazdauskas FC, Pittius C, Drohan WN (1992) High-level expression of a heterologous protein in the milk transgenic swine using the cDNA encoding human protein C. *Proc. Natl Acad. Sci. USA* **89**: 12003–12007.

Vercellotti GM (1995) Potential role of viruses in thrombosis and atherosclerosis. *Trends Cardiovasc. Med.* **5**: 128–133.

Verheijen JH, Chang GTG, Kluft C (1984) Evidence for the occurrence of a fast-acting inhibitor for tissue-type plasminogen activator in human plasma. *Thromb. Haemost.* **51**: 392–395.

Verheught FWA, ten Cate JW, Sturk A, Imandt L, Verhorst PMJ, van Eenige MJ, Verwey W, Roos JP (1987) Tissue plasminogen activator activity and inhibition in acute myocardial infarction and angiographically normal coronary arteries. *Am. J. Cardiol.* **59**: 1075–1079.

Verkerk AJMH, Pieretti M, Sutcliffe JS, Fu YH, Kuhl DPA, Pizzuti A, Reiner O, Richards S, Victoria MF, Zhang F, Eussen BE, Van Ommen GJB, Blonden LAJ, Riggins GJ, Chastain JL, Kunst CB, Galjaard H, Caskey CT, Nelson DL, Oostra BA, Warren ST (1991) Identification of a gene (FMR-1) containing a CGG repeat coincident with a breakpoint cluster region exhibiting length variation in fragile X syndrome. *Cell* **65**: 905–914.

Vermeer C (1990) γ-Carboxyglutamate-containing proteins and the vitamin K-dependent carboxylase. *Biochem. J.* **266** : 625–636.

Verstraete M (1993) The diagnosis and treatment of deep-vein thrombosis. *New Engl. J. Med.* **329**: 1418–1420.

Villa P, Aznar J, Jorquera J, Mira Y, Vaya A, Fernandez MA (1995) Inherited homozygous resistance to activated protein C. *Thromb. Haemost.* **74**: 793–810.

Villa P, Aznar J, Mira Y, Fernandez MA, Vaya A (1996) Third-generation oral contraceptives and low free protein S as a risk for venous thrombosis. *Lancet* **347**: 397.

Virchow R (1856) *Phlogose und Thrombose im Gefässystem. Gessammelte Abhandlungen zur Wissenschaftlichen Medecin.* Staatsdruckerei, Frankfurt.

Vogel F, Kopun M (1977) Higher frequencies of transitions among point mutations. *J. Mol. Evol.* **9**: 159–180.

Vogel F, Motulsky AG (1986) *Human Genetics: Problems and Approaches*, 2nd edn, Springer, Berlin.

Vogel F, Röhrborn G (1965) Mutationsvorgänge bei der Entstehung von Hämoglobinvarianten. *Humangenetik* **1**: 635–650.

Vogel F, Kopun M, Rathenburg R (1976) Mutation and molecular evolution. In: Goodman M, Tashian RE, Tashian JH (eds) *Molecular Anthropology*, pp.13–33. Plenum Press, New York.

Von der Ahe D, Nischan C, Kunz C, Otte J, Knies U, Oderwald H, Wasylyk B (1993) *Ets* transcription factor binding site is required for positive and TNFα-induced negative promoter regulation. *Nucleic Acids Res.* **21**: 5636–5643.

Voorberg J, Roelse J, Koopman R, Büller H, Berends F, ten Cate JW, Mertens K, van Mourik JA (1994) Association of idiopathic venous thromboembolism with single point-mutation at Arg[506] of factor V. *Lancet* **343**: 1535–1536.

Voskuilen M, Vermond A, Veeneman GH (1987) Fibrinogen lysine residue Aα 157 plays a crucial role in the fibrin-induced acceleration of plasminogen activation, catalyzed by tissue-type plasminogen activator. *J. Biol. Chem.* **262**: 5944–5948.

Wacey AI, Pemberton S, Cooper DN, Kakkar VV, Tuddenham EGD (1993) A molecular model of the serine protease domain of activated protein C: application to the study of missense mutations causing protein C deficiency. *Br. J. Haematol.* **84**: 290–300.

Wacey AI, Krawczak M, Kakkar VV, Cooper DN (1994) Determinants of the factor IX mutational spectrum in haemophilia B: an analysis of missense mutations using a multi-domain molecular model of the activated protein. *Hum. Genet.* **94**: 594–608.

Wada Y, Lord ST (1994) A correlation between thrombotic disease and a specific fibrinogen abnormality (Aa 554 Arg→Cys) in two unrelated kindred, Dusart and Chapel Hill III. *Blood* **84**: 3709–3714.

Wada H, Deguch K, Shirakawa S, Suzuki K (1993) Successful treatment of deep vein thrombosis in homozygous protein C deficiency with activated protein C. *Am. J. Hematol.* **44**: 218–219.

Wada H, Nakase T, Nakaya R (1994) Elevated plasma tissue factor antigen level in patients with disseminated intravascular coagulation. *Am. J. Hematol.* **45**: 232–236.

Wada H, Wakita Y, Shiku H (1995) Tissue factor expression in endothelial cells in health and disease. *Blood Coag. Fibrinol.* **6**: S26-S31.

Wade DP, Clarke JG, Lindahl GE, Liu AC, Zysow BR, Meer K, Schwartz K, Lawn RM (1993) 5′ control regions of the apolipoprotein (a) gene and members of the related plasminogen gene family. *Proc. Natl Acad. Sci. USA* **90**: 1369–1373.

Wagner OF, Binder BR (1986) Purification of an active plasminogen activator inhibitor immunologically related to the endothelial type plasminogen activator inhibitor from the conditioned media of a human melanoma cell line. *J. Biol. Chem.* **261**: 14474–14481.

Walker FJ (1980) Regulation of activated protein C by a new protein. A possible function for bovine protein S. *J. Biol. Chem.* **255**: 5521–5524.

Walker FJ (1981) Regulation of activated protein C by protein S. The role of phospholipid in factor Va inactivation. *J. Biol. Chem.* **256**: 11128–11131.

Walker FJ (1984a) Protein S and the regulation of activated protein C. *Sem. Thromb. Haemost.* **10**: 131–138.

Walker FJ (1984b) Regulation of vitamin K-dependent protein S. Inactivation with thrombin. *J. Biol. Chem.* **259**: 10335–10339.

Walker FJ (1986) Identification of a new protein involved in the regulation of the anticoagulant activity of activated protein C. *J. Biol. Chem.* **261**: 10941–10944.

Walker FJ (1989) Characterization of a synthetic peptide that inhibits the interaction between protein S and C4b-binding protein. *J. Biol. Chem.* **264**: 17645–17648.

Walker FJ, Fay PJ (1992) Regulation of blood coagulation by the protein C system. *FASEB J.* **6**: 2561–2567.

Walker FJ, Scandella D, Fay PJ (1990) Identification of the binding site for activated protein C on the light chain of factors V and VIII. *J. Biol. Chem.* **265**: 1484–1489.

Walsh JJ, Bonnar J, Wright FW (1974) A study of pulmonary embolism and deep vein thrombosis after major gynaecological surgery using labelled fibrinogen-phlebography and lung scanning. *J. Obstet. Gynaecol. Br. Commonw.* **81**: 311–316.

Walz DA, Bacon-Baguley T, Kendra-Francak S, DePoli P (1987) Binding of thrombospondin to immobilized ligands: specific interaction with fibrinogen, plasminogen, histidine-rich glycoprotein and fibronectin. *Sem. Thromb. Hemost.* **13**: 317–323.

Wang H, Riddell DC, Guinto ER, MacGillivray RTA (1988) Localization of the gene encoding human factor V to chromosome 1q21–25. *Genomics* **2**: 324–328.

Wang YZ, Patterson J, Gray JE, Yu C, Cottrell BA, Shimizu A, Graham D, Riley M, Doolittle RF (1989) Complete sequence of the lamprey fibrinogen α chain. *Biochemistry* **28**: 9801–9806.

Wardell MR, Abrahams J-P, Bruce D, Skinner R, Leslie AGW (1993) Crystallization and preliminary X-ray diffraction analysis of two conformations of intact human antithrombin. *J. Mol. Biol.* **234**: 1253–1258.

Wasley LC, Atha DH, Bauer KA, Kaufman RJ (1987) Expression and characterization of human antithrombin III synthesized in mammalian cells. *J. Biol. Chem.* **262**: 14766–14772.

Watanabe M, Osada J, Aratani Y, Kluckman K, Reddick R, Malinow MR, Maeda N (1995) Mice deficient in cystathionine β-synthase: animal models for mild and severe homocyst(e)inemia. *Proc. Natl Acad. Sci. USA* **92**: 1585–1589.

Watkins PC, Eddy R, Fukushima Y, Byers MG, Cohen EH, Dackowski WR, Wydro RM, Shows TB (1988) The gene for protein S maps near the centromere of human chromosome 3. *Blood* **71**: 238–241.

Watton J, Longstaff C, Lane DA, Barrowcliffe TW (1993) Heparin binding affinity of normal and genetically modified antithrombin III measured using a monoclonal antibody to the heparin binding site of antithrombin III. *Biochemistry.* **32**: 7286–7293.

Weaver DT, DePamphilis ML (1982) Specific sequences in native DNA that arrest synthesis by DNA polymerase α. *J. Biol. Chem.* **257**: 2075–2086.

Wei GM, Lemontt JF, Reddy VB, Hsiung N (1985) Cloning, sequencing and expression of human uterine tissue plasminogen activator cDNA. *DNA* **4**: 76.

Weiler-Guettler H, Yu K, Soff G, Gudas LJ, Rosenberg RD (1992) Thrombomodulin gene regulation by cAMP and retinoic acid in F9 embryonal carcinoma cells. *Proc. Natl Acad. Sci. USA* **89**: 2155–2159.

Weinmann EE, Salzman EW (1994) Deep-vein thrombosis. *New Engl. J. Med.* **331**: 1630–1641.

Weinstein RE, Walker FJ (1991) Species specificity of the fibrinolytic effects of activated protein C. *Thromb. Res.* **63**: 123–131.

Weisdorf DJ, Edson JR (1990) Recurrent venous thrombosis associated with inherited deficiency of heparin cofactor II. *Br. J. Haematol.* **77**: 125–126.

Weissbach L, Oddoux C, Procyk R, Grieninger G (1991) The β chain of chicken fibrinogen contains an atypical thrombin cleavage site. *Biochemistry* **30**: 3290–3294.

Welch WJ (1992) Mammalian stress response: cell physiology, structure/function of stress proteins, and implications for medicine and disease. *Physiol. Rev.* **72**: 1063–1081.

Wells PS, Blajchman MA, Henderson P, Wells MJ, Demers C, Bourque R, McAvoy A (1994) Prevalence of antithrombin deficiency in healthy blood donors: a cross-sectional study. *Am. J. Hematol.* **45**: 321–324.

Wells PS, Hirsh J, Anderson DR, Lensing AW, Foster G, Kearon C, Weitz J, D'Ovidio R, Cogo A, Prandoni P (1995) Accuracy of clinical assessment of deep-vein thrombosis. *Lancet* **345**: 1326–1330.

Wen D, Dittman WA, Ye RD, Deaven LL, Majerus PW, Sadler JE (1987) Human thrombomodulin: complete cDNA sequence and chromosome localization of the gene. *Biochemistry* **26**: 4350–4357.

White D, Abraham G, Carter C, Kakkar VV, Cooper DN (1992) A novel missense mutation in the antithrombin III gene (Ala 387→Val) causing recurrent venous thrombosis. *Hum. Genet.* **90**: 472–473.

Whitefleet-Smith J, Rosen E, McLinden J, Ploplis VA, Fraser MJ, Tomlinson JE, McLean JW, Castellino FJ (1989) Expression of human plasminogen cDNA in a baculovirus vector-infected insect cell system. *Arch. Biochem. Biophys.* **271**: 390–399.

Wilczynska M, Fu M, Ohlsson P-I, Ny T (1995) The inhibition mechanism of serpins. *J. Biol. Chem.* **270**: 29652–29655.

Wildgoose D, Klazin AL, Kisiel W (1990) The importance of residues 195–206 of human blood clotting factor VII in the interaction of factor VII with tissue factor. *Proc. Natl Acad. Sci. USA* **87**: 7290–7294.

Willems PJ (1994) Dynamic mutations hit double figures. *Nature Genetics* **8**: 213–215

Wiman B (1977) The primary structure of the B (light) chain of human plasmin. *Eur. J. Biochem.* **76**: 129–137.

Wiman B, Chmielewska J (1985) A novel fast inhibitor to tissue plasminogen activator in plasma, which may be of great pathophysiological significance. *Scand. J. Clin. Lab. Invest.* **45** (Suppl.): 43–47.

Wiman B, Hamsten A (1990) The fibrinolytic enzyme system and its role in the etiology of thromboembolic disease. *Sem. Thromb. Hemost.* **16**: 207–216.

Wiman B, Lijnen HR, Collen D (1979) On the specific interaction between the lysine-binding sites in plasmin and complementary sites in α2-antiplasmin and in fibrinogen. *Biochim. Biophys. Acta* **579**: 142–154.

Wiman B, Ljungberg B, Chmielewska J, Urden G, Blombäck M, Johnsson H (1985) The role of the fibrinolytic system in deep vein thrombosis. *J. Lab. Clin. Med.* **105**: 265–270.

Wiman B, Almquist A, Siguardardottir O, Lindahl T (1988) Plasminogen activator inhibitor (PAI) is bound to vitronectin in plasma. *FEBS Lett.* **242**: 125–128.

Winter JH, Bennett B, Watt JL, Brown T, San Roman C, Schinzel A, King J, Cook PJL (1982) Confirmation of linkage between antithrombin III and Duffy blood group and assignment of *AT3* to 1q22–q25. *Ann. Hum. Genet.* **46**: 29–34.

Winter PC, Scopes DA, Berg L-P, Millar DS, Kakkar VV, Mayne EE, Krawczak M, Cooper DN (1995) Functional analysis of an unusual length polymorphism in the human antithrombin III (*AT3*) gene promoter. *Blood Coag. Fibrinol.* **6**: 659–664.

Wion KL, Kelly D, Summerfield JA, Tuddenham EGD, Lawn RM (1985) Distribution of factor VIII mRNA and antigen in human liver and other tissues. *Nature* **317**: 726–729.

Witt I, Beck S, Seydewitz HH, Tasangil C, Schenck W. (1994) A novel homozygous missense mutation (Val325→Ala) in the protein C gene causing neonatal purpura fulminans. *Blood Coag. Fibrinol.* **5**: 651–653.

Wittman E, Walter J, Pabinger-Fasching I, Watzke HH (1994) Symptomatic hereditary type-II protein C deficiency caused by a missense mutation in exon IX of the protein C gene (Gly381 to Ser). *Ann. Hematol.* **68**: 255–259.

Wohl RC, Summaria L, Chediak J, Rosenfeld S, Robbins KC (1982) Human plasminogen variant Chicago III. *Thromb. Haemost.* **48**: 146–152.

Wojcik EGC, Simioni P, Berg M van den, Girolami A, Bertina RM (1996) Mutations which introduce free cysteine residues in the Gla-domain of vitamin K dependent proteins result in the formation of complexes with α1-microglobulin. *Thromb. Haemost.* **75**: 70–75.

Wolf M, Boyer-Neumann C, Molho-Sabatier P, Neumann C, Meyer D, Larrieu MJ (1990) Familial variant of antithrombin III (ATIII Bligny, 47 Arg to His) associated with protein C deficiency. *Thromb. Haemost.* **63**: 215–219.

Woodward SR, Weyward NJ, Bunnell M (1994) DNA sequence from Cretaceous period bone fragments. *Science* **266**: 1229–1232.

World Health Organization Collaborative Study of Cardivascular Disease and Steroid Hormone Contraception (1995) Venous thromboembolic disease and combined oral contraceptives: results of international multicentre case–control study. *Lancet* **346**: 1575–1582.

Wright HT, Blajchman MA (1994) Proteolytically cleaved mutant antithrombin-Hamilton has high stability to denaturation characteristic of wild type inhibitor serpins. *FEBS Lett.* **348**: 14–16.

Wu S, Seino S, Bell GI (1989) Human antithrombin III (*AT3*) gene length polymorphism revealed by the polymerase chain reaction. *Nucleic Acids Res.* **17**: 6433.

Wu HL, Chang BI, Wu DH (1990) Interaction of plasminogen and fibrin in plasminogen activation. *J. Biol. Chem.* **265**: 19658–19664.

Wu K, Urano T, Ihara H, Takada Y, Fuije M, Shikimori M, Hashimoto K, Takada A (1995) The cleavage and inactivation of plasminogen activator inhibitor type 1 by neutrophil elastase: the evaluation of its physiologic relevance in fibrinolysis. *Blood* **86**: 1056–1061.

Wuillemin WA, Huber I, Furlan M, Lämmle B (1991) Functional characterization of an abnormal factor XII molecule (FXII Bern). *Blood* **78**: 997–1004.

Wuillemin WA, Furlan M, Stricker H, Lämmle B (1992) Functional characterization of a variant factor XII (FXII Locarno) in a cross-reacting material positive FXII deficient plasma. *Thromb. Haemost.* **67**: 219–225.

Wun T-C, Kretzmer KK (1987) cDNA cloning and expression in *E. coli* of a plasminogen activator inhibitor (PAI) related to a PAI produced by HepG2 hepatoma cell. *FEBS Lett.* **210**: 11–16.

Wunderwald P, Schenk WJ, Port H (1982) Antithrombin-BM from human plasma an antithrombin binding moderately to heparin. *Thromb. Res.* **25**: 177–191.

Xu W-F, Chung DW, Davie EW (1996) The assembly of human fibrinogen. *J. Biol. Chem.* **271**: 27948–27953.

Xue J, Kalafatis M, Mann KG (1993) Determination of the disulfide bridges in factor Va light chain. *Biochemistry* **32**: 5917–5923.

Xue J, Kalafati M, Silveira JR, Kung C, Mann KG (1994) Determination of the disulfide bridges in factor Va heavy chain. *Biochemistry* **33**: 13109–13116.

Yamamoto K, Tamimoto M, Matsushita T, Kagami K, Sugiura I, Hamaguchi M, Takamatsu J, Saito H (1991a) Genotype establishments for protein C deficiency by use of a DNA polymorphism in the gene. *Blood* **77**: 2633–2636.

Yamamoto K, Takamatsu J, Saito H (1991b) Two novel sequence polymorphisms of the human protein C gene. *Nucleic Acids Res.* **19**: 6973.

Yamamoto K, Matsushita T, Sugiura I, Takamatsu J, Iwasaki E, Wada H, Deguchi K, Shirakawa S, Saito H (1992a) Homozygous protein C deficiency: identification of a novel missense mutation that causes impaired secretion of the mutant protein C. *J. Lab. Clin. Med.* **119**: 682–689.

Yamamoto K, Tanimoto M, Emi N, Matsushita T, Takamatsu J, Saito H (1992b) Impaired secretion of elongated mutant of protein C (Protein C Nagoya). *J. Clin. Invest.* **90**: 2439–2446.

Yamazaki T, Sugiura I, Matsushita T, Kojima T, Kagami K, Takamatsu J, Saito H (1993) A phenotypically neutral dimorphism of protein S: the substitution of Lys155 by Glu in the second EGF domain predicted by an A to G base exchange in the gene. *Thromb. Res.* **70**: 395–403.

Yamazaki T, Hamaguchi M, Katsumi A, Kagami K, Kojima T, Takamatsu J, Saito H (1995) A quantitative protein S deficiency associated with a novel nonsense mutation and markedly reduced levels of mutated mRNA. *Thromb. Haemost.* **74**: 590–595.

Yamazaki T, Katsumi A, Kagami K, Okamoto Y, Sugiura I, Hamaguchi M, Kojima T, Takamatsu J, Saito H (1996) Molecular basis of a hereditary type 1 protein S deficiency caused by a substitution of Cys for Arg474. *Blood* **87**: 4643–4650.

Yan SCB, Pazzano P, Chao YB, Walls JD, Berg DT, McClure DB, Grinnell BW (1990) Characterization and novel purification of recombinant human protein C from three mammalian cell lines. *Biotechnology* **8**: 655–661.

Yang X-J, Blajchman MA, Craven S, Smith LM, Anrari N, Ofosu F (1990) Activation of factor V during intrinsic and extrinsic coagulation. *Biochem. J.* **272**: 399–406.

Yang-Feng TL, Opdenakker G, Volckaert G, Francke U (1986) Human tissue-type plasminogen activator gene located near chromosomal breakpoint in myeloproliferative disorder. *Am. J. Hum. Genet.* **39**: 79–87.

Ye J, Esmon CT (1995) Factor Xa-factor Va complex assembles in two dimensions with unexpectedly high affinity: an experimental and theoretical analysis. *Biochemistry* **34**: 6448–6453.

Ye J, Esmon NL, Esmon CT, Johnson AE (1991) The active site of thrombin is altered upon binding to thrombomodulin. Two distinct structural changes are detected by fluorescence, but only one correlates with protein C activation. *J. Biol. Chem.* **266**: 23016–23021.

Ye J, Esmon CT, Johnson AE (1993) The chondroitin sulfate moiety of thrombomodulin binds a second molecule of thrombin. *J. Biol. Chem.* **268**: 2373–2379.

Ye S, Green FR, Scarabin PY, Nicaud V, Bara L, Dawson SJ, Humphries SE, Evans A, Luc G, Cambou JP, Arveiler D,Henney AM, Cambien F (1995) The 4G/5G genetic polymorphism in the promoter of the plasminogen activator inhibitor-1 (PAI-1) gene is associated with differences in plasma PAI-1 activity but not with risk of myocardial infarction in the ECTIM study. *Thromb. Haemost.* **74**: 837–841.

Yen PH, Li XM, Tsai SP, Johnson C, Mohandas T, Shapiro LJ (1990) Frequent deletions of the human X chromosome distal short arm result from recombination between low copy repetitive elements. *Cell* **61**: 603–610.

Yonegawa O, Voskuilen M, Nieuwenhuizen W (1992) Localization in the fibrinogen γ-chain of a new site that is involved in the acceleration of the tissue-type plasminogen activator-catalysed activation of plasminogen. *Biochem. J.* **283**: 187–191.

Yu S, Pritchard M, Kremer E, Lynch M, Nancarrow J, Baker E, Holman K, Mulley JC, Warren ST, Schlessinger D, Sutherland GR, Richards RI (1991) Fragile X genotype characterized by an unstable region of DNA. *Science* **252**: 1179–1181.

Yu K, Morioka H, Fritze LMS, Beeler DL, Jackman RW, Rosenberg RD (1992) Transcriptional regulation of the thrombomodulin gene. *J. Biol. Chem.* **267**: 23237–23247.

Yu S, Jhingan A, Christiansen WT, Castellino FJ (1994) Construction, expression and properties of a recombinant chimeric human protein C with replacement of its growth factor-like domains by those of human coagulation factor IX. *Biochemistry* **33**: 823–831.

Zammit A, Dawes J (1995) Fibrinogen inhibits the heparin cofactor II-mediated antithrombin activity of dermatan sulfate. *Blood* **85**: 720–726.

Zehnder JL, Jain M (1996) Recurrent thrombosis due to compound heterozygosity for factor V Leiden and factor V deficiency. *Blood Coag. Fibrinol.* **7**: 361–362.

Zhang L, Castellino FJ (1990) A γ-carboxyglutamic acid (γ) variant (γ⁶D, γ⁷D) of human activated protein C displays greatly reduced activity as an anticoagulant. *Biochemistry* **29**: 10828–10834.

Zhang L, Castellino FJ (1991) Role of hexapeptide disulfide loop present in the γ-carboxyglutamic acid domain of human protein C in its activation properties and in the *in vitro* anticoagulant activity of activated protein C. *Biochemistry* **30**: 6696–6704.

Zhang L, Castellino FJ (1992) Influence of specific γ-carboxyglutamic acid residues on the integrity of the calcium-dependent conformation of human protein C. *J. Biol. Chem.* **267**: 26078–26084.

Zhang L, Castellino FJ (1993) The contributions of individual γ-carboxyglutamic acid residues in the calcium-dependent binding of recombinant human protein C to acidic phospholipid vesicles. *J. Biol. Chem.* **268**: 12040–12045.

Zhang L, Castellino FJ (1994) The binding energy of human coagulation protein C to acidic phospholipid vesicles contains a major contribution from leucine 5 in the γ-carboxyglutamic acid domain. *J. Biol. Chem.* **269**: 3590–3595.

Zhang J-Z, Redman CM (1994) Role of interchain disulfide bonds on the assembly and secretion of human fibrinogen. *J. Biol. Chem.* **269**: 652–658.

Zhang L, Jhingan A, Castellino FJ (1992) Role of individual γ-carboxyglutamic acid residues of activated human protein C in defining its *in vitro* anticoagulant activity. *Blood* **80**: 942–952.

Zhang J-Z, Kudryk B, Redman CM (1993) Symmetrical disulfide bonds are not necessary for assembly and secretion of human fibrinogen. *J. Biol. Chem.* **268**: 11278–11282.

Zhang GS, Mehringer JH, Van Deerlin VMD, Kozak CA, Tollefsen DM (1994) Murine heparin cofactor II: purification, cDNA sequence, expression, and gene structure. *Biochemistry* **33**: 3632–3642.

Zhang Z, Fuentes NL, Fuller GM (1995) Characterization of the IL-6 responsive elements in the γ-fibrinogen gene promoter. *J. Biol. Chem.* **270**: 24287–24291.

Zheng Y-Z, Sakata T, Matsusue T, Umeyama H, Kato H, Miyata T (1994) Six missense mutations associated with type I and type II protein C deficiency and implications obtained from molecular modelling. *Blood Coag. Fibrinol.* 5: 687–696.

Zivelin A, Griffin JH, Xu X, Pabinger I, Samama M, Conard J, Brenner B, Eldor A, Seligsohn U (1997) A single genetic origin for a common Caucasian risk factor for venous thrombosis. *Blood* 89: 397–402.

Zöller B, Dahlbäck B (1994a) Linkage between inherited resistance to activated protein C and factor V gene mutation in venous thrombosis. *Lancet* 343: 1536–1538.

Zöller B, Svensson PJ, He X, Dahlbäck B (1994b) Identification of the same factor V gene mutation in 47 out of 50 thrombosis-prone families with inherited resistance to activated protein C. *J. Clin. Invest.* 94: 2521–2524.

Zöller B, He X, Dahlbäck B (1994) Homozygous APC-resistance combined with inherited type I protein S deficiency in a young boy with severe thrombotic disease. *Thromb. Haemost.* 73: 743–745.

Zöller B, Berntsdottir A, Garcia de Frutos P, Dahlbäck B (1995a) Resistance to activated protein C as an additional genetic risk factor in hereditary deficiency of protein S. *Blood* 85: 3518–3523.

Zöller B, Garcia de Frutos P, Dahlbäck B (1995b) Evaluation of the relationship between protein S and C4b-binding protein isoforms in hereditary protein S deficiency demonstrating type I and type III deficiencies to be phenotypic variants of the same genetic disease. *Blood* 85: 3524–3531.

Zöller B, Holm J, Svensson P, Dahlbäck B (1996) Elevated levels of prothrombin activation fragment 1+2 in plasma from patients with heterozygous Arg506 to Gln mutation in the factor V gene (APC-resistance) and/or inherited protein S deficiency. *Thromb. Haemost.* 75: 270–274.

Zotz RB, Maruhn-Debowski B, Scharf RE (1996) Mutation in the gene coding for coagulation factor V and resistance to activated protein C: detection of the genetic mutation by oligonucleotide ligation assay using a semi-automated system. *Thromb. Haemost.* 76: 53–55.

Zuber M, Toulon P, Marnet L, Mas J-L (1996) Factor V Leiden mutation in cerebral venous thrombosis. *Stroke* 27: 1721–1723.

Züger M, Demarmels Biasiutti F, Furlan M, Mannhalter C, Lämmle B (1996) Plasminogen deficiency: an additional risk factor for thrombosis in a family with factor V R506Q mutation? *Thromb. Haemost.* 76: 475–480.

Zushi M, Gomi Yamamoto S, Maruyama I, Hayashi T, Suzuki K (1989) The last three consecutive epidermal growth factor-like structures of human thrombomodulin comprise the minimum functional domain for protein C activating cofactor activity and anticoagulant activity. *J. Biol. Chem.* 264: 10351–10353.

Zushi M, Gomi K, Honda G, Kondo S, Yamamoto S, Hayashi T, Suzki K (1991) Aspartic acid 349 in the fourth epidermal growth factor-like structure of human thrombomodulin plays a role in its Ca^{2+}-mediated binding to protein C. *J. Biol. Chem.* 266: 19886–19889.

Appendix

Single base-pair substitutions causing venous thrombosis

This Appendix attempts to summarize *published* reports of single base-pair substitutions causing venous thrombosis which have occurred within the coding regions of a number of genes which encode key haemostatic proteins (table up to date as of 2 January 1997).

In terms of what they can teach us of protein structure and function relationships, missense mutations are the most important category of lesion. However, for the sake of completeness, and because they are relevant to any discussion of recurrent mutation, nonsense mutations are also included. We have not attempted to list other types of single base-pair substitution occurring in splice sites, or in promoter regions, or deletions and insertions. These categories of lesion are covered in the appropriate chapters. Further details of all types of lesion in these genes may be found either in the appropriate locus-specific databases (see text for references) or the Human Gene Mutation Database (http://www.cf.ac.uk/uwcm/mg/hgmd0.html).

Key

Gene: HUGO-recognized gene symbols are used throughout.

Deficiency type: types are as used in the text.

Activity/antigen: values are given when known and are for the propositus or proband. Where both anticoagulant and amidolytic activities have been determined, only the former is given. For the dysfibrinogenaemias, L (low), N (normal) and H (high) are used. For protein S deficiency, (F) and (T) denote free protein S antigen and total protein S antigen respectively.

Nucleotide substitution: *denotes a CG→TG or CG→CA substitution compatible with the mutational mechanism of methylation-mediated deamination of 5-methylcytosine (see Chapter 2).

Amino acid substitution: the three-letter code is used throughout.

Codon: numbering systems have adopted the conventions for each gene/protein. Negative numbers denote position in the leader peptide.

Comments: Name ascribed to dysfunctional protein. Venous thrombotic symptoms may be assumed except in a handful of cases of ATIII or protein C deficiency who had experienced arterial thrombotic symptoms or who, at the time of publication, were asymptomatic. Nonsense mutations and *de novo* mutations are specified. The biochemical basis of the protein dysfunction is given if known or hypothesized. Patients are heterozygous unless otherwise stated (i.e. double heterozygote or homozygote. In such cases, some indication of the other lesion is made).

Single base-pair substitutions causing venous thrombosis

Gene	Def. type	Act. (% norm)	Ant. (% norm)	Nucl. subst.	Amino acid subst.	Codon	Comments	Reference
AT3	I	<60	<60	TGC→TGA	Cys→Term	-4	Nonsense mutation	Saleun et al. (1996)
AT3	II	?	?	ATC→AAC	Ile→Asn	7	"ATIII Rouen III". New carbohydrate/heparin affinity	Bernnan et al. (1988), Perry and Carrell (1989)
AT3	II	50	92	ATG→ACG	Met→Thr	20	Double heterozygote; other allele is Tyr166→Cys	Lane et al. (1993)
AT3	II	56	105	CGC→TGC*	Arg→Cys	24	"ATIII Rouen IV". Reduced heparin binding	Perry and Carrell (1989), Borg et al. (1990)
AT3	II	47	104	CCG→CTG*	Pro→Leu	41		Chowdhury et al. (1995)
AT3	II	60	104	CCG→CTG*	Pro→Leu	41	"ATIII Basel". Reduced heparin confactor activity and loss of heparin binding. Asymptomatic	Chang and Tran (1986), Perry and Carrell (1989)
AT3	II	68	110	CCG→CTG*	Pro→Leu	41	"ATIII's Clichy 1, Clichy 2". Loss of heparin binding	Molho-Sabatier et al. (1989)
AT3	II	59	110	CCG→CTG*	Pro→Leu	41	"ATIII Franconville"	de Roux et al. (1990)
AT3	II	?	?	CGT→TGT*	Arg→Cys	47	"ATIII Alger". Loss of heparin binding. Homozygous	Brunel et al. (1987), Molho-Sabatier et al. (1989)
AT3	II	28	100	CGT→TGT*	Arg→Cys	47	"ATIII Kumamoto". Loss of heparin binding. Homozygous	Ueyama et al. (1990)
AT3	II	26	100	CGT→TGT*	Arg→Cys	47	"ATIII Toyama". Loss of heparin binding. Homozygous	Sakuragawa et al. (1983), Koide et al. (1984)
AT3	II	50	?	CGT→TGT*	Arg→Cys	47	"ATIII Tours". Loss of heparin binding. Asymptomatic	Duchange et al. (1986, 1987)
AT3	II	55	100	CGT→TGT*	Arg→Cys	47	"ATIII Amiens". Loss of heparin binding. Asymptomatic	Roussel et al. (1991)

Gene	Type			Codon	Substitution	Residue	Comments	Reference
AT3	II	60	100	CGT→TGT*	Arg→Cys	47	"ATIII Padua 2". Loss of heparin binding. Asymptomatic	Olds et al. (1990a)
AT3	II	54	127	CGT→TGT*	Arg→Cys	47	"ATIIIs Paris 1, Paris 2". Loss of heparin binding	Molho-Sabatier et al. (1989)
AT3	II	55	111	CGT→CAT*	Arg→His	47	"ATIII Rouen I". Loss of heparin binding. Asymptomatic	Owen et al. (1987), Perry and Carrell (1989)
AT3	II	60	100	CGT→CAT*	Arg→His	47	"ATIII Bligny". Reduced heparin cofactor activity. Also protein C deficient	Wolf et al. (1990)
AT3	II	64	102	CGT→AGT	Arg→Ser	47	"ATIII Rouen II". Loss of heparin binding. Asymptomatic	Borg et al. (1988)
AT3	II	?	?	CGT→TGT*	Arg→Cys	47	"ATIII Barcelona 2". Loss of heparin binding	Wells et al. (1994)
AT3	II	60	100	CGT→TGT*	Arg→Cys	47		Owen et al. (1989)
AT3	II	73	95	CGT→CAT*	Arg→His	47	"ATIII Padua I". Loss of heparin binding. Asymptomatic	Caso et al. (1990)
AT3	I	56	60	CGC→TGC*	Arg→Cys	57	Double heterozygote; other allele has 3 bp deletion (Asn 55)	Chowdhury et al. (1993)
AT3	I	55	45	TTT→TTG	Phe→Leu	58	"ATIII Budapest 6"	Lane et al. (1993)
AT3	I	36	42	TAT→TCT	Tyr→Ser	63		van Boven et al. (1994)
AT3	I	54	63	CCC→ACC	Pro→Thr	80	Pro80 may be required for formation of α-helix B. Mutation probably leads to improper folding	Millar et al. (1994a)
AT3	II	60–73	19–54	CTC→CCC	Leu→Pro	99	Two cases. Heparin binding defect. Homozygous. Early onset, severe arterial and venous thrombosis	Chowdhury et al. (1994)
AT3	II	?	77	CTC→TTC	Leu→Phe	99	"ATIII Budapest 3". Reduced heparin binding. Homozygous	Olds et al. (1992b), Lane et al. (1996b)

Single base-pair substitutions causing venous thrombosis (continued)

Gene	Def. type	Act. (% norm)	Ant. (% norm)	Nucl. subst.	Amino acid subst.	Codon	Comments	Reference
AT3	I	68	56	CTC→GTC	Leu→Val	99		Chowdhury et al. (1995)
AT3	I	58	?	CAA→AAA	Gln→Lys	101		Tait et al. (1994)
AT3	II	55	110	TCT→CCT	Ser→Pro	116	"ATIII Nagasaki"	Okajima et al. (1993)
AT3	II	52	104	CAG→CCG	Glu→Pro	118		Chowdhury et al. (1995)
AT3	I	69	62	CAC→TAC	His→Tyr	120	Predicted to lead to improper protein folding	Millar et al. (1994a)
AT3	I	<60	<60	CGA→TGA*	Arg→Term	129	Nonsense mutation	Olds et al. (1991a, b), Gandrille et al. (1991a)
AT3	I	50	49	CGA→TGA*	Arg→Term	129	Nonsense mutation	van Boven et al. (1994)
AT3	II	60	?	CGA→CAA*	Arg→Gln	129		Tait et al. (1994)
AT3	II	50	100	CGA→CAA*	Arg→Gln	129	"ATIII Geneva". Defective heparin cofactor activity	Gandrille et al. (1990)
AT3	I	68	47	TTA→TAA	Leu→Term	140	Nonsense mutation	Tomonari et al. (1992)
AT3	I	50	50	TAT→TGT	Tyr→Cys	166		Perry et al. (1995b)
AT3	I	48	47	TAT→TGT	Tyr→Cys	166	"ATIII Whitechapel". Double heterozygote; other allele is Met20→Thr	Chowdhury et al. (1993)
AT3	I	57	51	AAG→AAA	Lys→Lys (silent)	176	Intron 3a donor splice site mutation; leads to skipping of exon 3a in mature transcript. De novo	Berg et al. (1992)
AT3	I	70	73	AAC→AAA	Asn→Lys	187	Two unrelated patients, both asymptomatic	Tait et al. (1994), Perry et al. (1995a)
AT3	II	55	98	AAC→GAC	Asn→Asp	187	"ATIII Rouen VI". Reduction in progressive activity	Perry et al. (1991, 1995a), Bruce et al. (1994)
AT3	II	121	?	AAC→AAA	Asn→Lys	187	"ATIII Glasgow III"	Lane et al. (1991)
AT3	I	?	?	CGA→TGA*	Arg→Term	197	Nonsense mutation	Michiels et al. (1995)

Gene	Type			Codon	Substitution	Residue	Comment	Reference
AT3	I	52	68	CGA→TGA*	Arg→Term	197	Nonsense mutation	Michiels et al. (1995)
AT3	I	67	56	CGA→TGA*	Arg→Term	197	Nonsense mutation	Tomonari et al. (1992)
AT3	II	?	?	GAA→AAA	Glu→Lys	237	"ATIII Ituro"	Lane et al. (1993)
AT3	II	52	73	ATG→ATA	Met→Ile	251	May interfere with heparin binding or cocking with thrombin	Millar et al. (1994a)
AT3	I	51	?	CTT→CCT	Leu→Pro	270		Tait et al. (1994)
AT3	I	64	69	ATC→AAC	Ile→Asn	284		Lane et al. (1993)
AT3	II	40	60	TCC→CCC	Ser→Pro	349	"ATIII Hull I". Defective heparin binding	Grundy et al. (1992b)
AT3	I	55	49	CGA→TGA*	Arg→Term	359	Nonsense mutation	Chowdhury et al. (1993)
AT3	II	49	100	TTC→TCC	Phe→Ser	368	"ATIII Seoul". Defective heparin binding	Park et al. (1993)
AT3	II	56	100	GCA→ACA	Ala→Thr	382	"ATIII Glasgow II". Defective thrombin inhibition	Ireland et al. (1991a)
AT3	II	50	100	GCA→ACA	Ala→Thr	382	"ATIII Hamilton". Defective thrombin inhibition. Transformed into substrate	Devraj-Kizuk et al. (1988), Perry and Carrell (1989), Austin et al. (1991a)
AT3	II	65–80	?	GCA→TCA	Ala→Ser	384	10 unrelated patients	Tait et al. (1994)
AT3	II	75	103	GCA→TCA	Ala→Ser	384	"ATIII Cambridge II". Defective thrombin inhibition	Perry et al. (1991)
AT3	II	60	100	GCA→CCA	Ala→Pro	384	"ATIII Charleville". Defective thrombin inhibition	Molho-Sabatier et al. (1989)
AT3	II	56	119	GCA→CCA	Ala→Pro	384	"ATIII Sudbury". Defective thrombin inhibition	Pewarchuk et al. (1990)
AT3	II	62	105	GCA→CCA	Ala→Pro	384	"ATIII Vicenza". Defective thrombin inhibition	Caso et al. (1991)
AT3	II	54	95	GCA→CCA	Ala→Pro	384	"ATIII Cambridge". Defective thrombin inhibition	Perry et al. (1989)
AT3	I	50	60	GCT→GTT	Ala→Val	387		White et al. (1992)

Single base-pair substitutions causing venous thrombosis (continued)

Gene	Def. type	Act. (% norm)	Ant. (% norm)	Nucl. subst.	Amino acid subst.	Codon	Comments	Reference
AT3	II	53	111	GGC→GAC	Gly→Asp	392	"ATIII Stockholm". Defective thrombin inhibition	Blaichman et al. (1992a)
AT3	II	43	87	CGT→CAT*	Arg→His	393	"ATIII Glasgow I"	Thein and Lane (1988), Erdjument et al. (1988a), Owen et al. (1988)
AT3	II	?	?	CGT→TGT*	Arg→Cys	393	"ATIII Milano I". No thrombin inhibitory activity	Erdjument et al. (1988b)
AT3	II	60	100	CGT→TGT*	Arg→Cys	393	"ATIII Frankfurt I". No thrombin inhibitory activity	Ireland et al. (1991b)
AT3	II	61	100	CGT→CAT*	Arg→His	393	"ATIII Chicago". No thrombin inhibitory activity	Okajima et al. (1995)
AT3	II	37	96	CGT→CAT*	Arg→Cys	393		Erdjument et al. (1989)
AT3	II	62	100	CGT→CCT	Arg→Pro	393	"ATIII Pescara". No thrombin inhibitory activity	Lane et al. (1989b)
AT3	II	54	75	CGT→CAT*	Arg→His	393	"ATIII Sheffield". No thrombin inhibitory activity	Lane et al. (1989a)
AT3	II	59	100	CGT→CAT*	Arg→His	393	"ATIII Avranches". No thrombin inhibitory activity	Molho-Sabatier et al. (1989)
AT3	II	65	162	CGT→TGT*	Arg→Cys	393	"ATIII Northwick Park". No thrombin inhibitory activity	Thein and Lane (1988), Erdjument et al. (1988a)
AT3	II	54	92	TCG→TTG*	Ser→Leu	394	"ATIII Denver". Defective thrombin binding	Stephens et al. (1987)
AT3	II	66–71	100–123	TCG→TTG*	Ser→Leu	394	Four apparently unrelated patients, all of Danish origin. Defective thrombin binding predicted	Millar et al. (1994a)
AT3	II	?	?	TCG→TTG*	Ser→Leu	394	"ATIII Milano 2".	Olds et al. (1989)
AT3	II	?	?	TTC→TTA	Phe→Leu	402	Defective thrombin binding	Emmerich et al. (1994b)

Gene	Type			DNA	Protein	Codon	Description	Reference
AT3	I	52	70	TTC→TGC	Phe→Cys	402	"ATIII Rosny". Defective heparin binding	Lane et al. (1992b)
AT3	II	56	73	TTC→TTA	Phe→Leu	402	"ATIII Maisons Laffitte"	Lane et al. (1992b)
AT3	I	46	69	TTC→TCC	Phe→Ser	402	"ATIII Torino". Defective heparin binding	Lane et al. (1992b)
AT3	I	58	40	GCC→ACC	Ala→Thr	404	"ATIII Oslo". Defective thrombin inhibition	Bock et al. (1989), Lane et al. (1992b)
AT3	I	55	73	AAC→AAG	Asn→Lys	405	"ATIII La Rochelle"	Lane et al. (1991, 1992b)
AT3	I	56	58	AGG→ATG	Arg→Met	406	"ATIII Kyoto"	Tsuji et al. (1991), Nakagawa et al. (1991)
AT3	II	70	100	CCT→ACT	Pro→Thr	407	"ATIII Budapest 5"	Lane et al. (1991, 1992b)
AT3	I	50	50	CCT→CTT	Pro→Leu	407	"ATIII Utah"	Bock et al. (1985b, 1988)
AT3	I	45	46	ATC→ACC	Ile→Thr	421		van Boven et al. (1994)
AT3	I	68	68	ATC→ACC	Ile→Thr	421	Probably disturbs alignment of sheets 4B and 5B leading to adoption of an unstable conformation	Jochmans et al. (1994b)
AT3	I	59	48	GGC→CGC	Gly→Arg	424		Jochmans et al. (1994a)
AT3	I	48	64	GCC→GAC	Ala→Asp	427	"ATIII Hull II"	Millar et al. (1993c)
AT3	II	20	75	CCT→CTT	Pro→Leu	429	"ATIII Budapest"	Lane et al. (1991), Olds et al. (1992a)
AT3	I	55	60	TGT→TTT	Cys→Phe	430	Abolishes Cys247–Cys430 disulphide linkage	van Boven et al. (1994)
F5				CGA→CAA*	Arg→Gin	506	"Factor V Leiden". Underlies phenomenon of activated protein C resistance	Bertina et al. (1994), Greengard et al. (1994a), Zoller and Dahlbäck (1994)

Single base-pair substitutions causing venous thrombosis (continued)

Gene	Def. type	Act. (% norm)	Ant. (% norm)	Nucl. subst.	Amino acid subst.	Codon	Comments	Reference
FGA		L	N	CGT→TGT*	Arg→Cys	16	"Fibrinogen A α Hershey II". Delayed FpA release	Galanakis et al. (1990)
FGA		L	N	CGT→CAT*	Arg→His	16	"Fibrinogen A α New Albany". Delayed FpA release. Thrombosis	Henschen et al. (1981)
FGA		N	H	CGT→CAT*	Arg→His	16	"Fibrinogen A α Chapel Hill". Abnormal polymer formation. Bleeding/thrombosis	Carrell and McDonagh (1982)
FGA		N	N	AGG→GGG	Arg→Gly	19	"Fibrinogen A α Aarhus". Delayed polymerization. Thrombosis	Blombäck et al. (1988)
FGA		L	N	AAA→TAA	Lys→Term	461	"Fibrinogen A α Marburg 1". Bleeding/thrombosis Nonsense mutation. Defective monomer aggregation	Koopman et al. (1992b)
FGA		L	N	CGT→TGT*	Arg→Cys	554	"Fibrinogen A α Dusart (Paris V)". Defective fibrin polymerization	Koopman et al. (1993)
FGA		—	—	CGT→TGT*	Arg→Cys	554	Defective fibrin polymerization	Wada and Lord (1994)
FGB		L	N	CGT→TGT*	Arg→Cys	14	"Fibrinogen B β IJmuiden". Abnormal polymerization. Thrombosis	Koopman et al. (1992c)

Gene	Type	Reference	Description	Codon	Substitution	Codon change		
FGB		Koopman et al. (1992c)	"Fibrinogen Bβ Nijmegen". Abnormal polymerization. Thrombosis	44	Arg→Cys	CGT→TGT*	N	L
FGB		Koopman et al. (1992a)	"Fibrinogen Bβ Naples (Milano 2)". Defective thrombin binding. Thrombosis	68	Ala→Thr	GCT→ACT*	N	L
FGG		Borrell et al. (1995)	Abnormal polymerization. Normal tPA and plasminogen binding	275	Arg→His	CGC→CAC*	—	—
FGG		Reber et al. (1986)	"Fibrinogen Bergamo 2". Abnormal polymerization. Thrombosis	275	Arg→His	CGC→CAC*	N	L
FGG		Soria et al. (1987), Siebenlist et al. (1989)	"Fibrinogen Haifa". Defective fibrin polymerization	275	Arg→His	CGC→CAC*	N	L
FGG		Bantia et al. (1990)	"Fibrinogen Baltimore I". Delayed FpA release. Bleeding /thrombosis	292	Gly→Val	GGC→GTC	N	L
FGG		Grailhe et al. (1993)	Prolonged thrombin and Reptilase™ times	308	Asn→Lys	AAT→AAG	—	—
HCF2	II	Blinder et al. (1989), Borg et al. (1991), Toulon et al. (1991)	"Heparin cofactor II Oslo". Asymptomatic	189	Arg→His	CGC→CAC*	100	50
PLG	I	Ichinose et al. (1991)	"Plasminogen Nagoya I". Some relatives also doubly heterozygous for Ala601→Thr	355	Val→Phe	GTC→TTC	65	51
PLG	I	Azuma et al. (1993)	Occurs within activation cleavage region	572	Ser→Pro	TCC→CCC	54	54

Single base-pair substitutions causing venous thrombosis (continued)

Gene	Def. type	Act. (% norm)	Ant. (% norm)	Nucl. subst.	Amino acid subst.	Codon	Comments	Reference
PLG	II	69	100	GCT→ACT	Ala→Thr	601	"Plasminogen Tochigi I"	Aoki et al. (1978), Sakata and Aoki (1980)
PLG	II	8	98	GCT→ACT	Ala→Thr	601	"Plasminogen Kagoshima". Homozygous	Ichinose et al. (1991)
PLG	II	50	100	GCT→ACT	Ala→Thr	601	"Plasminogens Tochigi II, Nagoya I". Asymptomatic	Miyata et al. (1984)
PLG	I	60	60	GCT→ACT	Ala→Thr	675	Family members carrying lesion are essentially asymptomatic	Mima et al. (1996)
PLG	I	52	50	GAC→AAC	Asp→Asn	676		Tsutsumi et al. (1996)
PROC	I	71	66	ATG→ACG	Met→Thr	−42	Protein C Osaka 5: removes translational initiation codon	Zheng et al. (1994)
PROC	I	?	?	CAG→CCG	Gln→Pro	−40	May reduce stability of signal peptide and hinder transport of protein into endoplasmic reticulum	Gandrille and Aiach (1995)
PROC	I	?	?	CTG→CCG	Leu→Pro	−34	May destabilize signal peptide affecting translocation of protein across endoplasmic reticulum	Gandrille and Aiach (1995)
PROC	I	59	58	TGG→GGG	Trp→Gly	−29	Propeptide mutation	Poort et al. (1993), Hernandez et al. (1995)
PROC	I	38	42	TGG→TAG	Trp→Term	−29	Nonsense mutation	Millar et al. (1993a)
PROC	I	40	50	CGT→TGT*	Arg→Cys	−11	Propeptide mutation may affect carboxylase recognition. Two families affected	Ireland et al. (1996a)

PROC	I	?	?	CGG→TGG*	Arg→Trp	-5	Propeptide mutation	Gandrille and Aiach (1995)
PROC	II	40	70	CGG→TGG*	Arg→Trp	-5	Signal peptide; mutation impairs conformation of Gla domain but not γ-carboxylation	Gandrille et al. (1993a), Gaussem et al. (1994)
PROC	I	?	?	CGC→TGC*	Arg→Cys	-3	Propeptide mutation	Gandrille and Aiach (1995)
PROC	II	38	93	CGT→TGT*	Arg→Cys	-1	Propeptide cleavage site; novel inter-protein disulphide bonding?; impairs calcium binding	Gandrille et al. (1993a), Gaussem et al. (1994)
PROC	I/II	<10	75	CGT→CAT*	Arg→His	-1	Double heterozygote; other allele is Arg178→Gln	Alhenc-Gelas et al. (1995)
PROC	II	45	95	CGT→CAT*	Arg→His	-1	Propeptide cleavage site; mutation impairs γ-carboxylation	Gandrille et al. (1993a), Gaussem et al. (1994)
PROC	II	48	114	CGT→CAT*	Arg→His	-1	Propeptide cleavage site	Gandrille et al. (1993a)
PROC	II	30	107	CGT→TGT*	Arg→Cys	-1	Propeptide cleavage site. Reduced affinity for binding of Ca^{2+}; abolishes propeptide cleavage	Girolami et al. (1993)
PROC	II	45	91	CGT→AGT	Arg→Ser	-1	Protein C Osaka 10; normal propeptide cleavage but altered conformation of Gla domain	Miyata et al. (1995)
PROC	I	43	112	GAG→GAT	Glu→Asp	7	Modifies affinity of Gla domain for calcium ions	Gaussem et al. (1994)
PROC	II	4	47	CGT→TGT*	Arg→Cys	9		Sala et al. (1991), Tait et al. (1995)
PROC	II	60	102	AGC→TGC	Ser→Cys	12	Clinically asymptomatic; detected by population screening	Sala et al. (1991), Tait et al. (1995)
PROC	?	50	?	AGC→TGC	Ser→Cys	12		Ireland et al. (1996b)

Single base-pair substitutions causing venous thrombosis (continued)

Gene	Def. type	Act. (% norm)	Ant. (% norm)	Nucl. subst.	Amino acid subst.	Codon	Comments	Reference
PROC	II	62	111	AGC→TGC	Ser→Cys	12		Hernandez et al. (1995)
PROC	?	54	?	CTG→CCG	Leu→Pro	13		Ireland et al. (1996b)
PROC	II	52	72	CGG→TGG*	Arg→Trp	15		Gandrille et al. (1991b)
PROC	?	43	?	CGG→TGG*	Arg→Trp	15		Ireland et al. (1996b)
PROC	II	33	90	CGG→TGG*	Arg→Trp	15		Millar et al. (1993b)
PROC	I	?	?	CGG→CAG*	Arg→Gln	15		Gandrille and Aiach (1995)
PROC	II	44	83	CGG→GGG	Arg→Gly	15		Nakamura (1991)
PROC	I	52	55	CGG→TGG*	Arg→Trp	15		Lind et al. (1995)
PROC	I	60	58	CGG→CAG*	Arg→Gln	15		Gandrille et al. (1991b)
PROC	II	?	?	CGG→GGG	Arg→Gly	15	Abnormal conformation of Gla domain	Mimuro et al. (1993)
PROC	II	42	113	GAG→GCG	Glu→Ala	20	Gla residue. Double heterozygote; other allele is Val34→Met	Bovill et al. (1992), Lu et al. (1994a)
PROC	II	44	91	GAG→AAG*	Glu→Lys	25		Ireland et al. (1996b)
PROC	II	<5	20	GAG→AAG	Glu→Lys	26	Gla residue. Double heterozygote; other allele has frameshift mutation	Ido et al. (1993)
PROC	II	42	113	GTG→ATG	Val→Met	34	Double heterozygote; other allele is Glu20→Ala	Bovill et al. (1992), Lu et al. (1994)
PROC	I	?	?	GAT→GGT	Asp→Gly	35		Gandrille and Aiach (1995)
PROC	I	?	?	TTC→TTA	Phe→Leu	40		Gandrille and Aiach (1995)
PROC	I	?	?	GGT→TGT	Gly→Cys	47		Gandrille and Aiach (1995)
PROC	I	60	50	GGT→TGT	Gly→Cys	47	Introduction of Cys may disrupt disulphide linkages	Millar et al. (1995)
PROC	I	51	35	CAG→TAG	Gln→Term	49	Nonsense mutation	Soria et al. (1994a, 1996b)

Gene	Type			Base change	Substitution	Codon	Comment	Reference
PROC	I	?	?	TGC→CGC	Cys→Arg	50		Gandrille and Aiach (1995)
PROC	I	65	56	CCC→TCC	Pro→Ser	54		Soria et al. (1994a)
PROC	I	45	50	CCC→CTC	Pro→Leu	54		Ireland et al. (1996b)
PROC	II	46	60	CCC→TCC	Pro→Ser	54		Ireland et al. (1996b)
PROC	I	?	?	CCC→TCC	Pro→Ser	54		Gandrille and Aiach (1995)
PROC	?	?	?	CAC→AAC	His→Asn	66		Tsay et al. (1991)
PROC	I	?	?	GGC→GAC	Gly→Asp	67		Gandrille and Aiach (1995)
PROC	?	?	?	GGC→CGC	Gly→Arg	67		Reitsma et al. (1991)
PROC	I	42	56	GGC→CGC	Gly→Arg	72		Lind et al. (1995)
PROC	I	?	?	TTC→CTC	Phe→Leu	76		Gandrille and Aiach (1995)
PROC	I	74	64	TTC→CTC	Phe→Leu	76	Clinically asymptomatic; detected by population screening	Reitsma et al. (1991)
PROC	I	44	40	TGC→GGC	Cys→Gly	78		Tait et al. (1995)
PROC	I	45	45	TGC→TGA	Cys→Term	78	Nonsense mutation	Simioni et al. (1996c)
PROC	I	60	64	GGC→CGC	Gly→Arg	83	"Protein C Osaka 14"	Miyata et al. (1996)
PROC	I	43	46	CGC→CAC*	Arg→His	87		Ireland et al. (1996b)
PROC	II	26	52	CGC→CAC*	Arg→His	87	Doubly homozygous for this lesion and for a 7054 G→A transition on the other allele	Soria et al. (1996)
PROC	I	<5	<5	GAG→TAG	Glu→Term	92	Nonsense mutation	Reitsma et al. (1991)
PROC	I	46	39	GAG→TAG	Glu→Term	92	"Protein C Osaka 12". Nonsense mutation	Miyata et al. (1996)
PROC	I?	?	?	TCG→TAG	Ser→Term	99	Nonsense mutation	Ireland et al. (1996b)
PROC	I	47	50	GGC→CGC	Gly→Arg	103		Tsay et al. (1993)
PROC	I	48	36	TGC→TAC	Cys→Tyr	105		Reitsma et al. (1991)
PROC	I	34	35	TGC→TGA	Cys→Term	105	Nonsense mutation	Reitsma et al. (1991)
PROC	I	39	34	TGC→TAC	Cys→Tyr	105		Ireland et al. (1996b)
PROC	I	61	65	CAT→CCT	His→Pro	107	"Protein C Osaka U2"	Miyata et al. (1996)
PROC	I	51	40	TGG→TAG	Trp→Term	115	Nonsense mutation	Ireland et al. (1996b)
PROC	I	?	?	CGG→TGG*	Arg→Trp	116		Gandrille and Aiach (1995)

Single base-pair substitutions causing venous thrombosis (continued)

Gene	Def. type	Act. (% norm)	Ant. (% norm)	Nucl. subst.	Amino acid subst.	Codon	Comments	Reference
PROC	?	?	?	AGC→AGG	Ser→Arg	119		Reitsma et al. (1991)
PROC	I	?	?	TAC→TGC	Tyr→Cys	124		Gandrille and Aiach (1995)
PROC	I	?	<1%	TAC→TGC	Tyr→Cys	124	Double heterozygote; other allele is Gln132→Term	Soria et al. (1995b)
PROC	I	60	80	GGG→TGG	Gly→Trp	127		Ireland et al. (1996b)
PROC	I	19	23	CAG→TAG	Gln→Term	132	Nonsense mutation	Reitsma et al. (1991)
PROC	I	39	42	CAG→TAG	Gln→Term	132	Nonsense mutation	Ireland et al. (1996b)
PROC	I	?	<1%	CAG→TAG	Gln→Term	132	Compound heterozygote; other allele is Tyr124→Cys	Soria et al. (1995b, 1996b)
PROC	I	?	?	CAG→TAG	Gln→Term	132	Nonsense mutation	Gandrille and Aiach (1995)
PROC	I	1	1.3	GCA→CCA	Ala→Pro	136	Homozygous (neonatal purpura fulminans)	Long et al. (1994)
PROC	I	<1	?	GCA→CCA	Ala→Pro	136	Homozygous (neonatal purpura fulminans)	Dreyfus et al. (1991)
PROC	I	50–60	45–62	TTC→GTC	Phe→Val	139	"Protein C Osaka 11, Osaka 19, Osaka 27, Osaka 29"	Miyata et al. (1996)
PROC	I	56	67	TTC→CTC	Phe→Leu	139		Hernandez et al. (1995)
PROC	I	58, 61	72, 79	TTC→CTC	Phe→Leu	139	Two individuals. Both are clinically asymptomatic; detected by population screening	Tait et al. (1995)
PROC	I	56	57	TTC→CTC	Phe→Leu	139		Ireland et al. (1996b)
PROC	?	?	?	TGT→CGT	Cys→Arg	141		Reitsma et al. (1991)
PROC	I	31	48	TGT→TAT	Cys→Tyr	141		Hernandez et al. (1995)
PROC	I	52	58	TGG→TAG	Trp→Term	145	Nonsense mutation	Tsay et al. (1993)
PROC	?	44	?	CGG→TGG*	Arg→Trp	147		Ireland et al. (1996b)

Type						Comment	Reference
PROC ?	?	?	CGG→TGG*	Arg→Trp	147	Double heterozygote; other allele is frameshift mutation	Tsay et al. (1991)
PROC I	49	45	CGG→TGG*	Arg→Trp	147	Clinically asymptomatic; detected by population screening	Tait et al. (1995)
PROC II	58	85	CGC→TGC*	Arg→Cys	152		Gandrille et al. (1991b)
PROC I	60	58	CGA→TGA*	Arg→Term	157	Nonsense mutation	Poort et al. (1993)
PROC I	?	?	CGA→TGA*	Arg→Term	157	Nonsense mutation	Gandrille et al. (1992)
PROC I	65	54	CGA→TGA*	Arg→Term	157	Nonsense mutation	Gandrille et al. (1991b, 1992)
PROC I	?	?	CGA→TGA*	Arg→Term	157	Nonsense mutation	Gandrille and Aiach (1995)
PROC I	46	50	CGA→TGA*	Arg→Term	157	Patient experienced acute myocardial infarction but no venous thrombosis	Nakagawa et al. (1994)
PROC I	55	62	CCG→CTG*	Pro→Leu	168	Activation peptide	Tsay et al. (1993)
PROC I	10	?	CCG→CTG*	Pro→Leu	168	Homozygous	Conard et al. (1992)
PROC I	?	?	CCG→CTG*	Pro→Leu	168		Gandrille and Aiach (1995)
PROC I	60	59	CCG→CTG*	Pro→Leu	168		Gandrille et al. (1991b)
PROC I	23	36	CGG→CAG*	Arg→Gln	169	Thrombin cleavage site	Poort et al. (1993)
PROC I	?	?	CGG→TGG*	Arg→Trp	169	Thrombin cleavage site	Gandrille and Aiach (1995)
PROC I	49	60	CGG→TGG*	Arg→Trp	169	Thrombin cleavage site	Miyata et al. (1994)
PROC I	59	46	CGG→TGG*	Arg→Trp	169	Thrombin cleavage site	Tsay et al. (1993)
PROC II	30	40	CGG→TGG*	Arg→Trp	169	"Protein C Osaka 31"	Grundy et al. (1989)
PROC I	60	60	CGG→TGG*	Arg→Trp	169	Thrombin cleavage site	Miyata et al. (1996)
PROC I	29	41	CGG→TGG*	Arg→Trp	169	Thrombin cleavage site	Hernandez et al. (1995)
PROC I	?	?	CGG→TGG*	Arg→Trp	169	Thrombin cleavage site	Ohwada et al. (1992)
PROC II	<5	18	CGG→TGG*	Arg→Trp	169	Double heterozygote; other allele is deleted	Matsuda et al. (1988)

Single base-pair substitutions causing venous thrombosis (continued)

Gene	Def. type	Act. (% norm)	Ant. (% norm)	Nucl. subst.	Amino acid subst.	Codon	Comments	Reference
PROC	II	23	38	CGG→TGG*	Arg→Trp	169	Thrombin cleavage site	Ireland et al. (1996)
PROC	I	19	16	ACC→ATC	Thr→Ile	176	Residue 176 may play a role in pivoting the N terminus of protein C to fold into the oxyanion hole	Hallam et al. (1995b)
PROC	I	?	?	CGG→CAG*	Arg→Gln	178		Gandrille et al. (1992)
PROC	II	53	75	CGG→CCG	Arg→Pro	178		Millar et al. (1995)
PROC	I	33	50	CGG→CAG*	Arg→Gln	178		Millar et al. (1995)
PROC	I	?	?	CGG→CAG*	Arg→Gln	178		Soria et al. (1992)
PROC	I	?	?	CGG→CAG*	Arg→Gln	178		Soria et al. (1996c)
PROC	I	29	33	CGG→TGG*	Arg→Trp	178		Reitsma et al. (1991)
PROC	I	?	?	CGG→CAG*	Arg→Gln	178		Gandrille and Aiach (1995)
PROC	I	53	53	CGG→TGG*	Arg→Trp	178		Grundy et al. (1992a)
PROC	I	60	64	CGG→CAG*	Arg→Gln	178		Sala et al. (1991)
PROC	I	30	36	CGG→CAG*	Arg→Gln	178		Gandrille et al. (1992)
PROC	I/II	<10	75	CGG→CAG*	Arg→Gln	178	Double heterozygote; other allele is Arg1→His	Alhenc-Gelas et al. (1995)
PROC	I	46	46	CGG→CAG*	Arg→Gln	178		Poort et al. (1993)
PROC	I	?	<5	CGG→CAG*	Arg→Gln	178	Impaired secretion of defective protein. Double heterozygote; other allele is Cys331→Arg	Sugahara et al. (1994)
PROC	I	?	?	CGG→TGG*	Arg→Trp	178		Gandrille and Aiach (1995)
PROC	I	0	0	CAG→CAC	Gln→His	184	Neonatal purpura fulminans; homozygous; mutated base also position −1 of intron 7 donor splice site	Soria et al. (1994b)
PROC	?	44	?	TGC→TAC	Cys→Tyr	196		Ireland et al. (1996b)
PROC	I	?	?	GGG→GAG	Gly→Glu	197		Gandrille and Aiach (1995)

PROC ?	?	?	GGG→AGG*	Gly→Arg	197		Ireland et al. (1996b)
PROC I	45	49	ATC→ACC	Ile→Thr	201		Tsay et al. (1993)
PROC I	35	41	CAC→TAC	His→Tyr	202		Poort et al. (1993), Hernandez et al. (1995)
PROC I	?	?	GCG→GTG*	Ala→Val	209		Gandrille and Aiach (1995)
PROC II	43	146	CAC→CAA	His→Gln	211	Catalytic triad residue	Poort et al.(1993), Hernandez et al. (1995)
PROC I	?	?	CTT→CCT	Leu→Pro	220		Gandrille and Aiach (1995)
PROC I	?	?	CTT→TTT	Leu→Phe	223		Gandrille and Aiach (1995)
PROC I	26	16	CTT→TTT	Leu→Phe	223	Double heterozygote; other allele is Ile403→Met	Gandrille et al. (1993b)
PROC I	55	61	CTT→TTT	Leu→Phe	223		Reitsma et al. (1991)
PROC II	57	90	CGG→CAG*	Arg→Gln	229		Gandrille et al. (1993a)
PROC II	40	75	CGG→TGG*	Arg→Trp	229		Sala et al. (1991)
PROC II	57	90	CGG→CAG*	Arg→Gln	229		Gandrille et al. (1993a)
PROC I	59	69	CGC→TGC*	Arg→Cys	230		Reitsma et al. (1991), Hernandez et al. (1995)
PROC I	?	?	CGC→TGC*	Arg→Cys	230		Gandrille and Aiach (1995)
PROC I	?	?	GAG→AAG	Glu→Lys	232		Gandrille and Aiach (1995)
PROC I	6	<10	CCC→CTC	Pro→Leu	247	Homozygous	Grundy et al. (1992c)
PROC II	70	146	AGC→AAC	Ser→Asn	252	Creates consensus glycosylation sequence	Gandrille et al. (1993a)
PROC II	50	110	AGC→AAC	Ser→Asn	252	Creates consensus glycosylation sequence	Gandrille et al. (1993a)
PROC I	59	76	AAT→GAT	Asn→Asp	256		Allaart et al. (1993)
PROC I	55	57	GCA→ACA	Ala→Thr	259		Millar et al. (1995)
PROC I	<5	<5	GCA→GTA	Ala→Val	259	Homozygous	Grundy et al. (1991c)
PROC I	?	?	GCC→ACC*	Ala→Thr	267		Gandrille and Aiach (1995)
PROC I	20	20	GCC→ACC*	Ala→Thr	267	Homozygous	Conard et al. (1992)

Single base-pair substitutions causing venous thrombosis (continued)

Gene	Def. type	Act. (% norm)	Ant. (% norm)	Nucl. subst.	Amino acid subst.	Codon	Comments	Reference
PROC	I	?	?	TCG→CCG	Ser→Pro	270		Gandrille and Aiach (1995)
PROC	I	46	40	TCG→CCG	Ser→Pro	270	*De novo* mutation	Gandrille *et al.*(1994b)
PROC	I	27–36	36–45	CCG→CTG*	Pro→Leu	279	Two apparently unrelated families	Doig *et al.* (1994)
PROC	I	?	?	CCG→CTG*	Pro→Leu	279		Gandrille and Aiach (1995)
PROC	I	37	38	CCG→CTG*	Pro→Leu	279		Poort *et al.* (1993)
PROC	I	55, 62	50, 61	GGC→AGC*	Gly→Ser	282	Two individuals. Both are clinically asymptomatic; detected by population screening	Tait *et al.* (1995)
PROC	I	56	53	GGC→AGC*	Gly→Ser	282		Doig *et al.* (1994)
PROC	I	32	42	GGC→CGC*	Gly→Arg	282		Lind *et al.* (1995)
PROC	I	<1.3	<1.3	CGC→CAC*	Arg→His	286	Homozygous. Neonatal purpura fulminans	Long *et al.* (1994)
PROC	I	32	15	CGC→TGC*	Arg→Cys	286		Lind *et al.* (1995)
PROC	I	?	?	CGC→TGC*	Arg→Cys	286		Gandrille and Aiach (1995)
PROC	I	33	39	CGC→TGC*	Arg→Cys	286		Poort *et al.* (1993)
PROC	I	55–70	48–66	CGC→CAC*	Arg→His	286	"Protein C Osaka 18, Osaka 22"	Miyata *et al.* (1996)
PROC	I	52	48	GGC→AGC*	Gly→Ser	292		Doig *et al.* (1994)
PROC	I	?	?	GGC→AGC*	Gly→Ser	292		Gandrille and Aiach (1995)
PROC	?	?	28	GGC→AGC*	Gly→Ser	292	Homozygous (neonatal purpura fulminans)	Reitsma *et al.* (1992)
PROC	I	<1%	<1%	GGC→AGC*	Gly→Ser	292		Alessi *et al.* (1996)
PROC	I	42	37	GGC→AGC*	Gly→Ser	292		Ireland *et al.* (1996b)
PROC	I	<5	<5	GGC→AGC*	Gly→Ser	292	Homozygous	Yamamoto *et al.* (1992a)
PROC	I	?	?	CAG→CAC	Gln→His	293		Gandrille and Aiach (1995)

Gene			Codon	Substitution	Codon no.	Comments	Reference
PROC I	38–57	45–62	GTG→ATG*	Val→Met	297	"Protein C Osaka 17, Osaka 20, Osaka 24, Osaka 26, Osaka 30, Nara"	Miyata et al. (1996)
PROC I	?	?	GTG→ATG*	Val→Met	297		Gandrille and Aiach (1995)
PROC I	?	?	ACG→AAG	Thr→Lys	298		Gandrille and Aiach (1995)
PROC I	45–70	36–72	ACG→ATG*	Thr→Met	298	Only individuals with co-inherited factor V Leiden variant are clinically symptomatic	Brenner et al. (1996)
PROC I	20	20	ACG→ATG*	Thr→Met	298	"Protein C Vermont IIa". Double heterozygote; other allele is a frameshift insertion	Tomczak et al. (1994)
PROC I	36	33	ACG→ATG*	Thr→Met	298		Ireland et al. (1996)
PROC I	?	?	ACG→ATG*	Thr→Met	298		Gandrille and Aiach (1995)
PROC I	37	38	ACG→ATG*	Thr→Met	298		Tsay et al. (1993)
PROC I	40	47	GGC→GTC	Gly→Val	301	Clinically asymptomatic; detected by population screening	Tait et al. (1995)
PROC I	23	22	GGC→AGC	Gly→Ser	301	Homozygous	Conard et al. (1992)
PROC I	?	?	CGA→TGA*	Arg→Term	306	Nonsense mutation	Gandrille and Aiach (1995)
PROC I	67	72	CGA→TGA*	Arg→Term	306	Nonsense mutation	Grundy et al. (1992d)
PROC I	50	50	CGA→TGA*	Arg→Term	306	Nonsense mutation	Romeo et al. (1987)
PROC I	38	52	CGC→CAC*	Arg→His	314		Reitsma et al. (1991)
PROC I	?	?	GTC→TTC	Leu→Phe	318		Gandrille and Aiach (1995)
PROC I	?	?	ATT→TTT	Ile→Phe	323		Gandrille and Aiach (1995)
PROC I	?	?	GTG→GCG	Val→Ala	325		Gandrille and Aiach (1995)
PROC I	38	43	GTG→GCG	Val→Ala	325		Hernandez et al. (1995)

Single base-pair substitutions causing venous thrombosis (continued)

Gene	Def. type	Act. (% norm)	Ant. (% norm)	Nucl. subst.	Amino acid subst.	Codon	Comments	Reference
PROC	I	<2	<2	GTG→GCG	Val→Ala	325	Homozygous. Neonatal purpura fulminans	Witt et al. (1994)
PROC	I	<10	21	CCG→CTG*	Pro→Leu	327		Lind et al. (1995)
PROC	I	?	?	CCG→CTG*	Pro→Leu	327		Gandrille and Aiach (1995)
PROC	I	51	45	CCG→CTG	Pro→Leu	327	Protein C Osaka 6: may interfere with α helix formation	Zheng et al. (1994)
PROC	II	71	83	AAT→ACT	Asn→Thr	329	Glycosylation site. Reduced synthesis of β protein C variant with decreased functional activity	Simioni et al. (1996)
PROC	I	?	<5	TGC→CGC	Cys→Arg	331	Secretion of protein abolished due to misfolding as a consequence of loss of disulphide linkage	Miyata et al. (1996)
PROC	I	73	70	AGC→GGC	Ser→Gly	332		Ireland et al. (1996b)
PROC	I	?	?	ATG→ACG	Met→Thr	335		Gandrille and Aiach (1995)
PROC	I	15	40	ATG→ATA	Met→Ile	343	On oral anticoagulant therapy	Poort et al. (1993)
PROC	I	35	41	GCG→ACG	Ala→Thr	346	On oral anticoagulant therapy	Poort et al. (1993), Hernandez et al. (1995)
PROC	I	10	17	GCG→GTG*	Ala→Val	346	On oral anticoagulant therapy	Poort et al. (1993)
PROC	II	63	80	GGG→AGG*	Gly→Arg	350	Protein C Osaka 9: substitution occurs on surface loop with no obvious conformational disturbance	Zheng et al. (1994)

Gene	Type					Position	Notes	Reference
PROC	I	66	CGG→TGG*	73	Arg→Trp	352		Doig et al. (1994)
PROC	II	59	GAC→AAC*	91	Asp→Asn	359	Protein C Osaka 7 and 8: abolishes ionic interaction between residues 170 and 359 require for activation	Zheng et al. (1994)
PROC	?	?	GAC→AAC*	?	Asp→Asn	359		Tsay et al. (1991)
PROC	I	?	GGG→CGG	?	Gly→Arg	361		Gandrille and Aiach (1995)
PROC	?	?	GGG→GAG	40	Gly→Glu	362	"Protein C Osaka 21, Osaka 28"	Ireland et al. (1996b)
PROC	I	35–46	ATG→ATA	46–49	Met→Ile	364		Miyata et al. (1996)
PROC	I	50	ATG→ATA	50	Met→Ile	364		Miyata et al. (1994)
PROC	II	15	TTC→TCC	44	Phe→Ser	368	"Protein C Osaka 13"	Miyata et al. (1996)
PROC	I	45	ACC→GCC	42	Thr→Ala	371	Clinically asymptomatic; detected by population screening	Tait et al. (1995)
PROC	?	?	GGC→GAC	?	Gly→Asp	376	Double heterozygote; other allele is a frameshift	Sugahara et al. (1992)
PROC	II	49	TGG→TGC	95	Trp→Cys	380	"Protein C Osaka 15"	Miyata et al. (1996)
PROC	II	58	GGT→AGT	110	Gly→Ser	381		Wittman et al. (1994)
PROC	II	56	GGT→AGT	150	Gly→Ser	381	Substrate-binding pocket	Marchetti et al. (1993)
PROC	I	9	TGT→TAT	31	Cys→Tyr	384	On oral anticoagulant therapy	Tsay et al. (1993)
PROC	II	?	GGG→CGG	100	Gly→Arg	385		Miyata et al. (1994)
PROC	II	54	CTT→TTT	93	Leu→Phe	387	"Protein C Osaka 23"	Miyata et al. (1996)
PROC	I	?	AAC→AAA	?	Asn→Lys	389		Gandrille and Aiach (1995)
PROC	?	?	TAC→CAC	?	Tyr→His	393		Ireland et al. (1996b)
PROC	I	50	ACC→AAC	50	Thr→Asp	394		Ireland et al. (1996a)
PROC	I	54	CGC→CAC*	64	Arg→His	398	Clinically asymptomatic; detected by population screening	Tait et al. (1995)
PROC	I	52	TAC→CAC	61	Tyr→His	399	Protein C Osaka 4: may impair formation of inner hydrophobic core	Zheng et al. (1994)
PROC	I	50	TGG→TGC	50	Trp→Cys	402		Romeo et al. (1987)

Single base-pair substitutions causing venous thrombosis (continued)

Gene	Def. type	Act. (% norm)	Ant. (% norm)	Nucl. subst.	Amino acid subst.	Codon	Comments	Reference
PROC	I	?	?	ATC→ATG	Ile→Met	403		Gandrille and Aiach (1995)
PROC	I	71	70	ATC→CTC	Ile→Leu	403		Ireland et al. (1996b)
PROC	I	?	?	ATC→CTC	Ile→Leu	403		Gandrille and Aiach (1995)
PROC	I	26	16	ATC→ATG	Ile→Met	403	Double heterozygote; other allele is Leu223→Pro	Gandrille et al. (1993b)
PROC	I	53	45	ATC→ATG	Ile→Met	403		Allaart et al. (1993)
PROS	IIb	57	112(F), 87(T)	CGT→CTT	Arg→Leu	−2	Propeptide cleavage site; adversely affects γ-carboxylation	Gandrille et al. (1995)
PROS	IIb	64	101(F), 126(T)	CGT→CAT*	Arg→His	−1	Propeptide cleavage site	Gandrille et al. (1995)
PROS	II	24	62(F), 108(T)	AAA→GAA	Lys→Glu	9	May disrupt electrostatic interactions in Gla domain	Simmonds et al. (1996)
PROS	I	33	22(F), 57(T)	TGC→TGA	Cys→Term	22	Nonsense mutation	Gandrille et al. (1995)
PROS	I	27	45(F), 54(T)	GAA→GCA	Glu→Ala	26		Gandrille et al. (1995)
PROS	I	46	57(F), 52(T)	GAA→GCA	Glu→Ala	26		Simmonds et al.(1996)
PROS	I	45	72(F), 63(T)	TTT→TGT	Phe→Cys	31		Gandrille et al. (1995)
PROS	I	37	42(F), 60(T)	ACG→ATG*	Thr→Met	37		Gandrille et al. (1995)
PROS	I	?	60(F), 66(T)	CGC→CAC*	Arg→His	48	Does not co-segregate with disease phenotype in family pedigree but residue highly conserved	Gandrille et al. (1995)
PROS	I	42	38(F), 61(T)	GGG→GAG	Gly→Glu	54		Simmonds et al. (1996)
PROS	I	<5	15 (F), 41 (T)	TCA→TGA	Ser→Term	62	Nonsense mutation results in allelic exclusion. Detectable ectopic mRNA transcript lacked exon 4	Okamoto et al. (1996)

							Comment	Reference
PROS	IIb	46	80(F), 85(T)	ACT→AAT	Thr→Asn	103	Mutation in first EGF domain may affect function by steric hindrance of longer Asn side chain	Gandrille et al. (1995)
PROS	?	?	40(F)	TGT→TAT	Cys→Tyr	145	Disrupts disulphide linkage and may affect protein stability	Simmonds et al. (1996)
PROS	I	?	31(F), 31(T)	TGT→TAT	Cys→Tyr	145	Mutation abolishes disulphide link with Cys158	Beauchamp et al. (1996)
PROS	II	Low	normal (F & T)	AAG→GAG	Lys→Glu	155	Mutation in second EGF domain results in poor binding of MAb recognizing thrombin-sensitive domain	Hayashi et al. (1994)
PROS	I	?	36(F), 60(T)	TGT→TCT	Cys→Ser	200	Disrupts disulphide linkage and may affect protein stability	Simmonds et al. (1996)
PROS	I	?	45(F), ?(T)	GAT→GGT	Asp→Gly	204		Gandrille et al. (1995)
PROS	I	?	57(F), 46(T)	GAG→AAG	Glu→Lys	208		Gandrille et al. (1995)
PROS	III	38	45(F), 71(T)	AAT→AGT	Asn→Ser	217	May interfere with calcium binding	Formstone et al. (1995)
PROS	I	29	35(F), 55(T)	TGC→TGG	Cys→Trp	224		Gandrille et al. (1995)
PROS	I	16	15(F), 46(T)	TGC→CGC	Cys→Arg	224		Gandrille et al. (1995)
PROS	I	?	?	CCT→TTT	Leu→Pro	259		Duchemin et al. (1996)
PROS	?	?	23(F)	TCA→CCA	Ser→Pro	283		Simmonds et al. (1996)
PROS	I	20	2 (F), 26 (T)	CTG→CCG	Leu→Pro	310		Mustafa et al. (1995)
PROS	I		28(F), 50(T)	GAT→AAT	Asp→Asn	335		Gandrille et al. (1995)
PROS	II	8	7(F), 37(T)	GGT→GAT	Gly→Asp	340	Homozygous	Simmonds et al. (1996)
PROS	I		19(T)	GGT→GTT	Gly→Val	340		Gómez et al. (1995)
PROS	I	?	5(F), 25(T)	AAA→TAA	Lys→Term	368	Nonsense mutation	Simmonds et al. (1996)
PROS	I	?	20(F), 64(T)	TGT→TCT	Cys→Ser	408	Disrupts disulphide linkage and may affect protein stability	Simmonds et al. (1996)

Single base-pair substitutions causing venous thrombosis (continued)

Gene	Def. type	Act. (% norm)	Ant. (% norm)	Nucl. subst.	Amino acid subst.	Codon	Comments	Reference
PROS	I	48	5(F), 56(T)	CGA→TGA*	Arg→Term	410	Nonsense mutation	Andersen et al. (1996)
PROS	I		4(F), 23(T)	CGA→TGA*	Arg→Term	410	Nonsense mutation. Mutation also occurs in PROS pseudogene. Gene conversion?	Mustafa et al. (1995)
PROS	I		35(F), 64(T)	GGA→TGA	Gly→Term	448	Appears to be pathological lesion but defect does not co-segregate with disease state in family	Formstone et al. (1995)
PROS	II	?	(F)Red(T)Nor	TCC→CCC	Ser→Pro	460	Protein S Heerlen; occurs as polymorphism in general population, overrepresented in thrombotic patients	Duchemin et al. (1995)
PROS	?	?	40(F)	TCC→CCC	Ser→Pro	460		Simmonds et al. (1996)
PROS	I		32 (T)	GTA→GGA	Val→Gly	467		Gómez et al. (1995)
PROS	I	?	?	CGT→TGT*	Arg→Cys	474	Intracellular degradation and impaired secretion of in vitro expressed product	Yamazaki et al. (1996)
PROS	I	51	11(F), 44(T)	GCC→CCC	Ala→Pro	484	De novo mutation	Borgel et al. (1996)
PROS	I		2(F), 26(T)	TTA→TCA	Leu→Ser	511		Mustafa et al. (1995)
PROS	I	<10	14(F), 31(T)	CAG→TAG	Gln→Term	522	Nonsense mutation	Yamazaki et al. (1995)
PROS	I		30(F), 86(T)	CTG→CAG	Leu→Gln	543		Mustafa et al. (1995)
PROS	III		<10(F), 24(T)	ATG→ACG	Met→Thr	570	May affect protein S–C4bBP interaction	Formstone et al. (1995)
PROS	III	?	32(F), 66(T)	ATG→ACG	Met→Thr	570	May reduce C4bBP binding	Beauchamp et al. (1996)
PROS	I		2(F), 30(T)	GCC→CCC	Ala→Pro	575		Mustafa et al. (1995)
PROS	?	23	?	TAT→TAA	Tyr→Term	595	Nonsense mutation	Simmonds et al. (1996)
PROS	III	?	62(F), 81(T)	TCA→TTA	Ser→Leu	624	May reduce C4bBP binding	Beauchamp et al. (1996)

PROS	I	?	TGT→CGT	Cys→Arg	625		Duchemin et al. (1996)
PROS	I	13(F), 32(T)	TGT→CGT	Cys→Arg	625	Disrupts disulphide linkage and may affect protein stability	Simmonds et al. (1996)
PROS	I	24(T)	TAA→TAT	Term→Tyr	636	Abolishes normal Stop codon. Protein S molecule, elongated by 14 amino acids, predicted	Reitsma et al. (1994)
THBD	?	Low, variable	GAC→TAC	Asp→Tyr	468	May reduce expression of protein on membrane or mutant protein may undergo abnormal proteolysis	Öhlin and Marlar (1995)

Index

ORDERING DETAILS

Main address for orders

BIOS Scientific Publishers Ltd
9 Newtec Place, Magdalen Road,
Oxford OX4 1RE, UK
Tel: +44 1865 726286
Fax: +44 1865 246823

Australia and New Zealand
DA Information Services
648 Whitehorse Road, Mitcham, Victoria 3132, Australia
Tel: (03) 9210 7777
Fax: (03) 9210 7788

India
Viva Books Private Ltd
4325/3 Ansari Road, Daryaganj, New Delhi 110 002, India
Tel: 11 3283121
Fax: 11 3267224

Singapore and South East Asia
(Brunei, Hong Kong, Indonesia, Korea, Malaysia, the Philippines,
Singapore, Taiwan, and Thailand)
Toppan Company (S) PTE Ltd
38 Liu Fang Road, Jurong, Singapore 2262
Tel: (265) 6666
Fax: (261) 7875

USA and Canada
BIOS Scientific Publishers
PO Box 605, Herndon, VA 20172-0605, USA
Tel: (703) 661 1500
Fax: (703) 661 1501

Payment can be made by cheque or credit card (Visa/Mastercard, quoting number and expiry date). Alternatively, a *pro forma* invoice can be sent.

Prepaid orders must include £2.50/US$5.00 to cover postage and packing (two or more books sent post free)